Element	Symbol	Atomic Number	Atomic Weight	Element	Symbol	Atomic Number	Atomic Weight
Neodymium	Nd	60	144.24	Silicon	Si	14	28.086
Neon	Ne	10	20.179	Silver	Ag	47	107.868
Neptunium	Np	93	(237)	Sodium	Na	11	22.9898
Nickel	Ni			Strontium	Sr	38	87.62
Niobium	Nb			Sulfur	S	16	32.064
Nitrogen	N			Tantalum	Ta	73	180.948
Nobelium	No			Technetium	Tc	43	(99)
Osmium	Os			Tellurium	Te	52	127.60
Oxygen	O			Terbium	Tb	65	158.924
Palladium	Pd			Thallium	Tl	81	204.37
Phosphorus	P			Thorium	Th	90	232.038
Platinum	Pt			Thulium	Tm	69	168.934
Plutonium	Pu			Tin	Sn	50	118.69
Polonium	Po			Titanium	Ti	22	47.90
Potassium	K			Tungsten	W	74	183.85
Praseodymium	Pr			Uranium	U	92	238.03
Promethium	Pm			Vanadium	V	23	50.942
Protactinium	Pa			Xenon	Xe	54	131.30
Radium	Ra			Ytterbium	Yb	70	173.04
Radon	Rn			Yttrium	Y	39	88.905
Rhenium	Re			Zinc	Zn	30	65.37
Rhodium	Rh			Zirconium	Zr	40	91.22
Rubidium	Rb						(257)
Ruthenium	Ru						(260)
Samarium	Sm						
Scandium	Sc						
Selenium	Se						

QD 40 H 6   30386

Howald, Reed A.
The Science of Chemistry

DATE DUE

MAR 2 1 '91			
JUL 2 9 2003			

Waubonsee Community College

The Elements

THE SCIENCE OF CHEMISTRY:
Periodic Properties and Chemical Behavior

A Series of Books
in Chemistry
JAMES W. COBBLE,
Editor

The Science of Chemistry: Periodic Properties and Chemical Behavior

Reed A. Howald
Montana State University

Walter A. Manch
formerly of Montana State University

The Macmillan Company, New York / Collier-Macmillan Limited, London

Copyright © 1971,
The Macmillan Company

Printed in the
United States of America

All rights reserved. No part of
this book may be reproduced
or transmitted in any form or
by any means, electronic or
mechanical, including photo-
copying, recording, or any information storage and retrieval
system, without permission in
writing from the Publisher.

The Macmillan Company
866 Third Avenue,
New York, New York 10022

Collier-Macmillan Canada, Ltd.,
Toronto, Ontario

Library of Congress
catalog card number: 70-121679

First Printing

Preface

THE TRULY FASCINATING thing about chemistry is not its parts—it is rather the whole. Taken separately, parts such as stoichiometric calculations, the chemistry of vanadium or nylon, pollution control, and the structure of DNA remain simply parts.

A study of the unity of chemistry, of the interdependence of its parts and the relationships among the parts, although often difficult, can provide great insight into the nature of chemistry.

To emphasize the unity of chemistry it is necessary to recognize the major chemical themes. Chemistry is a vast subject, but there are continuous threads running through it. These are the threads that provide the story line. These are the threads that require emphasis in any first-year course. They provide the main themes upon which chemistry is based.

Our goal is to present the major chemical themes early and to expand upon them as the need arises. This means that periodicity, the electronic structure of atoms and molecules, elementary thermodynamics, and chemical bonding must be treated in the introductory chapters. Continued reference is then made to the material in these chapters, and each of these topics is amplified and developed in subsequent chapters. This approach makes the chapters interdependent. It also eliminates the illusion that chemistry consists of a group of independent topics to be treated and then forgotten. We do not feel that the idea of a separate chapter on thermodynamics, a chapter on molecular orbitals, etc., is a fruitful approach.

Periodicity, as emphasized in the subtitle, is one of the major themes of the book. Once introduced in Chapter 2, the concept of electronic structure is used in nearly every chapter. Similarly once thermodynamics is introduced, it is used throughout the text, and the periodicity of thermodynamic properties is apparent in all parts.

Preface

Two themes, electronic structure and thermodynamics, are used to organize much of the descriptive material in the book through properties such as polarizing abilities and bond energies.

The chapters on the molecular orbital method are separated in space to allow for an increasing sophistication with time. We are convinced that students are capable of using the qualitative picture that we have presented, but we feel it is best to spread it throughout the text. Molecular orbitals appear in Chapter 7, and a quantitative molecular orbital picture of the hydrogen halides is introduced in Chapter 11. The theory is amplified in Chapter 12, and the full ligand field treatment is given in Chapter 15.

Some of the standard topics will not be found in this book. The mathematics usually referred to as stoichiometry and molarity problems has been camouflaged. Other traditional topics have been left for the laboratory. In general, we have broken with the historical approach, although considerable effort has been made to emphasize the contributions of individuals. Thus the first chapter concentrates on the work of Mendeleev, and a number of Nobel prize winners are mentioned in Chapter 2.

In one sense this book is a very traditional one. In spite of our emphasis on theory we have included more descriptive material, not less. The most effective way to teach chemical theories is in terms of their application to actual chemical systems. We hope that students will learn from this book that there is a fundamental unity to chemistry, and that this unity is more fascinating than either the theoretical or descriptive parts.

We wish to acknowledge the assistance of our colleagues. Especially valuable comments were received from Dr. John Robbins, Dr. Arnold Craig, Dr. Charles Caughlan, and Dr. Richard Geer, all teachers who used the book in its preliminary edition. Also, a number of talented students have helped in the preparation of the manuscript.

We also wish to acknowledge the assistance of the many students we have had in our classes. Their desire for knowledge is responsible, in part, for the thought and effort that have gone into the development of the material presented here.

Finally we must acknowledge the contribution of our wives. They have worked with us for years with the conviction, but no assurance, that this book would become a reality. Our debt to them cannot really be adequately conveyed.

R. A. H.   W. A. M.

Contents

Chapter 1 Elements, Atoms, and the Periodic Arrangement 1

I. Elements 1
II. Atoms and Atomic Weights 2
III. The Periodic Law 4
    A. Some Consequences of Periodicity, Physical Properties of the Elements 6
        1. Melting Point and Boiling Point 8
        2. Electrical Conductivities 8
        3. Densities 12
        4. Thermal Conductance 12
    B. Compounds and the Periodic Law 15
        1. Formulas of Compounds 15
        2. Physical Properties of Some Simple Compounds of the Elements 15

Chapter 2 The Electronic Structure of Atoms 28

I. The Constituents of Atoms 28
II. The Nature of the Constituents of Atoms 29
    A. The Particle Nature of Electrons 29
    B. The Wave Nature of Electrons 30
    C. Particle versus Wave 35
    D. Some Consequences of Duality 38
III. Two Models of Atomic Structure 38
    A. The Particle Model 39
        1. Atomic Spectra 40
        2. The Stern-Gerlach Experiment 44
    B. The Wave Model 47
IV. The Electronic Structure of Atoms 50
    A. The Solutions to the Schrödinger Equation for Hydrogen 51
    B. The Physical Significance of ψ 53
    C. Wave Functions and Energy Level 56
    D. Electronic Structure and the Periodic Table 63

Contents

Chapter 3 Periodic Properties 67
- I. Ionization Potentials 68
  - A. X-ray Spectroscopy 74
  - B. First Ionization Potentials 77
- II. Atomic and Ionic Radii 79
  - A. Trends in Covalent and Metallic Radii 80
  - B. Ionic Radii 80
- III. Electronegativity 84
- IV. Oxidation States 86
- V. The Octet Rule 91

Chapter 4 The Energetics of Chemical Reactions 99
- I. Heat and Enthalpy 100
- II. Bond Dissociation Energies 107
  - A. Enthalpy Changes in Ionization 109
  - B. Bond Energies and Enthalpy of Formation 111
- III. Entropy 118
- IV. Activities and Equilibrium Constants 128

Chapter 5 Oxidation-Reduction Reactions 136
- I. Electrode Potentials 137
- II. Stable Oxidation States of the Elements 145

Chapter 6 The Alkali Metals and the Nature of Ionic Crystals 153
- I. The Chemistry of the Alkali Metals 153
- II. The Nature of Ionic Solids 157
  - A. The Closest Packing of Spheres 157
  - B. Some Typical Structures 162
    1. The Sodium Chloride Structure 162
    2. The Nickel-Arsenic (Ni-As) Structure 164
    3. The Fluorite Structure 164
    4. The Antifluorite Structure 165
    5. The Rutile Structure 165
    6. The Zinc Blende Structure 165
  - C. The Formation of Ionic Substances 165
  - D. The Lattice Energy 167
    1. The Born-Haber Cycle 170
    2. Variations in Lattice Energy 171

Chapter 7 Electrons in Molecules 176
- I. Bond Type and Electronegativity 177
- II. Ionic Bonding 178
- III. Covalent Bonding 181
  - A. The Molecular Orbital Method 183
    1. Bonding Orbitals 183
    2. Antibonding Molecular Orbitals 185
  - B. Molecular Orbital Diagrams 186

IV. Bond Type and Properties   191
V. Bond Strength and Electronegativity   192

I. The Valence Bond Method   196
II. The Gillespie-Nyholm Method   203
III. Applications   205
   A. Transition Metals and Coordination Compounds   205
   B. Multiple Bonding   207
   C. Bond Angles   209
   D. Exceptions to the Gillespie-Nyholm Method   211

Chapter 8   The Shapes of Molecules   196

I. Gas-Liquid Equilibria: Boiling Points   215
II. Ideal Gases   217
III. Kinetic Theory   222
IV. Boltzmann Constant   226
V. Heat Capacities   230
VI. Real Gases   236
   A. The Critical Point   237
   B. The Exact Treatment of Equilibria   239

Chapter 9   The Energy of Molecules in Gases, Liquids, and Solids   215

I. Isotopes of Hydrogen   247
II. Types of Hydrogen Compounds   248
   A. Saltlike Hydrides or Saline Hydrides   249
   B. The Metallic Hydrides   254
   C. Volatile Hydrides   254
      1. Group IVA Hydrides   259
      2. Group VA Hydrides   260
      3. Properties of Group VIA Hydrides   264
      4. Properties of the Group VIIA Hydrides   264
III. Hydrogen Bonding   266

Chapter 10   Hydrogen   247

I. The Relative Strength of Acids   272
   A. Conjugate Acid-Base Pairs   272
   B. Simple Hydrides   273
   C. The Oxyacids   274
   D. Molecular Orbitals and the Acidity of Simple Hydrogen Compounds   280
II. Acid-Base Equilibria   283
   A. The Relative Strength of Acids Revisited   286
   B. Equilibrium Calculations   286
   C. Buffers   292

Chapter 11   Acids and Bases   271

I. Introduction to Metals   295
II. The Band Theory of Bonding in Solids   308
III. Metals and Nonmetals   312

Chapter 12   Molecular Orbitals   295

Contents

Chapter 13 Polarizing Ability and Solubility 318

I. The Alkaline Earth Metals 318
II. The Solubility of Inorganic Salts 320
   A. Polarizing Ability of Cations and Ease of Polarization of Anions 320
   B. Measurement of Solubility 328
   C. The Solution Process 332
      1. The Solubility of Group IA Salts 335
      2. Solubility of Group IIA Halides 339
      3. The Solubility of Transition Metal Halides 340
      4. Solubility of Hydroxides 342
      5. Solubility of Sulfides 343
      6. The Solubility of Salts of Complex Anions 343
   D. Solubility Equilibrium 344

Chapter 14 Coordination Chemistry 351

I. Types of Coordination Compounds 353
II. Isomerism in Coordination Compounds 360
III. Nomenclature 361
IV. Coordinating Tendencies 362
   A. Complexes of Various Cations 364
   B. An Interpretation of Coordinating Tendencies with Simple Ligands 365
   C. Polarizing Ability and Coordination Number 367

Chapter 15 The Ligand Field 374

I. Introduction to Ligand Fields 374
   A. Molecular Orbitals for Beryllium Hydride (BeH_2) 376
   B. Molecular Orbitals for Boron Trifluoride (BF_3) 378
   C. Molecular Orbitals in Methane (CH_4) 383
   D. Degenerate States in Molecules 383
   E. Molecular Orbitals for Ammonia (NH_3) and Water (H_2O) 385
   F. Complexes Involving d Orbitals 388
II. The Strength of the Ligand Field 398

Chapter 16 Physical Methods in Chemistry 401

I. The Structure and Spectra of Transition Metal Complexes 401
II. The Application of Magnetic Susceptibility to Structure 409
III. The Application of Infrared Spectroscopy to Structure 410
IV. Dipole Moments and Structure 413
V. Mass Spectrometry 415
VI. Nuclear Magnetic Resonance 417

Chapter 17 The Common Metals 422

I. The Metallurgy of Copper 423
   A. Phase Diagrams 423
   B. The Phase Rule 424
   C. The Preparation of Pure Copper 426
   D. Copper Alloys 427
II. The Metallurgy of Iron 428

      A. Wrought Iron   428
      B. Steel   429
      C. Cast Iron and the Blast Furnace   430
      D. The Production of Steel   430
      E. Alloy Steels   431
III. More Active Metals   431

Chapter 18   Aluminum, Nickel, and Related Metals   436

  I. Chemical Properties of the Entire Group   437
      A. The Aluminum Subgroup   437
        1. Aluminum   438
        2. Beryllium   438
        3. Gallium   439
        4. Chromium   439
        5. Zinc   439
      B. The Nickel Subgroup   441
        1. The IIIB Family of Elements   441
          a. Occurrence and Discovery   441
          b. Chemistry of the +3 State   442
          c. Other Oxidation States   443
        2. Uranium and Thorium   444
        3. Families IVB and VB   444
        4. The 3*d* Transition Series   445
          a. Vanadium   446
          b. Manganese   447
          c. Iron   448
          d. Nickel   449

Chapter 19   The Structure of Nuclei   454

   I. The Formation of Atoms   454
  II. The Stability of Nuclei   460
      A. Energy Levels in Nuclei   460
      B. Beta Decay   462
 III. Rates of Radioactive Decay   465
 IV. Energy Requirements for Human Life   467

Chapter 20   Chemical Reactions   474

   I. Rates of Reaction and the Rate Law   474
  II. Reaction Mechanisms and the Rate Law   476
 III. The Theory of Activated Complexes   478
      A. The Activated Complex   478
      B. Entropy and Enthalpy of Activation   480
 IV. Complex Reaction Mechanisms   485
      A. Catalysis   485
      B. Successive Activated Complexes   487
      C. Dissociation of Diatomic Molecules and Recombination of Atoms   490
  V. $S_N 2$ Substitution   491
 VI. Heterogeneous Reactions   494

Contents

Chapter 21 The Robust Complexes 500

  I. The Rate of Ligand Exchange Reactions 500
     A. Robust and Labile Complexes 500
     B. The Effects of Size, Charge, and Electronic Structure on the Rate of Ligand Exchange 502
     C. The Amphoterism of Chromium (III) 505
  II. Mechanisms of Redox Reactions 507
     A. Higher Oxidation States of Chromium 509
     B. Cobalt 512

Chapter 22 The Chemistry of Nitrogen 516

  I. Elemental Nitrogen 516
  II. Compounds of Nitrogen 517
     A. Ionic Compounds of Nitrogen 517
     B. Covalent Nitrogen Compounds 519
        1. Ammonia 519
        2. Amino Acids and Proteins 520
        3. Oxides of Nitrogen 522
  III. Chemical Reactions 525
     A. Reactions of Nitrogen Oxides 525
     B. Nitration and Related Reactions 529

Chapter 23 The Chemistry of Phosphorus 533

  I. Elemental Phosphorus 533
  II. Compounds of Phosphorus 535
     A. Phosphides 535
     B. Phosphine 536
     C. Bond Energies in Phosphorus Compounds 537
     D. Uses of Phosphorus 544
     E. Phosphate Buffer Solutions 545

Chapter 24 Heavy Metals 550

  I. Arsenic 550
  II. Antimony 552
  III. Tin 554
  IV. Lead and Bismuth 555
     A. Acid-Base Behavior 556
     B. Colors and Charge Transfer Spectra 557
  V. A Measure of the Induced Polarizing Ability of a Cation 559
  VI. The Chemistry of the IB Cations 562
     A. Copper 562
     B. Silver 563
     C. Gold 565
  VII. Zinc, Cadmium, Mercury, and Thallium 566
  VIII. The Platinum Metals 567

Chapter 25 Nonmetals 572

  I. Elemental Sulfur 572
  II. Compounds and Reactions of Sulfur 575
     A. Sulfur (VI) 575
     B. Redox Reactions of Sulfur 576
     C. Sulfur Halides 581
  III. Chlorine Compounds 582

   IV. Heavier Elements in Groups VIA and VIIA 583
       A. Selenium 583
       B. Tellurium 584
       C. Iodine and Bromine 585
   V. The Electrostatic Valence Rule 586
  VI. Boron and Its Compounds 587
 VII. Silicon and Its Compounds 590
       A. Elemental Silicon 590
       B. Silanes 590
       C. Silicones 590
       D. Silicates 591
          1. The Structure of Orthosilicates 591
          2. The Structure of Metasilicates 592
          3. Amphiboles 593
          4. Layer Silicates 594

Chapter 26 Organic Chemistry 601

    I. Hydrocarbons 601
       A. Alkanes 602
       B. Cycloalkanes and Alkenes 604
   II. Reactions of C=C and C=O Bonds 607
       A. Polymerizations 607
       B. Hydrogenation 608
       C. Acid Catalyzed Additions 608
       D. Preparation of Aldehydes and Ketones 611
       E. Reactions of Aldehydes 612
  III. Optical Isomerism 614
  IV. Acids and Esters 615
   V. Organic Nitrogen Compounds 619
       A. The Carbon-Nitrogen Triple Bond 621
       B. Heterocyclic Compounds 624
       C. Dyes 625

Chapter 27 Biochemistry 632

    I. Iatrochemistry 632
   II. Proteins and Enzymes 634
  III. Adenosine Triphosphate and Glucose 635
       A. Glucose Formation and Oxidation 637
       B. Anaerobic Glycolysis–Fermentation 639
  IV. Deoxyribonucleic Acid (DNA) and Ribonucleic Acid (RNA) 641

Appendix I Wave Functions for the Hydrogen Atom 649

Appendix II Symbols and Abbreviations 650

Appendix III Fundamental Constants 652

Appendix IV Thermodynamic Properties of Selected Materials 653

Appendix V Nomenclature in Organic Chemistry 668

Appendix VI Answers to Problems II 675

Index 683

> For a true comprehension of the matter it is very important to see that all aspects of the distribution of the elements according to their atomic weights essentially expresses one and the same fundamental dependence—periodic properties.*
>
> DIMITRI IVANOVITCH MENDELEEV

Chapter 1 Elements, Atoms, and the Periodic Arrangement

I. Elements

The idea of an element has not changed since the time of Hippocrates and Aristotle. To the early Greeks all things were mixtures of elements that could not be reduced to simpler materials. The elements of physics (fire, water, air, earth) as defined by Aristotle, the elements of man (blood, phlegm, yellow bile, and black bile) as defined by Hippocrates, and the elements of man's nature as described by Galen (sanguine, phlegmatic, choleric, and melancholic) went into making the whole of matter and man.

Although the concept of an element has not changed the elements have, of course, been redefined according to the chemical techniques available at any specific time. If a substance could not be decomposed into simpler substances by the techniques of the day, it was considered an element. Robert Boyle, the man usually credited with the modern definition of an element, and Newton both believed that gold could be made from other elements. Lavoisier's list of 33 elements included oxygen and nitrogen (the supposed element air) as well as lime and

The meaning of the term element should become clear in the course of this chapter. The dictionary definition is "A substance which cannot be separated into substances different from itself by ordinary chemical means; one of a limited number of distinct varieties of matter which, singly or in combination, compose every material substance." Webster's Collegiate Dictionary (Springfield, Mass.: Merriam, 1948).

---

*The quotations from Mendeleev in this chapter are found in G. Holton and D. H. D. Roller, *Foundations of Modern Physical Science* (Reading, Mass.: Addison Wesley, 1958), and *Gateway to the Great Books*. Copyright, 1963, by Encyclopedia Britannica Inc.

magnesia. Lime was later separated into oxygen and a new element, calcium. Magnesia was found to be made up of oxygen and magnesium by Sir Humphrey Davy. Chlorine (from the Greek word for green), prepared by the reaction of hydrogen chloride (HCl) and oxygen (O_2) by Karl Wilhelm Scheele, was shown to be an element and was named by Davy. The list of elements accumulated through the centuries following Boyle until 65 were known by the time of Mendeleev (1869).

The fact that new elements were steadily being discovered presented a great challenge. A number of people believed that there should be some order, some fundamental relationship between the elements, that would allow them to be classified according to types or groups. Dimitri Ivanovitch Mendeleev provided the first clear and consistent classification.

With the additional chemical insight of the last 100 years we have introduced some small modifications into Mendeleev's classification, but it remains the fundamental basis for our understanding of the chemical and physical behavior of the elements.

II. Atoms and Atomic Weights

Mendeleev's classification was based primarily on the relative atomic weights of atoms and the proportions in which atoms combine. His work was directly dependent upon that of John Dalton who first made the idea of atoms, and particularly the idea that atoms had definite weights and combined in definite proportions, scientifically respectable. Dalton proposed that each element was composed of a particular kind of atom and that all atoms of a particular element were "perfectly alike in weight, figure, etc." but different from the atoms of any other element.

Like the idea of elements, the idea of atoms was proposed by the early Greeks, particularly Democritus in the fifth century B.C. The idea arose because of the belief that matter cannot be forever divisible. As obvious as this might seem at present, it is still amazing that such a deduction could be arrived at intuitively. The major objection voiced at that time concerned the influence that atoms must have upon one another across the void that is inherent in an atomic theory. This is the aspect of atomism that Plato and Aristotle emphasized and were unable to accept.

The same objection arises again much later with regard to Newton's idea of action at a distance. Leibniz objected to Newton's theory of gravitation on the grounds that the force was supposed to act directly across vast distances of space, and such a notion "must be either

Gottfried Wilhelm von Leibniz was a German scientist and mathematician contemporary with Newton. His name is often spelled Leibnitz.

magical or imaginary." So Aristotle and Plato argued against the voids in Democritus' atomic theory of matter and postulated that one must search for an explanation of matter in terms of inner "tension," "tones," "harmonies," and "patterns." Thus the two concepts of matter, as made up of aggregates of individual atoms or certain patterns maintained by inner "tensions" and "harmonies," began as rivals.

In reestablishing the atomic theory Dalton did not even try to answer these objections. He simply showed that, if one assumed that there were atoms of definite weights that combined in simple ratios, a large body of experimental data could be explained. With Dalton's theory, data on the combining ratios of elements in compounds could be interpreted.

Here we see a parallel between the theory of gravitation and the atomic theory. Granted his assumptions about a gravitational force, Newton was able to explain the observed geometrical forms of the planetary orbits. This was sufficient to establish the gravitational theory. It stands today, independent of all the models constructed around it. Similarly, Dalton established the atomic theory by showing how it explained the combining ratios of atoms. However, Dalton and even Democritus did not hesitate to propose more detailed models of atoms to explain other kinds of experimental data. It was tempting to say that atoms of liquids were smooth, in order to roll over one another freely, or that the atoms of acids combined with sharp points sticking out to prick the tongue. Such ideas strike us now as merely quaint, since they have been left so far back in the continuing development of better and more inclusive models. We have a problem, though, in making our modern explanations as vivid and as real as these.

From the modern viewpoint, there are two very basic errors in Dalton's theory. Atoms are *not* "perfectly alike in weight, figure, etc.," because isotopes of elements exist. In addition, atoms are not indestructible, and atomic bombs exist. (To be fair, however, Dalton did specify that "no new creation or destruction of matter is within the reach of *Chemical Agency*".) Hence, a distinction between chemical reactions or processes and nuclear reactions is necessary.

The existence of isotopes is now recognized and, although atomic weights are useful and important, they now refer to the average of the weights of the different isotopes of an element in the proportions that they are found in nature. The atomic weight of calcium (Ca) is 40.08 (the result of different amounts of stable isotopes with atomic weights of 40, 42, 43, 44, 46, and 48). The atomic weight of oxygen (O) is 15.9994 (the result of different amounts of stable isotopes with atomic weights of 16, 17, and 18) with the lightest of these ^{16}O having a weight of only 15.9949 on the atomic weight scale.

If a certain compound contains twice as many atoms of oxygen as

Atoms with the same nuclear charge which differ somewhat in atomic weight are not considered to be distinct elements but isotopes of a single element. The isotopes of an element differ in the number of neutrons in the nucleus. The word isotope comes from the Greek word meaning the same place, indicating that isotopes are assigned the same position in the periodic chart.

The unit on the atomic weight scale is $\frac{1}{12}$ of the weight of an atom of carbon-12.

Since the atomic weights of all individual isotopes are close to whole numbers, it is convenient to refer to isotopes by the integer closest to the atomic weight.

Table 1-1 *Analyses of Calcium-Phosphorus-Oxygen Compounds*

Formula	Calcium, %		Phosphorus, %		Oxygen, %	
	Calc.	Exptl.	Calc.	Exptl.	Calc.	Exptl.
CaO	71.47	72.23	—	—	28.53	28.77*
CaO_2	55.61	54.45	—	—	44.39	45.18
Ca_3P_2	66.00	66.18	34.00	33.82*	—	—
P_2O_3	—	—	56.34	56.47	43.66	43.53*
P_2O_5	—	—	43.64	43.59	56.36	56.41*
CaP_2O_6	20.24	20.29	31.28	31.08	48.48	48.63*
$Ca_2P_2O_7$	31.55	31.92	24.38	24.02	44.07	44.1*
$Ca_3P_2O_8$	38.76	38.4	19.97	20.2	41.27	41.4*

*Calculated by difference from analyses for the other elements.

calcium, the formula may be written CaO_2. Nevertheless, the compound is more than 50% calcium by weight because calcium atoms are heavier than oxygen atoms. The relative amounts of calcium and oxygen that combine can be expressed in any units: 40.08 tons of calcium are combined with 32.00 tons of oxygen (2×15.9994 rounded off), or 40.08 grams (g) of calcium are combined with 32.00 g of oxygen. In any case the percentage of calcium is $(100 \times 40.08)/(40.08 + 32.00) = 55.6\%$. Given also the atomic weight of phosphorus (P), the percent composition of P_2O_5 (43.6% P) and CaP_2O_6 (20.2% Ca, 31.3% P) can be calculated in a similar manner. In Table 1-1, the observed composition of compounds of these three elements are listed in addition to the formulas that correspond to the observations. The observed values given in this table are taken from the chemical literature. Note that there is considerable variation in the accuracy with which the various analyses were performed. Exceptionally pure materials and extreme care in the analysis is necessary to obtain agreement to four significant figures.

III. The Periodic Law

By the time of Mendeleev the atomic weight of about 50 of the 65 known elements had been determined with reasonable accuracy. The combining ratios of the elements and certain properties of the compounds were also known. Although Mendeleev could not understand the reason for the similarities and differences in the observed properties of elements and compounds, he did have the conviction that there must be a system that could be used to classify the elements and he set out to find it. He followed the largely ignored work of Newland (the octaves of Newland) and of deChancourtois (the telluric screw) which, according to Mendeleev, "merely wanted the boldness necessary to place the whole question at such a height that its reflection on the facts could be purely seen." Mendeleev had formulated the periodic law by 1870, using the atomic weights and the observed similarity between certain elements as a basis. "The researches of Marignac on niobium, and those of Roscoe on vanadium were of special moment. The striking analogies between vanadium and phosphorus on the one hand and between vanadium and chromium on the other, which became so apparent in the investigation connected with that element, naturally induced the comparison of $V = 51$ with $Cr = 52$, $Nb = 94$ with $Mo = 96$, and $Ta = 192$ with $W = 194$, while on the other hand $P = 31$ could be compared with $S = 32$, $As = 75$ with $Se = 79$ and $Sb = 120$ with $Te = 125$. From such approximations there remained but one step to the discovery of the law of periodicity."

This step was taken by Mendeleev. "The law of periodicity was thus a direct outcome of the stock of generalizations and established facts which had accumulated by the end of the decade 1860–1870. It is the embodiment of these data in a more or less systematic expression."

"The first step I took in this direction was the following: I selected the bodies with the smallest atomic weight and ordered them according to the magnitude of their atomic weights. Thereby it appeared that there exists a periodicity of properties and that even according to valence, one element follows the other in order of arithmetical sequence."

Valence is the term used in Mendeleev's time to describe the combining ratios of atoms.

Mendeleev's version of the law was preferable to those that had come before him because he was able to predict that elements of a certain atomic weight would be discovered. He recognized that titanium should appear under carbon and silicon and not in the third column under boron where it had been placed by others because it was the next highest atomic weight element then known after calcium. To complete his scheme he predicted the existence of an unsuspected element which he called ekaboron (Eb), having an atomic weight between calcium (40) and titanium (50), and a formula for the oxide, Eb_2O_3. He also predicted the existence of, and some of the properties of, ekaaluminum, ekasilicon, ekamanganese, ekatantalum, and dvimanganese. For example, we quote again from Mendeleev's summary of his own predictions in 1871.

"The element IV, 5 follows after IV, 3 ... the element silicon. I named this element ekasilicon and its symbol Es. The following are the properties which this element should have on the basis of the known properties of silicon, tin, zinc and arsenic. Its atomic weight is nearly 72, it forms a higher oxide EsO_2, a lower oxide EsO, compounds of the general form EsX_4 and chemically unstable lower compounds of EsX_2. Es gives volatile organometallic compounds; for instance $Es(CH_3)_4$, $Es(CH_3)_3Cl$ and $Es(C_2H_5)_4$ which boil at about 160°C...; also a volatile liquid chloride $EsCl_4$, boiling at about 90° and a density of 1.9...; the density of Es will be about 5.5 and EsO_2 will have a density of about 4.7...." These and similarly striking predictions for ekaboron and ekaaluminum were verified during Mendeleev's lifetime. The three elements are now called scandium (ekaboron), gallium (ekaaluminum), and germanium (ekasilicon). Rhenium (ekatantalum) was not discovered until 1925; technetium (ekamanganese) was the first element to be prepared by artificial means (in 1937), and promethium (dvimanganese) was not discovered until 1947 at an Oak Ridge research site. Neither technetium nor promethium occurs naturally.

Thus, granted periodicity Mendeleev could explain properties and make predictions. He succeeded because he recognized periodicity as the prime factor in the arrangement of elements and was willing to violate the initial assumption of arrangement by atomic weight in order to achieve the symmetrical classification provided by periodicity. There was, however, no hint of the why of periodicity. Sometime later Mendeleev added, "just as without knowing the cause of gravitation it is possible to make use of the law of gravitation, so it is possible to take advantage of the laws discovered by chemistry without being able to explain their causes;" and "above all I was interested to find out a general system of the elements."

A. SOME CONSEQUENCES OF PERIODICITY, PHYSICAL PROPERTIES OF THE ELEMENTS

The periodicity of which Mendeleev speaks is a feature which is nicely illustrated by a number of properties besides that of combining ratios. The regularity in atomic volume (the ratio of atomic weight to density) of the elements was first pointed out by Lothar Meyer only a short time after Mendeleev. Figure 1-1 is indeed a striking illustration of the periodic nature of the elements.

A less regular but still detectable periodicity is apparent in a number of other physical properties of the elements. These will be examined here, but before this is done it is desirable to classify the elements according to the modern form of the periodic table. The most convenient table, the so-called long form, is divided into A and B groups. The A elements are called main group or representative elements, whereas the B elements are called subgroup or transition elements. The reasons for choosing this classification will be examined later. Similarly an understanding of the trends observed in the properties to be discussed must be delayed until a later chapter. A glance at some of the observed trends indicates that some detailed knowledge of

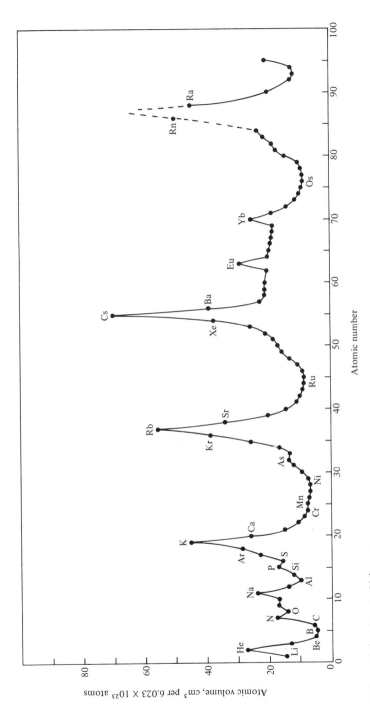

Figure 1-1 *Atomic Volumes.*

chemical bonding will be necessary to explain some of the features shown in the tables.

1. Melting Point and Boiling Point

The melting points and boiling points of the elements are shown in Tables 1-2 and 1-3. The periodicity is apparent but quite irregular. Low boiling points are found at both sides of the table and for one particular subgroup, group IIB.

Examination of the data shows that the only families with completely regular trends in the melting and boiling points are group VIIA and the inert gases. Irregularities occur in all of the other main groups.

Irregularities in the melting point (which represent a minimum) occur at magnesium (Mg), gallium (Ga), and tin (Sn). Other irregularities occur at antimony (Sb), bismuth (Bi), and polonium (Po). Of all the elements gallium has the largest liquid range, over 2200°C.

The trends in both melting point and boiling point for many of the transition metal groups (the B subgroups) are regular and straightforward. The boiling points and melting points increase with increasing atomic weight.

In general, across each row of representative elements (the A groups), there is an irregular rise in both melting point and boiling point up to a maximum in group IV or V, followed by a sharp decrease to the inert gases. In the first transition series [the elements between scandium (Sc) and zinc (Zn) inclusively] there is a steady rise in melting point up to vanadium (V), after which there is a general drop off toward zinc with manganese (Mn) appearing out of order. The elements between yttrium (Y) and cadmium (Cd) (the second transition series) show a similar pattern, with a maximum at molybdenum (Mo). There is a general decrease again to indium (In), with a minimum occurring at technetium (Tc). In the third transition series the maximum again occurs at group VIB. Note that mercury (Hg) has the lowest melting point of all the metals.

The terms family and group or subgroup are used almost interchangeably to describe vertical columns of the periodic table.

2. Electrical Conductivities

The distinction between most metals and nonmetals is immediately obvious from the electrical conductivities given in Table 1-4. In addition, the high values of the elements of group IB are immediately apparent. Thus, copper wire is our most common electrical conductor.

The exact division of elements into metals and nonmetals is quite arbitrary, since there are gradual changes in properties from the left to the right side of the periodic table. The conventional dividing line is shown as a heavy line in Table 1-4. Electrical conductance *alone* cannot be used to distinguish between metals and nonmetals, since

Sec. III The Periodic Law

Table 1-2 *Melting Points of the Elements, (°C)*

IA	IIA	IIIB	IVB	VB	VIB	VIIB	VIII			IB	IIB	IIIA	IVA	VA	VIA	VIIA	Inert Gases
H																	He
181 Li	1283 Be											2000 B	3727 C	−210 N	−219 O	−219 F	−249 Ne
98 Na	650 **Mg**											660 Al	1410 Si	44 P	119 S	−101 Cl	−189 Ar
64 K	850 Ca	1539 **Sc**	1668 Ti	1890 V	1875 Cr	1244 **Mn**	1535 Fe	1490 Co	1452 Ni	1084 Cu	420 **Zn**	30 **Ga**	936 Ge	817 As	217 Se	−7 Br	−157 Kr
39 Rb	770 Sr	1509 **Y**	1852 Zr	2468 Nb	2610 **Mo**	2140 Tc	2250 Ru	1960 Rh	1550 Pd	961 Ag	321 **Cd**	156 **In**	232 **Sn**	631 **Sb**	450 Te	114 I	−118 Xe
29 Cs	710 Ba	920 La	2230 Hf	2998 Ta	3410 W	3180 Re	3000 Os	2410 Ir	1769 Pt	1063 Au	−39 **Hg**	304 Tl	327 Pb	271 **Bi**	252 **Po**	At	−71 Rn

Ch. 1 Elements, Atoms, and the Periodic Arrangement

Table 1-3 *Boiling Points of the Elements (°C)*

IA	IIA	IIIB	IVB	VB	VIB	VIIB	VIII			IB	IIB	IIIA	IVA	VA	VIA	VIIA	Inert Gases
−253 H																	−269 He
1317 Li	2507 Be											2550 B	4200 C	−196 N	−183 O	−188 F	−246 Ne
883 Na	1110 Mg											2467 Al	2480 Si	280 P	445 S	−35 Cl	−186 Ar
760 K	1482 Ca	2727 Sc	3260 Ti	3000 V	2199 Cr	2097 Mn	3000 Fe	3100 Co	2700 Ni	2595 Cu	906 Zn	2403 **Ga**	2700 Ge	610 As	688 Se	58 Br	−153 Kr
688 Rb	1380 Sr	3200 Y	3580 Zr	4930 Nb	5560 Mo	Tc	4900 Ru	4500 Rh	3980 Pd	2212 Ag	765 Cd	2075 In	2270 Sn	1440 Sb	990 Te	185 I	−108 Xe
705 Cs	1500 Ba	3469 La	5400 Hf	6100 **Ta**	5900 W	5900 Re	5500 Os	5300 Ir	4530 Pt	2966 Au	357 Hg	1457 Tl	1737 Pb	1627 Bi	962 Po	(380) At	−62 Rn

Sec. III The Periodic Law

Table 1-4 *Electrical Conductivity* ($\mu ohm^{-1} cm^{-1}$)*

IA	IIA	IIIB	IVB	VB	VIB	VIIB	VIII			IB	IIB	IIIA	IVA	VA	VIA	VIIA	Inert Gases
0.108 H																	He
0.108 Li	0.25 Be											6×10^{-13} B	7×10^{-4} C	N	O	F	Ne
0.218 Na	0.224 Mg											0.382 Al	4×10^{-12} Si	10^{-17} P	10^{-23} S	Cl	Ar
0.143 K	0.218 Ca	0.016 Sc	0.024 Ti	0.04 V	0.078 Cr	0.054 Mn	0.103 Fe	0.160 Co	0.146 Ni	0.598 Cu	0.169 Zn	0.019 Ga	2.2×10^{-4} Ge	0.030 As	0.08 Se	10^{-18} Br	Kr
0.080 Rb	0.043 Sr	0.017 Y	0.025 Zr	0.08 Nb	0.18 Mo	Tc	0.13 Ru	0.22 Rh	0.093 Pd	0.63 Ag	0.146 Cd	0.12 In	0.091 0.003 Sn	0.026 Sb	2.3×10^{-6} Te	10^{-15} I	Xe
0.053 Cs	0.016 Ba	0.017 La	0.028 Hf	0.080 Ta	0.18 W	0.025 Re	0.11 Os	0.189 Ir	0.095 Pt	0.425 Au	0.011 Hg	0.055 Tl	0.046 Pb	0.0093 Bi	0.02 Po	At	Rn

*Most of the values are from the *Metals Handbook*, (Metals Park, Ohio: American Society for Metals, 1961).

one form of arsenic (As) is about as good a conductor as some of the transition elements.

Quantitative comparisons in conductivity are very difficult because this property is extremely dependent on the presence of traces of impurities. The conductivity of silicon (Si) has been variously reported in the range 10^{-10} to 10^{-11} μohm^{-1} cm^{-1}, but the purest silicon prepared (semiconductor grade) shows a value of only 4.4×10^{-12}. The common impurities in elemental titanium (Ti) have the opposite effect, and very pure titanium has a conductivity of 0.024 μohm^{-1} cm^{-1}, which is considerably higher than that of the commercially pure material, 0.021. Commercial titanium alloys have conductivities as low as 0.007. In compiling Table 1-4 we have tried to use values for the purest material available, but this means that the reported values are for samples of widely different purity.

An electrical conductor with a very high melting point is needed for the filament in an incandescent lamp. Referring back to Table 1-2 you should be led to consider tungsten (W). Edison would have been helped considerably in his search by a greater reliance on the periodic table.

3. Densities

The trends in density (Table 1-5) are much less complicated than the other properties listed thus far. The trend within most groups (groups IA and IIA are the only exceptions) is regular, and the density increases with atomic weight.

Across a row of representative elements the maximum density is observed for the group IIIA elements. Across each transition series there is a general increase followed by a decrease. In the third transition series the density reaches a maximum earlier than in the first and second. In addition there is a much larger difference between the second and third transition series than between the first and second.

4. Thermal Conductance

Thermal conductances for the elements are given in Table 1-6. The extremely low values for the elements in groups VIA, VIIA, and VIIIA and the high conductances of group IB metals are again most obvious. The other elements of the transition series are relatively poor conductors. Tungsten is one of the best.

By this time you should have a feeling for the kinds of relations that appear. It is clear that, in predicting properties of an element, knowledge of the corresponding data for other elements in the same family and also for other elements in the same row is essential. Thus, Mendeleev relied heavily on data for silicon, tin, zinc, and arsenic in making his predictions for germanium.

Table 1-5 *Densities of the Elements* ($g\ ml^{-1}$)

IA	IIA	IIIB	IVB	VB	VIB	VIIB	VIII			IB	IIB	IIIA	IVA	VA	VIA	VIIA	Inert Gases
H																	He
0.53 Li	1.85 Be											2.34 B	2.26 C	0.81 N	1.14 O	1.11 F	1.20 Ne
0.97 Na	1.74 Mg											2.70 Al	2.33 Si	1.82 P	2.07 S	1.56 Cl	1.40 Ar
0.86 K	1.55 Ca	3.0 Sc	4.51 Ti	6.1 V	7.19 Cr	7.43 Mn	7.86 Fe	8.9 Co	8.9 Ni	8.96 Cu	7.14 Zn	5.91 Ga	5.32 Ge	5.72 As	4.79 Se	3.12 Br	2.6 Kr
1.53 Rb	2.6 Sr	4.47 Y	6.49 Zr	8.4 Nb	10.2 Mo	11.5 Tc	12.2 Ru	12.4 Rh	12.0 Pd	10.5 Ag	8.65 Cd	7.31 In	7.30 Sn	6.62 Sb	6.24 Te	4.94 I	3.06 Xe
1.90 Cs	3.5 Ba	6.17 La	13.1 Hf	16.6 Ta	19.3 W	21.0 Re	22.6 Os	22.5 Ir	21.4 Pt	19.3 Au	13.6 Hg	11.85 Tl	11.4 Pb	9.8 Bi	(9.2) Po	At	Rn

Ch. 1 Elements, Atoms, and the Periodic Arrangement

Table 1-6 *Thermal Conductance* [Cal. (cm)$^{-2}$ (cm/°C) (sec.)$^{-1}$]

IA	IIA	IIIB	IVB	VB	VIB	VIIB	VIII			IB	IIB	IIIA	IVA	VA	VIA	VIIA	Inert Gases
4×10^{-4} H																	3×10^{-4} He
.17 Li	.38 Be											B	.057 C	6×10^{-5} N	6×10^{-4} O	F	10^{-4} Ne
.32 Na	.38 Mg											.50 Al	.20 Si	P	7×10^{-4} S	2×10^{-5} Cl	4×10^{-5} Ar
.23 K	.3 Ca	.015 Sc	Ti	V	.16 Cr	Mn	.18 Fe	.16 Co	.22 Ni	.94 Cu	.27 Zn	Ga	.14 Ge	As	10^{-5} Se	Br	2×10^{-5} Kr
Rb	Sr	.035 Y	Zr	.125 Nb	.35 Mo	Tc	Ru	.21 Rh	.17 Pd	.98 Ag	.22 Cd	.057 In	.16 Sn	.05 Sb	.014 Te	10^{-3} I	10^{-4} Xe
Cs	Ba	.033 La	.22 Hf	.13 Ta	.40 W	.17 Re	Os	.14 Ir	.17 Pt	.71 Au	.02 Hg	.093 Tl	.083 Pb	Bi	Po	At	Rn

A detailed examination of the values of any property would reveal trends similar to those examined here. Irregularities in the trends would also occur. The irregularities are often important clues to the understanding of the factors that influence a property.

B. COMPOUNDS AND THE PERIODIC LAW

1. Formulas of Compounds

Oxygen combines with nearly all of the elements. The composition of the oxide formed is often determined by the group number to which the element belongs. Similarly fluorine (F) and hydrogen (H) form fluorides and hydrides with simple ratios that are dependent upon the group number. For those cases where the group number is large (V, VI, VII), the combining ratios are not limited to a single oxide or fluoride, but in general a series of compounds may be formulated. The formulas of some oxides, fluorides and hydrides are listed in Tables 1-7a, b, 1-8a, b, and 1-9.

Many elements form several compounds with oxygen. The one with the greatest number of oxygen atoms per atom of the element concerned is the higher oxide.

There are definite limits placed upon the number and type of compounds that are normally stable. Although almost any combination is obtainable under special circumstances (for example, species such as N_2^+, O_2^+, O_2^-, and O have been detected in the upper atmosphere and prepared in the laboratory; CN has been detected in the coma surrounding the nucleus of comets, and chemists have studied the whole series CH, CH_2, and CH_3 as well as CH_4) only a few of the possible combinations are stable.

We will return to the question of predicting the formulas of stable molecules in Chapter 3.

2. Physical Properties of Some Simple Compounds of the Elements

In Tables 1-10 through 1-14 the periodicity of physical properties of some oxides, fluorides, and hydrides is illustrated. Transition metal compounds that were not included in the earlier tables have been included in these data where they are appropriate. Again it is clear that the periodic arrangement provides some obvious generalizations of the experimental data, but there is no need to go into the details at present. The important point is that essentially all properties of elements and compounds reflect periodicity in some way.

Periodicity is a fundamental concept of modern science. The arrangement of atoms in periodic fashion functions as a very useful framework for chemical knowledge and, as such, it emphasizes the fundamental unity of the subject. In 1970, 105 elements are known, one of which is Mendeleevium (101). The basis of periodicity in terms of the electronic structure of atoms is now also known.

Ch. 1 Elements, Atoms, and the Periodic Arrangement

The terms higher and lower used here refer to the relative numbers of oxygen or fluorine atoms, and so forth, combined with one atom of the other element.

Table 1-7a *Higher Oxides*

IA	IIA	IIIB	IVB	VB	VIB	VIIB	VIII			IB	IIB	IIIA	IVA	VA	VIA	VIIA	Inert Gases
												B_2O_3	CO_2	N_2O_5			
Li_2O	BeO											Al_2O_3	SiO_2	P_4O_{10}	SO_3	Cl_2O_7	
Na_2O	MgO											Ga_2O_3	GeO_2	As_2O_5	SeO_3	Br_2O_7	
K_2O	CaO											In_2O_3	SnO_2	Sb_2O_5	TeO_3		
Rb_2O	SrO											Tl_2O_3	PbO_2	Bi_2O_5			
Cs_2O	BaO																

Sec. III The Periodic Law

Table 1-7b *Lower Oxides*

IA	IIA	IIIB	IVB	VB	VIB	VIIB	VIII			IB	IIB	IIIA	IVA	VA	VIA	VIIA	Inert Gases
													CO	N$_2$O			
													SiO	P$_4$O$_6$	SO$_2$ SO	Cl$_2$O	
													GeO	As$_2$O$_3$	SeO$_2$	Br$_2$O	
													SnO	Sb$_2$O$_3$	TeO$_2$ TeO	I$_2$O$_5$	
													PbO	Bi$_2$O$_3$			

Ch. 1 Elements, Atoms, and the Periodic Arrangement

Table 1-8a *Higher Fluorides*

IA	IIA	IIIB	IVB	VB	VIB	VIIB	VIII			IB	IIB	IIIA	IVA	VA	VIA	VIIA	Inert Gases
LiF	BeF$_2$											BF$_3$	CF$_4$	NF$_3$	OF$_2$		
NaF	MgF$_2$											AlF$_3$	SiF$_4$	PF$_5$	SF$_6$		
KF	CaF$_2$											GaF$_3$	GeF$_4$	AsF$_5$	SeF$_6$		
RbF	SrF$_2$											InF$_3$	SnF$_4$	SbF$_5$	TeF$_6$	IF$_7$	
CsF	BaF$_2$											TlF$_3$	PbF$_4$				

18

Table 1-8b *Lower Fluorides*

IA	IIA	IIIB	IVB	VB	VIB	VIIB	VIII			IB	IIB	IIIA	IVA	VA	VIA	VIIA	Inert Gases
														NF_3	OF_2		
														PF_3	SF_4 SF_2	ClF_3 ClF	
														AsF_3	SeF_4 SeF_2	BrF_5 BrF_3 BrF	
													SnF_2	SbF_3	TeF_4 TeF_2	IF_5 IF_3 IF	
													PbF_2				

Ch. 1 Elements, Atoms, and the Periodic Arrangement

Table 1-9 *Hydrides*

IA	IIA	IIIB	IVB	VB	VIB	VIIB	VIII			IB	IIB	IIIA	IVA	VA	VIA	VIIA	Inert Gases
												B_2H_6	CH_4	NH_3	H_2O	HF	
LiH	BeH_2											AlH_3	SiH_4	PH_3	H_2S	HCl	
NaH	MgH_2											Ga_2H	GeH_4	AsH_3	H_2Se	HBr	
KH	CaH_2																
RbH	SrH_2											InH_3	SnH_4	SbH_3	H_2Te	HI	
CsH	BaH_2											TlH_3	PbH_4	BiH_3			

Table 1-10 *Densities of Hydrides* (g ml^{-1})

IA†	IIA†	IIIB	IVB	VB	VIB	VIIB	VIII			IB	IIB	IIIA	IVA	VA	VIA*	VIIA*	Inert Gases
0.8															.958	.991	
1.4															.993	1.187	
1.43	1.90														2.004	2.160	
2.59	3.27														2.65	2.799	
3.43	4.15																

† For solids.
* For liquids at boiling point.

Ch. 1 Elements, Atoms, and the Periodic Arrangement

Table 1-11 *Melting and Boiling Points of Hydrides**

IA	IIA	IIIB	IVB	VB	VIB	VIIB	VIII	IB	IIB	IIIA	IVA	VA	VIA	VIIA	Inert Gases
										−165.5 / −92.5	−183 / −162	−77.7 / −33.4	0 / 100	−83.1 / 19.9	
680	d									d	−185 / −111	−134 / −87.7	−85.5 / −60.3	−114 / −85.0	
d	d									−21.4 / 139	−166 / −84.4	−116 / −62.5	−65.7 / −41.3	−86.9 / −66.7	
d	d									d	−150 / −51.8	−88 / −17	−51 / −2.3	−50.8 / −35.4	
d	1200										−13	22			

*Melting point above boiling point in °C; single values are melting points; d means decomposes.

Table 1-12 *Melting and Boiling Points of Higher Fluorides**

IA	IIA	IIIB	IVB	VB	VIB	VIIB	VIII			IB	IIB	IIIA	IVA	VA	VIA	VIIA	Inert Gases
845 1681												−129 −99	−184 −182	−209 −129	−224 −145		
995 1704	1263 2227											1290 1257(sb)	−90.3	−84.5	−50.7 −63.7		
856 1502	1418										872	950	−15.0	−52.6	−34.6 −46.6(sb)		
775 1408	1400										1110	1170 1200		149	−37.7 −38.6	3.4(sb)	
682	1320 1799										645	550		550(sb)			

*Melting point above boiling point in °C; (sb) means sublimes. Single values are melting points.

Ch. 1 Elements, Atoms, and the Periodic Arrangement

Table 1-13 *Melting and Boiling Points of Higher Oxides**

IA	IIA	IIIB	IVB	VB	VIB	VIIB	VIII			IB	IIB	IIIA	IVA	VA	VIA	VIIA	Inert Gases
1727	2820											450	−56.5 −78.5	30 45		−91.5	
920	2802											2027	1710 2990	422	44.8		
>490	2587		1800	685	190	<20			1960	1230	2248	1727	1116 1200	d>400	118 d180		
>567	2457	2690	2677	1460	795	119.5	25			d>160	>1500		1927 1900(sb)	d	d>700		
490	1923	2590	2774	1880	1473	296	40			d>200	d	d<100	d752	d			

*Single values are melting points in °C; melting points above boiling points in °C when both are given; d means decomposes; (sb) means sublimes.

Table 1-14 *Physical Properties of Halides*

Compound	Melting point, °C	Boiling point, °C	Compound	Melting point, °C	Boiling point, °C
BF_3	−127	−101	AlF_3		1291 (subl)
BCl_3	−107	12	$AlCl_3$	192	180 (subl)
BBr_3	−46	91	$AlBr_3$	97	255
BI_3	43	210	AlI_3	180	381
$BeCl_2$	405	490			
$MgCl_2$	710	1400			
$CaCl_2$	780	1600			
PF_3	−160	95	PF_5	−83	−85
PCl_3	−112	76	PCl_5	148	163 (subl)
PBr_3	−40	173	PBr_5		decomp.
PI_3	61	decomp.			
SF_6	−51	−45	SCl_4	decomp.	
SF_4	−121	−40	SCl_2	−78	decomp.
S_2Br_2	−46	90	S_2Cl_2	−80	138

The purpose of this book is to study the periodic relationship in detail. Since periodicity is especially useful if the theoretical basis for it is understood, the next chapter deals with the structure of atoms.

QUESTIONS

1. How could you prove that air is a mixture and not an element?
2. How could you prove that water is a compound and not an element?
3. Estimate the melting point expected for the element astatine, atomic number 85.
4. Estimate the boiling point of technetium, atomic number 43.
5. Pick several elements with high electrical conductance and low melting point which might be considered for electrical fuses.
6. List all the metals with densities less than 5.0 g cm^{-3} and melting points above 300°C which thus might be considered for use in the construction of airplanes.
7. In what parts of the periodic table can you find elements that form gaseous oxides?
8. A handbook will list the formulas and properties of compounds of most of the elements with sulfur. Can you observe any periodicity in this data?
9. Estimate the boiling point of $CaCl_2$ given the values for $MgCl_2$, 1412; $SrCl_2$, 873; NaCl, 1413; KCl, 1500; and RbCl, 1390°C.
10. Using Table 1-7 predict the formulas for two oxides of polonium.

Ch. 1 Elements, Atoms, and the Periodic Arrangement

PROBLEMS I

1. Calculate the percent of platinum in PtO if the atomic weight of platinum is 195.
2. Calculate the percent of manganese (atomic weight 54.94) in the compound Mn_2O_7.
3. Calculate the percent composition of all the elements in calcium oxalate, CaC_2O_4.
4. The compound Fe_3O_4 is 72.4% iron by weight. Calculate the atomic weight of iron.
5. A sample of heavy water (mostly D_2O) is 81% oxygen by weight. What is the average atomic weight of the hydrogen present in this sample?
6. An oxide of nitrogen is 30.4% nitrogen (N). What is its formula?
7. Determine the formula of a compound which is 12.7% aluminum (Al), 67.4% oxygen (O), and 19.9% nitrogen.
8. The density of SiO_2 is 2.65 g cm^{-3}. Calculate the volume of SiO_2 which contains 16.0 g oxygen.
9. The density of In_2O_3 is 7.18 g cm^{-3}. Calculate the volume of In_2O_3 which contains 16.0 g oxygen.
10. If we accept the generalization that many solid oxides contain about 16 g oxygen in 12 ml, what value would we estimate for the density of Fe_3O_4?
11. Calculate the atomic weight of chlorine if 75.53% of the chlorine atoms have an atomic weight of 34.969 and the rest have the atomic weight 36.966.
12. What fraction of the carbon atoms in ordinary carbon have mass 13.00 if the atomic weight of carbon is 12.01 compared to the value 12.000 for carbon-12?
13. Calculate the percentage composition of the compound C_2N_2.
14. Another compound of carbon and nitrogen is 56.2% carbon. Calculate the empirical formula (simplest formula). What is the formula if there are actually four nitrogen atoms present in each molecule?

The symbol D is used for the isotope of hydrogen with an atomic weight of 2.014, ^2H.

PROBLEMS II

1. In the compound CO_2, 12 g carbon are combined with _____ g oxygen.
2. What weight of CO_2 contains 12.0 g carbon?
3. The molecular weight of CO_2 is _____.
4. What is the weight of 15.0 moles of CO_2?
5. What is the weight of 15.0 moles of SiO_2?
6. What is the molecular weight of H_3PO_4?
7. What is the molecular weight of $Ca_3(PO_4)_2$?
8. How many moles of $Ca_3(PO_4)_2$ are present in a 100-g sample?
9. How many moles of calcium atoms are present in 100 g $Ca_3(PO_4)_2$?
10. What is the percentage calcium (Ca) in $Ca_3(PO_4)_2$?
11. How many moles of calcium atoms will weigh 51.4 g?
12. How many moles of fluorine atoms will weigh 48.6 g?

13. What is the formula of a compound which is 51.4% calcium and 48.6% fluorine?
14. What is the formula of an oxide of manganese which is 72.4% manganese?
15. If 15.81 g of a compound contain 5.00 g phosphorus, 0.488 g hydrogen, and the rest is oxygen, what is its formula?

> Everything is simpler than you think and at the same time more complex than you imagine.
>
> GOETHE

Chapter 2 The Electronic Structure of Atoms

I. The Constituents of Atoms

All atoms are composed of three fundamental particles, the proton, the neutron, and the electron. The proton and the neutron each have roughly the same mass as the hydrogen atom whereas the electron has about 1/1840 of this mass. The neutron is electrically neutral, while the proton is positively charged and the electron negatively charged. Information concerning the arrangement of these particles in an atom, the approximate size of atoms, and another important quantity, the number of protons in an atom, was obtained from one of the first of a number of extremely important experiments performed on atoms. This experiment was performed by Rutherford in 1911.

Rutherford bombarded a thin metal foil about 1000 atoms thick with a stream of very high energy, double charged (positive) helium ions. A tremendous force would be required to deflect such a particle from its original path; yet Rutherford found that a small fraction of the particles were deflected at large angles (Fig. 2-1). Rutherford describes his amazement, "It was quite the most incredible event that ever happened to me in my life. It was almost as incredible as if you fired a 15-inch shell at a piece of tissue paper and it came back and hit

Atomic weights are really measurements of mass. Weight is a force proportional to mass that is also dependent on the strength of the gravitational field.

Doubly charged helium ions, He^{2+}, emitted from a nucleus are called alpha particles.

An ion is an atom (or group of atoms) with an excess or deficiency of electrons so that it has an electrical charge.

Sec. II The Nature of the Constituents of Atoms

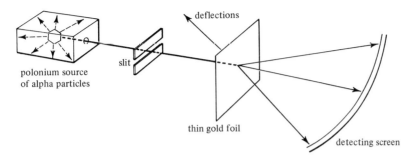

Figure 2-1 *Rutherford's apparatus.*

you!"* Rutherford soon found that the magnitude of the observed deflections could be accounted for by assuming that most of the mass of the atoms was contained in a minute, positively charged nucleus (Fig. 2-2). The nuclear charge is balanced by negatively charged electrons that surround the nucleus in some sort of electron configuration. This description of the atom is the basis of all modern theories of atomic structure.

The deflection of alpha particles observed by Rutherford requires a large force (hence a highly charged piece of the atom) exerted by something nearly as heavy as the atom.

The main subject of this chapter is the further description of the electron configuration of atoms. We will begin by considering the results of some experiments that may be performed on atoms, nuclei, and electrons.

II. The Nature of the Constituents of Atoms

A. THE PARTICLE NATURE OF ELECTRONS

The electron was discovered in 1897 by J. J. Thompson. The effects produced by electrons had been observed long before this, but until Thompson performed quantitative experiments it was not clear that different atoms gave the same negatively charged particles. Thompson was responsible for pioneer experiments leading to the determination of the charge of electrons and their charge-to-mass ratio.

Electrons exhibit a definite mass and charge; in this they exhibit the characteristics of particles. Additional confirmation of the particle nature of electrons is obtained using devices such as the Wilson cloud chamber. Figure 2-3 shows a Wilson cloud chamber and the results that are obtained when a particle enters the chamber. The lines shown represent the trail of a particle as it passes through the chamber.

The path of an electron in a cloud chamber is not quite straight since the electron can be scattered by nuclei more easily than alpha particles.

---

*E. Rutherford, "The Development of the Theory of Atomic Structure," in J. Needhorn and W. Pagel, eds., *Background to Modern Science*, (New York: Macmillan, 1938).

Ch. 2 The Electronic Structure of Atoms

The trail is formed when water vapor in the chamber condenses on ionized nitrogen molecules. The ions are produced when a high-energy particle, such as an electron or alpha particle, collides with a nitrogen molecule. The path of the particle is shown by the water droplets formed.

B. THE WAVE NATURE OF ELECTRONS

Probably one of the most important experiments ever performed using electrons was that of Davisson and Germer in 1927. In the course of experiments designed to study the scattering of electrons from the polycrystalline surface of a nickel sheet, the apparatus was broken and the carefully prepared nickel surface was destroyed. During the repair the nickel was unknowingly changed from many fine crystals into larger

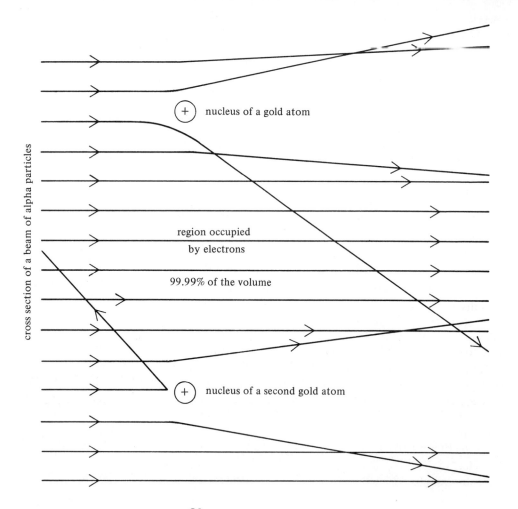

Figure 2-2 *Some typical paths of alpha particles in a small cross section of a beam.*

Sec. II The Nature of the Constituents of Atoms

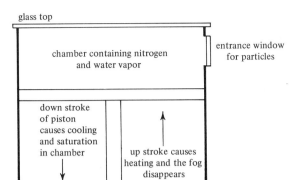

Figure 2-3 *Sketch showing the principal features of the Wilson cloud chamber and some typical cloud chamber tracks. (After G. D. Shaw, Sci. Am., 200; 173–182, June 1959.)*

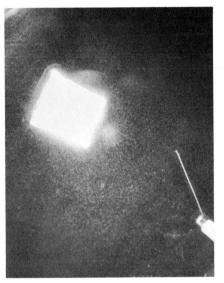

alpha track

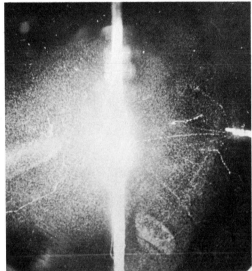

electron tracks

crystals. The results obtained by Davisson and Germer were very different from any scattering pattern previously observed for electrons. What Davisson and Germer actually observed was experimental verification that electrons can exhibit a wave character. They recognized the wave nature because the results they observed with electrons were exactly analogous to the diffraction patterns observed with X-rays. Figure 2-4 shows the characteristic patterns observed in X-ray diffraction. The diffraction of light from a crystal is the most convincing argument for the wave interpretation of the nature of light. If electrons give analogous patterns they must also be exhibiting a wave nature. Figure 2-5 is an electron diffraction pattern. Figure 2-6 shows other devices for demonstrating wave behavior and Fig. 2-7 shows the results

31

Ch. 2 The Electronic Structure of Atoms

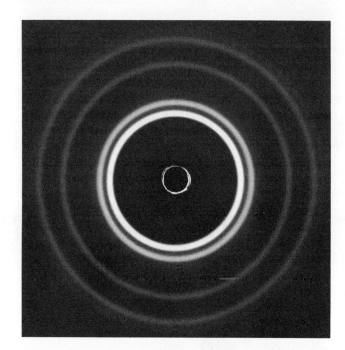

Figure 2-4 *X-ray diffraction pattern of aluminum (aluminum filings, 4×10^5 kcal mole^{-1} X-rays). (Courtesy of Film Studio, Education Development Center.)*

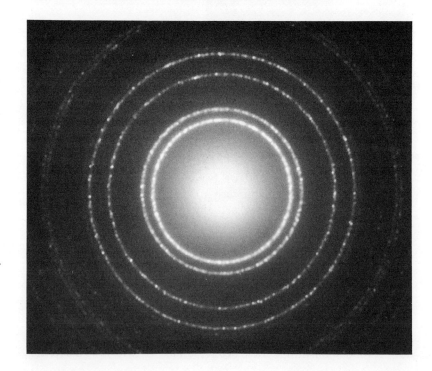

Figure 2-5 *Electron diffraction pattern of aluminum (aluminum foil, 1.4×10^4 kcal mole^{-1} electrons). (Courtesy of Film Studio, Education Development Center.)*

Sec. II The Nature of the Constituents of Atoms

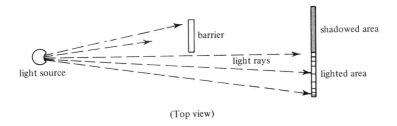

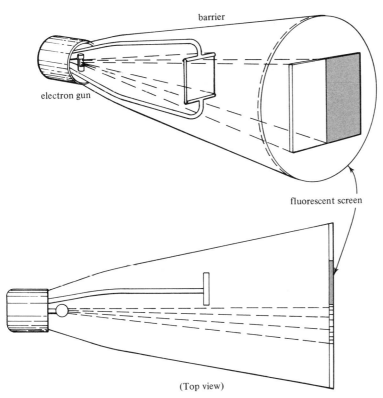

Figure 2-6 *Light diffraction and electron diffraction apparatus.*

for light and electrons. Similar patterns are obtained for light and electrons.

This type of pattern can only be explained simply if both electrons and X-rays exhibit a wave character. The patterns may be understood as follows. Two waves may reinforce when their maxima and minima coincide (the waves are said to be in phase). They will interfere with

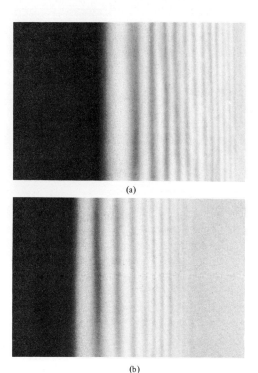

Figure 2-7 *Interference patterns. (a) Representation of interference pattern created by light passing a sharp-edged screen. (b) Representation of interference of electrons passing a sharp-edged screen. (After W. Glaser, Elektronen-und Ionenoptik, Handbuch der Physik, Editor Flügge, Berlin-Göttingen-Heidelberg; Springer, 1956, Vol. 33, p. 317.)*

each other when their maxima and minima do not coincide (the waves are out of phase). Reinforcement and interference are illustrated in Fig. 2-8. The amplitude (ψ) of the resultant wave is obtained from the algebraic sum of the amplitude of the two waves. When the waves are in phase the interference is constructive; when the waves are out of phase the interference is destructive.

In the diffraction of X-rays or electrons from a crystal (shown in Fig. 2-9) the diffracted waves will interfere constructively if the extra distance traveled (GE + EH) by the second wave equals one whole wavelength (or any integral number of wavelengths) of the incident wave. This condition for constructive interference is given by the Bragg equation

$$n\lambda = 2d \sin \theta \qquad (2.1)$$

In this equation λ is the wavelength, d the distance between layers of atoms in the crystal, n is an integer, and θ is the angle of diffraction. From the Bragg equation, electrons diffracted from a nickel crystal ($d = 2.5 \times 10^{-8}$ cm) at $\theta = 50°$ have a wavelength of 1.65×10^{-8} cm.

Sec. II The Nature of the Constituents of Atoms

In the same year of Davisson and Germer's experiment, G. P. Thompson also demonstrated the analogous behavior of electrons and X-rays when he obtained diffraction patterns on a photographic plate by passing electrons through a thin metal foil.

C. PARTICLE VERSUS WAVE

The experiments cited above demonstrate that the electrons exhibit the properties of a particle and the properties of a wave. Electrons

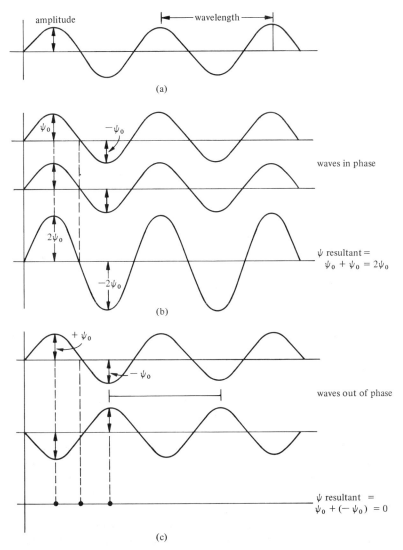

Figure 2-8 *Constructive and destructive interference of waves.*

Ch. 2 The Electronic Structure of Atoms

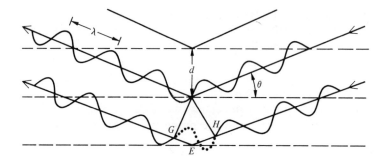

Figure 2-9 *Diffraction of X-rays from atoms or ions in a crystal.*

The wave nature of all particles was a postulate or hypothesis put forward by de Broglie in 1924. De Broglie's postulate was confirmed by Davisson and Germer.

exhibit a dual nature. Duality is not limited to electrons alone but, as de Broglie postulated, all material particles in motion exhibit a wave nature as well as a particle nature. De Broglie's postulate states that the wavelength of the wave associated with a particle is given by the expression

$$\lambda = \frac{h}{mv} \qquad (2.2)$$

where m is the mass of the particle, v its velocity, and h is Planck's constant (6.6252×10^{-27} erg sec). Thus, if the Davisson-Germer experiment is performed on neutrons or atoms, wave behavior may be detected. For heavy particles such as the atoms of mercury (Hg), cadmium (Cd), and arsenic (As), the velocity must be kept low or the wavelength will be too short to observe. The wavelength of the particle may be calculated by either the Bragg equation or the de Broglie equation. From the de Broglie relation the wavelength of 54 ev electrons is 1.67×10^{-8} cm. The wavelength obtained from diffraction experiments agrees well with that calculated from the de Broglie relation.

For heavy particles the Bragg equation must be modified slightly since the particles being scattered may be as heavy as the individual atoms of the crystal. This complication does not need to concern us here.

Duality is impossible to explain in any terms because nothing is quite comparable to it. The best we can do is to understand that it is observed and get used to it. The idea that duality is a part of nature is relatively new; however, the observation that light exhibits a particle and a wave nature is not.

The unit of energy most commonly used by chemists is kcal mole^{-1}. Other energy units (ergs, ev, and so forth) are used from time to time, but most energy values in this text will be in kcal mole^{-1}. The significance of the mole^{-1} part of the name will become clearer as we use this energy unit in subsequent chapters.

The wave nature of light was advanced in 1690 by Huygens, while the particle nature of light was developed by Newton. Both theories seemed equally capable of explaining the phenomena observed in optics at that time. However, the influence of Newton was so great that the wave theory was lightly regarded. In 1801, Thomas Young discovered interference and explained it simply by using the wave idea. In spite of this, the particle theory remained entrenched until methods were available for measuring the speed of light in a vacuum and in

The ev or electron-volt is a unit of energy; 1 ev is equivalent to 23.05 kcal mole^{-1}. Thus 54 ev electrons have an energy of 1250 kcal mole^{-1}.

a medium. Foucault demonstrated that the velocity of light was less in a medium than in a vacuum, verifying the predicted behavior using the wave theory.

The particle theory came back into fashion in 1905 when Einstein used it to explain the photoelectric effect. This effect was discovered by Heinrich Hertz in 1887. Hertz noted that electrons are ejected from a metal surface if the surface is illuminated by a light source. Subsequent experiments demonstrated that the energy of the ejected electrons was dependent upon the frequency of the light (Fig. 2-10) and not on its intensity. Below a certain frequency, electrons are not emitted regardless of the intensity. This is completely contradictory to the classical wave theory. According to this theory the surface will absorb energy from the impinging wavefront until a sufficient amount of energy has been received to cause ejection of photoelectrons from various points on the surface. If this were the case, the energy of the emitted electrons would depend upon the intensity of the light or a time lag would be expected to allow for the build up of a sufficient amount of energy. However, no time lag has ever been observed. The ejection of electrons is instantaneous even for very weak light.

In explaining these observations, Einstein suggested that light consisted of a stream of particles and called these particles photons. Each photon according to Einstein corresponded to a small packet (quantum) of energy ($h\nu$).

$$E = h\nu \tag{2.3}$$

In this expression h, Planck's constant, has the value 6.625×10^{-27} erg sec or 9.56×10^{-14} kcal $mole^{-1}$ sec.

The electron is emitted with a kinetic energy equal to $h\nu$ minus the energy needed to remove the electron from the metal, w.

$$E = h\nu - w \tag{2.4}$$

No emission can take place unless each quantum is equal to or greater than w, the energy needed to expel an electron.

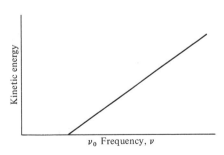

Figure 2-10 *Variations in maximum kinetic energy of photo electrons with frequency of the incident light.*

D. SOME CONSEQUENCES OF DUALITY

The only analogy that comes to mind to describe duality is an optical illusion. An optical illusion requires that the viewer recognize more than one object in the picture being viewed. For example, the object shown in Fig. 2-11 may be pictured as two faces or as a flower vase. It seems quite impossible to view both at the same time, and it is equally difficult to see the drawing for what it actually is (some lines on the paper) especially if the viewer's attention is not focussed on this third aspect. The observer attempts to see the picture in terms of something familiar. Once a familiar object is recognized, a definite effort must be made to recognize the second object, but the process becomes easier with practice. *This discussion does not imply that an electron is something other than a particle or wave. This is not known.* What is known is that at times a particle nature is observed whereas at other times a wave nature is observed. The behavior observed depends upon the apparatus used. The two can never be observed at the same time. An electron cannot behave as a particle (with a definitely specified position) and as a wave (with a specified momentum) simultaneously. If the position is definite, the electron is a particle. The photographic plate is a particle-detecting device. If the momentum is measured, the electron is a wave. The crystal or diffraction grating is a wave-detecting or momentum-measuring device, and the position is indefinite. In this instance the electron MAY APPEAR in any one of a large number of areas on the photographic plate.

The inability to know exactly the position and momentum of an electron at the same time is of extreme importance to the study of the electronic structure of atoms. It is known as the Heisenberg uncertainty principle after Werner Heisenberg, who first formulated the concept.

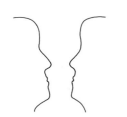

Figure 2-11 *An optical illusion.*

The fundamental connection between wavelength and momentum is expressed in de Broglie's equation (equation 2.2).

A single electron can only darken one spot on the photographic plate, but the diffraction pattern appears on a plate exposed to many electrons.

III. Two Models of Atomic Structure

We may now consider two models that may be used to describe the structure of atoms. In order to keep the models as simple as possible the hydrogen atom (H), with a proton as the nucleus and one associated electron, will be used. To further simplify the model we may assume that the proton is stationary with regard to any motion the electron might have. This assumption, known as the Born-Oppenheimer approximation, is a reasonable one in view of the relative mass of the proton and the electron. Because of the dual nature of electrons, the model constructed could use either the particle nature or the wave nature of the electron. The particle model is simplest and historically

A. THE PARTICLE MODEL

The simplest model of atomic structure is based on the particle nature of electrons. Electrostatic forces are assumed to be responsible for the attraction between the negatively charged electron and positively charged proton. The force of attraction is given by Coulomb's law

$$F_a = +\frac{Ze^2}{r^2} \tag{2.5}$$

where e is the charge on the electron, Z is the atomic number, Ze is therefore the charge on the nucleus, and r is the distance between the particles. Motion is necessary if the particles are to remain separate, so the force of attraction is exactly balanced by a centrifugal force given by

$$F_c = \frac{mv^2}{r} \tag{2.6}$$

where m is the mass of the electron, v its velocity, and r is the distance between the particles.

The potential energy of the electron is defined as the work needed to move the electron and proton to infinite separation. The work done when the electron is moved from a distance r_1 to a distance r_2 is proportional to the difference between $1/r_1$ and $1/r_2$, and it is equal to the change in potential energy.

$$PE_{r_2} - PE_{r_1} = w = -Ze^2 \left(\frac{1}{r_2} - \frac{1}{r_1} \right) \tag{2.7}$$

If the potential energy at r_2 is assumed to be zero when r_2 is equal to infinity then

$$w = PE = -\frac{Ze^2}{r} = Ze^2 \left(\frac{1}{r_1} - \frac{1}{r_\infty} \right) = 0 - PE_{r_1} \tag{2.8}$$

A negative potential energy represents a more stable system than one in which the proton and electron are separated by an "infinite distance." The total energy E, is the sum of the kinetic and potential energies

$$E = \frac{1}{2} mv^2 - \frac{Ze^2}{r} \tag{2.9}$$

where the first term is the kinetic energy. The total energy of the orbiting electron may also be written as

$$E = -\frac{1}{2}\frac{Ze^2}{r} \qquad (2.10)$$

$F_a = Ze^2/r^2$
$F_c = mv^2/r = F_a$
$mv^2/r = Ze^2/r^2$
$mv^2 = Ze^2/r$

This expression is obtained by substituting $(Ze^2)/r$ (obtained by equating F_a and F_c) for mv^2 in equation 2.9. Notice that the total energy in this system (a bound system because the electron is associated with the nucleus) is negative.

The type of orbit expected for an electron in the system described above is shown in Fig. 2-12. Actually, the orbit shown was not chosen for any special reason. Any circular orbit on the surface of a sphere of radius r is possible if only Coulombic forces are operating because Coulombic forces are independent of direction. Thus the orbit shown is only one of an infinite number with the same radius r that could have been selected.

We shall now examine some experimental observations and attempt to explain them in terms of the particle model.

1. Atomic Spectra

The bending or refraction of light by a prism is a diffraction phenomenon; however, because of the physical situation the light is only bent.

If a beam of white light is passed through a prism and the beam emerging is viewed on a screen (Fig. 2-13), a well-known phenomenon is exhibited. A prism can act as a dispersing device for light and, as such, can be used to separate light of various wavelengths because passage through the prism generally bends the short wavelength components of a light beam more than it does the longer wavelength components. The dispersion is observed on the screen as a continuous spectrum. If the source consists of atoms of an element such as hydrogen or sodium (using a gas discharge, for example), the pattern on the screen is quite different from that obtained with white light. In

An electrical discharge in a gas is a method of inducing the emission of light from gaseous atoms.

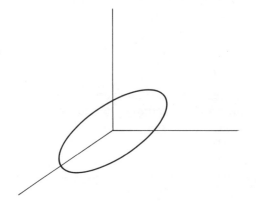

Figure 2-12 *Circular orbit of a particle electron.*

Sec. III Two Models of Atomic Structure

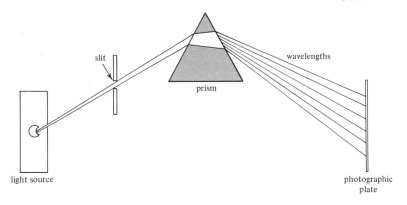

Figure 2-13 *Dispersion of light by a prism.*

the case of hydrogen (Fig. 2-14), a discrete number of lines are observed instead of a continuous spectrum. The discrete number of lines observed is a very important feature of atomic spectra.

When an atom or ion absorbs energy (that is, is excited in the source of a spectrograph), it is unstable and there will be a tendency to release all or part of the excess energy in the form of ultraviolet or visible light. The process may be considered as the excitation of a *single* electron in the atom. Light is then emitted when the electron releases the energy absorbed. It is clear that if there is no restriction on the energy absorbed, the light emitted would necessarily be continuous (all wavelengths would be observed). This is not the case for hydrogen. In order to explain the discrete number of lines observed, some restriction on the energy of light which can be absorbed and emitted must be operative; there must be some factor, not so far apparent from the particle model, that restricts the energy of light that can be absorbed and emitted by atoms. Examination of the emission spectrum of other elements and molecules shows that this situation is also observed for any other element or molecule.

A set of possible energy levels of the hydrogen atom that could give rise to the observed spectrum is shown schematically in Fig. 2-15a. The electron in the lowest lying level may be transferred to a higher

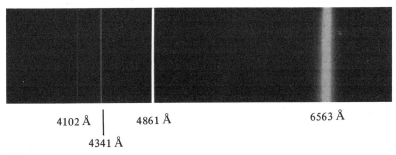

Figure 2-14 *Visible portion of the spectrum for hydrogen.*

41

Ch. 2 The Electronic Structure of Atoms

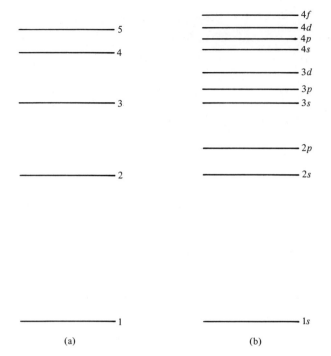

Figure 2-15 *Energy levels.*

energy level by absorption of a quantum of energy equal to the energy difference between the two levels. The energy of atomic systems is said to be *quantized* because only certain energies are allowed.

The energy of the levels in Fig. 2-15a is given by the formula

$$E_n = -\frac{2\pi^2 m e^4 Z^2}{n^2 h^2} \tag{2.11}$$

This equation may be derived for the particle model using one of Bohr's postulates. The derivation begins by equating the coulombic and centrifugal force of a particle electron circling the nucleus.

$$\frac{Ze^2}{r^2} = \frac{mv^2}{r} \tag{2.12}$$

Cancelling one power of r gives

$$\frac{Ze^2}{r} = mv^2 \tag{2.13}$$

Bohr postulated that the angular momentum mvr is quantized

$$mvr = n\frac{h}{2\pi} \quad \text{where } n = 1, 2, 3 \text{ etc.} \quad (2.14)$$

Solving for v we obtain

$$v = \frac{nh}{2\pi mr} \quad (2.15)$$

and substituting for v in equation 2.13, the relation

$$\frac{n^2 h^2}{m^2 r^2 (2\pi)^2} = \frac{Ze^2}{mr} \quad (2.16)$$

is obtained. Cancelling powers of r and m and solving for r gives

$$r = \frac{h^2 n^2}{(2\pi)^2 m Z e^2} \quad (2.17)$$

The total energy is thus

$$E = -\frac{Ze^2}{2r} = -\frac{2\pi^2 m Z^2 e^4}{h^2 n^2} \quad (2.18)$$

where m is the mass of the electron, e is the charge on the electron, Ze is the charge on the nucleus, h is Planck's constant, and n is an integer (1, 2, 3 etc.). Neglecting the constants this expression may be written

$$E \propto \frac{1}{n^2} \quad (2.19)$$

The symbol $\propto$ stands for "is proportional to."

The lowest possible energy level corresponds to n equal to 1; the next highest level corresponds to n equal to 2, etc.

The energy of a photon emitted by an atom is equal to the difference in the energy between the initial (E_2) and the final (E_1) states.

$$h\nu = (E_2 - E_1) = \frac{2\pi^2 m Z^2 e^4}{h^2}\left(\frac{1}{n_1^2} - \frac{1}{n_2^2}\right) \quad n_2 > n_1 \quad (2.20)$$

where n_2 refers to the initial state and n_1 refers to the final state. The frequency of the emitted light is

$$\nu = \frac{2\pi^2 m Z^2 e^4}{h^3}\left(\frac{1}{n_1^2} - \frac{1}{n_2^2}\right) \quad (2.21)$$

Ch. 2   The Electronic Structure of Atoms

A model capable of predicting quantized energy is required to explain the observed atomic spectra. The particle model (Bohr model) is capable of explaining the spectrum of the hydrogen atom and other one-electron systems (He^+, Li^{2+}) in a quantitative manner. The frequencies observed in the spectrum are those calculated and, furthermore, lines that had not been previously observed were correctly predicted by Bohr. However, in thinking about the particle model described above it is difficult to imagine a reason for quantized energies. Since the total energy of the system is dependent upon the distance separating the nucleus and the electron $[E = -(\frac{1}{2})(Ze^2/r)]$, a quantized energy requires that only certain distances are appropriate between a nucleus and an electron. Restricted distances are introduced (page 43) into the system by Bohr's postulate that the angular momentum (mvr) is quantized. There is, however, no known force that would require only a certain distance between two oppositely charged particles; nor is there a known force that would require a quantized angular momentum for a particle electron orbiting the nucleus. Certainly coulombic forces are not restricted in this way. Bohr had to make this restriction to make the particle model work. As we shall see there is an alternative model that accommodates quantized energies and quantized angular momenta.

2. The Stern-Gerlach Experiment

The Stern-Gerlach experiment is only one of several experiments that may be used to demonstrate quantized angular momentum.

A vector is a physical or mathematical quantity with both magnitude and direction.

While a quantized angular momentum is not justified in the particle model, it is a credit to Bohr's genius that the Stern-Gerlach experiment (performed long after Bohr made this assumption) shows that the angular momentum of an electron in an atom is indeed quantized. The Stern-Gerlach experiment may be used (1) to measure the z component of the angular momentum vector, (2) to measure the z component of the spin momentum vector, or (3) to demonstrate the shape of the atoms. We will come back to consider aspects (2) and (3) of this experiment later in this chapter.

The angular momentum of the particle electron orbiting the nucleus has been given (page 43) as mvr where mv is the linear momentum and r is the distance between electron and nucleus. The angular momentum of the electron has a magnitude and it is also associated with a direction (a line perpendicular to the orbit of the electron). Thus it may be represented by a vector (l) as shown in Fig. 2-16. We now designate specifically the z component of the angular momentum vector and call it m_l. This is also labeled in Fig. 2-16. The magnitude of m_l is the length of the projection of l along the z axis.

Any circular motion of electrical charge such as that associated with the angular momentum of electrons in atoms will create a magnetic field. An atom with electrons with angular momentum will behave like

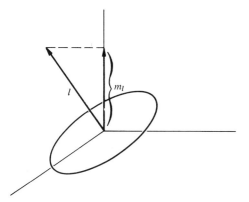

Figure 2-16 *The orbit, the angular momentum vector (l), and the z component of the angular momentum vector (m_l).*

a small magnet, and it may be deflected by a certain kind of magnetic field such as the one used in the Stern-Gerlach experiment, Fig. 2-17.

The Stern-Gerlach experiment can be used to measure the magnitude of m_l. In this experiment a beam of atoms is shot through an inhomogeneous magnetic field and detected on an appropriate screen. Atoms such as hydrogen pass straight through the magnet without deviation from the straight path. (This result is not strictly correct. Two spots are observed for hydrogen, but the splitting of the beam is due to the spin momentum of the electron which we are neglecting at present. In any case, the number of spots that are observed is not as important for the moment as the fact that a discrete number of spots are observed after a large number of atoms have hit the screen.) An atom such as boron (B) exhibits three spots (again neglecting the complication of spin). Helium (He), lithium (Li), and beryllium (Be) also produce one spot while aluminum (Al) gives three and vanadium

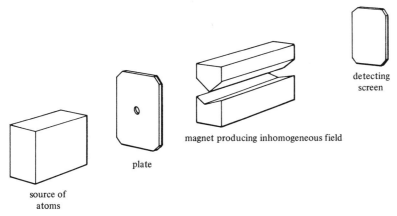

Figure 2-17 *The Stern-Gerlach apparatus.*

(V) five. The similarity of the patterns for hydrogen and lithium and for boron and aluminum might be expected from the position of these elements in the periodic table.

The significance of these results rests in the fact that the behavior of an atom in the magnetic field (in some cases a deviation from the straight-through path) depends upon the magnitude of m_l. For $m_l = 0$, no deflection is observed. For other values of m_l a deflection is observed. Thus the number of m_l values possible for the atom and the magnitude of the deflection are observables in this experiment. According to the results, the particle electron must occupy an orbit such that $m_l = 0$ in hydrogen, helium, lithium, and beryllium. This orbit must lie up and down the z axis so that l is in the xy plane as shown in Fig. 2-18. According to this experiment, this is the only possible orbit for the particle electron in the atoms that pass straight through the magnet. For boron (B) three values of m_l are observed, and three orbits are possible. These are shown in Fig. 2-19.

Since the values of m_l are restricted it is clear that we are again dealing with a quantized property. The particle model predicts an infinite number of spots because the forces involved are nondirectional. The infinite number of orientations of a circular particle orbit would give deflections for all values of m_l from $+l$ to $-l$, an infinite number of spots which would appear like a smeared out band. No atoms show this pattern in the Stern-Gerlach experiment; they all show a discrete number of spots.

To make the particle model work (that is, to obtain a quantized z component of the angular momentum), we must assume restricted orbits, which contradicts the model. Coulombic type forces would allow an infinite number of orbits and could not be responsible for the occupancy of only certain orbits. The simple model involving particle behavior of electrons is not capable of accommodating the quantized

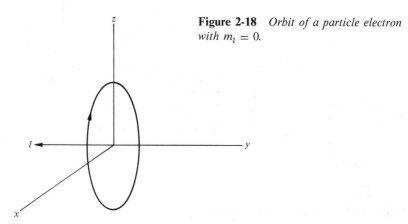

Figure 2-18 *Orbit of a particle electron with $m_l = 0$.*

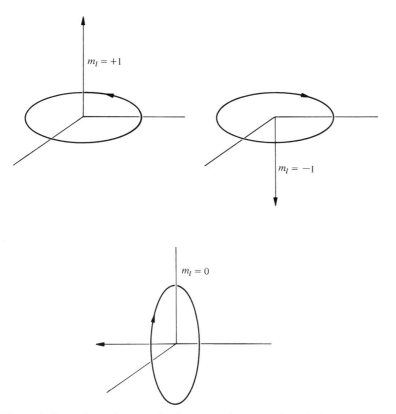

Figure 2-19 *Orbits of a particle electron with $m_l = +1, -1, 0$.*

properties observed in atomic systems. Further, it is not possible to continue to use the particle model while assuming some force of attraction other than coulombic that could result in a quantized energy and angular momentum because, as we shall see upon analyzing the results of the Stern-Gerlach experiment in more detail, even this could not explain the results. Besides, if a different type force were operative, it must be considerable and would probably have other observed effects. The problem does not lie in the forces assumed but rather in the model itself. For free particles, a wave behavior had to be introduced to understand experiments where momentum was measured. In the next section we shall attempt to explain the results of the atomic spectra experiment and the Stern-Gerlach experiment using the wave model.

B. THE WAVE MODEL

We shall begin the discussion of the wave model by considering an analogous situation. Imagine a piece of string suspended in air without

Ch. 2 The Electronic Structure of Atoms

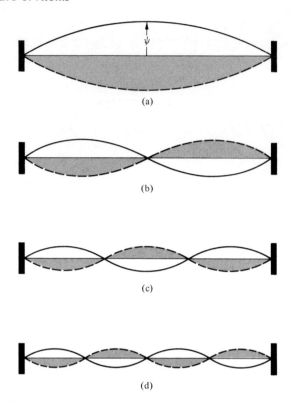

Figure 2-20 *Standing waves.*

The Greek letter psi (ψ, pronounced sī) is commonly used for wave functions.

supports and imagine what would happen if the string was disturbed by plucking it in the center. The string would collapse into an irregular form, and the resulting shape would not be predictable. If, however, the string was bound securely at both ends and then disturbed, the string would begin to vibrate in a regular manner. The results of several disturbances are illustrated in Fig. 2-20.

In this figure the string vibrates above and below the zero position (equilibrium position), and the amplitude (ψ) of the vibration (height of the string above the zero line) at any one point is equal on both sides of the zero line. The difference between a, b, and c in the figure is the frequency with which the string vibrates. Each portion of the string shown in Fig. 2-20b goes through one cycle of its motion faster than the string pictured in Fig. 2-20a. Frequencies that correspond to the vibrations shown in Fig. 2-21 are not possible because the wavelength is such that an integral number of loops do not fit into the available space.

The waves that can be exhibited by a vibrating string (standing waves because the crest always remains in the same position) involve one dimension. The wavelength of any wave is given in terms of the fundamental wavelength λ_0

$$\lambda = \frac{\lambda_0}{n} \qquad (2.22)$$

where n is a positive integer. In order to specify the waves in a two-dimensional system (a square drumhead), two numbers are necessary; three-dimensional vibrations require three numbers to specify a particular type of vibration of a three-dimensional system (an atom).

In an atomic system if the electron is to remain associated with the nucleus, there are certain places where the electron cannot go. Given enough energy an electron can be removed from an atom, but then it is free of the nucleus and it is no longer within the boundary of the atom (the energy of the electron becomes positive). Thus in an atomic system a boundary is placed on an electron. If an electron in an atom behaves as a wave, it must oscillate with discrete frequencies because only certain wavelengths will fit into the available space. In other words the energy of an electron in an atom is quantized because the electron wave oscillates with a discrete frequency and must do so because it is bound within a certain space. Thus the only requirement that must be placed on the atom for us to recognize that the energy must be quantized is that the electron wave must be bound by the atom. This restriction is a reasonable one, formally known as the boundary condition.

Several other conditions that follow from the analogy with the vibrating string may be mentioned now. First, ψ (the wave function) must be single-valued. The string cannot be on both sides of the zero line at one time. Secondly, ψ must be continuous (there can be no breaks in the string). These restrictions are important and will be met in further studies of wave mechanics. An examination of their impact on the theory is too difficult here; however, it might be pointed out that for ψ to be single-valued requires the angular momentum to be quantized. It is possible to understand the connection between wave

Figure 2-21 *Impossible oscillations of the string.*

character and angular momentum, and to do so we must examine the results of the Stern-Gerlach experiment in more detail.

The results of the Stern-Gerlach experiment for hydrogen indicate only one spot. What kind of electron configuration can give only one spot? The observation means that the electron configuration of each atom that enters the magnetic field is identical to every other atom entering the field. This would be very unlikely if the electron configuration were dependent upon angle (for example, a circular orbit) because, if this were true, it would be necessary for the source to emit each atom in exactly the same way. For example, with an electron in a circular orbit about the nucleus all atoms emitted by the source must come out of the source with exactly the same orientation of orbit to magnet in order for a discrete number of spots to be observed. With a random source, as the source in this experiment must be, only one kind of an electron configuration could give a single spot. That electron configuration is spherical. Any distribution other than spherical would result in an infinite number of values for m_l. (The possibility of a special force causing restricted orbits is inconsistent with the results.)

A particle electron cannot occupy a spherical configuration about the hydrogen nucleus, but a three-dimensional electron wave can. In the wave model, with a spherically symmetrical electron wave distributed about the nucleus, every atom is identical to every other, and the interaction of each hydrogen atom with the magnet is identical, and one spot should be observed. The observation for an atom such as boron, which shows three spots, is more difficult to interpret. In order to understand this observation we must first develop the wave model further.

IV. The Electronic Structure of Atoms

In order to express the wave behavior of electrons in atoms mathematically, a wave equation is necessary. This equation, first written by Erwin Schrödinger, is the basic equation of wave mechanics. The equation cannot be derived, but was obtained by Schrödinger from a series of substitutions into some of Hamilton's equations of motion. In this sense it is the first postulate of wave mechanics. It is, in essence, a hypothesis which will be kept as long as it works. We shall not write the equation out in full, but rather use it in the simplest possible form.

The Schrödinger equation may be written:

$$\bar{H}\psi = E\psi \qquad (2.23)$$

The Hamiltonian operator $\bar{H}$, may contain complex mathematical operations including differentiation, but in principle it is the same sort of thing as the operation of taking a square root which can be represented by the symbol $\sqrt{}$.

The symbols have the following significance: E represents the total

Sec. IV The Electronic Structure of Atoms

energy of the system. ψ is called a wave function; it describes the relationship between the nucleus and the electron or electrons in the system and contains everything that is known about the system. $\bar{H}$ is called the Hamiltonian operator. It is a mathematical representation of the total energy of the system.

The equation reads, if ψ is operated upon by $\bar{H}$ (the mathematical operations contained in $\bar{H}$ are performed), the same ψ is obtained multiplied by a number. This number is E the total energy of the system.

It is always possible to write the Hamiltonian for any system. It is sometimes difficult, but more often impossible, to write the exact wave function even for simple systems. This may be understood by considering the information contained in ψ. For the hydrogen atom the variables in ψ are the coordinates of the electron with respect to the position of the proton (again neglecting spin, a fourth coordinate). This in itself is a moderately complicated function, but consider the situation as the number of electrons increases. The total number of variables in the wave function for chromium (a 24-electron system) is 72.

A. THE SOLUTIONS TO THE SCHRÖDINGER EQUATION FOR HYDROGEN

We shall now examine the solutions to the Schrödinger equation for hydrogen. The choice of hydrogen is important and should not be forgotten.

When the Schrödinger equation is solved for hydrogen, a wave function is obtained for each of the energy levels given in Fig. 2-15a. Several of these functions are listed in Table 2-1 and in Appendix 1.

Table 2-1 *Some Wave Functions for the Hydrogen Atom**

n	l	m_l	$\Psi = [R_n^l(r)][\Theta_l^{m_l}(\theta)][\Phi^{m_l}(\varphi)]$
1	0	0	$\Psi = [2(Z/a_0)^{3/2}e^{-Zr/a_0}][(1/4\pi)^{1/2}][1]$
2	0	0	$\Psi = [(Z/2a_0)^{3/2}(2 - Zr/a_0)e^{-Zr/2a_0}][(1/4\pi)^{1/2}][1]$
2	1	0	$\Psi = [3^{-1/2}(Z/2a_0)^{3/2}(Zr/a_0)e^{-Zr/2a_0}][(3/4\pi)^{1/2}\cos\theta][1]$
2	1	± 1	$\Psi = [3^{-1/2}(Z/2a_0)^{3/2}(Zr/a_0)e^{-Zr/2a_0}][(3/4\pi)^{1/2}\sin\theta][\cos\varphi]$
			$\Psi = [3^{-1/2}(Z/2a_0)^{3/2}(Zr/a_0)e^{-Zr/2a_0}][(3/4\pi)^{1/2}\sin\theta][\sin\varphi]$
3	0	0	$\Psi = [(2/27)(Z/3a_0)^{3/2}(27 - 18Zr/a_0 + 2(Zr/a_0)^2)e^{-Zr/3a_0}][(1/4\pi)^{1/2}][1]$
3	1	0	$\Psi = [(1/81)3^{-1/2}(2Z/a_0)^{3/2}(6 - Zr/a_0)(Zr/a_0)e^{-Zr/3a_0}][(3/4\pi)^{1/2}\cos\theta][1]$
3	1	± 1	$\Psi = [(1/81)3^{-1/2}(2Z/a_0)^{3/2}(6 - Zr/a_0)(Zr/a_0)e^{-Zr/3a_0}][(3/4\pi)^{1/2}\sin\theta][\cos\varphi]$
			$\Psi = [(1/81)3^{-1/2}(2Z/a_0)^{3/2}(6 - Zr/a_0)(Zr/a_0)e^{-Zr/3a_0}][(3/4\pi)^{1/2}\sin\theta][\sin\varphi]$

*$a_0 = 0.529 \times 10^{-8}$ cm.

Ch. 2 The Electronic Structure of Atoms

The beauty of the theory is apparent when the solutions are written in generalized form.

$$\psi = R_n^l(r)\Theta_l^{m_l}(\theta)\Phi^{m_l}(\varphi) \tag{2.24}$$

The general form contains r, θ, and φ (the spherical coordinates, see Fig. 2-22) as variables and the quantities n, l, and m_l. These quantities are not new; they are the observables previously encountered in the atomic spectrum experiment and the Stern-Gerlach experiment. When values of these quantities (n, l, m_l) are placed into the general solution, the wave functions given in Table 2-1 are obtained. The theory incorporates the quantized properties that we have previously encountered as experimental quantities. They are the three parameters necessary in order to specify the type of vibration of a three-dimensional wave.

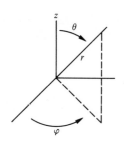

Figure 2-22 *Spherical polar coordinates.*

The possible values of n, l and m_l are given in Table 2-2. The number n, called the principal quantum number, can take values of 1, 2, 3, 4, and so forth. The quantum number l (the azimuthal quantum number)

Table 2-2 *Possible Values of the Quantum Numbers*

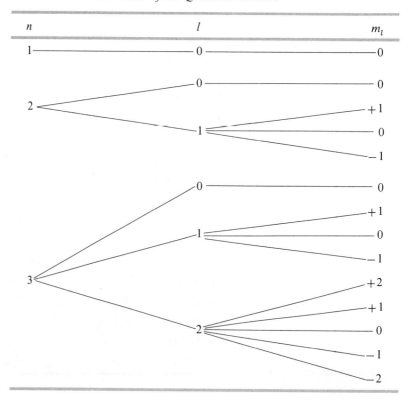

has values from 0 to $n-1$, while the quantum number m_l (the magnetic quantum number) takes all integral values between and including $+l$ and $-l$. Thus if $n = 1$, l may have only one value (0) and m_l may have only one value. If $n = 2$, there are two values for l, 0 and $n-1$ or 1. For each value of l, m_l has values from $+l$ to $-l$ (0 for $l = 0$ and $+1, 0, -1$ for $l = 1$). Note that l, the angular momentum vector can never be negative, but its projection on the z axis can be. For $n = 3$, l will have three values, 0, 1, and 2. The m_l values corresponding to each of these values of l may easily be determined and checked with the table.

B. THE PHYSICAL SIGNIFICANCE OF ψ

If the electron is characterized as a wave spread between the boundaries, it is not possible to describe the position of the electron. The electron will certainly not be evenly distributed through the whole of space because the forces of attraction decrease with increasing separation of charge. The electron is more likely to be close to the nucleus. This suggests that a probability interpretation might be necessary to describe the electron distribution in an atom; however, ψ by itself cannot represent a probability function because it is negative in some areas. For example, the value of the function $\cos\theta$ (from Table 2-1) is negative when θ is larger than $90°$. Instead ψ^2 is taken as being proportional to the probability. The total probability of finding the electron must be unity, and the probability of finding the electron in any small element is equal to ψ^2 times the volume element.

There are several ways of representing the electron distribution in an atom. Generally the functions are separated into the radial portion, $R(r)$ and the angular portion $Y(\theta, \varphi)$. The radial portion of the wave function is essentially dependent upon n; the angular part is controlled by l and m_l.

The most useful representation of the radial probability distribution is a plot of the probability that an electron will be in a small spherical shell of radius r, $4\pi r^2 [R_n^l(r)]^2$. These plots, given in Fig. 2-23, also show that the electron is usually farther from the nucleus for larger values of n (the greater the energy of the orbital). When $l = 0$ the probability distribution is spherically symmetrical. For $\psi^2_{1,0,0}$ where the subscripts indicate the values of $n, l,$ and m_l, the probability increases to a maximum and then decreases. There are two values of r where $\psi^2_{2,0,0}$ is a maximum, and three values where $\psi^2_{3,0,0}$ is a maximum. For both of these functions there is a distance where the probability is zero. A place where the probability is zero is called a node. In this case the nodes are radial or spherical. The number of spherical nodes in a wave function is given by $n - 1 - l$. The functions with $l = 0$ are called s functions.

Ch. 2 The Electronic Structure of Atoms

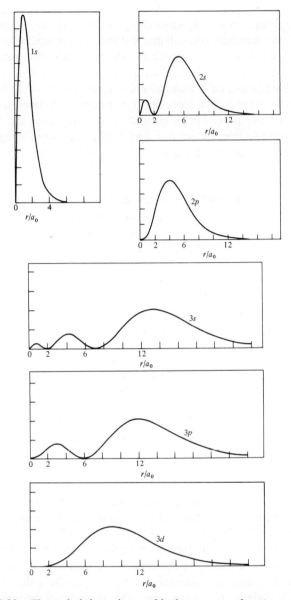

Figure 2-23 *The radial dependence of hydrogen wave functions.*

The letters s, p, d, and f are used for wave functions depending on the values of l according to the following table:

l	0	1	2	3
Designation	s	p	d	f

The angular portion of the wave function takes values other than 1 when the quantum number l has values other than 0. For $n = 2$, l may be equal to 0 or 1. When $l = 1$, there will be three wave functions: $\psi_{2,1,1}$, $\psi_{2,1,0}$, and $\psi_{2,1,-1}$. The angular dependence of these three functions is given in Fig. 2-24. These three functions are called p functions. The d functions ($l = 2$) are also given in Fig. 2-24.

Sec. IV The Electronic Structure of Atoms

To understand the variation in probability density of the angular portions, it is necessary to evaluate the angular portions of the wave functions for specific values of θ and φ. The value of the p_z function $[(3/4\pi)^{1/2} \cos \theta$, see Table 2-1] depends upon θ as follows: At $\theta = 0$, $\cos \theta = 1$; therefore, the maximum probability will be along the z axis.

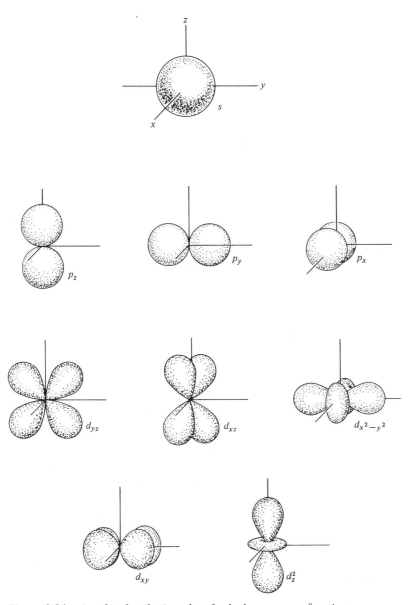

Figure 2-24 *Angular distribution plots for hydrogen wave functions.*

Ch. 2 The Electronic Structure of Atoms

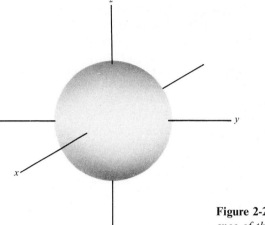

Figure 2-25 *The angular dependence of the p_z orbital.*

As θ increases toward 90°, the value of $\cos \theta$ decreases and the probability decreases until both are 0 at 90°. At $\theta = 90°$ we have a node everywhere in the xy plane. From 90 to 180° the value of $\cos \theta$ becomes more and more negative. The probability, ψ^2, increases until it is a maximum again along the minus z axis (at $\theta = 180°$). Thus at any particular radius the probability varies with θ from a maximum along the $+z$ axis to 0 in the xy plane. It then increases again to a maximum on the $-z$ axis. By looking at the angular dependence on the surface of a sphere, we obtain Fig. 2-25. The node in the xy plane cuts the sphere in half.

The variation in probability of the p_x function is similar to the p_z except the node is in the yz plane and the maximum probability is along the x axis.

With most of the d orbitals (for example, $d_{x^2-y^2} = \sin^2 \theta \cos 2\varphi$) there are two nodal planes (at $\varphi = 45°$ and $135°$ in this case). The number of nonspherical nodes is always given by the quantum number l.

C. WAVE FUNCTIONS AND ENERGY LEVELS

See equation 2.18

It may be remembered that the energy of the states of the hydrogen atom are proportional to $1/n^2$. From this the energy level diagram in Fig. 2-15a was drawn. Wave functions may now be assigned to the energy levels given in Fig. 2-15a according to the value of n. The function $\psi_{1,0,0}$ corresponds to the lowest energy state. This function is called the 1s function. The four functions with n equals 2 (the 2s and 2p functions) correspond to the next highest state. In the hydrogen atom the 2s and 2p functions have the same energy (the energy of the functions is independent of l for the particular case of hydrogen). The next highest state is assigned the 3s, 3p, and 3d functions.

Sec. IV The Electronic Structure of Atoms

Each of the wave functions under discussion could represent an electron in an atomic system. The question that arises now is how many electrons can each wave function accommodate? The answer to this question can also be obtained from the Stern-Gerlach experiment. The determination also represents the evaluation of the fourth quantum number suggested by Einstein's theory of relativity. The fourth quantum number represents the fourth number necessary to describe a four-dimensional wave.

For hydrogen the results actually obtained show two spots on the screen instead of one. The beam of atoms is not split because of angular momentum. The angular momentum vector (l) is 0 and the z component is 0. The beam is split because there are two possible values of the fourth quantum number m_s, which represents the z component of the *spin momentum vector*. The two values observed for hydrogen are assigned numerical values of $+\frac{1}{2}$ and $-\frac{1}{2}$. A complete description of a wave function then involves four quantum numbers: n, l, m_l and m_s. The fourth quantum number doubles the number of wave functions which must be placed onto the energy level diagram.

All of the functions described above are solutions to the Schrödinger equation *for hydrogen*. They are, of course, not strictly applicable to other atoms; however, it is necessary to use these functions when dealing with other atoms because solution of the Schrödinger equation for other atoms is nearly impossible. The approximation works out fairly well. The electronic structures of all 105 atoms may be described in terms of the solutions for hydrogen.

To obtain the electronic structure of any element, begin filling the lowest energy atomic orbitals first. Beginning with hydrogen, either $\psi_{1,0,0,+1/2}$ or $\psi_{1,0,0,-1/2}$ may be used. Helium uses both of these functions, $\psi_{1,0,0,+1/2}$ and $\psi_{1,0,0,-1/2}$. The electronic structure of helium may be represented as He $\frac{\cdot\cdot}{1s}$. The next element, lithium, has the electronic structure

$$\text{Li} \ \frac{\cdot\cdot}{1s}, \ \frac{\cdot}{2s}, \ \frac{\ }{\ } \ \frac{\ }{2p} \ \frac{\ }{\ }$$

The Stern-Gerlach experiment may be used to verify electronic structures. For example, another electronic structure which might be drawn for lithium is Li $\frac{\cdot\cdot}{1s}, \frac{\ }{2s}, \frac{\cdot}{\ } \frac{\ }{2p} \frac{\ }{\ }$. Since only one spot is obtained (two with spin), this structure is ruled out. The outside electron in lithium is not a p electron. Only at boron are the three spots corresponding to $l = 1$ or a $2p$ electron first observed. The electronic structures of the elements are given in Table 2-3. Note that in argon (Ar) all the functions through $\psi_{3,1,m_l,m_s}$ are occupied. For the next element, potassium, according to the diagram given earlier (Fig. 2-15b)

Ch. 2 The Electronic Structure of Atoms

Table 2-3 *Electronic Structures of Atoms*

Element	Atomic number	K	L		M			N				O				P				Q
		1s	2s	2p	3s	3p	3d	4s	4p	4d	4f	5s	5p	5d	5f	6s	6p	6d	6f	7s
H	1	1																		
He	2	2																		
Li	3	2	1																	
Be	4	2	2																	
B	5	2	2	1																
C	6	2	2	2																
N	7	2	2	3																
O	8	2	2	4																
F	9	2	2	5																
Ne	10	2	2	6																
Na	11	2	2	6	1															
Mg	12	2	2	6	2															
Al	13	2	2	6	2	1														
Si	14	2	2	6	2	2														
P	15	2	2	6	2	3														
S	16	2	2	6	2	4														
Cl	17	2	2	6	2	5														
Ar	18	2	2	6	2	6														
K	19	2	2	6	2	6		1												
Ca	20	2	2	6	2	6		2												
Sc	21	2	2	6	2	6	1	2												
Ti	22	2	2	6	2	6	2	2												
V	23	2	2	6	2	6	3	2												
Cr	24	2	2	6	2	6	5	1												
Mn	25	2	2	6	2	6	5	2												
Fe	26	2	2	6	2	6	6	2												
Co	27	2	2	6	2	6	7	2												
Ni	28	2	2	6	2	6	8	2												
Cu	29	2	2	6	2	6	10	1												
Zn	30	2	2	6	2	6	10	2												
Ga	31	2	2	6	2	6	10	2	1											
Ge	32	2	2	6	2	6	10	2	2											
As	33	2	2	6	2	6	10	2	3											
Se	34	2	2	6	2	6	10	2	4											
Br	35	2	2	6	2	6	10	2	5											
Kr	36	2	2	6	2	6	10	2	6											
Rb	37	2	2	6	2	6	10	2	6			1								
Sr	38	2	2	6	2	6	10	2	6			2								
Y	39	2	2	6	2	6	10	2	6	1		2								
Zr	40	2	2	6	2	6	10	2	6	2		2								
Nb	41	2	2	6	2	6	10	2	6	4		1								
Mo	42	2	2	6	2	6	10	2	6	5		1								

Table 2-3 (*Continued*)

Ele-ment	Atomic number	K	L		M			N				O				P				Q
		1s	2s	2p	3s	3p	3d	4s	4p	4d	4f	5s	5p	5d	5f	6s	6p	6d	6f	7s
Tc	43	2	2	6	2	6	10	2	6	(5)		(2)								
Ru	44	2	2	6	2	6	10	2	6	7		1								
Rh	45	2	2	6	2	6	10	2	6	8		1								
Pd	46	2	2	6	2	6	10	2	6	10										
Ag	47	2	2	6	2	6	10	2	6	10		1								
Cd	48	2	2	6	2	6	10	2	6	10		2								
In	49	2	2	6	2	6	10	2	6	10		2	1							
Sn	50	2	2	6	2	6	10	2	6	10		2	2							
Sb	51	2	2	6	2	6	10	2	6	10		2	3							
Te	52	2	2	6	2	6	10	2	6	10		2	4							
I	53	2	2	6	2	6	10	2	6	10		2	5							
Ne	54	2	2	6	2	6	10	2	6	10		2	6							
Cs	55	2	2	6	2	6	10	2	6	10		2	6			1				
Ba	56	2	2	6	2	6	10	2	6	10		2	6			2				
La	57	2	2	6	2	6	10	2	6	10		2	6	1		2				
Ce	58	2	2	6	2	6	10	2	6	10	(2)	2	6			(2)				
Pr	59	2	2	6	2	6	10	2	6	10	(3)	2	6			(2)				
Nd	60	2	2	6	2	6	10	2	6	10	(4)	2	6			(2)				
Pm	61	2	2	6	2	6	10	2	6	10	(5)	2	6			(2)				
Sm	62	2	2	6	2	6	10	2	6	10	6	2	6			2				
Eu	63	2	2	6	2	6	10	2	6	10	7	2	6			2				
Gd	64	2	2	6	2	6	10	2	6	10	(7)	2	6	(1)		2				
Tb	65	2	2	6	2	6	10	2	6	10	(8)	2	6	(1)		2				
Dy	66	2	2	6	2	6	10	2	6	10	(10)	2	6			(2)				
Ho	67	2	2	6	2	6	10	2	6	10	(11)	2	6			(2)				
Er	68	2	2	6	2	6	10	2	6	10	(12)	2	6			(2)				
Tm	69	2	2	6	2	6	10	2	6	10	13	2	6			2				
Yb	70	2	2	6	2	6	10	2	6	10	14	2	6			2				
Lu	71	2	2	6	2	6	10	2	6	10	14	2	6	1		2				
Hf	72	2	2	6	2	6	10	2	6	10	14	2	6	2		2				
Ta	73	2	2	6	2	6	10	2	6	10	14	2	6	3		2				
W	74	2	2	6	2	6	10	2	6	10	14	2	6	4		2				
Re	75	2	2	6	2	6	10	2	6	10	14	2	6	5		2				
Os	76	2	2	6	2	6	10	2	6	10	14	2	6	6		2				
Ir	77	2	2	6	2	6	10	2	6	10	14	2	6	7		2				
Pt	78	2	2	6	2	6	10	2	6	10	14	2	6	9		1				
Au	79	2	2	6	2	6	10	2	6	10	14	2	6	10		1				
Hg	80	2	2	6	2	6	10	2	6	10	14	2	6	10		2				
Tl	81	2	2	6	2	6	10	2	6	10	14	2	6	10		2	1			
Pb	82	2	2	6	2	6	10	2	6	10	14	2	6	10		2	2			
Bi	83	2	2	6	2	6	10	2	6	10	14	2	6	10		2	3			
Po	84	2	2	6	2	6	10	2	6	10	14	2	6	10		2	4			

Ch. 2 The Electronic Structure of Atoms

Table 2-3 (*Continued*)

Element	Atomic number	K	L		M			N				O				P			Q
		1s	2s	2p	3s	3p	3d	4s	4p	4d	4f	5s	5p	5d	5f	6s	6p	6d	6f	7s
At	85	2	2	6	2	6	10	2	6	10	14	2	6	10		2	5			
Rn	86	2	2	6	2	6	10	2	6	10	14	2	6	10		2	6			
Fr	87	2	2	6	2	6	10	2	6	10	14	2	6	10		2	6			1
Ra	88	2	2	6	2	6	10	2	6	10	14	2	6	10		2	6			2
Ac	89	2	2	6	2	6	10	2	6	10	14	2	6	10		2	6	(1)		(2)
Th	90	2	2	6	2	6	10	2	6	10	14	2	6	10		2	6	(2)		(2)
Pa	91	2	2	6	2	6	10	2	6	10	14	2	6	10	(2)	2	6	(1)		(2)
U	92	2	2	6	2	6	10	2	6	10	14	2	6	10	(3)	2	6	(1)		(2)
Np	93	2	2	6	2	6	10	2	6	10	14	2	6	10	(5)	2	6			(2)
Pu	94	2	2	6	2	6	10	2	6	10	14	2	6	10	(6)	2	6			(2)
Am	95	2	2	6	2	6	10	2	6	10	14	2	6	10	(7)	2	6			(2)
Cm	96	2	2	6	2	6	10	2	6	10	14	2	6	10	(7)	2	6	(1)		(2)
Bk	97	2	2	6	2	6	10	2	6	10	14	2	6	10	(8)	2	6	(1)		2
Cf	98	2	2	6	2	6	10	2	6	10	14	2	6	10	(10)| 2	6			2
Es	99	2	2	6	2	6	10	2	6	10	14	2	6	10	11	2	6			2
Fm	100	2	2	6	2	6	10	2	6	10	14	2	6	10	12	2	6			2
Md	101	2	2	6	2	6	10	2	6	10	14	2	6	10	13	2	6			2
No	102	2	2	6	2	6	10	2	6	10	14	2	6	10	14	2	6			2

we expect that the 3d levels will be used. The results then should show five values for m_l (simplifying again because for every value of m_l there are two values for m_s). This result is not obtained; instead a pattern similar to that of hydrogen is obtained. In view of this, l must equal 0 for the last electron in potassium. Thus the external electron in potassium must be an s electron. The electronic structure of potassium is:

$$K \frac{\cdot\cdot}{1s}, \frac{\cdot\cdot}{2s}, \frac{\cdot\cdot \;\; \cdot\cdot \;\; \cdot\cdot}{2p}, \frac{\cdot\cdot}{3s}, \frac{\cdot\cdot \;\; \cdot\cdot \;\; \cdot\cdot}{3p}, \frac{\cdot}{4s}$$

It is apparent from this discussion that n alone does not control the energy of the functions when more than one electron is involved. First for functions with the same principal quantum number the energy varies $s < p < d$. In addition the nd level usually has a higher energy than the $(n + 1)s$ level (for example $E_{4s} < E_{3d}$ and $E_{5s} < E_{4d}$). The electronic structures of atoms may be obtained by using Fig. 2-26 to determine the order of filling.

In order to explain the difference in energy between the s and p orbitals, it is necessary to look more closely at the radial plots. In Fig. 2-27 the probability density of the 2s and 2p functions for hydrogen are plotted on a graph showing the variation with r of the effective nuclear charge for the lithium ion. The effective nuclear charge is the charge on the nucleus minus the average amount of negative charge present between the nucleus and the point under consideration. The effective nuclear charge is very close to three at very small distances and drops off rapidly to one. The drop in the effective nuclear charge is due primarily to the presence of the 1s electrons close to the nucleus. Because the 2s function penetrates (has a substantial electron density) into the region where z is greater than 1, the electrostatic interaction between the nucleus and a 2s electron is much greater than the interaction with a 2p electron, which is nonpenetrating. The result is that the 2s orbital is, on the average, closer to the nucleus than the 2p orbital, and the energy of the 2s electron is lower than that of the 2p electron.

In the case of potassium (K), the 4s orbital penetrates the inner core of completed electron shells while the 3d electron does not penetrate.

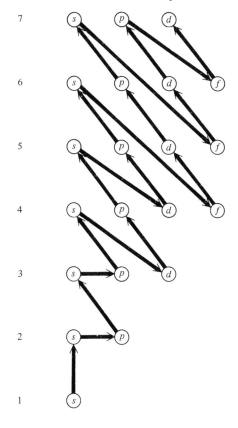

Figure 2-26 *The order of filling orbitals.*

Ch. 2 The Electronic Structure of Atoms

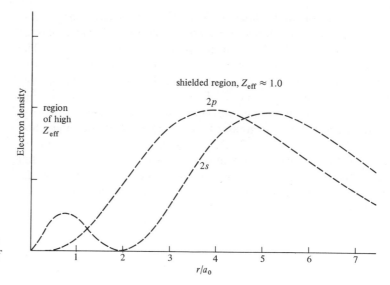

Figure 2-27 *A representation of the relative penetration of the 2s and 2p orbitals.*

Thus the energy of the 4s orbital is less than the energy of the 3d orbital. The same situation holds in all of the following cases: $E_{5s} < E_{4d}$, $E_{6s} < E_{5d}$, and $E_{6s} < E_{4f}$. Thus, for atoms other than hydrogen the quantum number l does influence the energy of the orbital. There is no such effect in hydrogen atoms because there is no inner core to be penetrated. In this case the electron feels the full positive charge anywhere inside the atom.

With only a few exceptions the build up of electronic structure follows a regular pattern as given in the energy sequence in Fig. 2-26. There are some small exceptions at chromium (Cr) and copper (Cu) and in some of the elements below them in the periodic table, but the exceptions are of only minor chemical significance.

In writing and thinking of electronic structures it is not necessary to continue to write them out completely. Only the outer level, the valence layer, must be shown. For example, the electronic structure K $\frac{\cdot}{4s}$, $- \frac{}{4p} -$ is sufficient for potassium, the argon core being implied in the symbol K.

The element symbols are used in several different ways: for the neutral atom, for the nucleus alone, and as here for the nucleus plus the filled core. The meaning is almost always clear from the context.

Two rules, Hund's rule and the Pauli exclusion principle, are used to arrive at the correct electronic structure for certain elements. Hund's rule states that for a particular atom the spins of the electrons will remain unpaired if possible. As a result the most stable electron configuration for nitrogen is

$$N \frac{\colon\colon}{2s}, \; \frac{\cdot \;\; \cdot \;\; \cdot}{2p} \quad \text{or} \quad \colon\colon, \; \frac{\uparrow \;\; \uparrow \;\; \uparrow}{} \tag{2.25}$$

where ↑ indicates a $+\frac{1}{2}$ spin and ↓ indicates a $-\frac{1}{2}$ spin.

The Pauli exclusion principle states that no two electrons in an atom can have the same four quantum numbers. Thus in oxygen the structure is

$$\text{O}\!:\!\!:, \quad \underset{+1}{\uparrow\downarrow}\ \underset{0}{\uparrow}\ \underset{-1}{\uparrow} \quad \text{instead of} \quad \text{O}\!:\!\!:, \quad \underset{+1}{\uparrow\uparrow}\ \underset{0}{\uparrow}\ \underset{-1}{\uparrow} \quad (2.26)$$

In this last structure the two electrons in the $p(+1)$ orbital have the same four quantum numbers n, l, m_l, m_s so the structure is not allowed.

D. ELECTRONIC STRUCTURE AND THE PERIODIC TABLE

Figure 2-28 shows the periodic table broken apart into groups of 2, 6, 10, and 14 elements. These are just the numbers of electrons which can occupy orbitals with $l = 0, 1, 2$, and 3 respectively. The shape of the periodic table is determined by the capacity and energy of the atomic orbitals. Periodicity is the result of the repeated filling of the s and p orbitals in the second and third rows of the table and of the s, d, p orbitals of rows 4, 5, and 6. After a p level is filled, for example, the $2p$ at neon (Ne), a new s level is used, for example, the $3s$ for sodium (Na). Each of the elements in a vertical group has an identical number of outer electrons, which generally occupy the same type of orbital.

There is a slight modification of Fig. 2-28 which will make it more useful chemically. Figure 2-26 shows correctly that the $4s$ orbital is filled before the $3d$, but the $4s$ orbital is not always lower in energy

Figure 2-28 *Sections of the periodic table according to the sets 2, 6, 10.*

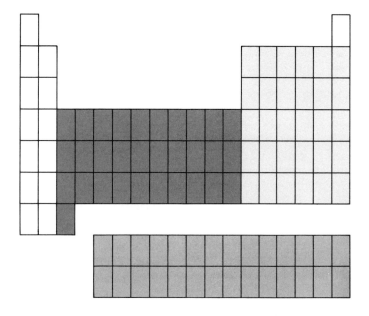

Ch. 2 The Electronic Structure of Atoms

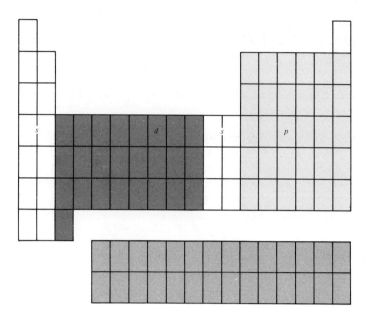

Figure 2-29 *Four sections of the periodic table.*

than the $3d$. The electronic structure of copper is $3d^{10}$, $4s^1$, and not $4s^2$, $3d^9$. This reversal is generally characteristic of the IB and IIB families, and it is an s orbital which is being filled for most of these elements. This change is shown in Fig. 2-29.

In the next chapter some of the properties that are periodic in nature will be discussed from the point of view of the electronic structure of atoms.

QUESTIONS

1. List the experiments important in the development of the theory of electronic structure of atoms.
2. List what each of these experiments measures.
3. What is the interpretation of ψ^2?
4. In what way is the particle model of the electronic structure of atoms inadequate?
5. For a particular atom the maximum value of m_l observed from experiment is $+2$. What are all the possible values of l, n, and m_l?
6. What condition must be imposed on an atomic system in order to explain quantized energies?
7. What information is available from atomic spectra?
8. List the information available from the Stern-Gerlach experiment.
9. Select from the list in Question 8 one aspect and explain the significance relative to the particle model.
10. If the particle model were valid, would the results of the Stern-Gerlach experiment depend upon the element used? Why?

Sec. IV The Electronic Structure of Atoms

11. Write the electronic structure (valence level only) of any element in the periodic table.
12. Determine the position of the nodes for the d_{xy} orbital.
13. How many different sets of quantum numbers are available for electrons with:
    (a) $n = 3$?
    (b) $n = 3, l = 2$?
    (c) $n = 2, l = 0$?
    (d) $n = 2, l = 6$?
14. What notation (s, p, d, f) is used to describe the electrons in Question 13(b), (c), and (d)?
15. Show that the angular parts of the $3d$ wave functions (for example, $\sin^2 \theta \cos 2\varphi$ for $3d_{x^2-y^2}$ and so forth) account for the diagrams shown in Fig. 2-24.
16. What are the shapes of the nodes for the $3d_{z^2}$ orbital?

PROBLEMS I

1. Calculate the wavelengths which would be associated with a nitrogen molecule at $300°C$ (velocity 7×10^4 cm sec^{-1}), a base ball (mass 100 g, velocity 4000 cm sec^{-1}), and a car traveling 30 miles per hour.
2. Show that the five d orbitals can produce a sphere (that is, add the electron densities for the five functions).
3. Radii of atoms are measured in angstroms (1 Å $= 10^{-8}$ cm) and the cross-sectional area of nuclei are measured in barns (1 barn $= 10^{-24}$ cm^2). What fraction of the volume of an atom 1 Å in radius is occupied by a nucleus with a cross section of 1 barn?
4. To get force in dynes from Coulomb's law, the charges should be in esu and distances in cm. The charge on an electron is 4.80×10^{-10} esu. Calculate the force of attraction between an electron and a helium nucleus at a distance of 1.00 Å and at 5.00 Å.

PROBLEMS II

1. What element has an atomic number of 10?
2. How many protons are present in a neon atom?
3. What isotope of neon contains 12 neutrons in its nucleus?
4. How many electrons are present in an atom of ^{20}Ne?
5. What element has only one electron?
6. What is the lowest value possible for the principal quantum number n?
7. What values of the quantum numbers l, m_l, and m_s are possible for an electron with $n = 1$?
8. List the four quantum numbers for the electron in a hydrogen atom in its ground state.
9. List the four quantum numbers for a $3p$ electron with $m_l = +1$.
10. What other values of m_l are possible for a $3p$ electron?
11. How many electrons can fit into a $3p$ subshell?
12. What element has the electronic structure $1s^2, 2s^2, 2p$?
13. Write the electronic structure of aluminum (Al) (just below boron (B) in the periodic table).

Ch. 2 The Electronic Structure of Atoms

14. Which orbital is lower in energy, the 3d or 4p?
15. Write the electronic structure for the element with atomic number 21 (scandium, Sc).
16. Write the electronic structure for iron (Fe, atomic number 26).
17. How many d electrons with parallel spins does an atom of iron (Fe) have?
18. How many valence electrons does a phosphorus (P) atom have?
19. What is the capacity of the 4f subshell?
20. What is the electronic structure of arsenic (As), atomic number 33?

But how is it that they suffer themselves to incline to and be swayed by probability, if they know not the truth itself?

Montaigne
Essays II 12

Chapter 3 Periodic Properties

Once the wave-particle duality exhibited by electrons is accepted, it can be used to understand and interpret the properties of atoms. This will be the subject of this chapter. Some of the properties with which we will be concerned in this chapter were introduced in Chapter 1. These are listed along with a large number of others in Table 3-1. All of the properties listed are dependent upon electronic structure; however, some are more obviously periodic and more important to understanding the nature of atoms than others. The importance of

Table 3-1 *Some Properties Dependent Upon Electronic Structure*

Atomic radius (and volume)	Ionic radius
Boiling point	Ionization potential
Coefficient of expansion	Ion mobility
Compressibility	Magnetic behavior
Density	Malleability
Electrical resistance	Melting point
Electron affinity	Optical spectrum
Electronegativity	Oxidation number or oxidation state
Hardness	
Heat of formation of a given type compound	Parachor
	Refractive index
Heat of solvation of ions	Standard redox potential
Heats of fusion, vaporization, and sublimation	Thermal conductivity

three of the properties listed cannot be overemphasized. These are ionization potential, electronegativity, and size. The ionization potential of an atom represents a direct method of obtaining information about orbital energies. Electronegativity is a property that also provides information on orbital energies. Although the size of an atom is a fairly elusive quantity because of the nature of atomic orbitals, it is very intimately connected with both ionization potential and electronegativity. In the first two sections of this chapter we shall discuss these three properties in detail. In the last two sections an example of the application of electronegativity to chemical bonding will be presented. Since ionization potentials are measured directly, we will begin our discussion with this property of atoms.

I. Ionization Potentials

The complete removal of an electron from an atom results in ionization, and the work necessary to remove completely the most loosely bound electron from the ground state of the atom is referred to as the ionization potential. If an atom contains several electrons, we may distinguish the first ionization potential, the second ionization potential, the third, and so on. In each case the most loosely bound electron is removed.

The first ionization potentials of the elements are given in Table 3-2; they are also plotted against atomic number in Fig. 3-1. The successive ionization potentials of oxygen are given in Table 3-3. The factors that influence the ionization potential may be determined by examining these data.

In Table 3-3 there is a general increase in ionization potential as the charge on the ion increases. The dependence upon nuclear charge is nicely illustrated by the magnitude of the eighth ionization potential. The removal of the electron from O^{7+} requires 20,094 kcal mole^{-1}, a quantity that is 64.06 or 8^2 times the energy required to remove the $1s$ electron from hydrogen (313.7 kcal mole^{-1}). However, in looking at some of the other potentials we see that there is not, in general, a simple relationship between the potential and the square of the nuclear charge. The potential required to remove an electron from O^{6+} is 17,049 kcal mole^{-1} or 7.4^2 times that required to remove the electron from hydrogen. The complicating feature here may be understood as follows. In O^{7+} the single electron is attracted by the full $+8$ charge on the nucleus. In O^{6+} the two electrons *repel* each other to some extent so that the combined attraction to the nucleus and repulsion between electrons *reduces* the net attraction between the nucleus and each electron. The result is that either electron is easier to remove than it

Sec. I Ionization Potentials

Table 3-2 *Ionization Potentials of the Elements*

Atomic number	Symbol	Valence configuration	Ionization potential, kcal mole^{-1}							
			1st	2nd	3rd	4th	5th	6th	7th	8th
1	H	$1s^1$	313.69							
2	He	$1s^2$	566.81	1,254.5						
3	Li	$2s^1$	124.29	1,743.8	2,823.0					
4	Be	$2s^2$	214.92	419.83	3,547.8	5,019.2				
5	B	$2s^22p^1$	191.31	579.94	874.44	5,979.4	7,843.3			
6	C	$2s^22p^2$	259.75	562.11	1,103.7	1,486.8	9,039.2	11,298		
7	N	$2s^22p^3$	335.29	682.69	1,093.6	1,786.0	2,256.7	12,727	15,377	
8	O	$2s^22p^4$	313.94	810.47	1,266.8	1,784.7	2,625.9	3,184.1	17,049	20,094
9	F	$2s^22p^5$	401.66	806.64	1,444.6	2,011.5	2,633.8	3,623.1	4,269.3	21,990
10	Ne	$2s^22p^6$	497.15	947.07	1,475.8	2,240.5	2,914.8	3,641.4		
11	Na	$3s^1$	118.48	1,090.5	1,652.2	2,280.2	3,196.1	3,974.6	4,806.6	6,091.4
12	Mg	$3s^2$	176.27	346.59	1,847.6	2,520.2	3,256.8	4,309	5,195.6	6,133.3
13	Al	$3s^23p^1$	137.99	434.06	655.83	2,766.3	3,545.9	4,391.1	5,578.9	6,575.1
14	Si	$3s^23p^2$	187.92	376.8	771.59	1,040.7	3,844.8	4,729.8	5,682.2	7,007.2
15	P	$3s^23p^3$	243.28	453.13	695.4	1,184.2	1,499.1	5,082.7	6,071.9	7,131.5
16	S	$3s^23p^4$	238.83	539.6	807.1	1,090.5	1,671.8	2,029.9	6,479.6	7,582.1
17	Cl	$3s^23p^5$	300.01	548.83	920.09	1,233.7	1,563.5	2,229.9	2,635.1	8,031.8
18	Ar	$3s^23p^6$	363.31	636.92	943.15	1,378.8	1,729.5	2,105.4	2,859.4	3,308.2
19	K	$4s^1$	100.06	733.54	1,060.8	1,404.4		2,299.1	2,721.1	3,574.3
20	Ca	$4s^2$	140.92	273.72	1,180.9	1,545	1,946		2,951.7	3,389.8
21	Sc	$3d^14s^2$	151.27	295.17	570.73	1,704.1	2,121.5	2,562	3,246.8	3,666.5
22	Ti	$3d^24s^2$	157.5	312.92	648.91	997.11	2,301.4	2,767.2	3,482.1	4,005.5
23	V	$3d^34s^2$	155.42	337.83	684.88	1,106.9	1,503.5	2,972.4	3,716	
24	Cr	$3d^54s^1$	155.98	380.26	714.86	1,162.2	1,678.8	2,082		
25	Mn	$3d^54s^2$	171.38	360.66	737.92	1,199.1	1,745.6			
26	Fe	$3d^64s^2$	182.17	373.11	703					
27	Co	$3d^74s^2$	181.25	393.17						
28	Ni	$3d^84s^2$	176.02	418.54						
29	Cu	$3d^{10}4s^1$	178.12	470.19	680.27					
30	Zn	$3d^{10}4s^2$	216.56	414.16	922.4					
31	Ga	$3d^{10}4s^24p^1$	138.36	472.96	705.64	1,471.2				

69

Table 3-2 (*Continued*)

Atomic Number	Symbol	Valence configuration	1st	2nd	3rd	4th	5th	6th	7th	8th
32	Ge	$3d^{10}4s^24p^2$	181.71	367.35	785.65	1,049.2	2,144.6			
33	As	$3d^{10}4s^24p^3$	226.22	465.81	645.68	1,150.7	1,441.2	2,943		
34	Se	$3d^{10}4s^24p^4$	224.83	495.79	781.73	985.12	1,678.8	1,877.1	2,837.30	
35	Br	$3d^{10}4s^24p^5$	273.03	498.1	592.64	1,153				
36	Kr	$3d^{10}4s^24p^6$	322.75	566.35	848.61	1,568.1				
37	Rb	$5s^1$	96.299	634.15	1,083.8	1,844.8				
38	Sr	$5s^2$	131.26	254.28						
39	Y	$4d^15s^2$	147.05	282.09	470.42					
40	Zr	$4d^25s^2$	157.62	297.84	553.44	779.43				
41	Nb	$4d^45s^1$	158.68	320.42	558.05					
42	Mo	$4d^55s^1$	164.44	362.5						
43	Tc	$4d^55s^2$	166.72	342.9						
44	Ru	$4d^75s^1$	169.84	382.73						
45	Rh	$4d^85s^1$	172.05	367.11						
46	Pd	$4d^{10}$	192.09	447.82						
47	Ag	$4d^{10}5s^1$	174.66	495.33	827.85					
48	Cd	$4d^{10}5s^2$	207.33	389.81	876.28					
49	In	$4d^{10}5s^25p^1$	133.4	434.17	643.37	1,332.9				
50	Sn	$4d^{10}5s^25p^2$	169.08	337.37	703.33	908.56	1,860.9			
51	Sb	$4d^{10}5s^25p^3$	199.22	438.14	569.58	1,014.6	1,279.8	2,471		
52	Te	$4d^{10}5s^25p^4$	207.77	495.79	703.33	869.36	1,383.6	1,660.3	3,165.4	
53	I	$4d^{10}5s^25p^5$	240.75	438.14						
54	Xe	$4d^{10}5s^25p^6$	279.65	489.1	737.92	1,060.8	1,752.6			
55	Cs	$6s^1$	89.773	578.81	807.1	1,176.1	1,337.5			
56	Ba	$6s^2$	120.14	230.62						
57	La	$5d^16s^2$	129.37	263.58	470.42					
58	Ce	$4f^26s^2$	159.34	341.29	463					
59	Pr	$4f^36s^2$	132.83							
60	Nd	$4f^46s^2$	145.28							
61	Pm	$4f^56s^2$								
62	Sm	$4f^66s^2$	129.14	258.27						

Sec. 1 Ionization Potentials

Table 3-2 (Continued)

Atomic number	Symbol	Valence configuration	Ionization potential, kcal mole^{-1}							
			1st	2nd	3rd	4th	5th	6th	7th	8th
63	Eu	$4f^76s^2$	130.75	259.19						
64	Gd	$4f^75d^16s^2$	142.05	276.72						
65	Tb	$4f^96s^2$	155.42							
66	Dy	$4f^{10}6s^2$	157.27							
67	Ho	$4f^{11}6s^2$								
68	Er	$4f^{12}6s^2$								
69	Tm	$4f^{13}6s^2$								
70	Yb	$4f^{14}6s^2$	143.43	279.03						
71	Lu	$4f^{14}5d^16s^2$	141.82	338.98						
72	Hf	$4f^{14}5d^26s^2$	126.83	343.59						
73	Ta	$5d^36s^2$	177.56							
74	W	$5d^46s^2$	184.02							
75	Re	$5d^56s^2$	181.48							
76	Os	$5d^66s^2$	200.62							
77	Ir	$5d^76s^2$	212.15							
78	Pt	$5d^96s^1$	207.54	427.99						
79	Au	$5d^{10}6s^1$	212.61	472.73						
80	Hg	$5d^{10}6s^2$	240.61	432.4	790.96	1,660.3	1,890.9			
81	Tl	$5d^{10}7s^26p^1$	140.8	470.89	684.88	1,164.5				
82	Pb	$5d^{10}6s^26p^2$	170.99	346.55	735.61	971.06	1,600.4			
83	Bi	$5d^{10}6s^26p^3$	168.04	445.06	586.19	1,040	1,284.4	2,180		
84	Po	$5d^{10}6s^26p^4$	194.4							
85	At	$5d^{10}6s^26p^5$								
86	Rn	$5d^{10}6s^26p^6$	247.78							
87	Fr	$7s^1$								
88	Ra	$7s^2$	121.69	233.83						
89	Ac	$6d^17s^2$								
90	Th	$6d^27s^2$			677.96					
91	Pa	$6d^37s^2$								
92	U	$5f^36d^17s^2$	92.24							

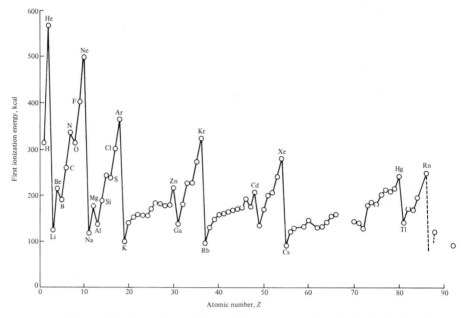

Figure 3-1 *Periodic variations with atomic number of the first ionization energies.*

would be if there were no electron-electron repulsions. The decreased effective nuclear charge acting on a particular electron may be considered to be due to the *shielding* of the nucleus by the other electron present. The shielding of an electron by other electrons in the same orbital is one factor that will affect the ionization potential of an atom.

A slightly different view of the concept of shielding was introduced in Chapter 2. An electron that is close to the nucleus is more effective at shielding the nuclear charge than an electron that is farther from

Table 3-3 *Successive Ionization Potentials of Oxygen*

	Reaction	Ionization potential, kcal mole^{-1}
1st	$O \longrightarrow O^+ + e^-$	313.94
2nd	$O^+ \longrightarrow O^{2+} + e^-$	810.47
3rd	$O^{2+} \longrightarrow O^{3+} + e^-$	1,266.8
4th	$O^{3+} \longrightarrow O^{4+} + e^-$	1,784.7
5th	$O^{4+} \longrightarrow O^{5+} + e^-$	2,625.9
6th	$O^{5+} \longrightarrow O^{6+} + e^-$	3,184.1
7th	$O^{6+} \longrightarrow O^{7+} + e^-$	17,049
8th	$O^{7+} \longrightarrow O^{8+} + e^-$	20,094

the nucleus. An electron that is close to the nucleus, a $1s$ electron for example, effectively reduces the nuclear charge by almost a full unit for an electron that is farther out. The $2s$ electron in lithium is quite well shielded by the $1s$ electrons present, and the ionization potential of lithium is actually smaller than that of hydrogen.

For the hydrogen atom and hydrogen-like ions (Li^{2+}, Be^{3+}), the $2s$ and $2p$ orbitals are equal in energy; however, in an atom with two or more electrons, the $2s$ orbital is always lower in energy than the $2p$. As shown in Fig. 2-27, the $2p$ orbital is more completely shielded by the filled $1s$ orbital than the $2s$ orbital. This is true because there is a fair probability that the $2s$ electron will be closer to the nucleus than the $1s$ electron, whereas the probability that the $2p$ electron will be closer to the nucleus than the $1s$ electron is quite small. These possibilities are apparent from Fig. 2-27. Notice in this figure that the $2s$ orbital has an inner bump of fairly high probability, whereas the $2p$ does not.

We say that the $2s$ orbital *penetrates* the inner core of $1s$ electrons to a greater extent than the $2p$ orbital. Thus on the average, the $2s$ electron of an oxygen atom will be close to the nucleus more often than a $2p$ electron, and the energy of the $2s$ will be lower than the energy of the $2p$. In this sense the two terms penetration and shielding are used interchangeably. It is only when we consider electrons in the same subshell that the term shielding must be used exclusively.

It is important to note that the radial distribution plots given in Fig. 2-24 and those used in Fig. 2-27 *do not* show that the $2s$ orbital is, on the average, closer to the nucleus than the $2p$ orbital. These plots are all for hydrogen. In hydrogen, the energy of the $2s$ orbital is the same as that of the $2p$; therefore the average distance for an electron in either orbital is the same. It is only in other atoms with completed cores that the energy of an orbital depends upon the quantum number l. In these orbitals the average distance from the nucleus does depend upon l because of the differing degree of penetration of the orbitals. In Fig. 3-2, the calculated radial dependence of the $2s$, $3s$, and $4s$ orbitals in lithium, sodium, and potassium are plotted. Notice that the position of maximum probability for these orbitals is quite different from that given in Fig. 2-24 for hydrogen orbitals.

In general the larger the angular momentum quantum number the less the penetration of that orbital. In other words an f electron penetrates the core less than a d electron; a d electron penetrates less than a p, and so on. As a result, for a given principal quantum number the s orbital will be lower in energy than a p orbital; a p orbital will be lower in energy than a d orbital, and so on. The importance of the concept of penetration cannot be overemphasized. The first application comes in interpreting the first ionization potentials of the ele-

Ch. 3 Periodic Properties

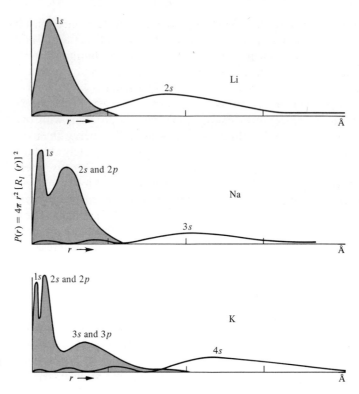

Figure 3-2 *Radial distribution functions in* Li, Na, *and* K.

ments. Before this is done, however, it is important to look briefly at an alternative method of obtaining reasonably accurate values of the orbital energies and the atomic numbers of the heavier elements. This method involves the use of X-rays.

A. X-RAY SPECTROSCOPY

Just as it is possible to study absorption and emission spectra using ultraviolet or visible light, absorption or emission may be studied in the X-ray region of the spectrum. To study absorption, a sample is arranged so that X-rays of varying energy or wavelength pass through it. Because of absorption, the intensity of the beam emerging from the sample is less than that of the incident beam. Figure 3-3 is a diagram illustrating the typical result obtained from this procedure. There is some absorption at all energies, and a general tendency toward decreased absorption as the energy increases. However as the energy of the impinging X-ray beam increases, there is a particular characteristic value of the energy at which the intensity of the emerging beam drops off very suddenly. The place where this sharp increase in absorption begins (the vertical line of Fig. 3-3) is called the absorption edge. In the ultraviolet or visible region, absorption at a particular wavelength may be considered to result in the excitation of an electron from one orbital to another of higher energy. In the X-ray region, with much larger energies involved, absorption corresponds to the complete removal of an electron from an atom. The energy required to remove a $1s$ or $2s$ (and so forth) electron from an atom may be measured in this way. The absorption edges are labeled according to the electron removed, thus a K absorption edge ($n = 1$), and L absorption edge ($n = 2$) and an M absorption edge ($n = 3$) may be observed depending upon the atom. Table 3-4 gives the absorption edges for a number of elements. From these data we see that the energy required to remove the $1s$ electron from an atom depends upon the nuclear charge of the atom. As the charge increases, the energy required also increases. In the element uranium the nuclear charge is so large that the K absorption edge is at 2.652×10^6 kcal mole^{-1}, an energy larger than that of many gamma rays. In addition, the number of observable edges

Gamma rays are photons emitted from the nucleus during a transition from a higher energy level to a lower energy level. The energy of the gamma ray corresponds to the difference in energy between the energy levels of the nucleus.

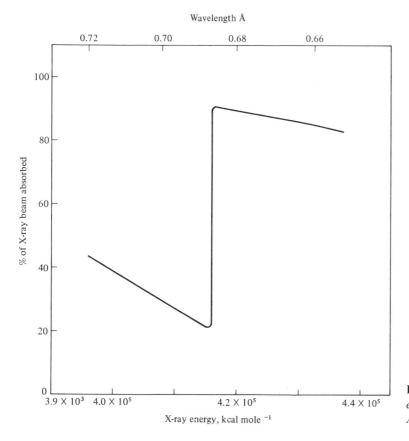

Figure 3-3 *X-ray absorption edge. (Data from H. W. Dunn, Anal. Chem., 34:116, 1962.)*

Table 3-4 *Critical K Absorption Edges*

	Energy in	
Element	kev	kcal mole^{-1}
Lithium	0.0547	1,261
Beryllium	0.112	2,580
Carbon	0.284	6,550
Sodium	1.08	24,900
Chlorine	2.822	65,050
Calcium	4.038	93,080
Scandium	4.496	1.036×10^5
Titanium	4.966	1.145×10^5
Vanadium	5.467	1.260×10^5
Chromium	5.988	1.380×10^5
Manganese	6.542	1.508×10^5
Iron	7.113	1.640×10^5
Cobalt	7.713	1.778×10^5
Nickel	8.331	1.920×10^5
Copper	8.982	2.070×10^5
Uranium	115.0	2.652×10^6

depends upon the total number of electrons in the atom. In an element such as oxygen only K and L edges are observed.

The experiment described here is much easier to perform than the determination of the ionization energy of a highly charged ion. If we compare the results of both approaches, we find additional evidence for the concept of shielding. The position of the K absorption edge in oxygen is 12,270 kcal mole^{-1}, over 70% of the energy required to remove an electron from O^{6+}. Thus the outer electrons, in this case the $2s$ and $2p$ electrons, are not very effective at shielding the $1s$ orbital from the nuclear charge.

Once an electron has been removed, the excited atom emits X-rays of characteristic frequencies. If a $1s$ electron has been removed, the electrons from higher levels may drop down to fill the $1s$ orbital, thereby emitting a quantum of energy equal to the difference between the two levels involved. The result is the K emission series. This is a series of lines of different intensity. The most intense line corresponds to the $2p$–$1s$ transition. If an L electron has been removed, the result is an L emission series. For atoms with a large number of electrons the emission spectrum will consist of a relatively large number of lines. To create an X-ray emission spectrum, the atom must contain at least one electron beyond the $1s$ level. Therefore, lithium is the first element for which a K emission series may be measured. This limitation is not as serious as the experimental problem of working with the low energy

X-rays from the light elements. The limitations of the method have been reduced in recent years and it is, in general, possible to study X-ray emission spectra for all the elements between carbon (C) and uranium (U). Very special equipment may extend this range slightly.

The first important study of this kind was carried out on 38 elements by Henry G. J. Moseley in 1914. In this study Moseley proved Rydberg's earlier prediction that the periodic classification depended upon a single "independent variable" and confirmed Mendeleev's reversal of the order of pairs of elements such as Co-Ni and Te-I, stating "the order of *atomic numbers* (as established by these X-ray studies) is the same as that of atomic weights except where the latter disagrees with the order of chemical properties." The relationship between the characteristic emission frequency and atomic number discovered by Moseley is shown in Fig. 3-4. The K emission lines of a number of elements are reproduced in Fig. 3-5. Note that there is obviously an element missing between Ca and Ti, Mendeleev's ekaboron.

B. FIRST IONIZATION POTENTIALS

In the preceding section we saw that the main factors influencing the ionization potential of an atom are the nuclear charge and the degree of penetration of the orbital from which the electron is removed. Shielding is involved as well, in so far as electrons of the same shell partially shield each other from the nucleus. These are the main factors that must be considered when discussing the main features of the plot shown in Fig. 3-1. Two other factors, atomic size and the effect of electron spin on electron-electron repulsion, must also be considered. This will be done later. Here we shall concentrate on the effects of changing nuclear charge and penetration.

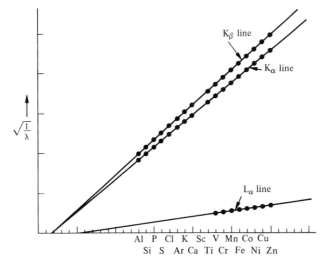

The K_α line corresponds to filling a 1s vacancy with a 2p electron, and similarly the K_β line corresponds to a $3p \to 1s$ transition.

Figure 3-4 *A plot of the X-ray emission data for some elements.*

Ch. 3 Periodic Properties

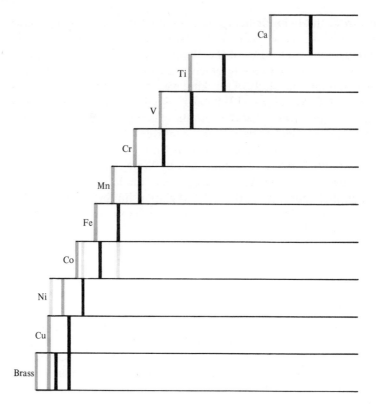

Moseley recognized the presence of iron and nickel in his cobalt sample. He also stated that the extra line in the nickel spectrum is due to manganese present as an impurity.

Figure 3-5 *X-ray lines as observed by Moseley.*

The effect of each factor taken separately is obvious. The greater the nuclear charge, the greater the ionization potential. As the penetration is increased, ionization potentials increase. Unfortunately the observed changes between elements often involve both factors at once, and only in those cases where one factor is predominant can the trends be easily interpreted. This is the case with the variations observed for group IA, the alkali metals. These atoms have a single ns electron beyond the very effective shielding inner core of electrons. The effective nuclear charge for the outermost electron in these elements is only slightly greater than $+1$. These elements have the smallest effective nuclear charge of any in the periodic table; thus they have the lowest ionization potentials. The decrease observed down group IA may be related to the increasing principal quantum number. The same general trend is observed in most families.

Across any row there is a general but irregular increase in ionization potential. The most important irregularities occur in group IIIA and group VIA. The first of these comes after the ns level has been filled. Ionization of boron (B) and aluminum (Al) results in the loss of a p

electron. In comparing beryllium (Be) with boron or magnesium (Mg) with aluminum, it is difficult to decide a priori, whether the increased nuclear charge (only partially compensated by the shielding of an additional outer electron) or the difference in energy between the s and p orbitals will predominate. It is obvious from the values given that the orbital penetration effect predominates, and the ionization potentials of the group IIIA elements are consistently lower than their group IIA or IIB neighbors.

The ionization potentials of the group IB and IIB elements are much larger than those of group IA and IIA. This may be understood from a consideration of the shielding capabilities of the orbitals involved. The effect of shielding of the inner levels is closely related to the penetration of the electron levels. A $3d$ level, for example, is more easily penetrated than a $3s$ or a $3p$ level (because the d level is nonpenetrating), thus the shielding provided by the $3d$ electrons is less than that provided by the $3p$ electrons.

The small irregularity observed at N-O, P-S, and As-Se has already been noted. The ionization potentials of oxygen, sulfur, and selenium are less than those of nitrogen, phosphorus, and arsenic in spite of the increased nuclear charge. These variations are connected with the stability of the half-filled shell occurring at nitrogen, phosphorus, and arsenic. A half-filled shell is more stable (requires more energy for ionization) than the corresponding p^4 system because the fourth p electron must occupy the same orbital as one of the other three electrons. This results in increased repulsion and less stability. Therefore the fourth p electron is easier to remove than the third.

The stability connected with a half-filled shell may be treated in a much more satisfactory manner in terms of spin correlation stabilization energy. When this is done in Chapter 4, we will have discussed all of the factors that affect orbital energies. To summarize, these factors are nuclear charge, penetration, the electrostatic effects of electron-electron repulsion, and the effect of the spin of electrons. The chemical consequences of these factors will be apparent throughout this book.

II. Atomic and Ionic Radii

Another apparently simple property of atoms that may be correlated with the periodic table is size. Because of the nature of atomic orbitals, the size of an atom cannot be determined unequivocally. In addition the electron distribution in an atom can be distorted by the presence of interacting atoms. The several different types of interactions observed (van der Waals, ionic bonds, covalent bonds, and metallic bonds) lead to different estimates of size, distinguished by the names van der Waal's radius, ionic radius, covalent radius, and metallic radius, respectively. The various types of interactions will be considered in more detail when bonding is discussed. Here we are interested in size variations and the periodic table.

A. TRENDS IN COVALENT AND METALLIC RADII

In a covalent bond a pair of electrons is shared equally.

The nonpolar covalent radius is defined as one half the distance between the nuclei of two like atoms bonded by a single covalent bond. The metallic radius is defined as one half the internuclear distance in a metallic crystal. Metallic radii are given in Table 3-5, and Table 3-6 shows the covalent radii for the nonmetals.

When considering the size variations in any period, two opposing factors are apparent from the model of the atom. Moving from left to right in a period the increasing nuclear charge tends to decrease the size while increasing electron repulsions tend to increase the size. An additional complication is the occupancy of a new subshell in group IIIA. The data in the tables show that the increasing nuclear charge predominates over both of these factors, and there is a general decrease from lithium to fluorine. The same tendency is shown by the elements from sodium to chlorine.

Examining the changes in detail we see that in the short periods (2 and 3) the largest decrease in size occurs at the beginning of the series with the addition of the second s electron. The second s electron would have an identical radial distribution as the first; however, now the effective (unshielded) nuclear charge has been approximately doubled and both electrons are attracted to the nucleus much more strongly, so the atom contracts. In group IIIA the first p electron is added. Since the p orbitals do not penetrate the inner core as does the s, the contraction effect is less for a p electron than for an s electron. Upon the addition of more p electrons each goes into the same level as the first; however, with an increased nuclear charge acting on all of the p electrons equally, a size decrease results. The size decrease becomes progressively smaller as p electrons are added. This decrease is observed in every row. The data in Table 3-5 also show that size increases with the occupancy of each new quantum level. Thus there is a general size increase down a family of elements.

B. IONIC RADII

The size differences listed above are helpful in explaining similarities and differences in the properties of the atoms. However, very much of the interesting inorganic chemistry takes place in solution and much of this chemistry involves ions. Therefore, a knowledge of ionic radii is also important. Table 3-7 gives some of the most common ions along

Sec. II Atomic and Ionic Radii

Table 3-5 *Metallic Radii of Metals*

IA	IIA	IIIB	IVB	VB	VIB	VIIB	VIII			IB	IIB	IIIA	IVA	VA	VIA	VIIA	Inert Gases
1.55 Li	1.12 Be																
1.90 Na	1.60 Mg											1.42 Al					
2.35 K	1.97 Ca	1.62 Sc	1.47 Ti	1.34 V	1.27 Cr	1.26 Mn	1.26 Fe	1.25 Co	1.24 Ni	1.28 Cu	1.40 Zn	1.41 Ga	1.37 Ge				
2.48 Rb	2.15 Sr	1.80 Y	1.60 Zr	1.46 Nb	1.39 Mo	1.36 Tc	1.34 Ru	1.34 Rh	1.37 Pd	1.44 Ag	1.54 Cd	1.66 In	1.62 Sn	1.59 Sb			
2.67 Cs	2.22 Ba	1.87 La	1.60 Hf	1.46 Ta	1.39 W	1.37 Re	1.35 Os	1.36 Ir	1.38 Pt	1.44 Au	1.57 Hg	1.71 Tl	1.75 Pb	1.70 Bi	1.76 Po		

Ch. 3 Periodic Properties

Table 3-6 *Nonpolar Covalent Radii of Nonmetals*

IA	IIA	IIIB	IVB	VB	VIB	VIIB	VIII			IB	IIB	IIIA	IVA	VA	VIA	VIIA	Inert gases
0.37 H																	
												0.82 B	0.77 C	0.75 N	0.73 O	0.72 F	
													1.11 Si	1.06 P	1.02 S	0.99 Cl	
														1.19 As	1.16 Se	1.14 Br	
															1.35 Te	1.33 I	

82

Sec. II Atomic and Ionic Radii

Table 3-7 *Ionic Radii*

IA	IIA	IIIB	IVB	VB	VIB	VIIB	VIII			IB	IIB	IIIA	IVA	VA	VIA	VIIA	Inert gases
												(+3) 0.20	(+4) 0.15	(−3) 1.71	(−2) 1.40	(−1) 1.36	
(+1) 0.60	(+2) 0.31											(+3) 0.50	(+4) 0.41	(+5) 0.34	(−2) 1.84 (+6) 0.29	(−1) 1.81 (+7) 0.26	
(+1) 0.95	(+2) 0.65											(+3) 0.62	(+4) 0.53	(+5) 0.47	(−2) 1.98	(−1) 1.95	
(+1) 1.33	(+2) 0.99	(+3) 0.81	(+2) 0.90 (+4) 0.68	(+3) 0.74 (+5) 0.59	(+3) 0.69 (+6) 0.52	(+2) 0.80 (+7) 0.46	(+2) 0.76 (+3) 0.64	(+2) 0.74 (+3) 0.63	(+2) 0.72	(+1) 0.96 (+2) 0.69	(+2) 0.74						
(+1) 1.48	(+2) 1.13	(+3) 0.93	(+4) 0.80	(+5) 0.70	(+4) 0.68 (+6) 0.62		(+3) 0.69 (+4) 0.67	(+2) 0.86	(+2) 0.86	(+1) 1.26	(+2) 0.97	(+3) 0.81	(+2) 1.12 (+4) 0.71	(+5) 0.62	(−2) 2.21	(−1) 2.16	
(+1) 1.69	(+2) 1.35	(+3) 1.15	(+4) 0.81	(+5) 0.73	(+4) 0.64 (+6) 0.68		(+4) 0.69	(+4) 0.66	(+2) 0.96	(+1) 1.37	(+2) 1.10	(+3) 0.95	(+2) 1.20 (+4) 0.84	(+3) 1.20 (+5) 0.74			

Ch. 3 Periodic Properties

The hydrated form of Cr(III) *that exists in aqueous* (water) *solutions is* $Cr(H_2O)_6^{3+}$. *It is usually written simply as* Cr(III), Cr^{3+}, *or* Cr^{3+}(aq).

with their radii. Ionic radii are extremely useful; however, we must always be aware of their significance. Only some of the ions (for example, Cr^{3+}) listed in Table 3-7 are capable of existing as hydrates in solution. Some (such as Cl^{7+}) exist only in the gas phase. The radius of these ions, of course, is not as significant as the radius of ions actually observed in solution.

The effect of increasing nuclear charge on the size in a series of ions with increasing nuclear charge and the same number of electrons (that is, O^{2-}, F^-, Na^+, Mg^{2+}, Al^{3+}) is particularly apparent. Note that the radii of the negative ions, found at the right hand side of the periodic table, are large. The generalization that size decreases across a period is only applicable to metallic or covalent radii. The ionic radii show the same general increases down a family as the metallic and covalent radii.

It must be understood that the values listed in Table 3-7 represent a consistent set of radii, but deviations from these values are sometimes observed. The Ag-I distance in silver iodide (AgI), for example, is not given by the sum of the silver and iodide ion radii listed in Table 3-7. The actual distance is much shorter, because the bonding in silver iodide cannot be classified as strictly ionic. In general the radii listed in the tables are additive only if the bonding is strictly of the type specified.

In almost all instances a knowledge of trends in size is sufficient, but in a few instances numerical values become important.

III. Electronegativity

The electronegativity of an element is defined as the ability of an atom, in a chemical compound, to attract electrons toward itself. It is essentially a qualitative concept that was first put on a quantitative basis by Pauling. Several other quantitative definitions are available now, including those of Mulliken, Allred-Rochow, and Sanderson. The Pauling definition is based upon the difference between observed and calculated bond energies for a certain group of elements. Mulliken's definition relates electronegativity to ionization potential. In a very real sense the electronegativity scale is *an attempt* to get a good picture of actual orbital energies by including the electrostatic effects that complicate the ionization potentials. It is essentially a measure of the unevenness of sharing of the electrons involved in chemical bonds. The values obtained by Pauling are given in Table 3-8.

Electronegativity is mainly dependent upon the combined effects of size and nuclear charge. Electronegativity decreases down the table

Table 3-8 *Periodic Relationships for the Electronegativity of the Elements*

IA	IIA	IIIB	IVB	VB	VIB	VIIB	VIII			IB	IIB	IIIA	IVA	VA	VIA	VIIA	Inert gases
H 2.1																	
Li 1.0	Be 1.5											B 2.0	C 2.5	N 3.0	O 3.5	F 4.0	
Na 0.9	Mg 1.2											Al 1.5	Si 1.8	P 2.1	S 2.5	Cl 3.0	
K 0.8	Ca 1.0	Sc 1.3	Ti 1.6	V 1.6	Cr 1.6	Mn 1.5	Fe 1.8	Co 1.8	Ni 1.8	Cu 1.9	Zn 1.6	Ga 1.6	Ge 1.8	As 2.0	Se 2.4	Br 2.8	
Rb 0.8	Sr 1.0	Y 1.2	Zr 1.4	Nb 1.6	Mo 1.8	Tc 1.9	Ru 2.2	Rh 2.2	Pd 2.2	Ag 1.9	Cd 1.7	In 1.7	Sn 1.8	Sb 1.9	Te 2.1	I 2.5	
Cs 0.7	Ba 0.9	La 1.1	Hf 1.3	Ta 1.5	W 1.7	Re 1.9	Os 2.2	Ir 2.2	Pt 2.2	Au 2.4	Hg 1.9	Tl 1.8	Pb 1.8	Bi 1.9	Po 2.0	At 2.2	

as a result of the size increase and increases across the table as a result of increasing nuclear charge and decreasing size.

Electronegativity is an important consideration in compound formation because the electron distribution of the bonding electrons (the bond type) depends upon the electronegativities of the atoms involved. This is described in simplified form in the next section and more completely in Chapter 6.

IV. Oxidation States

The simplest application of the electronegativity of atoms is the assignment of oxidation states to the atoms in chemical compounds. The formation of a chemical bond between two atoms often involves the sharing of a pair of electrons. If the atoms involved have identical electronegativities, the bonding electron pair is shared equally between the atoms and the electron distribution is symmetrical. The electron distribution will be distorted toward the more electronegative atom when the atoms involved are different. This simple idea will provide a basis for understanding the combining ratios discussed in Chapter 1.

In even the simplest molecules a full description of the electron distribution is difficult. There are simply too many factors involved. Because of this it is convenient when dealing with descriptive inorganic chemistry to apply some conventions or rules regarding the electron distribution in a molecule. These rules for obtaining the oxidation state or oxidation number of an atom in a molecule are not completely arbitrary, but another set would work equally well if applied consistently. In spite of the fact that oxidation states do not represent a true description of electron distribution, the concept proves to be a very useful one. In effect the scheme is simply a way of associating each electron in a molecule with a particular atom.

Oxygen is in group VI of the periodic table. The electron configuration is $2s^2 2p^4$.

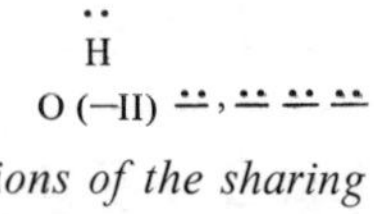

The oxidation state of an element in a molecule may be obtained by assigning the bonding electrons to the most electronegative element involved in the bond. Thus in water, Fig. 3-6, the oxygen atom is assigned eight valence electrons, or outer electrons, while the hydrogens have none. The oxidation state is then determined by counting the

$$: \overset{..}{\underset{..}{O}} : H$$
$$H$$
$$O\,(-II) \;\; \overset{..}{\underset{..}{\cdot\cdot}}, \overset{..}{\underset{..}{\cdot\cdot}} \; \overset{..}{\underset{..}{\cdot\cdot}} \; \overset{..}{\underset{..}{\cdot\cdot}}$$

Figure 3-6 *Two representations of the sharing of electrons in the water molecule, H_2O.*

electrons assigned to each atom. An atom that is assigned more electrons than the neutral atom is given a negative oxidation state. The oxygen in water has a -2 oxidation state and each hydrogen is $+1$. Although it is true that the oxygen is negative and the hydrogen is positive, the real charge distribution does not approach this magnitude. The term oxidation state is used instead of charge in order to emphasize that the charges are not real.

A few other examples will adequately illustrate the procedure. In oxygen fluoride (OF_2) the most electronegative element is fluorine. Each fluorine is -1 while the oxygen is $+2$. This is probably the only compound in which oxygen has an oxidation state of $+2$. In the perchlorate ion, ClO_4^-, each oxygen is -2, whereas the chlorine is $+7$ (the whole ion therefore carries one negative charge). This can be depicted schematically as follows. Chlorine(0) has seven valence electrons and chlorine(VII) has zero.

Fluorine is in group VIIA of the table. It has an electron configuration $2s^2 2p^5$.

Roman numerals are used to designate oxidation states.

$$Cl(O) \overset{..}{\underset{..}{}}, \; \overset{..}{\underset{..}{}} \; \overset{..}{\underset{..}{}} \; \overset{.}{}$$

$$Cl(VII) \text{---}, \; \text{---} \; \text{---} \; \text{---}$$

Each oxygen($-$II) contains eight valence electrons. A bond is formed using two of these eight electrons.

$$Cl(VII) \overset{..}{\underset{..}{:O:}}, \; \overset{..}{\underset{..}{:O:}} \; \overset{..}{\underset{..}{:O:}} \; \overset{..}{\underset{..}{:O:}}$$

Note that the sum of the positive and negative oxidation numbers of the atoms in a complex ion like ClO_4^- equals the net charge on the ion. A few other examples would illustrate that generally the *maximum positive oxidation state of an element is given by the group number of the element in the periodic table*. Since the group number gives the total number of electrons in the valence layer, it is not possible for an element to have an oxidation state higher than the group number. The loss of electrons from a completed core requires excessive amounts of energy (see Section I, Ionization Potentials, of this chapter).

The maximum negative oxidation state an atom can achieve is given by subtracting the group number from eight. Consequently nitrogen might form compounds in which it has an oxidation state of $+5$ as in the nitrate ion (NO_3^-) or -3 as in nitride (N^{3-}). In nitride ion the valence layer is filled.

$$N(-III) \overset{..}{\underset{..}{}}, \; \overset{..}{\underset{..}{}} \; \overset{..}{\underset{..}{}} \; \overset{..}{\underset{..}{}}$$

Three electrons represent the maximum number that nitrogen can gain. Additional electrons would have to occupy the next higher level ($3s$)

which is much too high in energy. A knowledge of the oxidation states to be expected from the electronic structure of an element can be used to predict stoichiometry. Consider the compounds of iodine (I) and fluorine (F) that are known to exist. They are IF_7, IF_5, IF_3, and IF. In this series the oxidation state of iodine changes by two, because molecules involving only representative elements are not usually stable if any unpaired electrons are present. The iodine atom in IF has six valence electrons which are not shared with fluorine.

The series ClO_4^-, ClO_3^-, ClO_2^-, ClO^-, Cl^- shows the successive increases in unshared electron pairs as the oxidation number decreases by two.

$$\text{Perchlorate ion } (ClO_4^-) \quad Cl(VII) \overset{xx}{\underset{O}{}}, \; \overset{xx}{\underset{O}{}} \; \overset{xx}{\underset{O}{}} \; \overset{xx}{\underset{O}{}}$$

$$\text{Chlorate ion } (ClO_3^-) \quad Cl(V) \overset{\cdot\cdot}{}, \; \overset{xx}{\underset{O}{}} \; \overset{xx}{\underset{O}{}} \; \overset{xx}{\underset{O}{}}$$

$$\text{Chlorite ion } (ClO_2^-) \quad Cl(III) \overset{\cdot\cdot}{}, \; \overset{\cdot\cdot}{} \; \overset{xx}{\underset{O}{}} \; \overset{xx}{\underset{O}{}}$$

$$\text{Hypochlorite ion } (ClO^-) \quad Cl(I) \overset{\cdot\cdot}{}, \; \overset{\cdot\cdot}{} \; \overset{\cdot\cdot}{} \; \overset{xx}{\underset{O}{}}$$

$$\text{Chloride ion } (Cl^-) \quad Cl(-I) \overset{\cdot\cdot}{}, \; \overset{\cdot\cdot}{} \; \overset{\cdot\cdot}{} \; \overset{\cdot\cdot}{}$$

Thus we may predict that sulfur can have a maximum oxidation state of VI, other positive oxidation states will be IV and II. Compounds formed between sulfur and fluorine will have the formulas SF_6, SF_4, SF_2. All of these compounds are known to exist. The common oxidation states of the elements are given in Table 3-9.

There are exceptions to many rules in chemistry; also, observable trends are not always followed completely. There are several exceptions to the above statements that may have been noticed. First, compounds such as chlorine dioxide (ClO_2) and nitrogen dioxide (NO_2) represent exceptions to the observation that oxidation states change by two. The oxidation state of chlorine in ClO_2 is $+4$, and that of nitrogen in NO_2 is also $+4$. Exceptions such as these are rare, however, because there is an odd number of electrons in each of these compounds. Although these molecules are quite reactive (ClO_2, for example, is highly explosive), they are stabilized by double bond formation and can be isolated.

Another difficulty must be mentioned here. Sulfur hexafluoride (SF_6) exists and could have been predicted; however, sulfur hexachloride (SCl_6) does not exist, nor do BrF_7, ClF_7, or $BrCl_7$. In addition phosphorus monofluoride (PF) can be made, but it is quite unstable. All

Table 3-9 *Common Oxidation States*

IA	IIA	IIIB	IVB	VB	VIB	VIIB	VIII	VIII	VIII	IB	IIB	IIIA	IVA	VA	VIA	VIIA	Inert gases
1																	
1	2											3	4	−3, 3, 5*	−2	−1	
1	2											3	4	3, 5*	−2, 2, 4, 6*	−1*, 1, 3, 5, 7	
1	2	3	2, 3, 4	3, 4, 5	2, 3*, 4, 6	2*, 3, 4, 7	2, 3*	2*, 3	2*	1*, 2	2	3	4	3*, 5	4, 6	−1*, 1, 3, 5	
1	2	3	3, 4	5	3, 4, 6		3, 5	3	2, 3	1*, 2	2, 3	3	2, 4	3*, 5	4, 6	−1*, 1, 3, 5	
1	2	3		5	6	4, 6	4, 6, 8	4, 6	6		1, 2	1, 3	2, 4	3*, 5			

*The oxidation state normally stable in aqueous solutions.

Sec. IV Oxidation States

of these compounds might have been predicted using the trends noted. The situation is obviously more complicated than a simple consideration of oxidation state would indicate, and other factors such as bond energies and relative sizes must be important.

Chlorine is larger than fluorine, and the repulsions between like atoms will be greater. Sulfur hexachloride (SCl_6) does not exist but sulfur tetrachloride (SCl_4) does. It is possible that the size factor is important here and that, although six chlorides may not fit around a sulfur, four will. The same relationship is observed for the interhalogen compounds; here iodine forms iodine heptafluoride (IF_7) whereas the highest fluoride of bromine is the pentafluoride (BrF_5), and chlorine forms only chlorine trifluoride (ClF_3). All of the 1:1 interhalogen compounds are known.

The examples used thus far have been taken from the representative elements at the right of the table. On the extreme left the charges inferred from oxidation states may be real. For example, the charge on the sodium ion in sodium chloride (NaCl) may be determined by X-ray methods, and it is found to be very close to +1.

The oxidation states of the transition metal ions are a great deal more complicated. The most common oxidation state in the first series is +2 with some ions also exhibiting +3. In general higher oxidation states are more common at the beginning of the series, while the lower states are more common toward the end of the series. Some of the common oxidation states of these ions are given in Table 3-9. The table also contains information regarding the stability of the lower states in water. The higher oxidation states are not stable in water as cations. Only the lower oxidation numbers have even a remote resemblance to real charges.

For almost all of the transition elements, additional oxidation states have been observed. Iron is Fe(VI) in FeO_4^{2-}, and Fe(0) in $Fe(CO)_5$. In this molecule carbon monoxide (CO) is a neutral molecule. Manganese (Mn) has an oxidation state of +7 in MnO_4^-, 0 in $Mn(CO)_5$, and −3 in $Mn(NO)_3CO$.

For compounds as complex as these, oxidation numbers are not sufficient to specify the stoichiometry.

In spite of the fact that the elements in later rows of the periodic table (rows 3, 4, 5) form compounds with more than four atoms attached to a central atom, there are a large number of compounds that are limited to only four attached groups. There is a special significance to this stoichiometry that we shall look into now.

Oxidation states alone can lead to an understanding of the formula of water (H_2O) the most stable compound of hydrogen and oxygen, and we can be certain that oxygen will not show an oxidation number of −6. We do not expect to find any compounds like H_6O. There are, however, other combining ratios possible for hydrogen and oxygen. Hydrogen peroxide (H_2O_2) is a reasonably stable compound, and OH and HO$_2$ have been studied even though they are extremely reactive. None of this information is directly contained in the concept of oxidation states. A theory of chemical bonding should be capable of predicting the properties of such molecules as well as those of formulations such as H$_6$O.

The simplest extension of the question of the relative stability and possible stoichiometry of molecules that goes beyond simple oxidation numbers was formulated in 1916 by G. N. Lewis. Stated most simply the argument is this: An oxygen atom has six valence electrons and tends to complete an octet by forming two chemical bonds. The water molecule may be represented by the electron dot formulas

$$O\!:\!\!:, \quad \overset{\cdot\cdot}{\underset{H\;\;H}{\cdot^{\times}\;\cdot^{\times}}} \quad \text{or} \quad \overset{\cdot\cdot}{\underset{H}{\overset{\cdot\cdot}{O}\!:\!H}}$$

Hence, the octet rule.

V. The Octet Rule

It is the purpose of this section to restate the octet rule in quantum mechanical language and to define its application. It should become obvious that, although the octet rule is not the last word on chemical bonding, it provides a powerful beginning. Many Lewis electron dot structures may be written easily, following the octet rule. The only absolute requirement is that the correct number of valence electrons be used. Octet rule structures were written in Section IV, Oxidation States, of this chapter for some oxyanions of chlorine. To write these structures, simply determine the oxidation state of the elements involved and arrange the electrons in groups of eight around each element. Octet rule structures for the sequence of ions ClO_4^-, ClO_3^-, ClO_2^-, and ClO^- are

$$\left(\begin{array}{c}:\ddot{O}:\\:\ddot{O}:\ddot{C}l:\ddot{O}:\\:\ddot{O}:\end{array}\right)^- \quad \left(\begin{array}{c}:\ddot{O}:\\:\ddot{C}l:\ddot{O}:\\:\ddot{O}:\end{array}\right)^- \quad \left(:\ddot{C}l:\ddot{O}:\right)^- \quad \left(:\ddot{C}l:\ddot{O}:\right)^-$$

The hydrides of the elements near oxygen may all be considered to contain H(I) and X($-m$) where m is eight minus the group number. In the sequence HF, H$_2$O, NH$_3$, CH$_4$, m equals 1, 2, 3, and 4 respectively. Octet rule structures for these molecules are

$$:\ddot{\underset{\cdot\cdot}{F}}:H \qquad :\ddot{\underset{H}{O}}:H \qquad :\overset{H}{\underset{H}{N}}:H \qquad H:\overset{H}{\underset{H}{C}}:H$$

Hydrogen atoms are given only two electrons because there are no 1p orbitals.

Of course, in any group the electron dot structures are identical for compounds of identical stoichiometry. The group VA hydrides may be represented

$$\begin{array}{ccccc} \text{H} & \text{H} & \text{H} & \text{H} & \text{H} \\ \text{:\ddot{N}:H} & \text{:\ddot{P}:H} & \text{:\ddot{A}s:H} & \text{:\ddot{S}b:H} & \text{:\ddot{B}i:H} \\ \text{H} & \text{H} & \text{H} & \text{H} & \text{H} \end{array}$$

The combination of ions in salts may also be written in terms of electron dot structures. To write only a few examples, consider sodium perchlorate ($NaClO_4$), sodium hypochlorite ($NaClO$), ammonium chloride (NH_4Cl), phosphonium bromide (PH_4Br), and hydronium perchlorate (H_3OClO_4).

$$Na^+ \begin{pmatrix} & :\ddot{O}: & \\ :\ddot{O}:\overset{..}{Cl}:\ddot{O}: \\ & :\ddot{O}: & \end{pmatrix}^- \quad Na^+ \begin{pmatrix} :\ddot{Cl}:\ddot{O}: \end{pmatrix}^-$$

sodium perchlorate sodium hypochlorite

$$\begin{pmatrix} \text{H} \\ \text{H:\ddot{N}:H} \\ \text{H} \end{pmatrix}^+ :\ddot{Cl}:^- \quad \begin{pmatrix} \text{H} \\ \text{H:\ddot{P}:H} \\ \text{H} \end{pmatrix}^+ :\ddot{Br}:^-$$

ammonium chloride phosphonium bromide

$$\begin{pmatrix} \text{H} \\ \text{H:\ddot{O}:} \\ \text{H} \end{pmatrix}^+ \begin{pmatrix} :\ddot{O}: \\ :\ddot{O}:\overset{..}{Cl}:\ddot{O}: \\ :\ddot{O}: \end{pmatrix}^-$$

hydronium perchlorate

The cations of the elements with very low electronegativities in families IA and IIA are generally written with no valence electrons in octet rule structures. Some other octet rule structures are given in Fig. 3-7.

Octet rule structures are extremely easy to write in cases where the octets are satisfied with a pair of electrons in each bond. However, there are many cases where this will not work. There may be a deficiency of electrons, as there is in lithium nitrate ($LiNO_3$). The nitrate ion (NO_3^-) has 24 valence electrons, but 26 are required for a structure with three electron pair bonds and an octet on each of the four atoms. In other molecules there appears to be an excess of electrons. For example, there are 40 valence electrons in phosphorus pentachloride (PCl_5). If each of the five bonds is assigned a pair of electrons there would be ten valence electrons around the phosphorus atom.

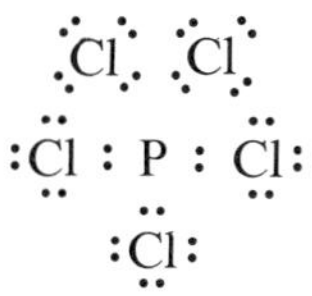

$$\begin{array}{c} \overset{..}{\underset{..}{Cl}}: \quad \overset{..}{\underset{..}{Cl}}: \\ :\overset{..}{Cl}:P:\overset{..}{Cl}: \\ :\overset{..}{\underset{..}{Cl}}: \end{array}$$

The concept of valence shell expansion may be employed for the phosphorus atom in PCl_5 using outside $3d$ orbitals to form the required number of bonds.

Valence shell expansion is discussed further in Chapter 8.

In the case of the nitrate ion (NO_3^-) the concept of resonance is

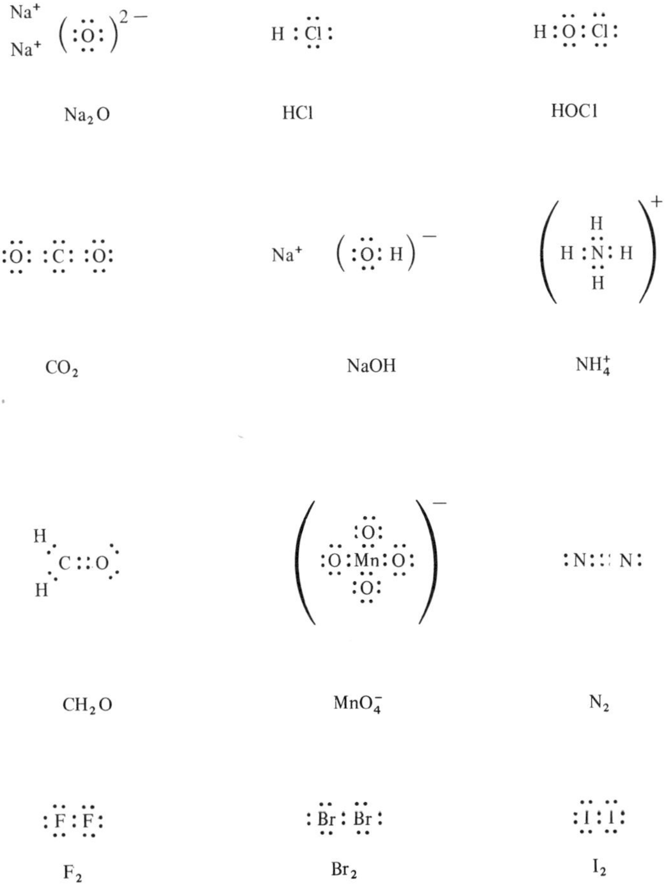

$$\begin{matrix} Na^+ \\ Na^+ \end{matrix} \left(:\overset{..}{\underset{..}{O}}: \right)^{2-}$$

Na₂O

$$H:\overset{..}{\underset{..}{Cl}}:$$

HCl

$$H:\overset{..}{\underset{..}{O}}:\overset{..}{\underset{..}{Cl}}:$$

HOCl

$$:\overset{..}{\underset{..}{O}}::\overset{..}{C}::\overset{..}{\underset{..}{O}}:$$

CO₂

$$Na^+ \left(:\overset{..}{\underset{..}{O}}:H \right)^-$$

NaOH

$$\left(\begin{matrix} H \\ H:\overset{..}{N}:H \\ H \end{matrix} \right)^+$$

NH₄⁺

$$\begin{matrix} H \\ C::\overset{..}{\underset{..}{O}}: \\ H \end{matrix}$$

CH₂O

$$\left(\begin{matrix} :\overset{..}{\underset{}{O}}: \\ :\overset{..}{\underset{..}{O}}:Mn:\overset{..}{\underset{..}{O}}: \\ :\overset{..}{\underset{..}{O}}: \end{matrix} \right)^-$$

MnO₄⁻

$$:N:::N:$$

N₂

$$:\overset{..}{\underset{..}{F}}:\overset{..}{\underset{..}{F}}:$$

F₂

$$:\overset{..}{\underset{..}{Br}}:\overset{..}{\underset{..}{Br}}:$$

Br₂

$$:\overset{..}{\underset{..}{I}}:\overset{..}{\underset{..}{I}}:$$

I₂

Figure 3-7 *Octet rule structures.*

usually employed after the structure is written, using one double bond and two single bonds.

$$\left(\begin{array}{c} :\ddot{\text{O}} \\ \ddot{} \\ :\ddot{\text{O}} \end{array} \!\!\! \diagup\!\!\!\text{N}\!=\!\ddot{\ddot{\text{O}}} \right)^{\!-}$$

However, the observed equality of the three bonds requires three structures. None of these structures represents the nitrate ion by itself. In resonance terminology they all contribute to the structure of nitrate. As a result all the N-O bonds are equivalent.

$$\left(\begin{array}{c}\text{O}\\ \text{O}\end{array}\!\!\!\!\diagup\!\!\!\text{N}=\text{O}\right)^{\!-}\longleftrightarrow \left(\begin{array}{c}\text{O}\\ \text{O}\end{array}\!\!\!\!\diagup\!\!\!\text{N}-\text{O}\right)^{\!-}\longleftrightarrow \left(\begin{array}{c}\text{O}\\ \text{O}\end{array}\!\!\!\!\diagup\!\!\!\text{N}-\text{O}\right)^{\!-}$$

It is possible to write octet rule structures for ions and molecules such as NO_3^- and PCl_5. All that is required is a distribution of the total number of electrons between the bonding positions and the nonbonding positions. To do this a fractional number of electrons must be used. A distaste for odd or fractional electrons is natural when electrons are considered as particles, but should electrons in molecules be considered as particles, or is a wave description more appropriate? There is no difficulty in accepting a probability distribution of electrons in a molecule, with a distribution of charge corresponding to this description. The symmetry of the molecule and the nature of the atoms involved determine the charge distribution, and any attempt to formulate a particle-like structure must also be considered as arbitrary. (The more commonly used oxidation state has been subject to the same criticism; in fact, in the view of some the term is meaningless.) But a concept or idea has merit if it is useful. The idea of atoms is useful, and the concept of oxidation state is useful. The reader may judge for himself the usefulness of the octet structures with fractional electrons. We believe that while the idea of fractional electrons may be arbitrary, the structures described below may be closer to a true description of the electronic structure of molecules than some of the more established methods.

The procedure for writing useful electron dot–octet rule structures for any molecule may be summarized as follows: (1) add up the total number of valence electrons (A) in the molecule; (2) determine the number of electrons (B) that will satisfy the octet rule without sharing any. To obtain this number (B) add two for each hydrogen atom and eight for every other atom, except the metals involved in ionic compounds.

The quantity $C = (B - A)$ gives the optimum number of electrons to be shared. Octet rule structures can now be written by placing the shared electrons into bonds as evenly as possible. The remaining electrons $(A - C)$ are distributed as nonbonding or unshared electrons to complete octets.

Figure 3-8 shows the octet rule structures for molecules that are usually described as resonance hybrids or with expanded octets. The

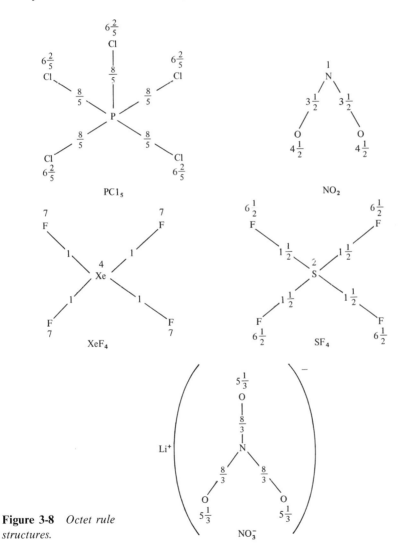

Figure 3-8 *Octet rule structures.*

Ch. 3 Periodic Properties

electrons are spread out as evenly as possible to correspond to the observed equivalence of the bonds. When used in this way the octet rule permits writing structures for inert gas compounds such as XeF_4, and the structure shows the weakness of the bonds involved (with only one electron per bond).

We will make use of these octet rule structures in later chapters to count the number of bonds present in molecules, to interpret bond energies, and to predict the shapes of molecules.

Quite often the octet rule cannot be satisfied, and some atoms must be left without a completed octet. This is necessarily true for single atoms of most elements and most free radicals. In the free hydroxyl radical (OH) the oxygen atom must be left with only seven electrons because hydrogen can only share two electrons. For all other pairs of atoms (excepting hydrogen) the maximum number of electrons that may be shared is six. In a free radical such as CN there are at least three bonds (six shared electrons) and three unshared electrons, two

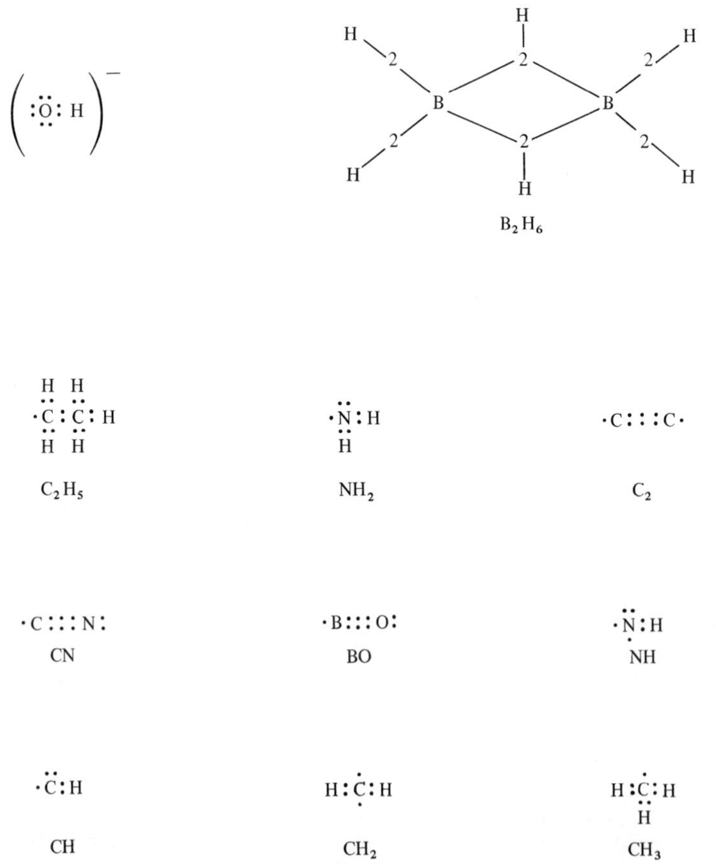

Figure 3-9 Octet rule structures of electron deficient compounds.

on the nitrogen and one on carbon. A number of other free radicals are shown in Fig. 3-9. From these structures highly reactive species can be easily recognized. Nearly all molecules with an odd number of valence electrons are highly reactive. For example, CH_3 (methyl radical) and CN rapidly dimerize to form the highly stable molecules C_2H_6, ethane, and $(CN)_2$, cyanogen. In this way the octet rule may be used to explain the difference in reactivity between OH and H_2O_2, hydrogen peroxide.

Other molecules such as B_2H_6, diborane, can be represented by octet rule structures in spite of a deficiency of electrons if there are no lone pairs and all electrons are used in bonding. The structures shown in Fig. 3-9 involve three center bonds. This type of bonding is rare, except for boron and aluminum compounds. The structure of compounds formed between boron and hydrogen, the boron hydrides, presented theoretical difficulties for a number of years. It was only after the molecular orbital method of treating chemical bonding was applied that the structures were understood. The octet rule structures introduced in this chapter are based on the molecular orbital method.

QUESTIONS

1. Which element in family IVA has the highest ionization potential?
2. Why is the $3d$ orbital higher in energy than the $3p$?
3. Is the $3d$ orbital always higher in energy than the $4s$?
4. Which is larger, an oxygen atom or the O^{2-} ion?
5. Which metal in row 3 of the periodic table has the smallest radius?
6. Determine the oxidation number of selenium in Na_2Se, SeO_2, SeF_6, and Na_2SeO_4.
7. Which element in group VA is the least electronegative?
8. Draw octet rule structures for the compounds in Question 6.
9. Write the formulas for the fluorides of Fe(III), P(V), Ca(II), and Br(V).
10. What is the difference in the electronic structure of GeF_4, SeF_4, and XeF_4?
11. Write the formulas for the sulfides of C(IV), As(III), Na(I), and Mn(II).
12. Write octet rule structures for NO, NO_2, and NO_3^-.
13. Calculate the oxidation number of phosphorus in H_3PO_4, $Na_4P_2O_7$, PH_4Cl, and PCl_3.
14. Draw octet rule structures for Na_2S, S_2Cl_2, SF_6, SO_3, and Na_2SO_3.
15. Which of the following bonds is the least polar: B-Al, B-Si, B-P, or B-S?

PROBLEMS I

1. Convert the X-ray absorption edges of carbon (284ev) and cobalt (7,713) to kcal mole^{-1}. How close are these values to being proportional to Z^2 where Z is the atomic number?
2. From the successive ionization potentials in Table 3-2, estimate the energy difference between the $2s$ and $2p$ orbitals of as many atoms as possible.

Ch. 3 Periodic Properties

3. Calculate the bond distances for each of the following bonds using metallic and/or covalent radii: Si-C, Al-H, Al-O, Na-Se, H-O, Se-Br, Li-Br, S-O, and Mo-S.
4. Repeat as many of the calculations in Problem 3 as possible using ionic radii.

PROBLEMS II

1. What is the energy required to remove a $3p$ electron from a chlorine atom?
2. What energy is required to remove a $1s$ electron from a chlorine atom?
3. What is the energy of the X-ray emitted when a vacancy in the $1s$ level of a chlorine atom is filled by a $3p$ electron?
4. What is the internuclear distance in Cl_2?
5. Estimate the C-Cl bond distance in CCl_4.
6. Estimate the Ca-Ca distance in metallic calcium.
7. Estimate the Ca-Cl distance in $CaCl_2$.
8. How does the observed Si-Cl bond distance in $SiCl_4$, 2.03 Å, compare with the bond distance predicted from covalent and ionic radii?
9. Arrange the following elements in order of electronegativity: Be, F, Ca, Ge, Br.
10. What is the oxidation number of bromine in Br_2?
11. What is the oxidation number of bromine in HBr, NaBr, and IBr?
12. What is the maximum oxidation state expected for bromine?
13. Write the formula of a fluoride of bromine containing bromine(V).
14. What is the oxidation number of oxygen in CaO_2?
15. What is the oxidation number of oxygen in Fe_3O_4?
16. What is the oxidation number of iron in Na_2FeO_4?
17. Write the formula of the oxide of iron(III).
18. How many valence electrons are there in the valence layer of phosphorus?
19. What is the highest oxide of phosphorus?
20. What other oxide of phosphorus is most likely to exist?
21. What is the most likely negative oxidation number of phosphorus?
22. Write the formulas for sodium phosphide and calcium phosphide.
23. What is the oxidation number of phosphorus in P_2H_4?
24. What phosphorus-hydrogen compound is more stable than P_2H_4?
25. Write the octet rule structure for PH_3.
26. How many valence electrons are present in a molecule of CS_2?
27. How many electrons must be shared in CS_2 to satisfy the octet rule?
28. How many electrons are shared in the octet rule structure for H_2SO_3?
29. How many electrons are shared in the octet rule structure for Na_2O?
30. How many electrons are shared in the octet rule structure for calcium peroxide, CaO_2?
31. How many electrons are shared in the SO_3^{2-} ion?

If someone points out to you that your pet theory of the universe is in disagreement with Maxwell's equations—then so much the worse for Maxwell's equations. If it is found to be contradicted by observation—well, these experimenters do bungle things sometimes. But if your theory is found to be against the second law of thermodynamics I can give you no hope; there is nothing for it but to collapse in deepest humiliation.*

SIR ARTHUR EDDINGTON

Chapter 4 The Energetics of Chemical Reactions

The subject matter in this chapter represents a different approach to chemistry than what we have seen thus far. Here we will be interested in macroscopic properties such as the energetics of chemical reactions rather than the microscopic properties of atoms and molecules.

The relationship between the two approaches that will become obvious in this chapter is the periodicity that is apparent in the macroscopic properties. Periodicity, of course, indicates a relationship to electronic structure. The details of this relationship may not be clear at this time; however, in order to accomplish our objective of integrating chemistry we must ultimately be able to discuss reactions and the energetics of reactions in terms of the nature of atoms, ions, and molecules.

There are two fundamental, almost intuitively obvious, features that characterize reactions in chemical systems. They are (1) the energy of the system changes, and (2) the disorder of the system changes. An understanding of chemical reactions and chemical reactivity may be obtained, at a certain level, from an examination of the energy and disorder terms. A more complete understanding of chemical reactivity

*Used by permission from Sir Arthur Eddington, *Nature of the Physical World* (Cambridge: Cambridge Univ. Press, 1953).

Ch. 4 The Energetics of Chemical Reactions

depends upon an understanding of these terms from an atomic or ionic point of view. This will be developed in detail in later chapters. In this chapter the energy and disorder terms will be discussed. Let us first consider the energy changes that are manifested as heat absorbed or heat evolved in a chemical reaction. This branch of thermodynamics is known as thermochemistry.

I. Heat and Enthalpy

The physical state of reactants and products, that is, solid (s), gas (g), or liquid (l), must be designated in any equation to be considered in thermochemistry.

The symbol Δ stands for a change. The symbol ΔH is used in a general sense to represent the enthalpy change of a reaction.

A mole of a substance contains 6.022×10^{23} molecules; 6.022×10^{23} is the conversion factor between atomic mass units (amu) and grams. The molecular weight of water is 18.016 amu, so 1 mole of water weighs 18.016 g. The number 6.022×10^{23} is known as Avogadro's number.

It is possible to measure the heat associated with a number of chemical reactions. One of the most common is the heat of combustion. For example, in the reaction of carbon with oxygen

$$C(s) + O_2(g) \longrightarrow CO_2(g)$$

heat is evolved. As a result we say that the heat content (enthalpy) of the product, given the symbol H, is less than the heat content (H) of the reactants. The change in heat content or enthalpy, ΔH, that accompanies the formation of 1 mole of gaseous carbon dioxide (CO_2) from 1 mole of gaseous oxygen (O_2) and 1 mole of solid carbon (C) is given in equation 4.1.

$$\Delta H = (H_{CO_2(g)} - (H_{C(s)} + H_{O_2(g)})) = -94 \text{ kcal mole}^{-1} \quad (4.1)$$

ΔH is a negative quantity. When heat is evolved during a chemical

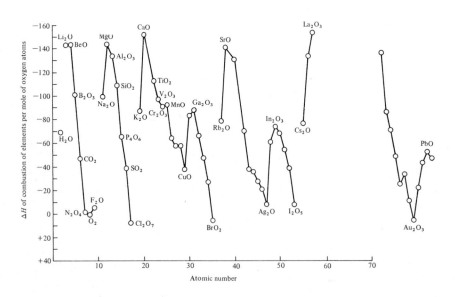

Figure 4-1 *Enthalpy of combustion.*

reaction, or equivalently if the change in enthalpy is negative, the reaction is termed exothermic.

The quantity of heat evolved is directly proportional to the amount of carbon reacting and to the amount of carbon dioxide produced. In the production of 3 moles of carbon dioxide (132.03 g) 3.00 moles × 94 kcal mole^{-1} or 282 kcal of heat is evolved.

In a reaction that absorbs heat the enthalpy change is positive. A positive enthalpy change corresponds to an endothermic reaction. In an endothermic reaction heat is absorbed from the surroundings because the heat content of the product is larger than the heat content of the reactants.

Combustion is a common reaction. All elements in the periodic table except some of the inert gases combine with oxygen. The values of the enthalpy changes for the combustion of a number of elements are shown in Fig. 4-1 and Table 4-1. Once again periodicity is apparent. In every row the maximum enthalpy change occurs in the early families of the table and the minimum change occurs in the late families. For elements of the first two groups there is a decrease in the heat evolved down the group. In the area of the transition metals there is an irregular change with the values generally becoming less negative from titanium (Ti) to copper (Cu).

The enthalpy changes that accompany the reaction of the elements with fluorine are shown in Table 4-2, and periodicity is again obvious.

The reaction of an element with oxygen is often termed combustion; however, a better term that is applicable to the formation of any compound is available. This term is the enthalpy of formation. The equation

$$S(s) + O_2(g) \longrightarrow SO_2(g)$$

represents the formation of SO_2 (sulfur dioxide) from the elements, and the equation

$$S(s) + F_2(g) \longrightarrow SF_2(g)$$

represents the formation of SF_2 (sulfur difluoride) from the elements. The enthalpy of formation of a substance is the enthalpy change (equal to the heat absorbed) when the substance is formed from the pure elements.

An important feature of the enthalpy of reaction may be illustrated by considering the combustion reaction of an element such as sulfur in which sulfur dioxide (SO_2) is produced.

$$S(s) + O_2(g) \longrightarrow SO_2(l)$$

Ch. 4 The Energetics of Chemical Reactions

Table 4-1 *Enthalpy of Formation of Oxides**†

IA	IIA	IIIB	IVB	VB	VIB	VIIB	VIII			IB	IIB	IIIA	IVA	VA	VIA	VIIA	Inert gases
$H_2O(l)$ −68.3																	
Li_2O −143.1	BeO −143.1											$BO(g)$ +6. B_2O_3 −304.2	$CO(g)$ −26.4 $CO_2(g)$ −94.1	$N_2O(g)$ 19.6 $N_2O_4(l)$ −4.66		$F_2O(g)$ −5.2	
Na_2O −99.4	MgO −143.7											Al_2O_3 −400.5	$SiO(g)$ −23.8 SiO_2 −217.7	P_4O_6 −392.0 P_4O_{10} −713.2	$SO_2(l)$ −76.6 $SO_3(s)$ −108.6	$Cl_2O(g)$ 19.2 $Cl_2O_7(l)$ 56.9	
K_2O −86.8	CaO −151.9		TiO_2 −225.5	V_2O_3 −290³ V_2O_5 −373	Cr_2O_3 −272 CrO_3 −140.9	MnO −92.1 MnO_2 −124.3	FeO −63.6 Fe_2O_3 −197.0	CoO −56.9 Co_3O_4 −213	NiO −57.3 Ni_2O_3 −117.0	Cu_2O −40.3 CuO −37.6	ZnO −83.2	Ga_2O −85 Ga_2O_3 −260.3	GeO −50.7 GeO_2 −131.7	As_2O_4 −189.7 As_2O_5 −221.1	SeO_2 −53.9 SeO_3 −39.9	BrO_2 11.6	
Rb_2O −78.9	SrO −141.1		ZrO_2 −262.3		MoO_2 −140.8 MoO_3 −178.1	Tc_2O_7 −266	RuO_2 −72.9 RuO_4 −57.2	Rh_2O_3 −82	PdO −20.4	Ag_2O −7.4	CdO −61.7	In_2O_3 −221.3	SnO −68.3 SnO_2 −138.8	Sb_2O_4 −216.9 Sb_2O_5 −232.3	TeO_2 −77.1	I_2O_5 −37.8	
Cs_2O −75.9	BaO −133.4	La_2O_3 −458	HfO_2 −271.5	Ta_2O_5 −499.9	WO_2 −140.9 WO_3 −201.4	ReO_3 −144.6 Re_2O_7 −296.4	OsO_4 −94.2	IrO_2 −65.5	Pt_3O_4 −39	Au_2O_3 19.3	HgO −21.7	Tl_2O −42.7	PbO −52.3 PbO_2 −66.3	Bi_2O_3 −137.2			

*Data are from Technical Note 270-3 (1968), Technical Note 270-4 (1969), Circular 500 (1952), (Washington, D.C.: U.S. National Bureau of Standards) and the JANAF Thermochemical Tables (Midland Mich.: Dow Chem. Co., 1960).
†All data are for solid crystalline oxides except where liquid or gas is indicated.

Table 4-2 Enthalpy of Formation of Fluorides*†

IA	IIA	IIIB	IVB	VB	VIB	VIIB	VIII			IB	IIB	IIIA	IVA	VA	VIA	VIIA	Inert gases
HF(g) −64.8																	
LiF −146.3																	
NaF −136.0	MgF$_2$ −263.5											BF(g) −29.2; BF$_3$(g) −271.8	CF$_3$(g) −114; CF$_4$(g) −221	NF$_3$(g) −29.8	OF$_2$ −5.2		
KF −134.5	CaF$_2$ −290.3		TiF$_2$ −198; TiF$_4$ −370		CrF$_2$ −181.0; CrF$_3$ −256.2	MnF$_2$ −189	CoF$_2$ −159; CoF$_3$ −187	NiF$_2$ −155.7		CuF$_2$ −129.7	ZnF$_2$ −182.7	AlF(g) −61.7; AlF$_3$(c) −359.5	SiF(g) 1.7; SiF$_4$(g) −386.0	PF$_3$(g) −219.6; PF$_5$(g) −381.4	SF$_4$(g) −185.2; SF$_6$(l) −289	ClF(g) −13.02; ClF$_3$(l) −45.3	
RbF −131.3	SrF$_2$ −290.3		ZrF$_2$ −230; ZrF$_4$ −445							AgF −48.9	CdF$_2$ −167.4	GaF(g) −60.2; GaF$_3$ −278	GeF(g) −7.97	AsF$_3$(l) −228.6	SeF$_6$(g) −267	BrF(g) −22.4; BrF$_3$(l) −71.9	
CsF −126.9	BaF$_2$ −286.9								ReF$_6$ −276								
							IrF$_6$(l) −130			AuF$_3$ −86.9		InF(g) −48.6		SbF(g) −11.3; SbF$_3$(s) −218.8	TeF$_6$(g) −315	IF(g) −22.9; IF$_5$(l) −206.7	XeF$_4$ −62.5
												TlF(g) −43.6; TlF(s) −77.6	PbF$_2$ −158.7; PbF$_4$ −225.1				

*Data are from Technical Note 270-3 (1968), Technical Note 270-4 (1969), and Circular 500 (1952), (Washington, D.C.: U.S. National Bureau of Standards) and the JANAF Thermochemical Tables (Midland, Mich.: Dow Chem. Co., 1960).
†All data are for solid crystalline fluorides except where liquid or gas is indicated.

The enthalpy change for this reaction is -76.6 kcal mole^{-1}. The enthalpy change for the production of sulfur trioxide (SO_3)

$$SO_2(l) + \tfrac{1}{2}O_2(g) \longrightarrow SO_3(l)$$

is -28.8 kcal mole^{-1}. The enthalpy change for the overall reaction

$$S(s) + \tfrac{3}{2}O_2(g) \longrightarrow SO_3(l)$$

is -105.4 kcal mole^{-1}, or the sum of the enthalpy changes of the separate reactions. This observation is not a fortuitous circumstance; the enthalpy changes that accompany reactions are additive. This fact was first noted by G. H. Hess in 1840 and it is now known as Hess's law. It is a corollary of the *first law of thermodynamics which states that energy is conserved.*

A few additional examples of Hess's law type calculations are given in Table 4-3. One of the important uses of Hess's law is that the ΔH of a reaction in which a compound of lower stoichiometry is formed can be obtained even in cases where it is not directly measurable. For example, it is impossible to obtain pure C_3O_2, SiO, or As_2O_4 by reacting the element with oxygen. In each case some of the compound with a higher stoichiometry will be formed, and measured heats do not represent the desired reaction. If the enthalpy of formation of the higher oxide is known and the enthalpy of the reaction of the lower oxide with oxygen is known, the enthalpy of formation of the lower oxide may be calculated. Some examples of this calculation are given in Table 4-4.

The enthalpy of formation should refer to the enthalpy of the reaction when the reactants and products are definitely specified

Table 4-3 *Enthalpies of Reaction for Lower and Higher Oxides and Fluorides (kcal mole^{-1})*

Reaction	ΔH
$Se(s) + O_2(g) \longrightarrow SeO_2(s)$	$\Delta H_1 = -53.9$
$SeO_2(s) + \tfrac{1}{2}O_2(g) \longrightarrow SeO_3(s)$	$\Delta H_2 = +14.0$
$Se(s) + \tfrac{3}{2}O_2 \longrightarrow SeO_3(s)$	$\Delta H = \Delta H_1 + (\Delta H_2) = -39.9$
$\tfrac{1}{2}I_2(s) + \tfrac{1}{2}F_2(g) \longrightarrow IF(g)$	$\Delta H_1 = -22.9$
$IF(g) + 2F_2(g) \longrightarrow IF_5(l)$	$\Delta H_2 = -183.8$
$\tfrac{1}{2}I_2(s) + \tfrac{5}{2}F_2(g) \longrightarrow IF_5(l)$	$\Delta H = \Delta H_1 + (\Delta H_2) = -206.7$
$Sb(s) + \tfrac{1}{2}F_2(g) \longrightarrow SbF(g)$	$\Delta H_1 = -11.3$
$SbF(g) + F_2(g) \longrightarrow SbF_3(s)$	$\Delta H_2 = -207.5$
$Sb(s) + \tfrac{3}{2}F_2(g) \longrightarrow SbF_3(s)$	$\Delta H = \Delta H_1 + \Delta H_2 = -218.8$

Table 4-4 *Application of Hess' Law (ΔH in kcal mole^{-1})*

$C(graphite) + O_2(g) \longrightarrow CO_2(g)$	$\Delta H_1 = -94.1$
$CO(g) + \frac{1}{2}O_2(g) \longrightarrow CO_2(g)$	$\Delta H_2 = -67.7$
$C(graphite) + \frac{1}{2}O_2(g) \longrightarrow CO(g)$	$\Delta H = \Delta H_1 + (-\Delta H_2) = -26.4$
$2As(s) + \frac{5}{2}O_2(g) \longrightarrow As_2O_5(s)$	$\Delta H_1 = -221.1$
$As_2O_4(s) + \frac{1}{2}O_2(g) \longrightarrow As_2O_5(s)$	$\Delta H_2 = -31.4$
$2As(s) + 2O_2(g) \longrightarrow As_2O_4(s)$	$\Delta H = \Delta H_1 + (-\Delta H_2) = -189.7$
$Si(s) + O_2(g) \longrightarrow SiO_2(s)$	$\Delta H_1 = -217.7$
$SiO(g) + \frac{1}{2}O_2(g) \longrightarrow SiO_2(s)$	$\Delta H_2 = -193.9$
$Si(s) + \frac{1}{2}O_2(g) \longrightarrow SiO(g)$	$\Delta H = \Delta H_1 + (-\Delta H_2) = -23.8$

according to physical state. That this is necessary can easily be seen from the reactions already considered. The ΔH of the reaction

$$S(g) + O_2(g) \longrightarrow SO_2(l)$$

is -143.2 kcal mole^{-1} if sulfur is in the gas phase instead of the solid phase. Thus, it is customary to designate a standard reference state for each substance. For gases, the standard state is taken as the ideal gas at 1 atm pressure, while the liquid state at 1 atm is used for liquids. The allotropic modification of the solid that is stable at room temperature and 1 atm pressure is used for solids. The temperature is usually specified as 25°C. If a standard enthalpy refers to any other temperature, this will be indicated by a subscript as in H^0_{1000K}. In addition to dependence upon physical state and temperature, the measured heat depends upon the quantity of substances involved. The term *standard enthalpy of formation* refers to the formation of 1 mole of product. The symbol H^0 is used to designate standard enthalpies of formation. These are the values shown in Tables 4-1 and 4-2.

Because the quantities of concern are differences, the actual magnitude of the enthalpy of a substance is of little significance. Any standard set could be used equally well, but it is conventional to assign the enthalpy of formation of the elements in their standard state a value of zero.

For those reactions that are suitable for direct measurement, the enthalpy change, ΔH^0, can be determined in a calorimeter. The reaction is allowed to take place in a reaction vessel surrounded by a weighed quantity of water. The product of the rise in temperature of the water and the heat necessary to raise the temperature of the water 1°C (the heat capacity) is equal to the heat evolved.

Since the heat of a reaction cannot always be measured, the standard

When two or more crystal structures are known for an element they are called allotropes.

H^0 is used only when referring to the enthalpy of formation of the substance in its standard state from the elements in their standard states. The substance may be an atom or a compound.

enthalpy of formation, H^0, takes on a special significance. Once known, H^0 can be used to calculate the enthalpy change for any reaction, according to the expression

The symbol Σ stands for a summation.

$$\Delta H = [\Sigma m H^0_{\text{prod}} - \Sigma n H^0_{\text{react}}] \quad (4.2)$$

This expression reads: the enthalpy of reaction is equal to the sum of the standard enthalpies of the products minus the sum of the standard enthalpies of the reactants. In this expression m and n are the coefficients in the balanced equation. The most extensive tabulation of standard enthalpies of formation is given in the National Bureau of Standards Circular 500, currently being revised in Technical Note 270-3 and Technical Note 270-4.

An equation written to show conservation of matter and electrical charge is said to be balanced. There must be the same number of atoms of each element on each side of a balanced chemical equation. Since matter and electrical charge are conserved, an unbalanced equation has no quantitative significance. The coefficients in a balanced chemical equation designate the relative numbers of moles of the various reactants and products involved.

A quantity of considerable significance to the discussion in the next section is the enthalpy of formation of gaseous atoms. This term refers to the energy required to produce 1 mole of gaseous atoms from the elements in their standard state. For example, the enthalpy of dissociation of hydrogen

$$H_2(g) \longrightarrow 2H(g)$$

is +104 kcal mole^{-1}. To form 1 mole of hydrogen atoms we have

$$\tfrac{1}{2}H_2(g) \longrightarrow H(g) \quad \Delta H^0 = +52 \text{ kcal mole}^{-1}$$

so the enthalpy of formation of 1 mole of gaseous hydrogen atoms is +52 kcal mole^{-1}. Similarly the enthalpy of formation of fluorine atoms is obtained from the dissociation reaction

$$F_2(g) \longrightarrow 2F(g) \quad \Delta H = +38 \text{ kcal mole}^{-1}$$

The enthalpy of formation of fluorine atoms is 19 kcal mole^{-1}.

The enthalpy of formation of gaseous atoms is a significant quantity because the bond dissociation energy of a molecule and the binding energy of solids are related to it. In the examples given the bond dissociation energy is twice the enthalpy of formation of the atoms, or just the dissociation energy of the molecule. The relationship illustrated by these two examples holds for diatomic molecules of elements such as hydrogen, fluorine, and chlorine, where only one bond is broken and the elements exist as gases in their standard states. If the element is a solid or if the bonding involved in the molecule is not a simple single bond, the enthalpy of formation is still related to the bond energy but the relationship is different. This will be clearer if we consider an example such as N_2. According to the octet rule the bonding in the nitrogen molecule may be represented as

$$:N\equiv N:$$

The dissociation reaction

$$N_2(g) \longrightarrow 2N(g) \quad \Delta H = 226 \text{ kcal mole}^{-1}$$

corresponds to breaking all three bonds, and the single nitrogen-nitrogen bond energy is not twice the value of H^0 for nitrogen atoms

$$\tfrac{1}{2}N_2(g) \longrightarrow N(g) \quad \Delta H = H^0(N) = 113 \text{ kcal mole}^{-1}$$

In addition the bonds are different in energy and the single bond energy cannot be obtained by dividing the enthalpy of dissociation by three. The relationship between bond energies and the enthalpy of formation of atoms will be discussed further later in this chapter and throughout the rest of the book.

Table 4-5 gives the value of the enthalpy of formation of atoms for all of the elements in the periodic table.

II. Bond Dissociation Energies

The energy released when a particular bond is formed is a quantity that is of primary importance to chemistry. It is very desirable to associate the dissociation of a particular bond with a particular quantity of energy. However, the enthalpy change involved in the formation of a given bond in one molecule may very well be different from the enthalpy change involved in the formation of the same bond in a different molecule. The molecular environment may have a considerable influence upon the strength of the bond. For example the energy required in each step of the dissociation of ammonia is different. The energy required to dissociate an N-H bond in ammonia is different from the energy required to dissociate an N-H bond in NH_2. The third dissociation also requires a different amount of energy.

$$NH_3(g) \longrightarrow NH_2(g) + H(g) \quad \Delta H = +114 \text{ kcal mole}^{-1}$$
$$NH_2(g) \longrightarrow NH(g) + H(g) \quad \Delta H = +90 \text{ kcal mole}^{-1}$$
$$NH(g) \longrightarrow N(g) + H(g) \quad \Delta H = +86 \text{ kcal mole}^{-1}$$

The difference between the energy required in each step of the dissociation of ammonia is due to the difference in the molecular

Ch. 4 The Energetics of Chemical Reactions

Table 4-5 H^0 of Atoms*

IA	IIA	IIIB	IVB	VB	VIB	VIIB	VIII			IB	IIB	IIIA	IVA	VA	VIA	VIIA	Inert gases
H 52.1																	He 0.0
Li 38.4	Be 78.3											B 134.5	C 171.3	N 113.0	O 59.6	F 18.9	Ne 0.0
Na 25.8	Mg 35.3											Al 78.0	Si 108.9	P 75.2	S 66.6	Cl 29.1	Ar 0.0
K 21.3	Ca 46.0	Sc 93	Ti 112.5	V 123	Cr 94.8	Mn 67.1	Fe 99.5	Co 101.5	Ni 102.7	Cu 80.9	Zn 31.2	Ga 66.2	Ge 90.0	As 72.3	Se 54.3	Br 26.7	Xe 0.0
Rb 20.5	Sr 39.2	Y 103	Zr 145.4	Nb 184	Mo 157.3	Tc 162	Ru 153.6	Rh 133.1	Pd 90.4	Ag 68.0	Cd 26.8	In 58.2	Sn 72.2	Sb 62.7	Te 47.0	I 25.5	Kr 0.0
Cs 18.8	Ba 42.0	La 100	Hf	Ta 186	W 203	Re 184	Os 189	Ir 159	Pt 135.1	Au 87.5	Hg 14.7	Tl 43.6	Pb 46.6	Bi 49.5			

*National Bureau of Standards Circular 500, Technical Note 270-3.

environment of the bond being broken. In general if the environment about a particular bond is changed, the energy of the bond will be changed.

It is the purpose of this section to introduce a method of calculating bond energies that does allow the association of a particular quantity of energy with a particular bond. Since the data needed for this system is obtained from ionization potential data we return briefly to the discussion of ionization potentials.

A. ENTHALPY CHANGES IN IONIZATION

In Chapter 3 one very important feature needed to interpret ionization potential data quantitatively was omitted. This has to do with the stability of the half-filled $2p$ subshell in nitrogen. Recall the idea that there is a special stability associated with a half-filled shell. It is now necessary to inquire further into the cause of this special stability.

In general the interaction between electrons in the same atom is termed electron correlation. This general term includes the repulsion between electrons because of charge (charge correlation) and the effect of electron spin on this repulsion. The latter is called spin correlation. It is the spin correlation that we are concerned with here.

According to the Pauli principle if the spins of two electrons are the same, one of the quantum numbers representing the space coordinates (n, l, or m_l) must be different. For electrons in the same subshell (that is, if n and l are the same) it is m_l that must be different. From the same principle if the spins of two electrons are different, n, l, and m_l may be the same.

In other words two electrons of the same spin tend to occupy different space. That is they tend to remain separated. On the other hand two electrons with different spins can occupy the same space (that is, the same orbital) and they will, for this reason be slightly closer together on the average.

The result of this is that electrons with different spins will repel each other more than electrons with the same spin. Thus the electronic structure in which electrons are unpaired as long as this is possible is more stable than a structure in which they are paired.

This is the explanation of Hund's rule. The tendency for electrons with opposed spins to come together results in a destabilization because of increased electron repulsion. Conversely the tendency of two electrons with the same spin to remain separate results in a stabilization. The stabilization is called the spin correlation stabilization.

Of the elements in row two, spin correlation stabilization will be important for carbon ($2s^2$, $2p^2$), nitrogen ($2s^2$, $2p^3$), and oxygen ($2s^2$, $2p^4$). Boron (B) has only one p electron; hence, electron repulsions within the $2p$ subshell are not important, and fluorine can also be treated as a one-electron system (in general a p^n system is equivalent to a p^{6-n} system). There is one stabilizing interaction in carbon atoms, three in nitrogen atoms, and oxygen is equivalent to carbon with only one stabilizing interaction (p^2 is equivalent to p^4).

The net effect of spin stabilization on the ionization potentials of the transition metals was made clear by C. K. Jørgensen. The method described by Jørgensen is simple and reliable and applicable (in slightly modified form) to the ionization energies of carbon, nitrogen, oxygen, and the elements below them in the periodic table. The stabilization energy resulting from spin correlation may be obtained from the expression $D(n-1)n$ where n is the number of unpaired electrons and D is an empirically obtained quantity set at 8 kcal mole^{-1} for $2p$ electrons. For carbon the stabilization energy due to spin correlation amounts to about 16 kcal mole^{-1}. When there are three interactions, as is the case with nitrogen, the stabilization energy is three times that of carbon or 48 kcal mole^{-1}. Oxygen has one spin correlation interaction and the spin correlation stabilization energy (SCSE) amounts to 16 kcal mole^{-1}.

The effect of spin correlation stabilization on the ionization energy of carbon, nitrogen, and oxygen may be understood by considering the ionization energy of *hypothetical atoms that are not stabilized by spin correlation*. These will be designated by using the symbol for the element with an asterisk. For example a hypothetical atom of carbon will be designated C*.

Since C* does not possess the extra stability due to spin correlation, the process

$$C^* \longrightarrow C^+ + e^-$$

is easier (requires less energy) than the ionization of real carbon atoms. The energy required for the ionization of C* is obtained by subtracting the spin correlation stabilization energy from the ionization potential of real carbon atoms. The value obtained is $(260 - 16) = 244$ kcal mole^{-1}.

Similarly the process

$$N^* \longrightarrow N^{+*} + e^-$$

would require less energy than the ionization of real nitrogen atoms. The difference in this case is 32 kcal mole^{-1}, so the ionization energy of N* is 303 kcal mole^{-1}. The value 303 kcal mole^{-1} is obtained from the ionization energy of N (335 kcal mole^{-1}) and the spin correlation stabilization energies of N* and N^{+*} which are 48 and 16 respectively.

On the other hand, the ionization $O^* \longrightarrow O^{+*} + e^-$ would require

32 kcal mole⁻¹ more energy than the ionization of real atoms. In this case the ion O^+ has SCSE of 48 kcal mole⁻¹ compared to 16 kcal mole⁻¹ for the atom. The ionization energy of O^* atoms is thus 346 kcal mole⁻¹ or 32 kcal mole⁻¹ greater than that of real oxygen atoms.

Since oxygen has a higher nuclear charge than nitrogen, we would expect from a simple consideration of the energy of the $2p$ orbital that the ionization potential of oxygen would be greater than that of nitrogen. For the actual atoms the spin correlation stabilization of N and O^+ is sufficient to reverse the order.

B. BOND ENERGIES AND ENTHALPY OF FORMATION

A consistent set of bond dissociation energies that are applicable to a large number of molecules cannot be calculated from the enthalpy of formation of atoms that are listed in Table 4-5 if the atom has a substantial spin correlation stabilization energy. Bond dissociation energies *that are* applicable in a general sense can only be obtained if the effects of spin stabilization are included. To do this we shall refer to the enthalpy of formation of hypothetical atoms. To obtain the enthalpy of formation of hypothetical atoms (H*) the spin correlation stabilization energy should be added to the observed enthalpy of formation (real atoms are more stable than the hypothetical atoms). The observed values were given previously in Table 4-5. Values of the calculated enthalpy of formation of hypothetical atoms (without spin correlation stabilization) are listed in Table 4-6. The spin correlation stabilization energy values needed to obtain the data in Table 4-6 are given in Table 4-7. They are based on the estimates of the D values for different subshells listed at the top of Table 4-7. The enthalpy of formation of the hypothetical atoms from boron to neon are summarized in Table 4-8. Also included in Table 4-8 are the *calculated* ionization potentials of the hypothetical atoms. Notice that these ionization potentials increase smoothly from boron to neon. Figures 4-2 and 4-3 show the relation between ionization energies of real and hypothetical atoms and the relation between H^0 and H^*.

Since our objective is to treat the chemistry of the elements of the periodic table as a whole, we shall now examine the enthalpy of formation of other hypothetical valence state atoms. The value $D = 8$ may be assigned to correlate with the measured C-H and O-H bond energies and the ionization potentials of carbon, nitrogen, and oxygen. The problem in extending this idea to the elements in other rows in the periodic table is that satisfactory values of bond dissociation energies are not available; therefore, the spin correlation stabilization energy must be evaluated from ionization potentials alone. Rough estimates of D give $D = 4$ kcal mole⁻¹ for elements containing electrons in the $3p$ and $3d$ subshells, $D = 3$ for electrons in $4p$ and $4d$

The ionization potential of an atom is defined as the work needed to remove an electron from the atom completely. The justification for using ionization potentials in a calculation involving enthalpies rests in the fact that ionization potentials are enthalpies at 0 K. The values at 25°C are only slightly different.

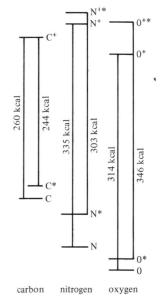

Figure 4-2 *Ionization energies of O, N, C, O^*, N^*, and C^*.*

Table 4-6 H^* of Atoms

IA	IIA	IIIB	IVB	VB	VIB	VIIB	VIII			IB	IIB	IIIA	IVA	VA	VIA	VIIA	Inert gases
													C 187	N 161	O 76		
													Si 117	P 99	S 75		
			Ti 120	V 147	Cr 150	Mn 147	Fe 148	Co 126	Ni 111				Ge 96	As 90	Se 60		
			Zr 151	Nb 211	Mo 205	Tc 222	Ru 181	Rh 145	Pd 90				Sn 76	Sb 75	Te 51		
			Hf —	Ta 198	W 227	Re 224	Os 213	Ir 171	Pt 135				Pb —	Bi —			

Sec. II Bond Dissociation Energies

Table 4-7 Spin correlation stabilization energy $= D(n-1)n$

$D = 8$ for $2p$
$D = 4$ for $3p$ and $3d$
$D = 3$ for $4p$ and $4d$
$D = 2$ for $5p$ and $5d$

IA	IIA	IIIB	IVB	VB	VIB	VIIB	VIII	VIII	VIII	IB	IIB	IIIA	IVA	VA	VIA	VIIA	Inert gases
													16	48	16		
			8	24	55*	80	48	24	8				8	24	8		
			6	27*	48*	60	27*	12*	0				6	18	6		
			4	12	24	40	24	12	0				4	12	4		

*Estimates for atoms with more unpaired electrons than expected.

Ch. 4 The Energetics of Chemical Reactions

Table 4-8 *Enthalpies of Formation and Ionization Energies of Real and Hypothetical Atoms (in kcal mole⁻¹)*

Neutral species	H^0	H^*	Ion	H^0	H^*	Ionization potential (exptl.)	Ionization potential (calc. hypothetical atoms and ions)
B	134		B+	325		191	191
C	171		C+	431		260	
C*		187	C+	431			244
N	113		N+	448		335	
N*		161	N+*		464		303
O	60		O+	374		314	
O*		76	O+*		422		346
F	19		F+	421		402	
F	19		F+*		437		418
Ne	0		Ne+	497		497	497

subshells, and $D = 2$ for electrons in $5p$ subshells. Using the appropriate value of D, the H^*'s of the atoms for the remaining elements in the periodic table may be calculated. They are listed in Table 4-6.

The concept of spin correlation stabilization energy and the concept of hypothetical atoms have been introduced because they are important

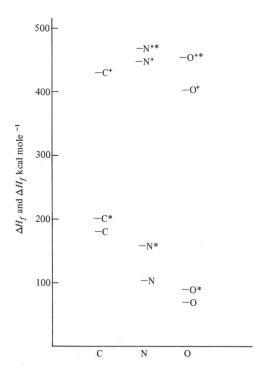

Figure 4-3 H^0 and H^* for O, N, and C.

in calculating bond dissociation energies. Using the concept of hypothetical atoms, bond dissociation energies that apply to an entire series of molecules may be calculated. For example, Pauling has used these same arguments to explain the difference between the observed dissociation energy of the O-H bonds in water (H_2O). The first O-H bond requires 119 kcal mole^{-1} for dissociation, whereas the second O-H bond requires only 102 kcal mole^{-1}. In the first dissociation reaction

$$H_2O \longrightarrow OH + H$$

there is only one odd electron on the OH radical; therefore, spin correlation effects are not important. However, in the second step

$$OH \longrightarrow O + H$$

oxygen atoms are produced in which there are two unpaired spins. The second reaction is thereby promoted by the production of a species containing electrons with unpaired spins, and less energy is required to break the second O-H bond. For the reaction

$$OH \longrightarrow O^* + H$$

the bond dissociation energy is calculated from the dissociation energy (102 kcal mole^{-1}) plus the spin correlation stabilization energy for oxygen (16 kcal mole^{-1}). Bond dissociation energy = 102 + 16 = 118 kcal mole^{-1}. If bond energies are referred to hypothetical valence state atoms, both of the O-H bonds in water have a similar energy (118 and 119 kcal mole^{-1}). This is the value of the concept of hypothetical atoms. It is very desirable to associate a particular quantity of energy with a particular bond. Without using hypothetical valence state atoms this is impossible. Additional examples of the utility of these ideas are in order here.

As another example, consider the energy required to dissociate methane (CH_4) in a stepwise manner. The measured bond dissociation energy for the reaction

$$CH_4 \longrightarrow CH_3 + H$$

is 103 ±1 kcal mole^{-1}. The total bond dissociation energy required to dissociate all four bonds in methane in the reaction

$$CH_4(g) \longrightarrow C(g) + 4H(g)$$

is 398 kcal mole^{-1}. If this value is used to calculate the C-H bond

The bond energy is the energy released when a bond is formed. In this case the total bond energy is +398 kcal mole^{-1}.

dissociation energy, each C-H bond has an average energy of 398/4 = 99.5 kcal mole^{-1}.

If C* is used in place of C, the enthalpy of the reaction

$$CH_4(g) \longrightarrow C^*(g) + 4H(g)$$

is 398 + 16 or +414 kcal mole^{-1}. This value corresponds to a single bond dissociation energy of 414/4 = 103.5 kcal mole^{-1}. When hypothetical valence state atoms are used the C-H bond dissociation energy agrees with the measured bond energy of 103 ± 1 kcal mole^{-1}.

By using the enthalpy of formation of C* (187 kcal mole^{-1}) *and the value of 103 kcal mole^{-1} for C-H bond dissociation energy, the enthalpy of formation of any molecule* CH_x *may be calculated*. The only other information required is a knowledge of the number of bonds involved in the molecule under consideration. This may be obtained from electron dot–octet rule structures as previously discussed.

This calculation is important and it will be illustrated here by giving several examples. The molecules that will be considered are CH_3, CH_2, and CH. Lewis electron dot structures for these molecules are given in Fig. 4-4. In Fig. 4-5 a convenient scheme for setting up the calculation is shown for CH_3. This scheme is called a Born-Haber cycle. It illustrates the energy terms involved in the formation of CH_3 by two different but energetically equivalent paths. One path, the direct reaction of C(s) and H_2(g) to produce CH_3 involves the enthalpy of formation of CH_3. The second path involves the formation of 3H atoms and one hypothetical carbon atom, C*, first, and then the formation of three C-H bonds. Since this is written in the cycle *as the formation of 3C-H bonds*, the sign of the bond dissociation energy term must be reversed for this step. Knowledge of all energy terms but one permits the calculation of the remaining term. Here we are interested in calculating the enthalpy of formation of CH_3. The remaining terms in the cycle are as follows. The total bond dissociation energy is three times the bond dissociation energy (103 kcal mole^{-1}), H^0 of C* is 187 kcal mole^{-1}, and H^0 of hydrogen atoms is 52 kcal mole^{-1}. Thus

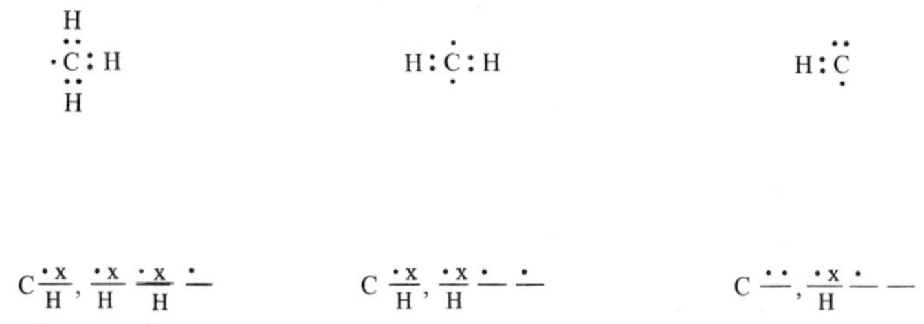

Figure 4-4 *Electron dot structures.*

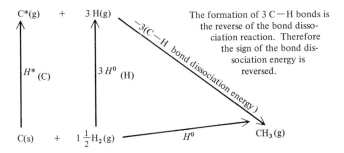

$H*(C*) = 187$ kcal mole^{-1}
$H^0(H) = 52$ kcal mole^{-1}
Bond dissociation energy = 103 kcal mole^{-1}
$H^0 = -3(C-H \text{ bond energy}) + H*(C*) + 3H^0(H)$
$H^0 = -3(103) + 187 + 156 = 34$ kcal mole^{-1}

Figure 4-5 *Born-Haber cycle.*

$H^0(CH_3) = 187 + (3)(52) + (3)(-103) = 34$ kcal mole^{-1}. The best current experimental value is 33 ± 1 kcal mole^{-1}.

In CH there is one bonding electron pair, one lone pair of electrons, and one unpaired electron. Using the Born-Haber cycle the enthalpy of formation may be calculated *from the C-H bond dissociation energy* and the enthalpy of formation of C* and H in a fashion similar to the calculation illustrated in Fig. 4-5. H^0 of CH is found to be $187 + 52 + (-103) = 136$ kcal mole^{-1}. The experimental value is 142 kcal mole^{-1}. In CH_2 there are two bond pairs and two unpaired electrons. Thus a spin correlation stabilization should be applied to the molecule as well as the gaseous carbon atom. The calculated H^0 for $(CH_2)*$ is 85 kcal mole^{-1}.

$$H* = 187 + (2)(52) + (2)(-103) = 85$$

However, this is the enthalpy of formation of the hypothetical molecule. Spin correlation stabilization amounting to 16 kcal mole^{-1} for one spin-spin interaction must be subtracted from 85 to give a value of 69 kcal mole^{-1} for the real molecule. A number of other similar calculations are summarized in Table 4-9. Except for the value of CH_2 the calculated values show good agreement with the experimental values.

The only substances listed in Table 4-9 that have negative values of H^0 are water (H_2O), ammonia (NH_3), and methane (CH_4). These substances are the only ones listed in the table that are stable under ordinary conditions.

A negative enthalpy of formation results when the bonding in the product is stronger than the bonding in the elements. However, a negative enthalpy alone is not a sufficient criterion for predicting whether a chemical reaction will occur. In a large number of cases

Table 4-9 *Examples of the Calculation of H^0 (kcal mole^{-1}) for Simple Molecules*

Species	H^0 or H^*	SCSE	Species	H^0 (calc.)	H^0 (exptl.)
N*	161	48	N	113	113
NH*	104	16	NH	88	79
NH_2	46	—	NH_2	46	41
NH_3	−11	—	NH_3	−11	−11
O*	76	16	O	60	60
OH	9	—	OH	9	9
H_2O	−58	—	$H_2O(g)$	−58	−58
C*	188	16	C	172	171
CH	136	—	CH	136	136
CH_2*	84	16	CH_2	68	84
CH_3	34	—	CH_3	34	33
CH_4	−18	—	CH_4	−18	−18

an exothermic reaction will occur spontaneously, but in many instances endothermic reactions occur spontaneously as well. As an example, we may consider one of the simplest possible changes, freezing and melting. The enthalpy of ice is less than that of water so the freezing of water is an exothermic reaction that occurs spontaneously. However, the reverse reaction, the melting of ice, is also spontaneous at temperatures above 0°C in spite of the fact that the reaction is endothermic. The stability of a phase or the chemical stability of a set of reagents is not determined entirely by the heat content of the substances involved. To give an example of a chemical change that is endothermic but spontaneous, we need only look at the solution of certain substances in water. These substances are soluble but the reactions are endothermic. Ammonium chloride (NH_4Cl) is a substance in this category. To get a measure of whether a reaction will occur spontaneously we must look at the degree of disorder of the system in addition to the enthalpy. The degree of disorder of a system is the entropy of the system.

III. Entropy

In the example of melting ice the degree of disorder in water is greater than the disorder in the ice crystal. In water the molecules are free to assume a large number of possible orientations, but in ice crystals the number of possible orientations is greatly decreased. Ice, like all crystals, has a highly ordered structure.

In general, the greater the number of possible orientations of a system, the greater the entropy of the system and the greater the probability that the system will be in this state rather than in some other state of lower entropy. Thus it is much more likely for a piece of chalk to break when dropped than for two pieces to join together, and it is much more likely that an individual will age rather than become younger. In each case the change corresponding to the highest probability is the spontaneous and irreversible change, and it is the one in which entropy increases. We may write an expression that involves entropy and the concept of probability. It is

$$S = k \ln K \qquad (4.3)$$

where K is the number of equivalent ways (distinguishable in principle but indistinguishable in practice) in which a system may be arranged; k is a proportionality constant; and S is the entropy. From this expression, we see that an increase in K results in an increase in entropy. To further illustrate equation 4.3, think about the entropy of the following systems: (1) a deck of playing cards before and after it is shuffled; (2) a single ace of spades; (3) the molecules of a vapor compared to the same molecules in the liquid phase.

The symbol ln represents a logarithm to the base e. In order to convert from base 10 to base e it is necessary to multiply by 2.303. The expression in equation 4.3 then becomes $S = 2.303\,k \log K$. The value of e is 2.718.

The importance of the concept of entropy to a consideration of chemical reactions is obvious. A chemical system will tend to change in a manner consistent with an increase in entropy. This is the substance of the second law of thermodynamics. This law was stated by Rudolf Clausius as: "Die Entropie der Welt strebt einem Maximum zu." The entropy of the universe tends toward a maximum. All spontaneous, irreversible reactions are accompanied by an increase in the entropy of the universe. A decrease may occur in one portion of the universe, for example, in human cells during the production of proteins, but the net change must be balanced by an equal or slightly larger positive entropy change in some other portion of the universe. Life originated and is possible in spite of the negative entropy changes involved because of the tremendous energy changes that involve an entropy increase in the surroundings. The second law of thermodynamics is different from the first law in this sense: the first law deals with the possible and the impossible, whereas the second law deals with the probable and the improbable. The concept of entropy and spontaneity will be illustrated using the simple physical change referred to previously, the melting of ice and the freezing of water.

During the reaction

$$H_2O(l) \longrightarrow H_2O(s) \qquad \Delta H_{0°C} = -1440 \text{ cal mole}^{-1}$$

Ch. 4 The Energetics of Chemical Reactions

1440 cal mole^{-1} is liberated, and the reaction is spontaneous; however, as indicated above the reverse reaction, an endothermic one, is also spontaneous, for ice will melt if the temperature is high enough.

The heat evolved or absorbed in the ice-water reaction stems entirely from the fact that the enthalpy of ice is less than that of water. Upon the addition of energy to ice, it will melt; upon the removal of energy from water, it will freeze. Thus the ice $\longrightarrow$ water endothermic reaction is spontaneous if the temperature is increased whereas the exothermic reaction is spontaneous at low temperatures. The temperature T, ΔH, and ΔS affect stability, and the direction in which the reaction proceeds is dependent upon all three quantities. The relationship between ΔS, ΔH, and T must be such that, at a certain temperature, the entropy change and the enthalpy change balance each other. The enthalpy change favors the water $\longrightarrow$ ice reaction whereas the entropy change favors the ice $\longrightarrow$ water reaction. If and when the two factors, ΔH and ΔS, balance, the reaction can proceed in both directions and ice and water will be in equilibrium with each other. The entropy difference between ice and water may be obtained from the data given in Table 4-10. The melting of ice results in an increase in entropy of 5.27 cal mole^{-1} deg^{-1}, whereas the freezing of water results in a decrease in entropy of 5.27 cal mole^{-1} deg^{-1}. (Later the more convenient shorthand symbol eu for entropy units will be used.) To balance at a certain temperature, the relationship between the entropy change and enthalpy change must be

$$\Delta S = \frac{1}{T}(\Delta H) \quad (4.4)$$

Since ΔH is 1440 cal mole^{-1} and ΔS is 5.27 cal mole^{-1} deg^{-1}, the temperature at which ΔS and ΔH will balance is 273 K or 0°C, the melting point of ice. At 0°C, 1440 cal must be added to the ice to melt 1 mole of it. The result is an increase in the entropy of the system by +5.27 cal mole^{-1} deg^{-1}. On the other hand, 1440 cal must be removed from 1 mole of water to freeze it. The entropy of the ice-water system then decreases by 5.27 and $\Delta S = -5.27$ cal mole^{-1} deg^{-1}.

When water freezes, is this truly a decrease in entropy? No, it is not. The second law states that the entropy of the universe tends to increase. Thus, the decrease in entropy of the ice-water system must be balanced by an increase in entropy of the surroundings. When a mole of water freezes, 1440 cal are liberated and added to the surroundings and the heat content of the surroundings is increased by 1440 cal. With the addition of this energy to the surroundings, the molecules of the surroundings become more randomly oriented due to increased thermal motion; therefore, the entropy of the surroundings

Table 4-10 *Standard Entropies*

Species	S^0, eu
$H_2O(g)$	45.11
$H_2O(l)$	16.72
$H_2O(s)$	11.45
$H_2O_2(g)$	55.65
$H_2O_2(l)$	26.2
$OH(g)$	43.89

The effect of temperature on the motion of molecules will be considered in detail in Chapter 9.

increases. The entropy change of the surroundings may be calculated from

$$\Delta S_{surr} = \frac{\Delta H_{surr}}{T} = \frac{1}{273 \text{ K}}(1440) = +5.27 \text{ eu}$$

During freezing the entropy decrease suffered by the ice-water system is exactly compensated for by the entropy increase of the surroundings if the temperature of the surroundings is 273 K.

At $-10°C$ ice will not melt because the removal of 1440 cal from the surroundings at $-10°C$ results in an entropy change for the surroundings

$$\Delta S_{surr} = \frac{1}{263 \text{ K}}(-1440)$$

of -5.48 eu. At this temperature the surroundings decrease in entropy more than the ice-water system increases, and the melting process is not spontaneous. In this case

$$\Delta S_{total} = \Delta S_{sys} + \Delta S_{surr}$$
$$\Delta S_{total} = -0.21 \text{ eu}$$

On the other hand, at $+10°C$ the change in entropy of the surroundings is -5.09 eu, and $\Delta S_{total} = +0.18$ eu. With a positive value for ΔS_{total} the melting process is spontaneous.

While we are discussing the ice-water system because it is simple, a very important generalization follows from the above discussion. Equilibrium is possible in this system because at $0°C$ the entropy and enthalpy changes are opposed and they balance each other. In general, equilibrium will be possible when the entropy change and the enthalpy change balance at a specific temperature, *and if they are opposed*. In a simple phase change, the entropy change of the system is balanced by the entropy change of the surroundings at equilibrium. This will be true for all reactions, including chemical reactions; therefore, the criterion for equilibrium in chemical systems is just the same. For equilibrium in a chemical reaction the *entropy change of the surroundings*, $\Delta H_{surr}/T_{surr}$, must balance the *entropy change of the system*. Equilibrium is possible in a reaction if the entropy change of the surroundings is the same in magnitude but is opposite in sign to ΔS_{sys}. If

$$\Delta S_{sys} = \frac{-\Delta H_{surr}}{T} = \frac{\Delta H_{sys}}{T} \qquad (4.5)$$

Ch. 4 The Energetics of Chemical Reactions

equilibrium will exist. Then both the forward and backward reaction will be spontaneous because

$$\Delta S_{total} = \Delta S_{surr} + \Delta S_{sys} = 0 \qquad (4.6)$$

This type of reaction is called reversible. Since it is usually possible to balance the entropy change in a reaction with that of the surroundings, all chemical reactions can, in principle, be considered to be equilibrium reactions. In practice reactions will tend to go to completion if they are highly exothermic; they will not proceed to any appreciable degree if they are highly endothermic.

All of this may be expressed in a different form. If a new symbol is introduced to represent the difference between the enthalpy and entropy terms, we may choose the form

$$\Delta G = \Delta H - T\Delta S \qquad (4.7)$$

$H_2O(s) \longrightarrow H_2O(l); \Delta G = +.054$ kcal mole^{-1} at 263 K, and $\Delta G = -0.051$ kcal mole^{-1} at 283 K. Note that the units of G are the same (kcal mole^{-1}) as those used for H.

When the difference between the enthalpy of the reaction and the product of temperature times entropy change for the reaction is calculated for the ice-water system, we find that at 263 K this difference is positive. At 283 K the difference is negative. This difference we are calculating is called the Gibbs free energy after Josiah Willard Gibbs, and the symbol ΔG is used to represent it. Here we have another way of stating the criterion for spontaneity. *If ΔG is negative the reaction is spontaneous.* Stated another way

$$\Delta H = \Delta G + T\Delta S$$

Any way the equation is written, for a reaction to be spontaneous ΔG must be negative. Thus a negative enthalpy term and a positive entropy term are favorable, and the reaction tends to proceed. If the enthalpy term is favorable and the entropy term is unfavorable, or vice versa, the value of ΔG depends upon the relative magnitude of the opposing terms. If the free energy change is equal to zero, as is the case for the ice-water system at 0°C, the system is in equilibrium and the reaction may proceed in either direction.

The Gibbs free energy relationship is one of extreme importance. To understand the relationship between ΔH and ΔS (that is, ΔG) better, we shall examine these functions in just a bit more detail.

A chemical reaction that is exothermic is often accompanied by an unfavorable entropy change. In such a reaction the energy of the products is lower than the energy of the reactants. This implies tighter bonding between atoms and molecules and, therefore, a more ordered system results. The energy released to the surroundings goes to increase the entropy of the surroundings. In an endothermic reaction the products have a greater energy than the reactants, implying decreased interaction and less order, or more disorder; therefore, an endothermic reaction is often accompanied by an increase in entropy. In an endothermic reaction the surroundings decrease in entropy, since energy must be removed if the reaction is to proceed. If a forward reaction is primarily dependent upon a favorable enthalpy change, the reverse reaction will be dependent upon a favorable entropy change.

In practice a reaction is reversible if the energy requirements can be met by the surroundings. In an exothermic reaction the surroundings increase in entropy but, if there is an entropy decrease in the system, equilibrium will be reached at some temperature.

To make a very endothermic reaction go, the entropy decrease of the surroundings must be balanced or outweighed by an entropy increase in the system. The balance depends upon the temperature of the surroundings since the equation

$$\Delta S_{surr} = \Delta H / T$$

is applicable. At high temperatures highly endothermic reactions can proceed and very exothermic reactions may become reversible.

To emphasize this point and to illustrate that ΔG is not something new, we can rewrite the equation

$$\Delta G = \Delta H_{sys} - T \Delta S_{sys} \qquad (4.8)$$

in terms of the energy change of the surroundings after first dividing by T.

$$\frac{\Delta G}{T} = \frac{\Delta H_{sys}}{T} - \Delta S_{sys} \qquad (4.9)$$

Since the first law of thermodynamics requires that energy be conserved

$$\Delta H_{sys} = -\Delta H_{surr}$$

and since

$$\Delta S_{surr} = \frac{1}{T}(-\Delta H_{sys}) \qquad (4.10)$$

$$\frac{\Delta G}{T} = -\Delta S_{surr} - \Delta S_{sys}$$

$$\frac{\Delta G}{T} = -(\Delta S_{surr} + \Delta S_{sys}) = -\Delta S_{total}$$

For reactions of this type, $-\Delta G/T$ represents the change in total entropy. Thus the sign of ΔG is an indication of whether or not a reaction will proceed; in addition, the magnitude of ΔG provides a quantitative measure of the equilibrium. The more negative ΔG, the greater the probability that the reaction will be complete and the less likely the reverse reaction. This concept is customarily written in the form

$$\Delta G^0 = -RT \ln K \quad (4.11)$$

where K is called an equilibrium constant. Since

$$\frac{\Delta G^0}{T} = -\Delta S_{total}$$

we see that this is another way of writing

$$\Delta S_{total} = k \ln K$$

In equation 4.11, R is a constant and ΔG^0 is the standard free energy change. This new symbol is used to represent the free energy change for reactions that involve compounds in their standard state. ΔH^0 and ΔS^0 have a similar meaning.

Free energy, G, has one important advantage over the entropy S in that it can be used without being concerned with the changes that take place in the surroundings. This is included in the way G is defined. If the enthalpy change and the entropy change are known, the free energy change can be calculated. Like H^0, G^0 is additive and from a listing of G^0 values the ΔG^0 of any reaction may be calculated. Also like H^0, the free energy of formation of substances shows periodicity. The free energy of formation of oxides and fluorides is shown in Tables 4-11 and 4-12.

Equation 4.11 can be written in the form

$$K = e^{-\Delta G^0/RT} \quad (4.12)$$

substituting

$$\Delta G^0 = \Delta H^0 - T\Delta S^0$$

we obtain

$$K = e^{-\Delta H^0/RT} e^{\Delta S^0/R}$$
$$K = 10^{-\Delta H^0/2.303RT} 10^{\Delta S^0/2.303R} \quad (4.13)$$

Table 4-11 G^0 of Oxides

I	II	IV	VI	VII	VIII	VIII	VIII	IB	IIB	III	IV	V	VI	VII
H$_2$O(g) −54.6; H$_2$O$_2$(g) −25.2														F$_2$O(g) −1.1
Li$_2$O −134.3	BeO −136.1									BO −1	CO(g) −32.8; CO$_2$(g) −94.3	N$_2$O(g) 24.9; N$_2$O$_4$(l) 23.3		
Na$_2$O −90.1	MgO −136									Al$_2$O$_3$ −378.2	SiO(g) −30.2; SiO$_2$ −204.8	P$_4$O$_{10}$(c) −644.8	SO$_3$(c) −88.2	Cl$_2$O(g) 23.4
	CaO −144.4	TiO −118.3; TiO$_2$ −212.6	Cr$_2$O$_3$ −252.9	MnO −86.7; MnO$_2$ −111.2	FeO −58.4; Fe$_2$O$_3$ −177.4	CoO −51.2	NiO −50.6	Cu$_2$O −34.9; CuO −31.0	ZnO −76.1	Ga$_2$O$_3$ −238.6	GeO(c) −56.7; GeO$_2$(c) −118.8	As$_2$O$_5$(c) −187.0		
	SrO −133.8	ZrO$_2$ −248.5	MoO$_2$ −127.4; MoO$_3$ −159.7	ReO$_2$ −88	RuO$_2$ −51			Ag$_2$O −2.7	CdO −54.6	In$_2$O$_3$ −198.6	SnO(c) −61.4; SnO$_2$(c) −124.2	Sb$_2$O$_4$(c) −190.2; Sb$_2$O$_5$(c) −198	TeO$_2$(c) −64.6	
	BaO −126.3		WO$_3$ −182.6		OsO$_4$ −72.9				HgO −14.0	Tl$_2$O −35.2	PbO −44.9; PbO$_2$ −52.0	Bi$_2$O$_3$ −118.0		

Table 4-12 G^0 of Fluorides

1	2	3–11 (transition)	13	14	15	16	17
HF(g) −65.3							
LiF −139.6	BeF$_2$ −234.7		BF$_3$(g) −267.8	CF$_4$(g) −210	NF$_3$(g) −19.9		ClF(g) −13.37
NaF −129.9	MgF$_2$ −256.0		AlF$_3$ −340.6; AlF(g) −67.8	SiF(g) −5.8; SiF$_4$(g) −375.7	PF$_3$(g) −214.5	SF$_4$(g) −174.8; SF$_6$(g) −264.2	BrF(g) −26.1; BrF$_5$(l) −84.1
KF −128.5	CaF$_2$ −277.7	TiF$_4$ −372.7; CrF$_3$ −260; MnF$_2$ −179; CoF$_2$ −154.7; NiF$_2$ −144.4; ZnF$_2$ −107.5	GaF$_3$ −259.4		AsF$_3$(l) −217.3	SeF$_6$(g) −243	IF(g) −28.3
		ZrF$_2$ −219; ZrF$_4$ −432.6; MoF$_6$(l) −352.1; AgF −44.2; CdF$_2$ −154.8					
	BaF$_2$ −274.5	WF$_6$(l) −389.9; IrF$_6$ −110.3		PbF$_2$ −147.5			

The equilibrium constant K, like ΔG^0, expresses the degree to which a reaction will proceed. In Fig. 4-6 the relationship between the value of e^x and the value of x is given. Using these figures we can understand the relationship between K, ΔH^0, and ΔS^0 for different possible values of ΔH^0 and ΔS^0.

If ΔH^0 is very highly negative, the value of K will be large. Reactions that have a highly negative enthalpy change tend to go to completion unless ΔS^0 is also highly negative. For these reactions small positive and negative values of ΔS^0 have relatively little effect on the value of the equilibrium constant.

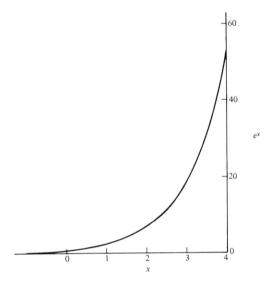

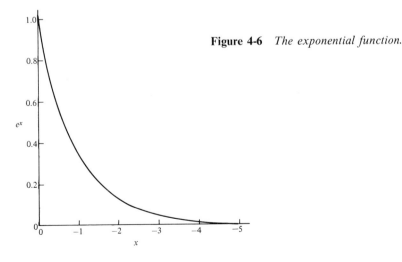

Figure 4-6 *The exponential function.*

If ΔH^0 is small the equilibrium constant approaches 1. In reactions of this type equilibrium is likely unless ΔS^0 is very highly positive.

If ΔH^0 is highly positive K is much less than 1. Reactions of this type will not proceed until or unless ΔS^0 is positive.

If ΔH^0 is slightly positive the reaction may proceed if ΔS^0 is also positive. The magnitude of the equilibrium constant will be essentially determined by ΔS^0.

Temperature will have an effect upon the equilibrium. For highly exothermic reactions increasing temperature promotes equilibrium or decreases the equilibrium constant. Similarly, increasing the temperature increases the equilibrium constant for endothermic reactions.

IV. Activities and Equilibrium Constants

The magnitude of the equilibrium constant referred to above is dependent upon the magnitude of the free energy change for the reaction when the reactants and products are in their standard states. In specifying the standard state of a substance a definite quantity is involved. The standard free energy of formation refers to the formation of 1 mole of the substance. The net result is that ΔG^0 does not vary with the amount of substance. Consequently K, the equilibrium constant, is a constant at any given temperature.

Equilibrium constants can be calculated exactly from thermodynamic data. The magnitude of the constant reflects the magnitude of ΔG^0. The larger the value of K, the greater the tendency for the reaction to proceed. A negative free energy change and a value of K greater than 1 indicate that the free energy of the products is less than the free energy of the reactants. In the change from reactants to products, the free energy of the materials decreases. However the change is not ΔG^0. Once the reaction begins the reactants are no longer in their standard states, and there are, in fact, no products at the beginning. The symbol that is applicable when discussing the free energy change of the reaction is not ΔG^0 but ΔG. When equilibrium is established the value of ΔG is zero, but the value of the equilibrium constant, K, can still be calculated from ΔG^0.

For a particular reaction the equilibrium constant is calculated from data on the substances in their standard states. This means that the intermolecular or interionic forces are exactly those of the substance in its standard state. The free energy of a substance that is not in its standard state is not represented by G^0. In other words there is a difference in the substance that may lead to an increased or decreased tendency to react. There is an obvious difference between solid and liquid water. The difference is usually referred to in terms of the activity of the substance. The activity of a substance in its standard state has some particular value. By changing the conditions of a reaction the activities of the substances taking part in the reaction are changed.

The free energy of material that is not in its standard state, G, is related to the free energy of the substance in its standard state, G^0, by the equation

$$G = G^0 + RT \ln a$$

in which a is the activity of the material. For any general reaction

$$mA \longrightarrow nB$$

where m moles of A react to produce n moles of B, the free energy change may be written as the difference between the free energy of the products and the reactants.

$$\Delta G = [nG_B - mG_A]$$

By substituting

$$G = G^0 + RT \ln a$$

for the free energy of the products and reactants we obtain

$$\Delta G = [n(G_B^0 + RT \ln a_B) - m(G_A^0 + RT \ln a_A)]$$

where a_B refers to the activity of B and a_A refers to the activity of A. At equilibrium $\Delta G = 0$, so

$$[n(G_B^0 + RT \ln a_B) - m(G_A^0 + RT \ln a_A)] = 0$$

or

$$[(nG_B^0 + RT \ln a_B^n) - (mG_A^0 + RT \ln a_A^m)] = 0$$

In this expression the activity of B is raised to the power n and the activity of A is raised to the power m. By rearranging terms we obtain

$$nG_B^0 - mG_A^0 = -RT \ln a_B^n + RT \ln a_A^m$$

$$nG_B^0 - mG_A^0 = -RT \ln \frac{a_B^n}{a_A^m}$$

$nG_B^0 - mG_A^0$ is the standard free energy change, ΔG^0, so

$$\Delta G^0 = -RT \ln \frac{a_B^n}{a_A^m}$$

The term involving activities

$$\frac{a_B^n}{a_A^m}$$

is equal to the equilibrium constant K

$$K = \frac{a_B^n}{a_A^m}$$

In this expression the activity of the product appears in the numerator and the activity of the reactant appears in the denominator. Also each activity is raised to a power corresponding to the coefficient in the balanced equation.

The activity of a substance can be determined experimentally; however, in practice, activities are usually estimated. In very dilute solutions, concentrations are equal to activities and the equilibrium constants calculated using concentrations are exact and independent of the concentration. Since this is the case it is customary to use concentrations as an approximation of activities when calculating equilibrium constants or when using equilibrium constants in a calculation. Three ways of expressing the concentration of a substance will be used. The mole fraction is used to represent the activity of a solid or liquid. Molarity is used as an approximation for the activity of a dissolved substance. Pressure in atmospheres is used for gases.

One equilibrium constant expression that is often used is that for the dissociation of a weak acid in water. Substances such as hydrofluoric acid (HF) and acetic acid ($CH_3-\overset{\overset{O}{\|}}{C}-OH$ or HAc) are weak acids in water. When dissolved in water they react as follows

$$HF + H_2O \rightleftharpoons H_3O^+ + F^-$$

H_3O^+ is called the hydronium ion. The equilibrium constant for this equation is

$$K_a = \frac{(a_{H_3O^+})(a_{F^-})}{(a_{HF})(a_{H_2O})}$$

In dilute solutions the mole fraction of water will be very nearly one, so the expression simplifies to

$$K_a = \frac{(a_{H_3O^+})(a_{F^-})}{(a_{HF})}$$

This may then be rewritten in terms of concentrations as

$$K_a = \frac{[H_3O^+][F^-]}{[HF]}$$

For HF the value of K_a is 6.75×10^{-4}. The magnitude of the constant indicates that the dissociation of HF is far from complete.

The equation that represents the dissociation of HAc is

$$HAc + H_2O \rightleftharpoons H_3O^+ + Ac^-$$

The equilibrium constant expression is

$$K_a = \frac{[H_3O^+][Ac^-]}{[HAc]}$$

The chemical formula of a molecule or ion enclosed in square brackets is used to signify the concentration of that species in solution in units of moles liter^{-1}. Thus in a 2.00 molar (M) solution of NaCl [Na$^+$] = 2.00.

The value of the constant is 1.8×10^{-5}, again indicating that HAc dissociates only very slightly.

A pH scale has been devised as a means of expressing the activity of H_3O^+ present in solution. By definition

$$pH = -\log a_{H_3O^+}$$

or

$$a_{H_3O^+} = 10^{-pH}$$

If the activity of the hydronium ion in a solution is 2.0×10^{-1}, the pH of the solution may be calculated from

$$pH = -\log a_{H_3O^+}$$
$$pH = -\log (2.0 \times 10^{-1})$$
$$pH = -\log 2.0 - \log 10^{-1}$$
$$pH = 1 - \log 2.0$$
$$pH = 1 - 0.30$$
$$pH = 0.70$$

If the activity of hydronium ion is 3.4×10^{-4}, the pH is

$$pH = -\log a_{H_3O^+}$$
$$pH = -\log (3.4 \times 10^{-4})$$
$$pH = -\log 3.4 - \log 10^{-4}$$
$$pH = 4 - \log 3.4$$
$$pH = 4 - 0.53$$
$$pH = 3.47$$

QUESTIONS

1. What is the difference between the heat of a reaction and the enthalpy of a reaction? An endothermic and an exothermic reaction?
2. Examine the data in Table 4-1 to determine which groups show irregularities in H^0 of oxides.
3. Are there similar irregularities in the data given in Table 4-2?
4. Plot the data in Table 4-2 for the elements in row 4 as they are plotted in Fig. 4-1.
5. From the data in this chapter which element, oxygen or fluorine, would be a better oxidizer for use as a rocket fuel with hydrogen, lithium, carbon, magnesium?
6. What is the first law of thermodynamics?
7. What is spin correlation?
8. Explain the origin of spin correlation.
9. How are Hund's rule of maximum multiplicity and the Pauli exclusion principle related?
10. Define or state briefly the meaning of the term entropy.
11. Decide whether the following processes are likely to lead to an increase or decrease in the entropy of the system.
    (a) the synthesis of a new compound
    (b) the formulation of a new theory
    (c) digestion
    (d) a thunderstorm
    (e) a party
    (f) the end of a party
12. Do any of the above processes mean that the entropy of the universe decreases? Why?
13. How are free energy and entropy related?
14. Define the term activity.

PROBLEMS I

1. Calculate the enthalpy changes that accompany the following reactions:

$$BF(g) + F_2 \longrightarrow BF_3(g)$$
$$AlF_3(c) \longrightarrow AlF(g) + F_2(g)$$
$$Li_2O(s) + 2KF(s) \longrightarrow 2LiF(s) + K_2O$$
$$2LiF(s) + CdO(s) \longrightarrow CdF_2(s) + Li_2O(s)$$

2. Which of the above reactions is exothermic? Which is endothermic? Which absorbs heat? Which evolves heat?
3. Assuming that H^0 of H^+(aq) is 0, calculate H^0 for Cl^-, Br^-, Na^+, K^+ from the equations given below.

$\frac{1}{2}H_2(g) + \frac{1}{2}Cl_2(g) \longrightarrow H^+(aq) + Cl^-(aq)$
$\Delta H^0 = -40.0$ kcal mole^{-1}
$\frac{1}{2}H_2(g) + \frac{1}{2}Br_2(l) \longrightarrow H^+(aq) + Br^-(aq)$
$\Delta H^0 = -29.0$ kcal mole^{-1}
$Na(s) + \frac{1}{2}Cl_2(g) \longrightarrow Na^+(aq) + Cl^-(aq)$
$\Delta H^0 = -97.3$ kcal mole^{-1}
$K(s) + \frac{1}{2}Cl_2(g) \longrightarrow K^+(aq) + Cl^-(aq)$
$\Delta H^0 = -100.3$ kcal mole^{-1}

4. Assuming the D values given in this chapter, calculate the spin correlation stabilization energy of N^{2+}, Mn^+, P^+.
5. Calculate the ionization energy and H^* of P^*, As^*, Mn^*, Cu^*, and Cl^*.
6. From the H^0 values given in Chapter 10 calculate the average M-H bond energy for the hydrides of group IV. Check your answers with the M-H bond energies given in Table 10-4.
7. Hydrogen peroxide (H_2O_2) boils at 152.1°C. From the data in Table 4-10 calculate the enthalpy of vaporization of H_2O_2.
8. Calculate the enthalpy of vaporization of H_2O.
9. From the data in Appendix IV calculate ΔG^0 for

$Na + \frac{1}{2}Cl_2 \longrightarrow NaCl$
$N_2 + 3H_2 \longrightarrow 2NH_3$

10. Using G^0 values given in Appendix IV calculate the equilibrium constants for the formation of the hydrides of the group IVA elements.

PROBLEMS II

1. In the reaction $S(s) + O_2(g) \longrightarrow SO_2(g)$, 71 kcal mole^{-1} of heat is released. What is ΔH for this reaction?
2. An exothermic reaction will always have a _____ value for ΔH.
3. A reaction for which ΔH is positive is called _____.
4. $\Delta H = -800$ kcal mole^{-1} for the reaction $4Al(s) + 3O_2(g) \longrightarrow 2Al_2O_3(s)$. How much heat is liberated when 3 moles of oxygen react with aluminum?
5. How much heat is liberated when 101.96 g of Al_2O_3 are formed from the elements?
6. What is the standard enthalpy, H^0, of $O_2(g)$?
7. What is the standard state for aluminum?
8. Calculate H^0 for Al_2O_3.
9. $H^0 = -267$ kcal mole^{-1} for solid Fe_3O_4. How much heat is liberated when 55.85 g of iron react according to the equation $3Fe(s) + 2O_2(g) \longrightarrow Fe_3O_4(s)$?

Ch. 4 The Energetics of Chemical Reactions

10. Calculate ΔH^0 for the reaction $3Fe_3O_4(s) + 8Al(s) \longrightarrow 9Fe(s) + 4Al_2O_3(s)$.
11. H^0 for $N_2O(g)$ is $+19.6$ kcal mole^{-1}. The reaction $2N_2(g) + O_2(g) \longrightarrow 2N_2O(g)$ is _____.
12. Calculate ΔH^0 for the reaction $2Al(s) + 3N_2O(g) \longrightarrow Al_2O_3(s) + 3N_2(g)$ _____.
13. How much heat is liberated when 14.0 g N_2O react with aluminum?
14. How much Al(s) must be combined with $O_2(g)$ to supply 1200 kcal of heat?
15. H^0 of Na_2O is -99.4 kcal mole^{-1} and that of Na_2SO_3 is -260.6 kcal mole^{-1}. Calculate ΔH^0 for the reaction $Na_2O(s) + SO_2(g) \longrightarrow Na_2SO_3(s)$.
16. What weight of SO_2 will react with 3.1 g Na_2O?
17. How much heat is released when 3.1 g Na_2O react with SO_2?
18. How much Na_2O must react with SO_2 to supply enough heat to warm 50 g H_2O by $15°C$?
19. When 22.98 g of sodium reacts with oxygen to form Na_2O_2, 60.3 kcal of heat is liberated. Calculate H^0 for Na_2O_2.
20. Calculate ΔH^0 for the reaction $2Na_2O_2(s) \longrightarrow 2Na_2O + O_2$.
21. Calculate ΔH^0 for the reaction $2\,Na_2O_2(s) + S(s) \longrightarrow Na_2SO_3 + Na_2O$.
22. H^0 for $H(g)$ is $+52$ kcal mole^{-1}. How much heat is required to break an H-H bond?
23. ΔH^0 for the reaction $H(g) + Cl_2(g) \longrightarrow HCl(g) + Cl(g)$ is -45 kcal mole^{-1}. Which bond is stronger Cl-Cl or H-Cl?
24. To dissociate Cl_2 into Cl atoms 58 kcal mole^{-1} are required. What is the H-Cl bond dissociation energy?
25. Calculate H^0 for HCl.
26. H^0 for $H_2O(g)$ is -57.8 kcal mole^{-1}. Calculate ΔH^0 for the reaction $4H_2(g) + Fe_3O_4 \longrightarrow 4H_2O(g) + 3Fe$.
27. H^0 of oxygen atoms is 59.6 kcal mole^{-1}. What energy is required to break the double bond in O_2?
28. ΔH^0 for the reaction $H_2O(g) \longrightarrow H(g) + OH(g)$ is $+119$ kcal mole^{-1}. Calculate H^0 for $OH(g)$.
29. Calculate ΔH^0 for the reaction $H(g) + O(g) \longrightarrow OH(g)$.
30. H^* for O^* is $+76$ kcal mole^{-1}. Calculate ΔH^0 for the reaction $H(g) + O^* \longrightarrow OH(g)$.
31. What is the O-H bond dissociation energy?
32. H^0 for $H_2O_2(g)$ is -32.6 kcal mole^{-1}. Calculate ΔH^0 for the formation of $H_2O_2(g)$ from $2H(g) + 2O^*$.
33. Calculate the O-O single bond dissociation energy in H_2O_2.
34. If the Cl-O single bond dissociation energy is 58 kcal mole^{-1}, estimate H^0 for $Cl_2O(g)$.
35. Estimate H^0 for HClO(g).
36. Using this estimate calculate ΔH^0 for the reaction $H_2O(g) + Cl_2(g) \longrightarrow HOCl(g) + HCl(g)$.
37. Given $H^* = 99$ kcal mole^{-1} for P^* and $H^0 = 1.3$ kcal mole^{-1} for $PH_3(g)$ calculate the P-H bond energy.

38. The P-Cl bond energy is 85 kcal mole^{-1}. Estimate H^0 for the hypothetical molecule

$$\text{H}-\overset{..}{\underset{\underset{\text{H}}{|}}{\text{P}}}-\overset{..}{\underset{..}{\text{Cl}}}:$$

39. Using the results of Problem 38, estimate ΔH^0 for the reaction $PH_2Cl(g)$ $\longrightarrow HCl(g) + \frac{1}{3}PH_3(g) + \frac{1}{6}P_4(s)$.

SUGGESTED READINGS

Bent, H. A., *The Second Law*. New York: Oxford University Press, 1965

Harvey, K. B., and G. B. Porter, *Introduction to Physical Inorganic Chemistry*. Reading, Mass.: Addison-Wesley, 1963, Chapter 8

Klotz, I. M., *Chemical Thermodynamics*. New York: W. A. Benjamin, 1964

Mahan, B. H., *College Chemistry*. Reading, Mass.: Addison-Wesley, 1966, Chapter 10

Mahan, B. H., *Elementary Thermodynamics*. New York: W. A. Benjamin, 1963

> When atoms like the electronegative ones in which the corpuscles are very stable are mixed with atoms like the electropositive ones, in which the corpuscles are not nearly so firmly held, the forces to which the corpuscles are subject by the action of the atoms upon each other, may result in the detachment of corpuscles from the electropositive atom and their transference to the electronegative.*
>
> J. J. THOMPSON

Chapter 5 Oxidation-Reduction Reactions

In addition to predicting formulas, the oxidation state of an element may be used to predict roughly the tendency of a material to act as an oxidizing agent or a reducing agent. An oxidizing agent is a substance that is capable of accepting electrons. A reducing agent releases electrons. The oxidation state of one of the atoms in an oxidizing agent or reducing agent changes accordingly. One or more of the atoms composing an oxidizing agent decreases in oxidation state (to a less positive value), whereas one or more of the atoms in a reducing agent increases in oxidation state.

The oxidation state of any element is zero. In forming sulfur dioxide (SO_2), the oxidation state of sulfur has increased; sulfur has acted as the reducing agent. Each oxygen atom, on the other hand, has gained two electrons in functioning as an oxidizing agent. In any such reaction the increase in oxidation state should be balanced by the decrease in oxidation state. In the above example the changes do balance.

Several other examples of oxidation-reduction reactions are given in Table 5-1. In reaction (6) we see that sulfur is involved as an oxidizing agent, whereas in the reaction considered above sulfur was a reducing agent. This type of behavior is possible for many elements and any compound with one element in an intermediate oxidation

---

Phil. Mag., **7**:237 (1904).

Table 5-1 *Typical oxidation-reduction reactions*

(1) $2Fe^{3+}(aq) + 3I^-(aq) \rightleftharpoons 2Fe(aq)^{2+} + I_3^-(aq)$
(2) $H_2SO_3(aq) + H_2O_2(aq) + H_2O \rightleftharpoons SO_4^{2-}(aq) + 2H_3O^+(aq)$
(3) $2Na + 2H_2O \rightleftharpoons 2NaOH + H_2$
(4) $Cd + 2Ag^+ \rightleftharpoons Cd^{2+} + 2Ag$
(5) $H_2(g) + Cl_2(g) \rightleftharpoons 2HCl(g)$
(6) $H_2(g) + S(s) \rightarrow H_2S(g)$

state. Only compounds that contain elements in their maximum positive or maximum negative oxidation states cannot behave as both an oxidizing agent and reducing agent. The atoms in these compounds either have a full complement of electrons or they have lost as many electrons as possible. Thus, S^{2-} can only be a reducing agent, and SO_4^{2-} can only behave as an oxidizing agent.

From this we see that there is a general relationship between the magnitude of the oxidation state and the oxidizing power of the material. Some caution must be used in this relation, however, because inversions often occur. Although nitric acid (HNO_3) is a better oxidizing agent than nitrous acid (HNO_2) and sulfuric acid (H_2SO_4) is a better oxidizing agent than sulfurous acid (H_2SO_3), perchloric acid ($HClO_4$) is not a better oxidizing agent than chloric acid ($HClO_3$). This indicates that oxidation state alone is not an adequate measure of oxidation capacity. There are a large number of factors which significantly affect the oxidizing and reducing power of elements and compounds. In order to carry this discussion further, a quantitative measure of oxidizing power is needed.

When SO_4^{2-} is treated with an EXTREMELY strong oxidizing agent the sulfur atom cannot be oxidized, but it is possible to oxidize the oxygen atoms present.

I. Electrode Potentials

The oxidizing power or reducing power of an element or compound can be determined experimentally by means of an electrolytic cell such as that shown in Fig. 5-1. The particular cell shown consists of zinc (Zn) and copper (Cu) strips immersed in solutions of zinc and copper sulfate. The metal strips are connected as shown, and the solutions are connected by a salt bridge. (The tube filled with a concentrated salt solution is called a salt bridge.) If the metal strips are weighed after a suitable length of time, the copper strip will be heavier and the zinc strip lighter than they were originally. Copper ions from solution have been deposited onto the copper strip while zinc has dissolved as zinc ions. Electrical neutrality is maintained when anions

Ch. 5 Oxidation-Reduction Reactions

migrate through the salt bridge. The net reaction is

$$Zn + Cu^{2+} \longrightarrow Zn^{2+} + Cu$$

The reaction occurs with the spontaneous passage of electrons through the wire from the zinc to the copper. The metal strips are called electrodes. The zinc is the negative electrode whereas the copper is the positive electrode. The reaction occurring at the zinc electrode is

$$Zn \longrightarrow Zn^{2+}(aq) + 2e^-$$

in which zinc behaves as a reducing agent and is *oxidized*. Similarly, the reaction occurring at the copper electrode is

$$Cu^{2+}(aq) + 2e^- \longrightarrow Cu(s)$$

In this reaction $Cu^{2+}(aq)$ is an oxidizing agent and it is therefore reduced. The process as a whole is the sum of both reactions. The processes occurring at each electrode are called *half-reactions,* and the two parts of the cell (where the two half-reactions occur) are called half-cells. The half-cell in which reduction takes place is called the cathode, and the half-cell in which oxidation takes place is called the anode. In the reaction given above, the zinc half-cell is the anode and the copper half-cell is the cathode. The reaction takes place spontaneously because one atom or ion attracts electrons more strongly than another atom or ion. The electron-attracting ability of a half-cell is called its potential. The difference in potential between two half-cells is the potential from an external source which exactly balances the emf of the cell so that there is no net flow of current in the wire. In the above example, the emf of the cell is 1.13 volts.

The two terms, anode and cathode, imply oxidation and reduction respectively. Note that the two words beginning with vowels go together.

The abbreviation emf stands for the electromotive force of a cell. It is used as a synonym for potential. Both terms are very similar to voltage, but they are equal to voltage only when there is no current flowing in the cell.

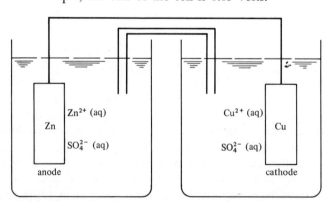

Figure 5-1 *An electrolytic cell.*

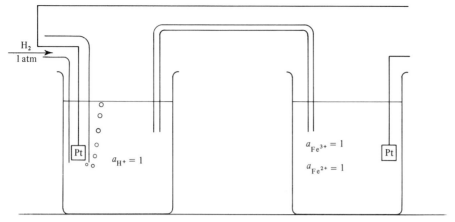

Figure 5-2 *The standard cell and the $Fe^{3+}|Fe^{2+}$ couple.*

To obtain a scale of relative redox potentials, it is customary to set the potential of a standard half-cell at 0.00 volts. Then the difference between any half-cell and the standard is the cell potential of the half-cell in question. If a cell attracts electrons more strongly than the standard half-cell, the assigned potential is positive. If it attracts electrons less strongly, the assigned potential is negative.

Since the measured potentials will depend upon the conditions under which they are measured, it is necessary to specify the standard conditions to which the potentials apply. The standard conditions are a temperature of 25°C, unit activity for all dissolved species, 1 atm pressure for gases, and all solids in their most stable form at 25°C. When standard conditions are met, the cell potential is the standard cell potential, designated as E^0. The half-cell chosen as the reference standard is represented by the equation

$$2H_3O^+(aq) + 2e^- = H_2(g) + 2H_2O$$

In Fig. 5-2 the standard cell using platinum (Pt) electrodes is shown in conjunction with the $Fe^{3+} | Fe^{2+}$ couple. The cell may be designated as

$$Pt \,|\, H_2(1\text{ atm}) \,|\, H_3O^+(a=1) \,\|\, Fe^{+3}(a=1), Fe^{+2}(a=1) \,|\, Pt$$

In this diagrammatic representation of a cell the electrodes are separated from the solutions by | and the half-cells are separated by ||. The half-cell containing the anode is represented first. The measured potential of the cell shown in Fig. 5-2 is +0.77 volt. That of the cell

$$Zn \,|\, Zn^{2+}(a=1) \,\|\, H_3O^+(a=1) \,|\, H_2(1\text{ atm}) \,|\, Pt$$

is -0.76 volt, while that of the cell

$$\text{Pt} \mid \text{H}_2(1\text{atm}) \mid \text{H}_3\text{O}^+(a = 1) \parallel \text{Cu}^{2+}(a = 1) \mid \text{Cu}$$

is $+0.37$ volt. Some additional examples of standard redox potentials are given in Tables 5-2 and 5-3. Note that each redox potential applies to an oxidizing agent and a reducing agent paired together. Many of the half-reactions in these tables are easily balanced by inspection.

The potential of any reaction may be obtained from the half-cell potentials listed in Tables 5-2 and 5-3 in the following manner. In general the reduced form of a couple with a large negative potential will be a good reducing agent and will reduce the oxidized form of any substance with a more positive potential. Thus, given

$$2\text{H}_3\text{O}^+ + \text{S} + 2e^- \rightleftharpoons \text{H}_2\text{S} + 2\text{H}_2\text{O} \qquad E^0 = 0.17 \text{ volt}$$

and

$$\text{Cr}_2\text{O}_7^{2-} + 14\text{H}_3\text{O}^+ + 6e^- \rightleftharpoons 2\text{Cr}^{3+} + 21\text{H}_2\text{O} \qquad E^0 = 1.33 \text{ volt}$$

we see that hydrogen sulfide will reduce dichromate ($\text{Cr}_2\text{O}_7^{2-}$) in acid solution. To obtain the balanced equation, the couple with the more negative potential is reversed. The equations are added after multiplying by suitable factors to balance the number of electrons appearing in the two half-reactions. To equate the number of electrons in the previous example, multiplication of the H_2S reaction by 3 is necessary. This is the couple that should be reversed.

The half-cell with the more negative potential is the poorer electron acceptor. In this case H_2S is a reducing agent. This is indicated by reversing the reaction.

$$6\text{H}^+ + 3\text{S} + 6e^- \longrightarrow 3\text{H}_2\text{S} \qquad\qquad E^0 = +0.17 \text{ volt}$$

$$3\text{H}_2\text{S} \rightleftharpoons 6\text{H}^+ + 3\text{S} + 6e^- \qquad\qquad E^0 = -0.17 \text{ volt}$$

$$\text{Cr}_2\text{O}_7^{2-} + 14\text{H}^+ + 6e^- \rightleftharpoons 2\text{Cr}^{3+} + 7\text{H}_2\text{O} \qquad E^0 = 1.33 \text{ volts}$$

$$\text{Cr}_2\text{O}_7^{2-} + 3\text{H}_2\text{S} + 8\text{H}^+ \rightleftharpoons$$
$$\qquad\qquad 2\text{Cr}^{3+} + 3\text{S} + 7\text{H}_2\text{O} \qquad E^0 = +1.16 \text{ volts}$$

When the H_2S reaction is reversed the sign of the potential should be changed. This is one way to keep track of the signs.

The E^0 value that may be obtained for any combination of half-cells is related to the standard free energy change for the reaction according to the expression

$$-\Delta G^0 = n\mathcal{F} E^0$$

where n is the number of electrons involved in the cell reaction and

Table 5-2 *Redox Couples in Acid Solution*

Couple	$E°$, volts
$F_2 + 2H_3O^+ + 2e^- = 2HF + 2H_2O$	3.317
$Ag^{2+} + e^- = Ag^+$	1.98
$PbO_2 + SO_4^{2-} + 4H_3O^+ + 2e^- = PbSO_4 + 6H_2O$	1.678
$Cr_2O_7^{2-} + 14H_3O^+ + 6e^- = 2Cr^{3+} + 21H_2O$	1.33
$O_2 + 4H_3O^+ + 4e^- = 6H_2O$	1.229
$Br_2 + 2e^- = 2Br^-$	1.077
$Ag^+ + e^- = Ag$	0.800
$Hg_2^{2+} + 2e^- = 2Hg$	0.796
$Fe^{3+} + e^- = Fe^{2+}$	0.770
$O_2 + 2H_3O^+ + 2e^- = H_2O_2 + 2H_2O$	0.695
$MnO_4^- + e^- = MnO_4^{2-}$	0.555
$I_2(s) + 2e^- = 2I^-$	0.535
$Cu^+ + e^- = Cu$	0.520
$Cu^{2+} + 2e^- = Cu$	0.340
$Hg_2Cl_2 + 2e^- = 2Hg + 2Cl^-$	0.268
$AgCl + e^- = Ag + Cl^-$	0.222
$2H_3O^+ + S + 2e^- = H_2S + 2H_2O$	0.174
$2H_3O^+ + 2e^- = H_2 + 2H_2O$	0.000
$Pb^{2+} + 2e^- = Pb$	−0.126
$Sn^{2+} + 2e^- = Sn$	−0.141
$PbSO_4 + 2e^- = Pb + SO_4^{2-}$	−0.355
$Cd^{2+} + 2e^- = Cd$	−0.402
$Fe^{2+} + 2e^- = Fe$	−0.409
$Zn^{2+} + 2e^- = Zn$	−0.762
$Mn^{2+} + 2e^- = Mn$	−1.182
$Al^{3+} + 3e^- = Al$	−1.68
$H_2 + 2e^- = 2H^-$	−2.25
$Mg^{2+} + 2e^- = Mg$	−2.363
$Na^+ + e^- = Na$	−2.714
$Ca^{2+} + 2e^- = Ca$	−2.866
$Ba^{2+} + 2e^- = Ba$	−2.905
$Cs^+ + e^- = Cs$	−2.923
$Rb^+ + e^- = Rb$	−2.925
$K^+ + e^- = K$	−2.936
$Li^+ + e^- = Li$	−3.045

Table 5-3 Redox Couples in Basic Solution

Couple	$E°$, volts
$F_2 + 2e^- = 2F^-$	2.890
$BrO_3^- + 3H_2O + 6e^- = Br^- + 6OH^-$	0.584
$ClO_4^- + 8e^- + 4H_2O = Cl^- + 8OH^-$	0.560
$O_2 + 4e^- + 2H_2O = 4OH^-$	0.401
$PbO_2 + 2H_2O + 2e^- = Pb(OH)_3^- + OH^-$	0.214
$O_2 + H_2O + 2e^- = HO_2^- + OH^-$	0.065
$Pb(OH)_3^- + 2e^- = Pb + 3OH^-$	-0.538
$Fe(OH)_3 + e^- = Fe(OH)_2 + OH^-$	-0.547
$Cd(OH)_2 + 2e^- = Cd + 2OH^-$	-0.824
$2H_2O + 2e^- = H_2 + 2OH^-$	-0.828
$Zn(OH)_4^{2-} + 2e^- = Zn + 4OH^-$	-1.189
$Mn(OH)_2 + 2e^- = Mn + 2OH^-$	-1.533
$Mg(OH)_2 + 2e^- = Mg + 2OH^-$	-2.690
$Ca(OH)_2 + 2e^- = Ca + 2OH^-$	-3.017

$\mathcal{F}$ is a constant called the Faraday. This relationship provides a direct method of measuring the standard free energy changes for redox reactions. If some of the materials are not in their standard states, the equation should be written as

$$-\Delta G = n\mathcal{F}E$$

If a cell reaction is already at equilibrium

$$\Delta G = 0$$

and there will be no cell voltage available to do electrical work. The equation is just a statement of the relationship between the free energy function and work. For the reaction

$$Zn + Cu^{2+} \rightleftharpoons Cu + Zn^{2+}$$

the free energy change is given by

$$\Delta G = \Delta G^0 + RT \ln \frac{a_{Zn^{2+}}}{a_{Cu^{2+}}}$$

From this, it follows that

$$E = E^0 - \frac{(RT)}{n\mathcal{F}} \ln \frac{a_{Zn^{2+}}}{a_{Cu^{2+}}}$$

In general this expression may be written

$$E = E^0 - \frac{RT}{n\mathcal{F}} \ln Q$$

where Q is a ratio of activities.

This equation is called the Nernst equation. It may be used to relate the half-cell potential at nonstandard conditions to: (1) the standard half-cell potential, (2) the number of moles of electrons (n) in the equation for the half-cell reaction, and (3) the activity quotient Q. The values of the constants used in the equation are $R = 8.32$ joule deg^{-1}, and $\mathcal{F} = 96{,}500$ coulombs. To convert from $\ln K$ to $\log K$ the conversion factor is 2.303. Alternative units for the constants in the Nernst equation are 2.303 $R = 4.5757$ cal mole^{-1} deg^{-1} and $\mathcal{F} = 23.05$ kcal mole^{-1} volt^{-1}.

With the constant evaluated at a temperature, T, of 298 K (25°C), the Nernst equation becomes

$$E = E^0 - \frac{0.059}{n} \log Q$$

The activity quotient Q contains the product of the activities of the products in the numerator and the product of the activities of the reactants in the denominator, just like the ratio in the equilibrium constant expression used in Chapter 4. For example, for the cell reaction

$$H_2 + 2H_2O + Cu^{2+} = Cu + 2H_3O^+$$

the Nernst equation is

$$\Delta E = 0 = \Delta E^0 - \frac{0.059}{n} \log Q = 0.34 - 0.030 \log \frac{(a_{H_3O^+})^2(a_{Cu})}{(a_{Cu^{2+}})(a_{H_2})(a_{H_2O})^2}$$

The Nernst equation may be used to calculate the equilibrium constant for the H_2S, $Cr_2O_7^{2-}$ reaction used as an example above. At

Ch. 5 Oxidation-Reduction Reactions

equilibrium, where $E = 0$, the equation for the H_2S, $Cr_2O_7^{2-}$ cell is

$$\Delta E° - \frac{0.059}{6} \log Q = 1.16 - \frac{0.059}{6} \log \frac{(a_{Cr^{3+}})^2(a_S)^3(a_{H_2O})^7}{(a_{Cr_2O_7^{2-}})(a_{H_2S})^3(a_{H_3O^+})^8}$$

$$0 = 1.16 - \frac{0.059}{6} \log K_{eq}$$

$$1.16 = \frac{0.059}{6} \log K_{eq}$$

$$\log K_{eq} = \frac{1.16}{0.0098} = 118$$

$$K_{eq} = 10^{118}$$

The value of this equilibrium constant shows that the reaction, like many oxidation-reduction reactions, goes nearly to completion. At equilibrium the activity of one of the species in the denominator must be very close to zero. The method allows the determination of equilibrium constants that could not be obtained from the analysis of an equilibrium mixture.

Another important application of the Nernst equation is the calculation of the effect of concentration changes on the potential. The standard reference cell is

$$2H_3O^+(aq) + 2e^- = H_2(g) + 2H_2O$$

If the activity of the hydronium ion (H_3O^+) is reduced to 10^{-7} without a change in the pressure of gaseous hydrogen, the resulting potential may be calculated.

Pressure is used to represent the activity of a gas.

$$E = E° - \frac{0.059}{n} \log \frac{P_{H_2}}{(a_{H_3O^+})^2}$$

$$E = 0 - \frac{0.059}{2} \log \frac{1}{(10^{-7})^2} = -0.059 \frac{14}{2}$$

$$E = -0.413 \text{ volt}$$

Continued reduction of the activity of H_3O^+ causes the potential to become even more negative. Similarly the potential of most couples that depend upon the acidity of the solution will be more negative in basic solution than in acidic solution.

The most frequent error made when using the Nernst equation is one of sign. In this book we will use the convention that assigns positive electrode potentials to couples that are more strongly oxidizing than the standard couple. With this in mind any errors made in writing the Nernst equation may easily be found. An increase in the activity or concentration of the oxidizing agent will result in an increase in the electrode potential. In the example above increasing $a_{H_3O^+}$ would make E more positive.

II. Stable Oxidation States of the Elements

There is a relationship between the stability of an oxidation state and the redox potential which is nicely illustrated by diagrams called E_h-pH diagrams or oxidation state maps. In these diagrams the cell potentials are plotted against pH. The Nernst equation is used to calculate the effect of a pH change on the cell potential.

Oxidation state maps are available over the entire periodic table. The construction of a few of these maps will be outlined in detail here. The general principles of construction are the same regardless of the element in question. The potential of the $F_2 \mid F^-$ half-reaction

$$F_2(g) + 2H_3O^+(aq) + 2e^- \rightleftharpoons 2H_2O + 2HF(aq)$$

is $+3.08$ volts in acid solution. The reaction in basic solution is

$$F_2(g) + 2e^- \rightleftharpoons 2F^-(aq)$$

and the potential is $+2.89$ volts. The effect of pH on the electrode potentials is apparent from these values. Because of the formation of the weak acid, hydrofluoric acid, fluorine is a better oxidizing agent in acid solution than it is in basic solution. In addition the potential of the cell is independent of pH in base, but the potential in acid will be highly dependent upon $H_3O^+(aq)$. The dependency can be understood using the Nernst equation

$$E = E^0 - \frac{0.059}{2} \log \frac{(a_{F^-})^2}{a_{F_2}}$$

The potential of the $F_2 \mid F^-$ half-cell will increase with increasing acidity because the activity of F^- will decrease due to the production of HF. In a plot of potential against pH, the activity of other species (except H_3O^+ and OH^-) is kept at unit activity. From the first of the two fluorine half-reactions we have

$$E = E^0 - \frac{0.059}{2} \log \frac{(a_{HF})^2}{(a_{F_2})(a_{H_3O^+})^2}$$

Ch. 5 Oxidation-Reduction Reactions

With the activity of HF and F_2 set at unity, the expression becomes

$$E = E^0 - \frac{0.059}{e} \log \frac{1}{(a_{H_3O^+})^2} = 3.06 - \frac{0.059}{2} \log (a_{H_3O^+})^{-2}$$

$$E = 3.06 + 0.059 \log a_{H_3O^+}$$

$$= 3.06 - 0.059 \text{ pH}$$

In Fig. 5-3 the plot of E against pH shows a straight line with slope -0.059. In base, the half-cell is not affected by pH; therefore a straight line at $E = E^0 = 2.87$ volts is drawn. *Lines 1 and 2 must meet at the place where the activity of HF equals the activity of F^-.* From

$$K_a = \frac{(a_{H_3O^+})(a_{F^-})}{(a_{HF})}$$

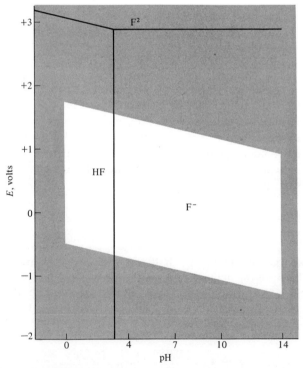

Figure 5-3 E_h-pH diagram for F_2.

if $a_{HF} = a_{F^-}$ the equilibrium constant for HF is equal to the activity of H_3O^+

$$K_a = a_{H_3O^+} = 6.9 \times 10^{-4}$$

or

$$pK_a = pH = 3.16$$

Line 3 is drawn at the pK_a of HF.

Each area of the map is labeled according to the fluorine species that is thermodynamically most stable under the conditions in that area. F_2 exists at very positive potentials above lines 1 and 2. HF and F^- are stable at potentials below lines 1 and 2 with HF more important to the left of line 3 and F^- to the right of line 3.

From the diagram we can obtain all of the following information. If the half-cell is coupled with a cell that has a potential below line 1 in acid media, HF would be the product. F^- would be produced from F_2 if the half-cell were coupled with a half-cell that had a potential below line 2 at a pH greater than 3.16. F_2 cannot exist at equilibrium in contact with water in either acid or basic solution since the potential of the half-cell

$$O_2 + 4H_3O^+(aq) + 4e^- \longrightarrow 6H_2O \qquad E^0 = +1.23 \text{ volts}$$

is well below the line (line 1) representing the oxidizing strength of F_2 in acid. Therefore, the reaction

$$2F_2(g) + 2H_2O \longrightarrow O_2(g) + 4HF$$

takes place with the evolution of oxygen gas.

The upper limit of stability in aqueous acid solution is 1.23 volts; the lower limit of stability in aqueous acid media, 0.00 volt, is determined by the reaction

$$2H_3O^+(aq) + 2e^- = H_2(g) + 2H_2O$$

The variation of the limits of stability with pH can be calculated using the Nernst equation. We shall extend both the upper and lower limit by an extra 0.5 volt because of a phenomenon (called overvoltage) related to the fact that water usually reacts quite slowly with oxidizing and reducing agents. This, in effect, means the voltage range over which water is reasonably stable toward oxidation and reduction is 1 volt larger than that calculated from the standard cell potentials.

Ch. 5 Oxidation-Reduction Reactions

This region is outlined in tint in E_h-pH diagrams such as Fig. 5-3.

The principles used to produce the diagram shown in Fig. 5-3 may be used to prepare similar diagrams for all the elements in the periodic table. The diagrams for the halogens are presented in Figs. 5-3 and 5-4. From these figures the relative oxidation-reduction strength of the elements in various oxidation states is apparent at a glance. For example, from these diagrams it is apparent that Cl_2 is a poorer oxidizing agent than F_2, and that I_2 is the poorest of all the halogens. In the same way I^- is a better reducing agent than any of the other halide ions.

Considering the chlorine diagram (in acid solution) we see that Cl_2 is an oxidizing agent when placed with a couple that has a potential below 1.36 volts, and the product is Cl^-. For any couple with a half-cell potential above 1.40 volts, Cl_2 is a reducing agent, and the product is ClO_4^-. *The product is ClO_4^- if equilibrium is reached. Actually equilibrium in the Cl^--ClO_4^- couple is almost never reached and most strong oxidizing agents oxidize Cl^- only as far as ClO_3^-.* The diagram is very simple in spite of the complexity of oxidation states of chlorine. Like F_2, Cl_2 is a stronger oxidizing agent than O_2, but the difference here is somewhat smaller. The consequences of the slight difference in oxidizing capacity of O_2 and Cl_2 are important. Cl_2 in acid solutions is another example in which the phenomenon of overvoltage (or rate of reactions) is important. The voltage 1.36 is larger than 1.23, so the reaction

$$2Cl_2 + 6H_2O \longrightarrow 4Cl^- + 4H_3O^+ + O_2$$

is spontaneous as written. The equilibrium constant for this reaction is given by

$$K = \frac{(P_{O_2})(a_{H_3O^+})^4(a_{Cl^-})^4}{(P_{Cl_2})^2} = 10^{(0.13)(4/0.059)} = 10^{8.8} = 6 \times 10^8$$

In spite of this Cl_2 gas can be bubbled through a solution of HCl with only a slight amount of reaction occurring. This is due to an inherent slowness in the reactions producing O_2. As a result an oxidizing agent such as manganese dioxide (MnO_2) which is capable of oxidizing both Cl^- and H_2O will always produce more Cl_2 than O_2 on reacting with HCl solutions.

The construction of an E_h-pH diagram is an exercise in the use of the Nernst equation and illustrates the application of thermodynamics to an important property. As such, an E_h-pH diagram is a convenient summary of much of the descriptive chemistry of an element.

Sec. II Stable Oxidation States of the Elements

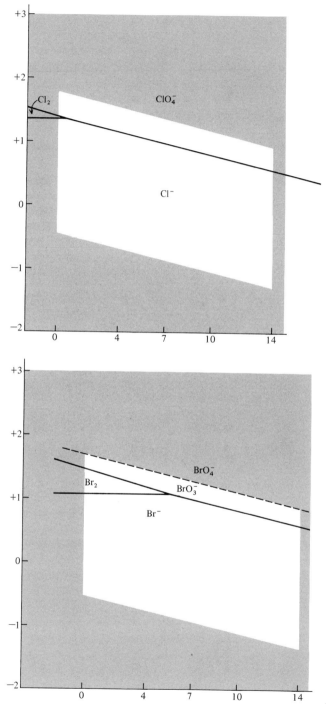

Figure 5-4 E_h-pH diagrams for the halogens.

Ch. 5 Oxidation-Reduction Reactions

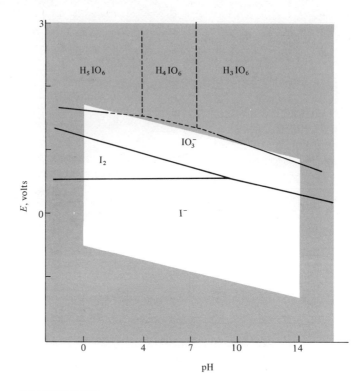

Figure 5-4 (*Continued*)

QUESTIONS

1. What are the oxidation numbers of B in BO_3^{3-}; S in $S_2O_3^{2-}$; C in CO_3^{2-}; and P in $H_2PO_4^-$?
2. Interpret the trends in ionization potential shown by the data for P, 242 kcal; S, 239 kcal; and S^+, 540 kcal.
3. In each group of five species below, pick the strongest oxidizing agent:
   (a) Ni^{2+}, Cu, Cu^{2+}, Zn, Zn^{2+}
   (b) Si, P, S, Cl_2, Ar
   (c) O_2, N_2, P_4, S_8, H_2O
4. Write chemical formulas for: the fluoride of Cr(III); the oxide of Ti(IV); the sulfide of Al(III); a sulfate of K; calcium chloride.
5. Why cannot FeI_3 exist?
6. What is the effect of pH on the strength of H_2SO_4 as an oxidizing agent?

PROBLEMS I

1. Complete and balance the following half-reactions.
   $CO = CO_3^{2-}$
   $Fe^{2+} = Fe^{3+}$
   $Cr^{3+} = Cr_2O_7^{2-}$
   $S_4O_6^{2-} = SO_4^{2-}$
   $NiS = Ni^{2+} + SO_4^{2-}$

2. Combine two half-reactions to obtain a balanced chemical equation for the reaction of Fe^{2+} with $Cr_2O_7^{2-}$.

3. In outlining the area of stability of sulfur on an E_h-pH diagram, one must calculate the point where the areas for S, H_2S, and HS^- meet. Calculate the electrode potential and pH at which S, H_2S, and HS^- all have unit activity.
K_{ioniz} for H_2S is 1.0×10^{-7} and E^0 for $S + 2H_3O^+ + 2e^- = H_2S + 2H_2O$ is 0.141 volt.

4. Using electrode potential data calculate the values of the equilibrium constants for the following reactions.

$2Na + Zn^{2+} = Zn + 2Na^+$
$2Hg + 2Fe^{3+} = Hg_2^{2+} + 2Fe^{2+}$
$2Br_2 + 6H_2O = 4Br^- + 4H_3O^+ + O_2$
$Ag^+ + Cl^- = AgCl$

5. A copper electrode in a certain $CuSO_4$ solution has a potential (against the standard hydrogen electrode) of 0.300 volt. Calculate the activity of Cu^{2+} in this solution.

6. What pressure of H_2S is required to bring the potential of the H_2S-S couple to 0.138 at a pH of 2?

7. Antimony (Sb) has oxidation states of $-3, 0, +3$, and $+5$. Sb_2O_5 dissolves in basic solutions with the formation of $Sb(OH)_6^-$ ions. Label the five regions (a, b, c, d, and e) on the E_h-pH diagram for antimony with the chemical formula of the predominate species in each region. Write a balanced half-reaction for the reduction of $Sb(OH)_6^-$ in basic solutions.

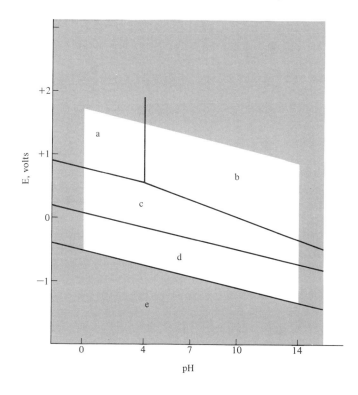

Ch. 5 Oxidation-Reduction Reactions

PROBLEMS II
1. Balance the half-reaction for Cl_2 going to Cl^-.
2. How many Cl^- ions must be formed per ClO_3^- ion in the half-reaction
   $\_ClO_3^- + \_H_3O^+ + \_e^- \longrightarrow \_H_2O + \_Cl^-$?
3. How many electrons are required to reduce the oxidation number of chlorine from $+5$ to -1?
4. How many H_3O^+ ions are needed in the half-reaction in Problem 2 so that it is balanced with respect to charge?
5. Finish balancing the half-reaction in Problem 2.
6. Does the number of electrons equal the change in oxidation number for the chlorine atom in Problem 2? Are both hydrogen and oxygen balanced?
7. Balance the half reaction $\_PbCl_2 + \_e^- \longrightarrow \_Pb + \_Cl^-$
8. Write a balanced chemical equation for the oxidation of lead by ClO_3^- in the presence of Cl^-.
9. Will H_2 react with Pb^{2+} at standard conditions? $E^0 = -0.13$ for $Pb^{2+} + 2e^- = Pb$.
10. Write a balanced equation for the reaction in Problem 9.
11. Write the equilibrium constant expression in Problem 9.
12. Calculate the value of the equilibrium constant in Problem 9.
13. What must $a_{Pb^{2+}}$ be reduced to in order to shift the Pb–Pb^{2+} potential to -0.27 volt?
14. The standard potential for the half-reaction $PbCl_2 + 2e^- \longrightarrow Pb + 2Cl^-$ is -0.27 volt. Calculate K for the reaction $PbCl_2 \longrightarrow Pb^{2+} + 2Cl^-$.
15. E^0 for $O_2 + 4H^+ + 4e^- \longrightarrow 2H_2O$ is 1.23 volts. What is the potential of this half-reaction in pure water (pH $= 7.0$) that is exposed to air if the pressure of O_2 is 0.20 atm?
16. At what pH will the reaction $Pb + 2H^+ \longrightarrow Pb^{2+} + H_2$ be at equilibrium with $P_{H_2} = 1$ atm and $(Pb^{2+}) = 0.10$ M?

By 1887 the electrolytic dissociation theory had been sufficiently completed for a careful exposition which appeared in the first volume of the *Zeitschrift für physikalische Chemie*. It is said that this publication ranks in importance with Madame Curie's announcement of her discovery of radium; yet her's was accorded instant approval and praise by the scientific world, whereas that of Arrhenius met with violent opposition and ridicule.*

H. N. ALYEA

Chapter 6 The Alkali Metals and the Nature of Ionic Crystals

I. The Chemistry of the Alkali Metals

The E_h-pH diagram for sodium shown in Fig. 6-1 is exceptionally simple. Sodium is such a strong reducing agent ($E^0 = -2.71$ volts) that, even with the overvoltage allowance, metallic sodium cannot exist in the presence of water. The other alkali metals are even stronger reducing agents than sodium, with E^0 values ranging from -3.045 to -2.923 volts. The reaction of each of these metals with water

$$2M + 2H_2O \longrightarrow 2M^+ + 2OH^- + H_2$$

where M is any alkali metal, is rapid and highly exothermic. The spontaneity and exothermicity (enough heat is produced to boil water, melt the metals, and in some cases to ignite the hydrogen produced) of the reactions with water (and other oxidizing agents) is a necessary consequence of the position of the alkali metals at the extreme bottom of the E_h-pH diagrams, the position for extremely strong reducing agents.

The alkali metal ions are soluble over the entire pH range as shown

---

*Reprinted with permission from Volume 2 of *Collier's Encyclopedia* © 1967 Crowell-Collier Educational Corporation.

Ch. 6 The Alkali Metals and the Nature of Ionic Crystals

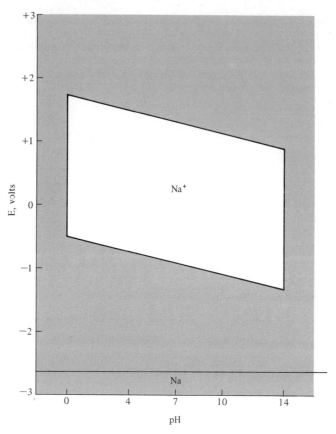

Figure 6-1 E_h-pH diagram for sodium.

in Fig. 6-1 for sodium ion (Na⁺). No insoluble hydroxides are formed. In fact, with only a few exceptions (for example, LiF, KClO₄) the simple salts of the alkali metals are all quite soluble in water. We will discuss the reasons for the solubility and insolubility of salts in later chapters.

In the presence of water the alkali metals will be found as simple ions in solution. This is reflected in the natural occurrence of the alkali metals. The metals do not exist as such in nature because they are such strong reducing agents. There are significant quantities of lithium (Li), sodium (Na), and potassium (K) in the +1 oxidation state in almost all rocks; however, the major sources of the alkali metals are the oceans and deposits left from the evaporation of former oceans. Sodium is the most abundant of the alkali metals, and sodium ion is the principal cation in sea water. Some biological fluids, such as blood and lymph, are similar to sea water in salt composition. Plants and animals are very efficient at differentiating between sodium and the potassium they need for growth. Some sea weeds contain ten times as much potassium as sodium.

A soil may become depleted in potassium; however, the deposits of potassium chloride (KCl), potassium sulfate (K_2SO_4), and so forth, left from the evaporation of ancient seas are so substantial that there appears to be no danger of running out before the year 3650. The use of potassium as a fertilizer in the form of potassium nitrate was reported in 300 B.C.

One of the major difficulties in understanding the function of sodium and potassium in biological systems and the mechanism by which potassium ion (K^+) is concentrated in the cell is the fact that the ions are so similar chemically. The alkali metals are so similar that the easiest method of distinguishing them uses the color of the light emitted from the atoms in a flame.

In Table 6-1 the ionization potentials of the alkali metals are compared with the energy required for the removal of an electron with the same quantum number from hydrogen. The difference in potential represents the amount the ns electron of the alkali metal is stabilized because of penetration of the inner core. The ionization potentials of the alkali metals decrease in a regular manner down the table. Qualitatively, the larger the atom, the lower the ionization potential. Some of the decreases from one row to the next are smaller than others, partly because of the difference between the number of elements in a row and partly because of the difference in electron configuration of the inner cores. In lithium the inner core is a $1s^2$ configuration, but in sodium the inner core is a $1s^2$, $2s^2$, $2p^6$ configuration.

The ionization potential of the np electrons in each of the excited atoms is also listed in Table 6-1. The difference in penetrating ability of the ns and the np electron is apparent from these data. The energy difference between the ns and np levels corresponds to the principal lines in the alkali metal emission spectra. The difference for sodium accounts for the characteristic yellow color observed at 5893 Å in the flame test for sodium. The wavelength of the observed line in the other alkali metals and their corresponding colors are given in Table 6-2. The electron transition observable with the naked eye in potassium (K), rubidium (Rb), and cesium (Cs) is not the $np \longrightarrow ns$ but the $(n + 1)p \longrightarrow ns$. This accounts for the difference in sensitivity of the flame tests for sodium and potassium. The most probable transition in each case is the $np \longrightarrow ns$ because this is the transition that requires the least amount of energy. The intensity of the observed light is directly related to the probability that a transition will occur, therefore the sodium light is more intense than that of potassium.

An energy level diagram showing the electron transitions for potassium is given in Fig. 6-2. Similar diagrams may be obtained for all

*Some typical data on K^+/Na^+ ratios in marine algae are summarized in R. W. Eppley, J. Gen. Physiol., **41**:901 (1958).*

Table 6-1 *Ionization Potentials of Hydrogen and the Alkali Metals, kcal $mole^{-1}$*

Element	s	p
Li	124	81.4
H n = 2	78.4	78.4
Na	118	69.5
H n = 3	34.8	34.8
K	100.0	63.0
H n = 4	19.6	19.6
Rb	96.4	60
H n = 5	12.5	12.5
Cs	89.7	58
H n = 6	8.5	8.5

The relation between energy in kcal $mole^{-1}$ and wavelength of light in Å is given by $E = 2860/\lambda$.

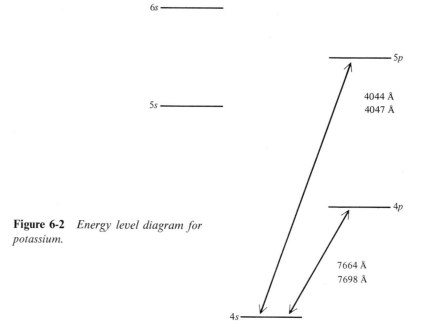

Figure 6-2 *Energy level diagram for potassium.*

the remaining alkali metals. Note the position of the $np \longrightarrow ns$ line in potassium. It is just out of the visible region.

The +1 cations of the alkali metals are colorless. Electronic transitions in the alkali metal ions are observed only at very short wavelengths, since this amounts to breaking into a completed core of electrons. The ions will have a spectrum that is similar to that of the inert gas of the preceding row, but the position of the lines will be shifted toward shorter wavelengths because of the larger nuclear charge on the ions. Both argon (Ar) and potassium ion are colorless. The chloride ion (Cl$^-$) has the same electron configuration and should also be

Table 6-2 *Position of Electronic Transitions and Colors of the Alkali Metals (in a flame)*

Atom	Wavelength in Å of the $np \longrightarrow ns$ Transitions	$(n+1) p \longrightarrow ns$	Color
Li	6708		red
Na	5890		yellow
K	7664, 7698	4047	violet[a]
Rb	7800, 7947	4201	indigo[a]
Cs	8521, 8943	4593	blue[a]

[a] Color given results from the $(n+1) p \longrightarrow ns$ transition.

colorless. Most of the properties of a salt like potassium chloride (KCl) are those of the ions involved, so we expect the solid potassium chloride to be colorless.

In spite of the similarity in electronic configuration of potassium ion, chloride ion, and argon, the physical properties of solid potassium chloride are quite different from those of solid argon. Solid potassium chloride is held together by strong electrostatic forces, and the material is very high-melting. Before discussing the chemistry of group I cations further, we shall look at the nature of ionic solids.

II. The Nature of Ionic Solids

The ionic bond is by far the simpler of the two extreme bond types because its treatment is uncomplicated by wave character. Ionic substances such as sodium chloride may be considered to be composed of positive and negatively charged ions held together by electrostatic forces. In the remainder of this chapter the structure of ionic solids and the energetics of their formation will be discussed. A discussion of bonding in general terms will be given in Chapter 7.

A. THE CLOSEST PACKING OF SPHERES

The structure of many inorganic substances can be related to the close packing of spheres. The ions of the compound, for example, (Na^+ and Cl^-) are taken as spheres and are packed together as efficiently as possible (so that the volume is a minimum and oppositely charged ions are as close as possible while similarly charged ions are as far apart as possible). One of the ions (for example, the anion) is taken as the basic layer shown in Fig. 6-3. The remainder of the structure

Ions that have a negative charge are anions; those with a positive charge are cations.

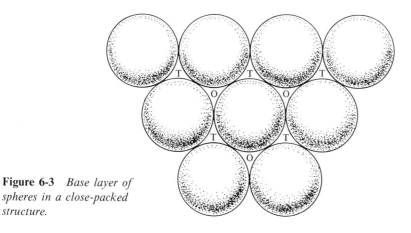

Figure 6-3 *Base layer of spheres in a close-packed structure.*

may be built by adding a second layer of anions, placing them over the holes marked T. There are two alternatives for adding the third layer of spheres. First, the spheres of the third layer may be arranged so that they are directly over the spheres of the first layer. This arrangement is known as hexagonal close packed (hcp) and the sequence of layers is represented as ABABAB . . . . In this case the total symmetry is hexagonal (the symmetry of the individual layers) as shown in Fig. 6-4, and each sphere has 12 nearest neighbors. In the alternative packing arrangement, called cubic close packed (ccp) or face centered cubic (fcc), shown in Fig. 6-5, the spheres of the third layer are placed over holes (marked O in Fig. 6-3) in the first layer. In this case the stacking sequence is ABCABC . . . . Each sphere again has 12 nearest neighbors.

These two structures (stacking sequences of ABAB . . . , and ABCABC . . .) account for the structures of a large number of inorganic crystals, metals, and inert gases as well as solids such as methane (CH_4), carbon monoxide (CO), and nitrogen (N_2). Thus, it appears that bond type has little effect, if any, on the stability of a stacking sequence.

It is really quite amazing that any stacking sequence is maintained as well as it is, considering the fact that the difference between the two sequences appears to be so minor. In most metals (for example,

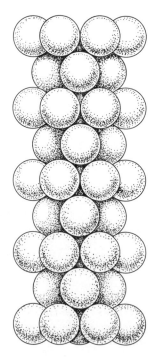

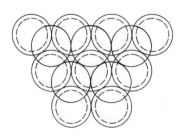

Figure 6-4 *Hexagonal close packing of spheres. The third layer of spheres is directly over the first layer of spheres.*

Sec. II The Nature of Ionic Solids

Figure 6-5 *Cubic close packing of spheres. The packing repeats after three layers.*

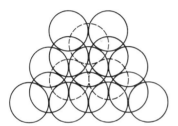

Ca, Ni, Sc, La, Ce) the packing sequence is rigorously maintained throughout the whole crystal. Only one case of a mixed cubic close packing and hexagonal close packing is known for the elements. This type of defect occurs in metallic cobalt.

As we shall see, many salts (and metals) exist in more than one distinct form. For example, zinc-sulfide (ZnS) occurs with either the zinc blende (ccp) or wurtzite (hcp) structure.

Examination of Fig. 6-6 reveals that there are two types of holes

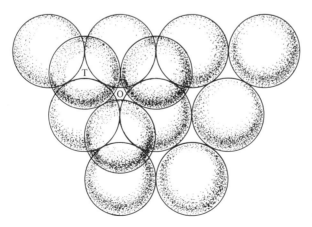

Figure 6-6 *Holes in a close-packed structure.*

159

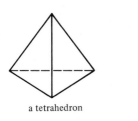

a tetrahedron

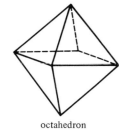

octahedron

The coordination number of an ion refers to the number of anions around a cation or the number of cations around an anion.

in a close packed structure. The holes marked T are called tetrahedral holes because they are associated with four spheres at the vertices of a tetrahedron. Those marked O are surrounded by six spheres, so these are octahedral holes. The structures of ionic solids are related to close packed spheres by filling the holes left in the packing of one kind of ion with the oppositely charged ions. In both structures (hcp and ccp) there are twice as many tetrahedral holes as octahedral, and the number of octahedral holes equals the total number of spheres. This relationship between the number of spheres and the number of holes is important because the stoichiometry of a crystal is obviously dependent upon it. The occupancy of all the octahedral holes provides a 1:1 stoichiometry (for example, NaCl) whereas full occupancy of the tetrahedral holes allows 2:1 (for example, Li_2O).

A compound crystallizes in the crystal arrangement that provides the greatest stability, or the arrangement in which the attractive forces are the greatest and the repulsive forces are the least. The most stable arrangement is the one that provides the maximum coordination number for the metal and the smallest number of anion-anion or cation-cation contacts. It is apparent from this discussion that the size of the cation (for a given anion) will be of extreme importance. Figure 6-7 shows the arrangement of ions in the cube face layer of alkali halide crystals with the sodium chloride structure. In this figure the circles represent ions with radii corresponding to ionic radii and they are drawn to scale with the observed interionic distance. In lithium chloride (LiCl), lithium bromide (LiBr), and lithium iodide (LiI) the anions are in mutual contact. Of all of the lithium halides, only in lithium fluoride (LiF) is the cation approaching contact with the anions and only in LiF are the relative sizes of the cation and anion such that the anions are not in mutual contact. To achieve cation-anion contact in LiCl, LiBr, and LiI, the anions would have to be squeezed together, thereby increasing the repulsive forces. The difficulties related to anion-anion contact are reflected in an exceptionally low bond energy and low melting point for LiI. It is apparent that the ratio of the radius of the anion and the radius of the cation is a significant quantity in the stability of crystals.

The size of the holes in a close packed lattice depends upon the anions (or cations) that make up the lattice. Some range in size of the cation occupying the holes is allowable, of course. However, if the cation is very small, it is possible that a different structure, with the cations in smaller holes, would be more stable. These considerations led Pauling to the so-called radius ratio effect. If eight negatively charged ions of radius r^- are placed around a positively charged ion of radius r^+ so that the negative ions are just touching each other and the anions are in contact with cations, the shortest distance (x) between two anions is given by

$$x = \frac{2}{\sqrt{3}}(r^+ + r^-) = 2r^-$$

and

$$\frac{r^-}{r^+} = 1.37$$

This calculation is illustrated in Fig. 6-8. If the anion is too large or the cation too small, either cation-anion contact is lost or the anions must be squeezed together. Either choice results in a loss of stability. In either of these situations, when the radius ratio is greater than 1.37, a change to a lower coordination number (for example, NaCl structure) will be favored. To maintain a structure with eight coordination, the radius ratio should be 1.37 or smaller.

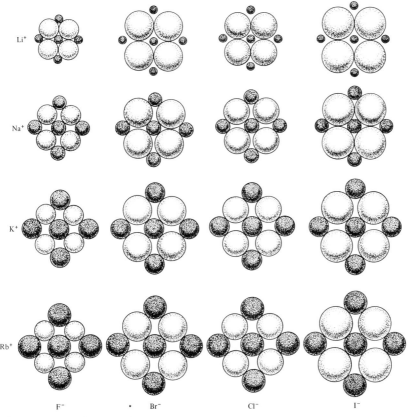

Figure 6-7 *Packing in alkali halide crystals.*

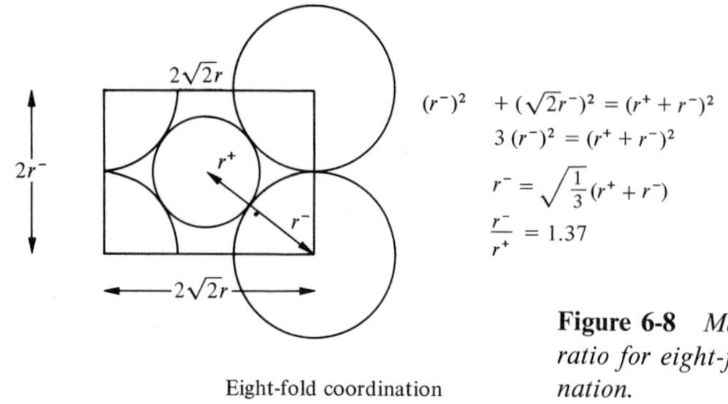

Ch. 6 The Alkali Metals and the Nature of Ionic Crystals

$$(r^-)^2 + (\sqrt{2}r^-)^2 = (r^+ + r^-)^2$$
$$3(r^-)^2 = (r^+ + r^-)^2$$
$$r^- = \sqrt{\frac{1}{3}}(r^+ + r^-)$$
$$\frac{r^-}{r^+} = 1.37$$

Eight-fold coordination

Figure 6-8 *Maximum radius ratio for eight-fold coordination.*

For the NaCl structure (6-6 coordination) the maximum radius ratio is 2.44. The calculation used to obtain this value is illustrated in Fig. 6-9. If the ratio exceeds this value, the zinc blende structure in which the metal ions occupy tetrahedral holes is favored. The zinc blende structure is satisfactory for radius ratios up to 4.55.

There are two main structures for 2:1 stoichiometry, with coordination numbers of 8-4 and 6-3. The fluorite structure (8-4 coordination) is expected if the radius ratio is less than 1.37, while the rutile structure (6-3 coordination) is expected if the ratio is greater than 1.37. These crystal structures with their coordination numbers are listed in Table 6-3 along with some typical crystals of each type.

B. SOME TYPICAL STRUCTURES

1. THE SODIUM CHLORIDE STRUCTURE (FIG. 6-10)

The sodium chloride structure involves a face centered cubic array of anions with the cations occupying octahedral holes. This structure may be represented as two interpenetrating face centered cubic lattices with each ion surrounded octahedrally by ions of opposite charge. As

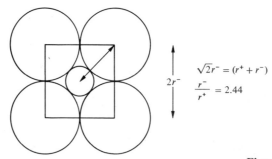

$$\sqrt{2}r^- = (r^+ + r^-)$$
$$\frac{r^-}{r^+} = 2.44$$

Six-fold coordination

Figure 6-9 *Maximum radius ratio for six-fold coordination.*

Sec. II The Nature of Ionic Solids

Table 6-3 *Examples of Some Common Crystal Structures*

Structure	Examples	Coordination numbers
CsCl	CsCl, CsBr, CsI	8-8
NaCl (Fig. 6-10)	Alkali metal halides except CsCl, CsBr, CsI	6-6
	Alkaline earth metal oxides except BeO	
	Other salts including AgCl, FeO, MnO, NiO	
Ni-As (Fig. 6-11)	Sulfides of divalent transition metal ions. Selenides (Se^{2-}) and tellurides (Te^{2-}) of divalent transition metal ions	6-6
Zinc blende (Fig. 6-14)	ZnO, BeO, ZnS, BeS, MnS	4-4
Fluorite (Fig. 6-12)	CaF_2, SrF_2, $SrCl_2$, BaF_2	8-4
Antifluorite	Mg_2Si, Li_2O, Na_2O, K_2O, Rb_2O	4-8
Rutile (Fig. 6-13)	TiO_2, MgF_2, MnF_2, FeF_2, CoF_2, NiF_2, ZnF_2	6-3
$CdCl_2$	$CdCl_2$, $MgCl_2$, $MnCl_2$, $FeCl_2$, $CoCl_2$, $NiCl_2$	6-3
CdI_2	Bromides and iodides of Mg^{2+}, Mn^{2+}, Fe^{2+}, Co^{2+}, and Ni^{2+}	6-3

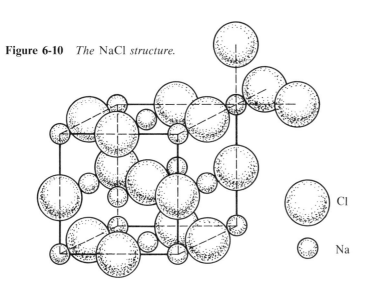

Figure 6-10 *The NaCl structure.*

Ch. 6 The Alkali Metals and the Nature of Ionic Crystals

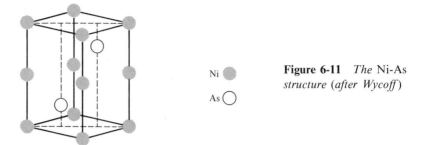

Figure 6-11 *The Ni-As structure (after Wycoff)*

noted in Table 6-3 this structure is observed for the halides of group IA, (except for CsCl, CsBr, and CsI). It is also found for the halides of silver (Ag^+) and ammonium ions (NH_4^+), and 1:1 transition metal oxides, and the oxides, sulfides, selenides, and tellurides of group IIA (except beryllium oxide and sulfide).

2. THE NICKEL-ARSENIC (Ni-As) STRUCTURE (FIG. 6-11)

The Ni-As structure involves a hexagonal close packed array of arsenic ions with the nickel ions occupying all the octahedral holes. Each nickel is surrounded by six arsenic ions but there are, in addition, two close nickel ions. Each arsenic is surrounded octahedrally by six nickel ions. Substances that crystallize in this structure include the 1:1 sulfides of the divalent transition metal ions, many selenides (CrSe, FeSe, CoSe, NiSe) tellurides (CrTe, MnTe, FeTe, CoTe, NiTe, RhTe), arsenides (MnAs, NiAs), antimonides (CrSb, MnSb, FeSb, CoSb, NiSb, PdSb, PtSb), bismuthides (MnBi, NiBi, PtBi), and stannides (MnSn, FeSn, NiSn, CuSn, PtSn). Since this structure puts the cations closer together than necessary it is not found for any compound in which the bonding is very ionic.

3. THE FLUORITE STRUCTURE (FIG. 6-12)

In the fluorite structure the cations are arranged in a face centered

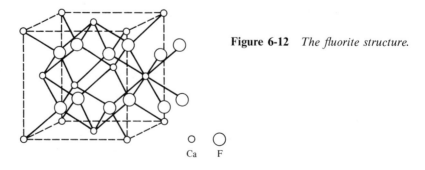

Figure 6-12 *The fluorite structure.*

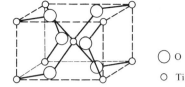

Figure 6-13 *The rutile structure.*

cubic array with the anions occupying the tetrahedral holes. The resulting stoichiometry is 1:2. Examples of this structure include CaF$_2$, BaF$_2$, CdF$_2$, HgF$_2$, and SnF$_2$. In CaF$_2$, for example, each calcium ion is surrounded by eight fluoride ions at the corners of a cube. Each F$^-$ is surrounded by four Ca^{2+} ions at the corners of a tetrahedron.

4. The Antifluorite Structure

This structure allows a 2:1 stoichiometry through a cubic close packed array of anions with metal ions occupying all of the tetrahedral holes. Compounds crystallizing with this structure include Li$_2$O, Na$_2$O, and K$_2$O.

5. The Rutile Structure (Fig. 6-13)

The fluorides of magnesium and the divalent transition metal ions crystallize in the rutile structure. These cations are smaller than those that crystallize in the fluorite structure, again illustrating the radius ratio effect. The fluoride ions are in a very distorted hexagonal close packed structure and the metal ions occupy one half of the octahedral holes. Thus, the stoichiometry is 1:2. The coordination numbers are 6 and 3 respectively. Each metal is the center of an octahedron while each fluorine is the center of a plane triangle of cations.

6. The Zinc Blende Structure (Fig. 6-14)

This structure can be considered to be a cubic close packed array of sulfur ions, with one half of the tetrahedral holes occupied by zinc ions. Examples of this structure include ZnS, BeS, and MnS. Zinc sulfide also crystallizes in the wurtzite structure, a hexagonal close packed array of sulfur with zinc occupying one half of the tetrahedral holes. In fact, many substances crystallize in both structures.

C. THE FORMATION OF IONIC SUBSTANCES

The distribution of electrons in the bond that results when two atoms interact to form a compound depends upon the electronegativity of the atoms. In the event that the electronegativities of the reacting elements are quite different, essentially complete transfer of an electron may take place, resulting, of course, in the formation of ions. The ions are held together by electrostatic forces and the bond is called an

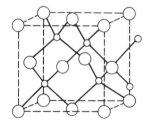

Figure 6-14 *Zinc blende structure.*

electrovalent or ionic bond. The simplest way to visualize the formation of an ionic substance is to require the metal to lose an electron and the nonmetal to gain an electron.

$$M(g) \longrightarrow M^+(g) + e^-$$

$$A(g) + e^- \longrightarrow A^-$$

The formation of a cation requires energy (the ionization potential) but adding an electron in forming the anion usually releases energy (electron affinity). The energy involved in the overall reaction

The electron affinity of a nonmetal refers to the energy released when the nonmetal gains an electron. The electron affinities of the halogen atoms are as follows. F, -82.1; Cl, -86.4; Br, -81.2; and I, -71.0 kcal mole^{-1}. These values are calculated from the data in Technical Note 270-3.

$$Na(g) + Cl(g) \longrightarrow Na^+(g) + Cl^-(g)$$

will depend on the relative magnitude of the ionization energy and the electron affinity. If the electron affinity is large enough to compensate for the ionization potential, the reaction will proceed and may give off heat. If the reverse is true, the reaction cannot take place spontaneously.

In the example used, the formation of sodium chloride, the ionization potential of sodium is $+118.5$ kcal mole^{-1} and the formation of chloride ion releases 86.4 kcal mole^{-1}; therefore, the reaction will not proceed. As a matter of fact, examination of the ionization energies and electron affinities shows that only cesium has an ionization potential so low that it is almost balanced by the electron affinity of chlorine (the atom with the largest electron affinity). In all cases the free atoms are energetically favored over the gaseous ions. However, there is another important effect to consider. This is the attractive force that exists between the products Na$^+$(g) and Cl$^-$(g). This attraction provides us with another term, the energy released when the gaseous ions come together

$$Na^+(g) + Cl^-(g) \longrightarrow NaCl(g) \longrightarrow NaCl(s)$$

To illustrate the effect of molecule formation, we shall consider the joining of one sodium ion and one chloride ion.

It was shown previously (Chapter 2) that the potential energy of two oppositely charged particles is $-e^2/r$ where r is the distance between the particles and e is the fundamental unit of charge. In Fig. 6-15 the potential energy is plotted against r. Notice that as r decreases the potential energy becomes more negative. The equilibrium interionic distance (where further approach stops) will be that distance at which the potential energy is a minimum. On the same graph a repulsion term is also plotted. This term arises because of the repulsion that takes place between the filled electron shells of the ions and the repulsion of the nuclei. The sum of the attractive and repulsive contri-

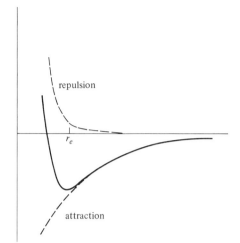

Figure 6-15 *The variation in potential energy of two ions with distance (r).*

butions gives the net potential energy of the two ions shown in Fig. 6-15. For ionic bonds the net potential energy is approximated to within 10-20% by the attractive contribution, $-e^2/r_0$, alone. The interionic distance in gaseous sodium chloride is known to be 2.38 Å. Using this distance for r and 4.8×10^{-10} esu for the value of e, the coulomb potential energy is

$$-\frac{e^2}{r} = -\frac{(4.8 \times 10^{-10} \text{ esu})^2}{2.38 \times 10^{-8} \text{ cm}} = -9.68 \times 10^{-12} \text{ erg}$$

$$= -139.5 \text{ kcal mole}^{-1}$$

1 erg = 1.439 × 10¹³ kcal mole⁻¹.

One mole of diatomic molecules is 139.5 kcal more stable than the separate ions. Thus, even though some energy must be invested to form the ions, this is more than compensated for by the energy due to the mutual coulombic attraction.

D. THE LATTICE ENERGY

The ionic crystal lattice energy is a very important concept because it is directly related to the stability of the crystal. Some other properties directly related to crystal lattice energy are the enthalpy of fusion, enthalpy of sublimation, melting point, boiling point, and solubility. In the previous section $-e^2/r_0$ was used to calculate the mutual coulombic attraction between two ions. In more general terms this might be written as an attractive term

$$\left(\frac{Z_1 Z_2 e^2}{r_0}\right)$$

where Z_1 and Z_2 are the charges on the cation and anion (neglected above because they were unity) and the repulsion term

$$+\frac{be^2}{r^n}$$

where n is called the Born exponent. The Born exponent depends upon the electron configuration of the ion ($n = 5$ for He configuration, 7 for Ne, 9 for Ar or Cu^+, 10 for Kr or Ag^+). The quantity b is a repulsion coefficient. The charge of an electron, e, has its usual significance in these equations.

In a crystal of sodium chloride, each sodium ion is surrounded by six chlorine ions at a distance r_0. Additional distances for next nearest neighbors are as follows (see Fig. 6-16)

$$12Na^+ \text{ at } \sqrt{2}r_0$$
$$8Cl^- \text{ at } \sqrt{3}r_0$$
$$6Na^+ \text{ at } \sqrt{4}r_0 = 2r_0$$
$$24Cl^- \text{ at } \sqrt{5}r_0$$
$$24Na^+ \text{ at } \sqrt{6}r_0$$

If all the electrostatic interactions are summed, the attractive potential energy is obtained as follows if Z_1 and Z_2 are unity

$$PE = -\frac{6e^2}{r_0} + \frac{12e^2}{\sqrt{2}r_0} - \frac{8e^2}{\sqrt{3}r_0} + \frac{6e^2}{\sqrt{4}r_0} - \frac{24e^2}{\sqrt{5}r_0} + \frac{24e^2}{\sqrt{6}r_0} \cdots$$

$$PE = -\frac{e^2}{r_0}\left(6 - \frac{12}{\sqrt{2}} + \frac{8}{\sqrt{3}} - \frac{6}{\sqrt{4}} + \frac{24}{\sqrt{5}} - \frac{24}{\sqrt{6}} \cdots\right)$$

It takes a great many terms to evaluate the series in brackets accurately,

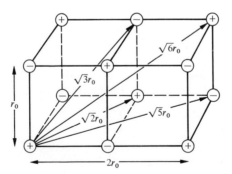

Figure 6-16 *Nearest neighbors in NaCl (after Douglas and McDaniel).*

Table 6-4 *Madelung Constants*

Structure	Coordination numbers	Madelung constant
CsCl	8-8	1.763
NaCl	6-6	1.748
Zinc blende	4-4	1.638
Wurtzite	4-4	1.641

but when this is done it is found to have a value of 1.747558. This number is known as the Madelung constant, A. The Madelung constant may be evaluated for any crystal structure. (The electrostatic attractive potential energy of the structure is given by $-Ae^2/r_0$.) Table 6-4 gives the Madelung constants for four substances with different crystal structures and a 1:1 stoichiometry. Note that there is a general trend to larger values of the Madelung constant for higher coordination numbers. The cesium chloride structure is the most favored structure if the cations are large enough for eight-fold coordination without anion-anion contact. The two structures with 4-4 coordination are very similar in electrostatic energy, and it is not surprising that many substances, like zinc sulfide (ZnS), crystallize in either the zinc blende or the wurtzite structure.

The repulsion term given previously is dependent upon r^n. Because the repulsive effects fall off very rapidly, it is necessary to consider only nearest neighbors in the repulsion term. If a quantity B is used to represent the product of the number of nearest neighbors (six for NaCl) times the parameter b, the potential energy of an ion in a crystal may be written as

$$\text{PE} = -\frac{Ae^2Z^2}{r} + \frac{Be^2}{r}$$

If B is evaluated when the potential energy is a minimum (when $r = r_0$) the potential energy becomes

$$\text{PE} = -\frac{Ae^2Z^2}{r_0}\left(1 - \frac{1}{n}\right)$$

The total energy of a mole of solid crystal, the lattice energy, is

$$U_0 = N^0(\text{PE}) = -\frac{N^0 Ae^2Z^2}{r_0}\left(1 - \frac{1}{n}\right)$$

where N^0 is Avogadro's number. U_0 is negative.

A direct determination of lattice energies (U_0 or ΔH_u) has been carried out in only a few cases. Some experimental and calculated lattice energies are compared in Table 6-5. Clearly the electrostatic calculation gives reasonably accurate values. Because of the problems involved in calculating lattice energies most are obtained from Born-Haber cycles.

1. THE BORN-HABER CYCLE

The Born-Haber cycle may be used to evaluate lattice energies by relating them to other thermochemical quantities. In the cycle in Fig. 6-17 the formation of a solid is considered by two different paths. In one pathway the cycle again represents the various steps into which the overall reaction

$$M(s) + \tfrac{1}{2}X_2(g) \longrightarrow MX(s)$$

may be broken. In the first two steps energy is required to vaporize the metal and dissociate the nonmetal. The corresponding enthalpy changes, ΔH_v and ΔH_D, are positive. In the second step of the cycle energy is required to ionize the metal (positive enthalpy change ΔH_{IP}), and energy is released when the nonmetal gains an electron (a negative enthalpy change ΔH_{EA}). Finally energy is released when solid MX forms from the gaseous ions. The large negative enthalpy change corresponding to this step (ΔH_u) is equal to the lattice energy (U_0). In the second pathway only one enthalpy term appears; that is the standard enthalpy of formation H^0.

To make the problem of keeping track of signs easier we wish to list the enthalpy change, ΔH, in each step. Thus we use the notation ΔH_{IP} for ionization potentials in place of IP as introduced in Chapter 3. Similarly ΔH_{EA} is the enthalpy change when the gaseous electronegative atom acquires an electron, or the electron affinity ($\Delta H_{EA} = -86.4$ kcal mole^{-1} for $Cl + e^- \longrightarrow Cl^-$). Ionization potentials are defined in terms of work at 0 K, but this is exactly equal to the enthalpy change at this temperature.

The total energy involved in both paths of the cycle must be equal so

$$H^0 = \Delta H_v + \tfrac{1}{2}\Delta H_D + \Delta H_{IP} + \Delta H_{EA} + \Delta H_u$$

or

$$\Delta H_u = +H^0 - \Delta H_v - \tfrac{1}{2}\Delta H_D - \Delta H_{IP} - \Delta H_{EA}$$

Table 6-5 *Comparison of Experimental and Calculated Lattice Energies (note lattice energies are negative)*

Substance	$-U_0$ (exptl.), kcal mole^{-1}	$-U_0$ (calc.), kcal mole^{-1}
KI	155.2	147.8
CsI	143.6	134.9
RbBr	154.3	151.3
NaCl	185.3	179.2
NaI	167.6	159.6
NaBr	169.9	170.5
KBr	160.0	156.6

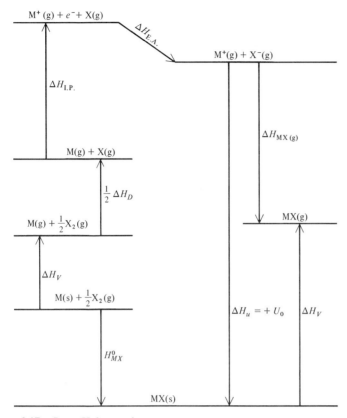

Figure 6-17 *Born-Haber cycle.*

If all the quantities on the right are known, $\Delta H_u = U_0$ may be calculated. A large number of values calculated using this cycle are given in Table 6-6.

Since Madelung constants for ionic lattices are considerably greater than 1.00, much more energy is released when the crystal is formed than when two gaseous ions combine. As shown in Fig. 6-17 the difference between these two energies is the enthalpy of vaporization (ΔH_v) of the compound. Ionic substances, in general, will have a high enthalpy of vaporization, which in turn leads to a high boiling point. The enthalpy of vaporization of solid sodium chloride is 40.8 kcal mole^{-1}. Sodium chloride boils at 1413°C.

2. Variations in Lattice Energy

The effects of size and charge on lattice energy are apparent from Table 6-6. The lattice energies of the alkali metal compounds decrease with increasing size of the cation or anion. The lattice energies of the group IIA compounds are considerably larger than those of group IA.

Table 6-6 Lattice Energies of Halides* (kcal mole^{-1}) (Born-Haber Cycle Values)

LiF	245.8	LiCl	203.0	LiBr	191.1	LiI	181.8
NaF	218.2	NaCl	185.3	NaBr	169.9	NaI	167.6
KF	193.8	KCl	168.5	KBr	160.0	KI	155.2
RbF	184.9	RbCl	162.4	RbBr	154.3	RbI	149.8
CsF	172.3	CsCl	154.8	CsBr	147.4	CsI	143.6
MgF$_2$	700.5	MgCl$_2$	597.0	MgBr$_2$	571	MgI$_2$	553.2
CaF$_2$	624.5	CaCl$_2$	536.0	CaBr$_2$	511.9	CaI$_2$	497.4
SrF$_2$	588.7	SrCl$_2$	508.2	SrBr$_2$	485.9	SrI$_2$	469.3
BaF$_2$	553.2	BaCl$_2$	483.7	BaBr$_2$	461.3	BaI$_2$	445.7
AgF	228.4	AgCl	215.8	AgBr	211.2	AgI	212.0
		TiCl$_2$	591.3	TiBr$_2$	568	TiI$_2$	558
		VCl$_2$	612	VBr$_2$	588	VI$_2$	578
CrF$_2$	690	CrCl$_2$	611.0	CrBr$_2$	593.3	CrI$_2$	577.6
MnF$_2$	663	MnCl$_2$	599.6	MnBr$_2$	581.2	MnI$_2$	567.5
FeF$_2$	696	FeCl$_2$	621.9	FeBr$_2$	604.5	FeI$_2$	590.8
CoF$_2$	714.9	CoCl$_2$	636.0	CoBr$_2$	618.7	CoI$_2$	606.1
NiF$_2$	726.5	NiCl$_2$	655.6	NiBr$_2$	637.9	NiI$_2$	624.9
CuF$_2$	700.5	CuCl$_2$	635.2	CuBr$_2$	621.1		
ZnF$_2$	718.3	ZnCl$_2$	646.6	ZnBr$_2$	630.6	ZnI$_2$	620.7

*H^0 values for the halides used in the calculation of these lattice energies may be found in Appendix IV, the references cited in Appendix IV, or L. Brewer, G. R. Somayajulu, and E. Brackett, *Chem. Rev.*, **63**:111 (1963). Ionization potentials and H^0 values for atoms are from Tables 3-2 and 4-5, respectively.

Table 6-7 Comparison of Calculated and Born-Haber Cycle Values of the Lattice Energy

Substance	Cycle value, $-\Delta H_w$, kcal mole^{-1}	Δ*
MgF$_2$	700	+3
CaF$_2$	624	+6
MnF$_2$	663	+7
NiF$_2$	726	+29
CuF$_2$	700	+10

*Δ is the difference between cycle values and calculated values.

The effect of size is again noted in group IIA by the continued downward trend.

An additional factor in the magnitude of the lattice energy is the radius ratio effect discussed earlier. In a crystalline array the size of the holes determines the type of ion that fits best. If the cation is too small, the attractive forces are reduced because the cation and anion are not in contact. To achieve cation-anion contacts, the anions must overlap; this increases the repulsive forces. For a given anion or cation there is an optimum size that insures the largest lattice energies, and deviation from this size results in a decrease in stability. This, of course, does not mean that the anion and cation must be the same size. This is usually impossible because the largest cation, Cs^+ (excepting Fr^+) is approximately the same size as the smallest of the halide anions, F^-.

The influence of electron configuration is noted for the ions of the transition elements where the d orbitals are ineffective at shielding. The result is a tendency toward larger lattice energies than the size and charge factors alone would indicate. Comparison of the lattice energies of the transition metal compounds listed in Table 6-6 with appropriate compounds of the representative elements illustrates this effect. To make this comparison, pairs of ions such as Mg^{2+}—Cu^{2+} or Na^+—Ag^+, which are about the same size should be used.

The deviation of the calculated values (based on size and charge only) from the experimental values, as listed in Table 6-7 also shows the effect for the ions Ni^{2+} and Cu^{2+}.

QUESTIONS
1. Arrange the following five ions in order of size: Si^{4+}, Ca^{2+}, Ti^{4+}, S^{2-}, and O^{2-}.
2. In each group of five species below, pick the one best described at the left.
    (a) strongest oxidizing agent: Na^+, Na, Cl^-, K^+, Br^-
    (b) strongest reducing agent: K, Ca, Sc, Ti, V
    (c) strongest oxidizing agent: Li^+, Na^+, K^+, Rb^+, Cs^+
    (d) largest ion: Li^+, Na^+, K^+, Rb^+, Cs^+
3. Explain the following sequence of melting points: LiI, 450; NaI, 651; KI, 686; RbI, 642°C.
4. Predict the melting point of CsI.
5. In each of the following sets of five compounds or ions, pick the one specified at the left.
    (a) largest ion: Li^+, Be^{2+}, Na^+, Cs^+, Ba^{2+}
    (b) strongest reducing agent: F_2, Cu, Cu^{2+}, Na, Na^+
    (c) lowest melting point: LiF, NaF, KF, RbF, CsF
    (d) most electronegative: Li, Na, K, Rb, Cs

(e) smallest ion: O^{2-}, F^-, Na^+, S^{2-}, Mg^{2+}
(f) lowest electronegativity: Na, K, Cl, Br, O
(g) smallest first ionization potential: O, F, Ne, Na, Mg
(h) highest energy K absorption edge: Na, Rb, F, Cl, I
(i) largest coordination number for Cl^-: Li^+, Na^+, K^+, Rb^+, Cs^+
(j) the zinc blende crystal structure: Li_2O, BeO, CaO, BeF_2, LiF

6. How many valence electrons will be around the lithium atom in the octet rule structure of $LiNO_2$? Why?
7. Contrast the NaCl and NiAs crystal structures. Which will have the higher Madelung constant and why?

PROBLEMS I

1. Calculate the radius of the largest sphere which will fit in the tetrahedral hole between four touching spheres of 1.00 in. radius.
2. Calculate radius ratios for the salts LiI, CsCl, CaF_2, and MgO.
3. Draw the octet rule structure for K_2CO_3.
4. Using the data in Tables 3-2 and 4-5, calculate H^0 for Na^+, Mg^{2+}, and Mn^{2+}.
5. H^0 for bromine atoms is 26.7 kcal mole^{-1}, and the electron affinity of bromine is -82.2 kcal mole^{-1}. Calculate H^0 for $Br^-(g)$.
6. H^0 for $MgBr_2(s)$ is -124 kcal mole^{-1}. Calculate ΔH^0 for the reaction $Mg^{2+}(g) + 2Br^-(g) \longrightarrow MgBr_2(s)$.
7. Using ionic radii to estimate r, calculate the lattice energy for NaBr.
8. Estimate the heat of vaporization of NaBr.
9. Using the result of Problem 7, estimate H^0 for NaBr.
10. Estimate the lattice energy for MgBr, and from this estimate H^0 for MgBr and ΔH^0 for the reaction $2MgBr \longrightarrow Mg + MgBr_2$.

PROBLEMS II

1. What is the percentage of sulfur in Na_2S?
2. Write the electronic structure of rubidium.
3. What is the oxidation state of rubidium in Rb_2O_2?
4. How many neutrons, protons, and electrons are present in a $^{85}Rb^+$ ion?
5. Arrange the following in order of size: Rb, Rb^+, K^+, Cs.
6. Arrange the following ions in order of size: Se^{2-}, Br^-, Rb^+, K^+, Te^{2-}.
7. In which of the following salts is there the greatest discrepancy between the sizes of cations and anions: LiF, LiI, CsF, CsI?
8. Write the octet rule structure of KN_3.
9. H^0 for KI is -78 kcal mole^{-1}. Estimate the H^0 value for LiI.
10. H^0 for LiI is -65 kcal mole^{-1}. Calculate ΔH^0 for the reaction $Li(s) + KI(s) \longrightarrow LiI(s) + K(s)$.
11. Assuming ΔS^0 for the reaction in Problem 10 is -2 eu, calculate the equilibrium constant for the reaction.
12. Calculate K for the reaction $2\,NaH(s) \longrightarrow 2Na(s) + H_2$. $G^0 = -8.02$ kcal mole^{-1} for NaH(s).
13. What pressure of H_2 will be in equilibrium with the pure solids NaH and Na?

14. Given the standard electrode potentials below, list the three alkali metals in order of strength as reducing agents.

$Na^+ + e^- = Na$ -2.71 volts
$Rb^+ + e^- = Rb$ -2.92 volts
$Li^+ + e^- = Li$ -3.04 volts

15. What pressure of H_2 would be in equilibrium with pure Li, pure H_2O, and the ions H^+ and Li^+ with $a_{Li^+} = 1.00$ and pH $= 14$?

> They (atoms) move in the void and catching each other up jostle together and some recoil in any direction that may chance, and others become entangled with one another in various degrees according to the symmetry of their shapes and sizes and positions and order, and they remain together and thus the coming into being of composite things is effected.*
>
> SIMPLICIUS

Chapter 7 Electrons in Molecules

In Chapter 3, an elementary approach to the electronic structure of molecules was presented in terms of the octet rule. Atoms complete octets by sharing pairs of electrons, or in some cases fractional numbers of electrons. In this chapter we shall examine the bonding in molecules in terms of molecular orbitals.

It should be emphasized again that a complete description of the electronic structure of molecules is difficult or impossible for even reasonably simple molecules. The difficulty, of course, is due to the wave nature of electrons in atoms. Many aspects of ionic bonding can be treated without considering the wave nature of electrons explicitly; however, the wave nature of electrons is basic to the description of the covalent bond. The two models used to describe chemical bonds, covalent and ionic, are extremes, and the bonds in most molecules are intermediate between them. There is, in general, a gradual transition between the two extremes that is related to the electronegativity of the atoms involved.

*Used by permission from Cyril Bailey, *The Greek Atomists and Epicurus, A Study*, [1928] (New York: Russell & Russell, 1964).

I. Bond Type and Electronegativity

The distribution of the bonding electron pair in a molecule is determined by the electronegativity of the atoms involved. In a homonuclear diatomic molecule the electrons are symmetrically distributed in the bonding orbital. In a heteronuclear diatomic molecule, as the electronegativity difference between atoms increases, the distribution becomes distorted toward the more electronegative atom until, in a molecule with very different atoms, both electrons of the bond are essentially localized about one atom. When the electron transfer to the more electronegative atom is essentially complete, the bond is called ionic. Bonds involving the intermediate situation are called partially ionic or partially covalent. The term polar covalent is also useful for a bond in the intermediate range. This distinction between bond types is important because different properties are associated with the different types of molecules.

The term homonuclear means that the atoms involved are the same. Heteronuclear means that the atoms are different.

Partially ionic bonds in a molecule will result in a dipole moment if the molecule is not structurally symmetrical. A linear molecule made up of two different atoms will have a dipole moment that is equal to the product of the charge on the atoms and the distance separating them. The molecule is termed a polar molecule. Water has a dipole moment because the bonds are partially ionic and the molecule is bent. In contrast, the dipole moment of carbon dioxide (CO_2), a linear molecule, is zero in spite of the polar nature of the bonds.

Dipole moments were initially used by Pauling to calculate ionic character as it relates to electronegativity difference. The partial ionic character of the bonds in the hydrogen halides is obtained from the ratio of the observed dipole moment to the dipole moment calculated for the 100% ionic bond. The observed dipole moment for hydrogen fluoride (HF) is 1.82 debye (1 debye = 10^{-18} cm-esu). The dipole moment expected for a 100% ionic bond in hydrogen fluoride is 4.4 debye [$(4.8 \times 10^{-10}$ esu$)(0.917 \times 10^{-8}$ cm$)$]. The bond in hydrogen fluoride is thus 41% ionic. Figure 7-1 gives the relationship between ionic character and electronegativity difference. The scattering of points in this figure indicates that there is only a general correlation between the percent ionic character calculated from dipole moments and the electronegativity difference.

The charge on one electron is 4.8×10^{-10} esu and the bond length in HF is 0.917×10^{-8} cm.

In spite of the fact that most bonds are of the intermediate type, the terms ionic and covalent are convenient to use in order to obtain simple pictures of the orbitals involved in a bond. This was done in the introduction to ionic bonding in the previous chapter. In the next section we will build on this discussion of ionic systems before discussing covalent systems.

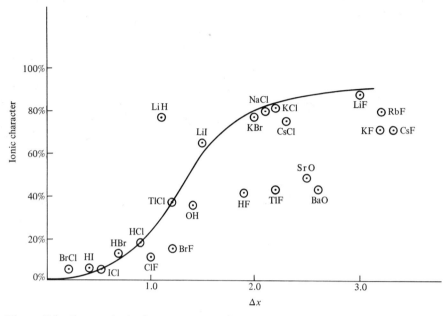

Figure 7-1 *Percent ionic character versus electronegativity difference ($\Delta\chi$).*

II. Ionic Bonding

The valence electrons in sodium chloride are shared between the sodium and chlorine, but the sodium has such a small share that it is reasonable to assume that the electron transfer is complete. It is possible to describe the bonding fairly accurately in terms of the transfer of electrons from the high energy $3s$ orbital of sodium to the lower energy $3p$ orbital of chlorine. This means that the bonding molecular orbital is principally the valence orbital of the chlorine atom. It is relatively easy to estimate the energy of the molecular orbitals for bonds of this type since the molecular orbital is fundamentally (with some increased electron repulsion) just the atomic orbital of the most electronegative atom. A method that may be used to estimate the energy of molecular orbitals will be discussed later. At present we are primarily concerned with energy differences.

The bond dissociation energy of sodium chloride

$$NaCl(s) \longrightarrow Na(g) + Cl(g)$$

$$\Delta H = + \text{ bond dissociation energy} = 153 \text{ kcal mole}^{-1}$$

is a measure of the difference in orbital energies because it is essentially the energy required to transfer one electron from the chlorine ion to the sodium ion. If electronegativity is a useful guide to orbital energies, we might expect the bond dissociation energy to increase linearly with the difference in electronegativity between the two elements. Figure 7-2 shows that there is some tendency in this direction; however, it illustrates even more that electronegativity alone is not a completely satisfactory measure of the strength of ionic bonds or of orbital energies. Two additional factors are important.

The major additional factor is a matter of electrostatics. Energy is required to produce ions. Where this energy requirement is exceptionally large, the bonding is weaker. This dependence is apparent from Fig. 7-3 where the bond dissociation energy (BDE) is shown to be

$$\text{BDE} = -(\Delta H_u + \Delta H_{\text{IP}} + \Delta H_{\text{EA}})$$

since

$$\text{BDE} + \Delta H_{\text{IP}} + \Delta H_{\text{EA}} + \Delta H_u = 0$$

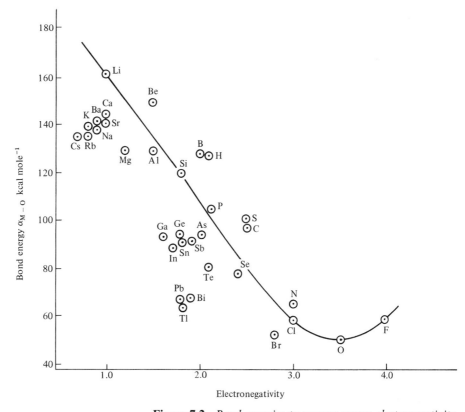

Figure 7-2 *Bond energies to oxygen versus electronegativity.*

In this figure ΔH_u is the lattice enthalpy, ΔH_{IP} is the ionization potential, ΔH_{EA} is the electron affinity, and ΔH_v is the energy required to vaporize solid NaCl. As ΔH_{IP} and ΔH_{EA} become more positive (ΔH_{EA} is negative) the bond energy should decrease.

The energy required to form a doubly charged oxygen ion results in a bond that is weaker than might be expected from the large electronegativity difference. As a result the M-O bond energies do not fall on the line shown in Fig. 7-2. The Na-O bond is much stronger in NaOH than it is in Na_2O so the reaction $Na_2O + H_2O \longrightarrow 2NaOH$ is very exothermic ($\Delta H = -36.3$ kcal mole^{-1}). In fact Na-O bonds reach their full strength only if the negative charge is spread over more than one oxygen atom in salts like $NaNO_3$ and Na_2SiO_3.

The influence of the lattice energy, U_0, on bond energies is shown in the data for the alkali halides listed in Table 7-1. the Cs-Cl bond (152 kcal) is weaker than the K-Cl bond in spite of the fact that cesium is less electronegative than potassium because the internuclear distance is larger in CsCl than it is in KCl. A shorter bond distance leads to a more negative value for U_0 and this implies a greater bond energy (see Fig. 7-3).

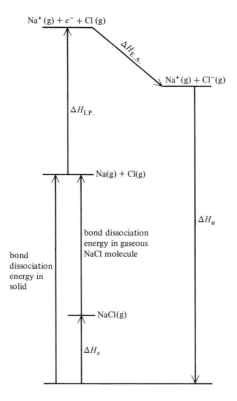

Figure 7-3 Born-Haber cycle.

Table 7-1 *Bond Energies and Bond Lengths of Some Alkali Metal Salts*

Bond	Bond length, Å	Bond energy, Kcal mole^{-1}
Li-F	2.01	202
Na-F	2.31	181
K-F	2.67	174
Rb-F	2.82	170
Cs-F	3.00	165
Li-Cl	2.56	164
Na-Cl	2.81	153
K-Cl	3.15	154
Rb-Cl	3.29, 3.24	152
Cs-Cl	3.51, 3.57	152
Li-Br	2.75	148
Na-Br	2.98	139
K-Br	3.30	142
Rb-Br	3.43	140
Cs-Br	3.72	140
Li-I	3.00	128
Na-I	3.24	121
K-I	3.53	125
Rb-I	3.67	124
Cs-I	3.94	125

III. Covalent Bonding

It is apparent from orbital energies or from electronegativity considerations why an electron is transferred from a sodium atom to a chlorine atom, and it is clear that the resulting ions will have an electrostatic attraction for each other. It is not as easy to see why sharing a pair of electrons will hold together molecules such as H_2 and Cl_2. Some insight into the bonding in these molecules may be gained from a further consideration of shielding.

It is apparent from the discussion of shielding in Chapter 3 that, at small values of r, a hydrogen atom will exhibit some attraction for an additional electron because the hydrogen nucleus is not fully shielded by the electron in the $1s$ level. Neutral atoms do not behave as neutral particles at small distances because an effective nuclear charge is operative. As a result, at small distances a force of attraction will be exerted on negatively charged particles, such as an electron

Ch. 7 Electrons in Molecules

Van der Waal's forces will be discussed in Chapter 9.

from another atom, and a force of repulsion is exerted on another nucleus. Figure 7-4 shows a plot of the potential energy as a function of distance for the hydrogen molecule.

At large distances there is a slight attraction between two atoms (picture hydrogen atoms for simplicity) that is called the van der Waal's attraction. The overlap of the 1s functions is small, and shielding of the nuclei is nearly complete. As the distance between the atoms decreases the shielding decreases, and the effective nuclear charge on each atom increases as the orbitals overlap more. Both the attractive and repulsive forces increase; however, in a hydrogen molecule there are four attractive and two repulsive interactions (electron-electron and nucleus-nucleus repulsions). If the charges are properly placed, there will be a net attractive force. It is reasonable to expect that the original atomic orbitals distort from the completely spherical distribution as the electrostatic attractions and repulsions increase. As the atomic orbitals merge and the nuclei come closer, the effective nuclear charge becomes larger and the electron distribution becomes more concentrated between the nuclei. Eventually, as r continues to decrease the forces balance and an equilibrium distance is reached. At this point the electron density of the electron pair bond is substantially distorted from the original atomic orbitals. The electron density is very high between the atoms, the repulsive forces equal the attractive forces, and the equilibrium distance (0.74 Å) is established. The probability distribution plot shown in Fig. 7-5 indicates that the maximum electron density is on the atoms, but the probability that the electrons will be between the nuclei is very high. If this enhanced electron density between the nuclei is permitted by the Pauli exclusion principle, a chemical bond results.

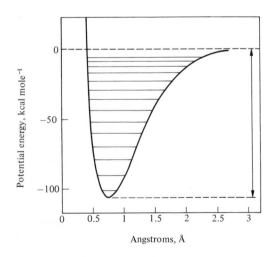

Figure 7-4 *Potential energy versus internuclear distance for the hydrogen molecule. (After W. J. Moore, Physical Chemistry, Englewood Cliffs, N.J.: Prentice-Hall, 1950.)*

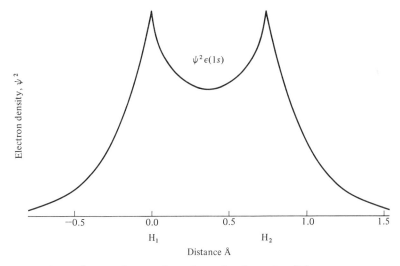

Figure 7-5 *Electron density along the molecular axis of H_2.*

The two criteria for covalent chemical bond formation are apparent here. One is that the atomic orbitals must overlap in space. The atoms must be sufficiently close so that the shielding is not complete. The second requirement is that the two electrons must not have all four coordinates (quantum numbers) the same. Since the electrons must occupy the same space (the orbitals overlap) they must have different spin coordinates.

We are not interested here in a detailed mathematical description of bonding, but it is important to distinguish between two major methods of treating chemical bonds. One, the most powerful method, is the molecular orbital method. We shall introduce some of the terminology here and take up the method again in Chapter 12. The other method, the valence bond method, is used primarily to predict the structure of covalent molecules. It will be presented together with the Gillespie-Nyholm method in Chapter 8.

A. THE MOLECULAR ORBITAL METHOD

1. Bonding Orbitals

The simplest description of the bonding function for the hydrogen molecule can be written as

$$\psi_{MO} = \sqrt{\tfrac{1}{2}}(\psi_{1s(H_1)} + \psi_{1s(H_2)})$$

The function ψ_{MO} is called a linear combination of atomic orbitals (LCAO). The normalization constant ($\sqrt{\tfrac{1}{2}}$) assures that the total prob-

Ch. 7 Electrons in Molecules

ability of finding the electron will be 1. The combination implies that the atomic orbitals interact, or overlap to form a new orbital. By definition, this is a bonding molecular orbital. A chemical bond is just such an orbital filled with two electrons.

One of the criteria for the effective overlap between two orbitals is the symmetry of the combining orbitals with respect to the molecular axis. To form an effective overlap the symmetry of the two orbitals must be the same. Some combinations that have appropriate symmetry are shown in Fig. 7-6. An example of a pair of orbitals that do not overlap effectively is shown in Fig. 7-7.

If the bonding function has cylindrical symmetry about the axis connecting the two atoms, it is termed a σ (sigma) bonding function whether the atomic orbitals involved are s or p orbitals. A bonding function that has a single planar node containing the molecular axis is called a π (pi) function; π functions do not have cylindrical sym-

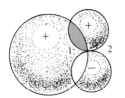

Figure 7-7 *Ineffective overlap of an s and p function.*

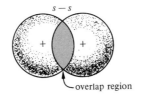

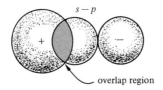

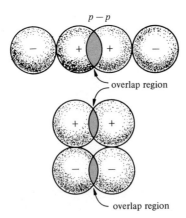

Figure 7-6 *Effective overlaps.*

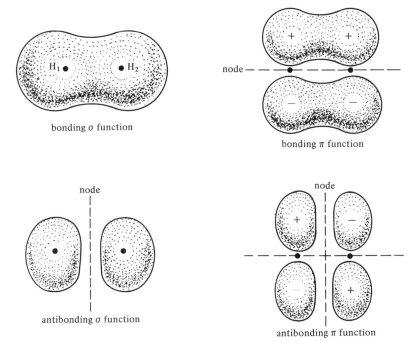

Figure 7-8 *Bonding and antibonding functions.*

metry. The electron distribution in a σ bond and a π bond is shown in Fig. 7-8.

2. Antibonding Molecular Orbitals

The bonding functions described above were obtained by taking the positive combinations. It is interesting and important to see what happens if the negative combination is taken

$$\psi_{MO} = \sqrt{\tfrac{1}{2}}(\psi_{1s(H_1)} - \psi_{1s(H_2)})$$

In this function the negative sign changes the sign of the function on hydrogen (2). As a result ψ_{MO} changes sign between the two atoms. In other words this combination places a node between the atoms. A node always represents zero probability because the function has a positive value on one side and a negative value on the other. Combinations of this type are called antibonding functions. They are labelled as antibonding states with an asterisk, as ψ^*, σ^*, or π^*. They are less stable (higher in energy) than bonding orbitals and even less stable than the original atomic orbitals because the electron concentration between the atoms is low (resulting in less shielding and increased nuclear repulsion). The antibonding function that corresponds to each bonding function is shown in Fig. 7-8.

B. MOLECULAR ORBITAL DIAGRAMS

The relative energies of the bonding and antibonding levels of hydrogen (H_2) may be represented by the molecular orbital diagram shown in Fig. 7-9. In this diagram the atomic orbitals are written on the left and right and the molecular orbitals are drawn in the center. This diagram illustrates the principle that when molecule formation takes place the total number of orbitals is conserved. We start with two atomic orbitals and we obtain two molecular orbitals. Another point illustrated by Fig. 7-9 is that full occupancy of both bonding and antibonding orbitals (by four electrons, two in each orbital) will result in no net bonding (for example, He_2 is unstable). This explains qualitatively why filled subshells do not ordinarily contribute to bonding.

It is especially easy to write molecular orbital diagrams for diatomic molecules. This will be done for a few examples now. The same ideas with only minor modifications are involved in writing molecular orbital diagrams and molecular orbitals for any molecule.

The bonding in oxygen (O_2) may be visualized from the orbital diagrams shown in Fig. 7-10. In this figure a cartesian coordinate system is used to represent each atom of the diatomic molecule. The systems are arranged so that one of the axes is common to both atoms, while the other axes are kept parallel. The labeling of the axes is arbitrary; however, by convention the z axis is taken as the molecular axis. If the p_z axis is taken as the molecular axis, the p_z function will be involved in σ bonding while the p_x and p_y orbitals will form π bonds. The overlapping orbitals are: s_1 with s_2; p_{z_1} with p_{z_2}; p_{y_1} with p_{y_2}; p_{x_1} with p_{x_2} where the numbers designate the atom involved.

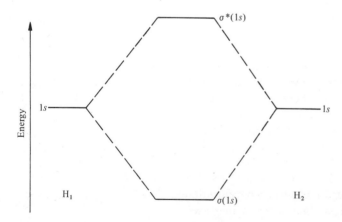

Figure 7-9 *Molecular orbital diagram for* H_2.

Sec. III Covalent Bonding

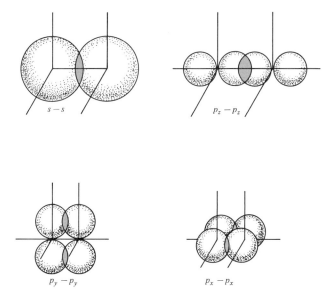

Figure 7-10 *Overlaps in* O_2.

A molecular orbital diagram may be written for the molecule if the relative energies of the orbitals are known. In general, σ bonding orbitals (σ bonding states) are lower in energy than π bonding states, and bonding states are lower in energy than antibonding states of the same symmetry (that is, σ or π). The energy of many of the levels may be determined experimentally for a specific molecule, but for many purposes it is adequate to order the levels intuitively by assigning the lowest energy to the bonding orbital with the largest overlap. A complete molecular orbital diagram for oxygen is given in Fig. 7-11. We shall discuss this figure in detail.

In Fig. 7-11 the lines on the extreme left and right represent the atomic orbitals of the separate atoms. The states in the center of the diagram represent the molecular energy levels. The lines connecting the atomic and molecular levels indicate the origin of the molecular orbitals. This general diagram may be applicable to a number of diatomic molecules. To apply it to a specific case, such as oxygen, an appropriate number of electrons must be added. Starting with the lowest lying level, two electrons are added to each σ state and four electrons are added to each π state. This has been done in Fig. 7-12. It should be noted that the two electrons occupying the antibonding π state (designated as the π* state in Fig. 7-11 and Fig. 7-12) are written in separate orbitals because the energy of these orbitals is the same. Hund's rule is applicable whenever orbitals for electrons have the same energy, whether the orbitals are in an atom or a molecule.

187

Ch. 7 Electrons in Molecules

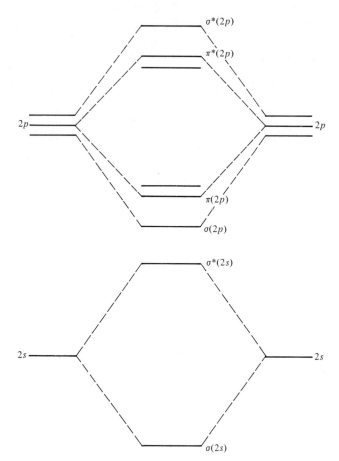

Figure 7-11 *Molecular orbital diagram for O_2. (The 1s orbitals are omitted.)*

In Chapter 4 we saw that two electrons with the same spin are more stable than two electrons with opposite spin. Whenever the orbitals have the same energy, no advantage can be gained by pairing; therefore the electrons remain unpaired. The presence of two unpaired electrons in oxygen can be verified experimentally. This is a feature that comes out of the molecular orbital method that none of the other methods of treating chemical bonding can handle convincingly.

Electrons that are placed in bonding orbitals give the molecule stability over and above the uncombined atoms. In the oxygen molecule there are a total of eight bonding electrons and four antibonding electrons. Very roughly the stabilization gained from the occupancy of a bonding level equals the destabilization resulting from the occupancy of the antibonding level of the same symmetry. In oxygen the net result is a bond energy equivalent to four bonding electrons or two bonds. The experimental bond energy observed for the oxygen molecule supports this argument. In fact, the ability of oxygen to sustain life supports this idea. The great difference in chemical reactivity between oxygen molecules and nitrogen molecules is apparent and results from the presence of antibonding electrons in oxygen. The nitrogen molecule has ten valence electrons, just enough to fill all four bonding molecular orbitals in Fig. 7-11 and only one antibonding orbital. This gives a net of six bonding electrons, or three bonds for nitrogen.

The molecular orbital diagrams appropriate for the diatomic molecules formed by some of the other elements of the first short row are given in Fig. 7-13. The levels in this figure have been ordered according to the correct ground state configuration of the molecule. The ground state configurations are:

A ground state is the state of lowest energy for a molecule.

$$B_2 \; [\sigma(2s)]^2, \; [\sigma(2s)^*]^2, \; [\pi(2p)]^2$$

$$C_2 \; [\sigma(2s)]^2, \; [\sigma(2s)^*]^2, \; [\pi(2p)]^4$$

$$N_2 \; [\sigma(2s)]^2, \; [\sigma(2s)^*]^2, \; [\pi(2p)]^4, \; [\sigma(2p)]^2$$

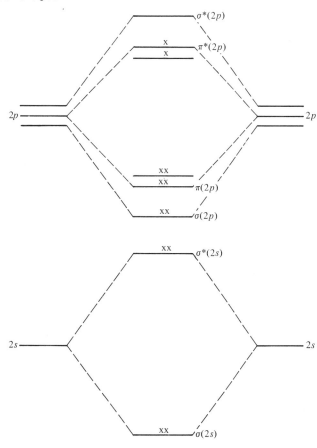

Figure 7-12 *Molecular orbital diagram for O_2.*

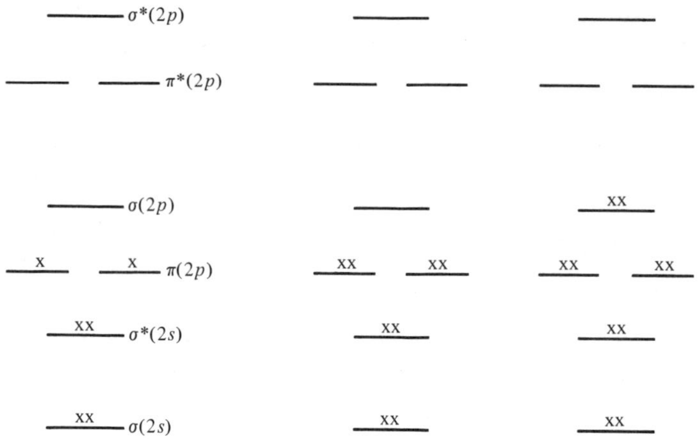

Ch. 7 Electrons in Molecules

Energy level diagram (left to right, for B₂, C₂, N₂):

- $\sigma^*(2p)$: — / — / —
- $\pi^*(2p)$: — — / — — / — —
- $\sigma(2p)$: — / — / xx
- $\pi(2p)$: x x / xx xx / xx xx
- $\sigma^*(2s)$: xx / xx / xx
- $\sigma(2s)$: xx / xx / xx

Figure 7-13 Energy levels for B_2, C_2, and N_2. (The 1s orbitals are omitted in each case.)

According to these configurations B_2 has two unpaired electrons, whereas the electrons in C_2 and N_2 are all paired. The ordering of levels in these molecules is somewhat different than that given for O_2. The difference amounts to an inversion in order of the σ (2p) and π (2p) bonding levels. This inversion is not expected from an intuitive ordering of the levels. It results from a slight complication in the method that is still to be introduced. To obtain the diagrams in Fig. 7-13, the simple method presented here must be extended to include the concept of mixing, or alternatively, of hybridization. These are the names given to two mathematically different but equivalent methods of extending the simple method presented above. Hybridization is introduced in the next chapter. In Chapter 11 some additional molecules are treated by the molecular orbital method, and we shall return to the question of the actual ordering in Chapter 12.

Depending upon the molecule, one of the two alternative diagrams (Fig. 7-11 or 7-13) will be appropriate; however, in any situation where the order given in Fig. 7-11 is likely, the ordering is actually immaterial because in these cases a sufficient number of electrons will be present so that the levels will be nearly filled. Oxygen and fluorine are examples. Homonuclear molecules formed by atoms to the left of oxygen and to the right of beryllium in the first row as well as nearly all heavier atoms will require the order shown in Fig. 7-13.

As a final illustration we shall consider one other example. In neon (Ne_2) each neon atom supplies eight valence electrons to the molecule. The total number of valence electrons is sixteen or just exactly the number required to fill all the levels in the diagram. As a result the diatomic neon molecule is not stable.

IV. Bond Type and Properties

Most chemical bonds are intermediate between the covalent and ionic type. However, an ionic bond with a small amount of covalent character (10%) can differ a great deal from a purely ionic bond; whereas a bond with only a small amount of ionic character generally behaves like a completely covalent bond.

Many substances that involve covalent bonds are low melting and low boiling. They are generally nonconducting and insoluble in water. Substances that involve ionic bonds are usually high melting and high boiling, and they are often soluble in water. The difference in properties is not related to the strength of the bonds involved in covalent and ionic substances, but to the type of solids that usually result. *Ionic substances* consist of a three-dimensional crystalline array in which all ions are held by strong electrostatic forces. Considerable energy is required to rupture an ionic crystal and separate the ions.

It is necessary to distinguish two classes of solids containing covalent bonds. If all the covalent bonding is within certain aggregates (molecules) (carbon tetrachloride, CCl_4; carbon dioxide, CO_2; and so forth), the substance is called a molecular solid. *Molecular solids* generally have low melting and boiling points since the molecules are held together only by relatively weak forces (van der Waal's forces, see Chapter 9). On the other hand in many cases the covalent bonding extends throughout the crystal, and separate molecules cannot be distinguished. This is the case in diamond crystals and in silicon dioxide (SiO_2). These so-called *covalent solids* are very hard and have high melting points and high boiling points. The distinction between covalent and molecular solids accounts for the drastic changes in properties between solid carbon and gaseous nitrogen (N_2) and between carbon dioxide (CO_2, sublimes at $-80°C$) and silicon dioxide (SiO_2, melts at $1710°C$, and does not boil without decomposition).

It should be noted that if a substance forms molecular crystals, the melting point will be low regardless of the type of bond involved in the molecule. For example, the bonds in silicon tetrafluoride (SiF_4) are about 50% ionic; yet it exists as a gas at room temperature. The radius ratio in silicon tetrafluoride is 3.3. This is in the range for which a tetrahedral coordination is expected. Since the four ligands are symmetrically arranged, silicon tetrafluoride has no net dipole moment in spite of the polarity of the individual bonds.

On the other hand, the physical properties of aluminum trifluoride (AlF_3) (which also contains bonds that are about 50% ionic) are those of a typical ionic compound. It sublimes without melting at $1257°C$. The bonds in aluminum trifluoride are not very different from those in silicon tetrafluoride, but the difference in stoichiometry and an increase in coordination number from four to six are sufficient to account for the change from molecular crystals for silicon tetrafluoride to ionic crystals for aluminum trifluoride.

V. Bond Strength and Electronegativity

At the beginning of this chapter we considered the effect of the difference in electronegativity of the bonded atoms on the strength of ionic bonds. We saw in Fig. 7-2 that there is a general but irregular increase in bond strength as the electronegativity difference increases. When there is an exceptionally large electronegativity difference this may be overbalanced by the electrostatic energy requirement for a highly unsymmetrical charge distribution. Thus the curve of bond strength versus $\Delta\chi$ bends down when $\Delta\chi$ is large. There must also be considerable curvature for small values of the electronegativity difference since, as we have seen, the bond energy does not go to zero when the electronegativity difference is zero. The strength of a purely covalent bond will obviously depend upon the amount of overlap of the valence orbitals and the repulsion from the filled inner orbitals.

$\Delta\chi$ represents the electronegativity difference.

Pauling has shown that it is possible to estimate empirically the purely covalent part of the bond energy for any single bond, A-B, as the geometric mean of the A-A and B-B single bond energies. We can illustrate this relationship with data for pairs of elements of very nearly the same electronegativity. The bond energy in H_2 is 104 kcal mole^{-1} and the P-P single bond energy is about 67 kcal mole^{-1}. The geometric mean of these values is $(104 \times 67)^{1/2} = 83.5$. This is very close to the measured P-H bond energy of 85 kcal mole^{-1}. Similarly the N-Cl bond energy is 63 kcal mole^{-1} which can be compared with $(\alpha_{N-N}\alpha_{Cl-Cl})^{1/2} = (67 \times 58)^{1/2} = 62$. The C-I bond energy (55 kcal) is closely approximated by $(\alpha_{C-C}\alpha_{I-I})^{1/2} = (91 \times 37)^{1/2} = 58$. It is true that there are cases where the agreement is not this good. Two examples are the C-S bond (calc., 81; exptl., 73 kcal mole^{-1}) and the H-Te bond (calc., 72; exptl., 65 kcal mole^{-1}). In fact within the accuracy of bond energy calculations a simple average can often be used as an estimate of the geometric mean. Nevertheless Pauling's relationship is sufficiently accurate to give meaningful results, and it is the basis for the electronegativity values commonly listed for nonmetals.

The symbol α_{M-M} is used to designate the bond energy for a single M-M bond.

In a molecule like hydrochloric acid (HCl) the bonding electron pair is centered more on the chlorine than on the hydrogen, and the bond energy is greater than that expected for a purely covalent bond. This difference represents a stabilization that is related to the electronegativity difference. To relate this stabilization to electronegativity,

Pauling proposed the relationship

$$\alpha_{A-B} - (\alpha_{A-A}\alpha_{B-B})^{1/2} = 30(\Delta\chi)^2$$

which appears to be reasonably accurate for values of $\Delta\chi$ up to about 1.5.

With this relationship the experimental H-Cl bond energy (103 kcal mole^{-1}) can be used to calculate the electronegativity of chlorine if we assume that the electronegativity of hydrogen is 2.1.

$$\alpha_{H-Cl} - (\alpha_{H-H}\alpha_{Cl-Cl})^{1/2} = 30(\Delta\chi)^2$$
$$103 - (104 \times 58)^{1/2} = 103 - 78 = 25 = 30(\Delta\chi)^2$$
$$\Delta\chi^2 = 0.83$$

From this $\Delta\chi$ is equal to 0.9, and the electronegativity of chlorine is 3.0. The reverse calculation can be used to estimate the bond energy. For Cl-S

$$\alpha_{Cl-S} = (72 \times 58)^{1/2} + 30(3.0 - 2.5)^2 = 64.5 + 7.5 = 72$$

The experimental value for the sulfur dichloride (SCl$_2$) molecule is $\alpha_{Cl-S} = 69$ kcal mole^{-1}.

Experimental bond energies cover a wide range of values but, other things being equal, the greater the electronegativity difference, the greater the bond energy. This generalization accounts for the highly exothermic reactions of metals with nonmetals and for the displacement reactions

$$2AlCl_3 + 3F_2 \longrightarrow 2AlF_3 + 3Cl_2$$

and

$$AlCl_3 + 3Na \longrightarrow Al + 3NaCl$$

since both the Al-F and Na-Cl bonds should be stronger than those in AlCl$_3$.

QUESTIONS

1. Define or state briefly the meaning of the terms ionic bond, covalent bond, partially ionic bond, and partially covalent bond.
2. Designate the bond type in the following molecules: LiCl, LiF, H$_2$S, H$_2$O, F$_2$, Cl$_2$.

3. Estimate the percent ionic character of the bond in LiCl, H_2S, and H_2O.
4. Explain the effects of decreasing electron affinity and decreasing ionization potential on the bond energy of a solid.
5. Explain the importance of shielding in chemical bonding.
6. Define what is meant by a bonding and an antibonding orbital.
7. Diagram the positions of the nodes in a σ^*, and π^* orbital.
8. What is an effective overlap?
9. Construct a molecular orbital diagram for Ne_2. Why is Ne_2 unstable?
10. Construct a molecular orbital diagram for Be_2. From this diagram do you expect that Be_2 would be stable?
11. Draw the overlaps and the molecular orbital diagram for F_2.
12. Explain the meaning of the terms triplet and singlet.
13. Why are the π bonds in N_2 different in energy from the σ bond?
14. How many of the valence electrons in N_2 are bonding electrons? Antibonding electrons? What is the bond order?
15. Compare the bonding in C_2, N_2, and O_2.
16. Draw the octet rule structure for CN. Why cannot seven valence electrons be shared in this molecule?
17. Classify the following materials as covalent, molecular, or ionic. Which ones have boiling points below 100°C? $AlCl_3$, F_2, AlF_3, Cl_2, Na, NaCl, NaH, HF, H_2, Al_2O_3, B, C, N_2.

PROBLEMS I

1. From the covalent radii given in Chapter 3 and Fig. 7-1 estimate the dipole moment of HBr, HI, and IF assuming 100% ionic bonding.
2. Using the data in Chapters 3 through 6, estimate the bond energy in $MnF_2(s)$, and NiF_2.
3. The dipole moment of HCl is 1.04 debye. Given the interatomic distance, 1.28 Å, estimate the percent ionic character of the bond.
4. Using Pauling's relationship between bond energy and electronegativity, estimate the P-Cl, H-Cl, N-I, and H-C bond energy.
5. From the calculation in Problem 4 and the H^0 values for the atoms, estimate H^0 for PCl_3 and CH_4.
6. Given the H^0 values below, calculate ΔH_u for these four solids.

Compound	H^0
$MgCl_2$	−153.4
$CaCl_2$	−190.0
$SrCl_2$	−198.0
$BaCl_2$	−205.6

7. The Madelung constant for the CaF_2 structure is 1.68. Using the ionic radii from Table 3-7 calculate ΔH_u for CaF_2 and $CaCl_2$.

PROBLEMS II

1. Estimate the bond length in BrF.

2. Calculate the dipole moment for charges of $\pm 4.80 \times 10^{-10}$ esu at a distance of 1.86 Å.
3. The measured dipole moment of BrF is 1.29 debye. Estimate the percentage ionic character in the Br-F bond.
4. Classify BrF as covalent, molecular, or ionic.
5. How many of the valence electrons in F_2 are in bonding molecular orbitals?
6. How many antibonding electrons are there in F_2?
7. Which is easier to dissociate, F_2 or N_2? Why?
8. Why is the ionization potential of F_2 less than that of fluorine atoms?
9. H^0 for fluorine atoms is $+19$. Calculate the F-F bond energy.
10. The Cl-Cl bond dissociation energy is 58 kcal mole^{-1}; estimate the Cl-F bond energy.
11. Estimate H^0 for ClF.

SUGGESTED READINGS

ANDER, P., and A. J. SONNESSA, *Principles of Chemistry*. New York: The Macmillan Co., 1965, Chapter 3

DRAGO, R. S., *Prerequisites for College Chemistry*. New York: Harcourt Brace and World, Inc., 1966, Chapter 6

GRAY, H. B., *Electrons and Chemical Bonding*. New York: W. A. Benjamin Inc., 1965, Chapter 2

MAHAN, B., *College Chemistry*. Reading, Mass.: Addison-Wesley, 1966, Chapter 11

ORCHIN, M., and JAFFE, H. H., *The Importance of Antibonding Orbitals*. Boston: Houghton Mifflin Co., 1967

> The forms of things unknown, the
> poet's pen
> Turns them to shapes, and gives
> to airy nothing
> A local habitation and a name.
>
> WILLIAM SHAKESPEARE

Chapter 8 The Shapes of Molecules

I. The Valence Bond Method

In introducing the valence bond method, it is instructive to consider bond formation using atomic orbitals. To simplify matters it is best to start with a covalent bond where each atom contributes one electron to the bond. A convenient molecule to use is methane (CH_4).

The electronic structure of carbon is

$$C \; \frac{\cdot\cdot}{1s}, \; \frac{\cdot\cdot}{2s}, \; \frac{\cdot}{} \; \frac{\cdot}{2p} \; \frac{}{}$$

By restricting the possible bonding to covalent bonds, the number of bonds has been defined as equal to the number of unpaired electrons on the carbon. The expected stoichiometry of the compound formed between carbon and hydrogen is 1 : 2. A diagram such as that in (8.1) may be used to illustrate the compound expected. In this diagram the carbon contributes one electron ($\cdot$) and each hydrogen contributes one electron (x) to the bond. The carbon $2p$ and hydrogen $1s$ orbital overlap is designated by placing two electrons in one orbital ($\cdot$x).

$$C \; \frac{\cdot\cdot}{1s}, \; \frac{\cdot\cdot}{2s}, \; \frac{\overset{H}{\cdot\text{x}}}{} \; \frac{\overset{H}{\cdot\text{x}}}{2p} \; \frac{}{} \qquad (8.1)$$

Sec. 1 The Valence Bond Method

If atomic orbitals were suitable for bonding, we would expect a compound with the formula CH_2. The fact that this is not the only observed stoichiometry indicates that other factors must be important.

In the most stable compound four bonds are present. If we retain the concept of one electron from each atom for every bond, the only way to designate four bonds to carbon is to rearrange the electronic structure to

$$C\frac{\cdot\cdot}{1s}, \frac{\cdot}{2s}, \frac{\cdot}{} \frac{\cdot}{2p} \frac{\cdot}{}$$

Such a rearrangement requires an amount of energy equal to the energy difference between the $2s$ level and the $2p$ level. In carbon this amounts to 97 kcal mole^{-1}. This energy is referred to as the promotion energy.

With the stoichiometry correct we may look at the structure of CH_4, using the atomic orbitals of carbon. There are four possible bonds that may be represented by the C_{2s}-H_{1s}, C_{2p_x}-H_{1s}, C_{2p_y}-H_{1s}, C_{2p_z}-H_{1s} orbital overlaps. If we visualize each of the orbital overlaps as shown in Fig. 8-1, we get three bonds at right angles to each other (from the C_{2p}-H_{1s} overlaps) and a fourth bond which may be placed anywhere since the carbon $2s$ orbital is spherical. Another difficulty with using atomic orbitals for bonding is apparent if we consider the strength of the C_{2p}-H_{1s} and C_{2s}-H_{1s} bonds. The bond energy is dependent upon the amount of overlap. Because different orbitals are involved, the C_{2p}-H_{1s} overlap will most likely be different from the C_{2s}-H_{1s} overlap; therefore the bonds should be different in energy.

It is apparent from experiment that all of the bonds in methane (CH_4) are equivalent, and the angle between the bonds is not 90° but 109°. The hydrogen atoms lie at the apices of a tetrahedron surrounding the carbon atom.

The above discussion illustrates some of the difficulties involved when an attempt is made to use atomic orbitals in the prediction of structure. For the most part *unmodified* atomic orbitals are not suitable for a description of the chemical bonds in molecules.

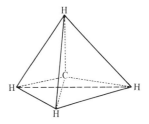

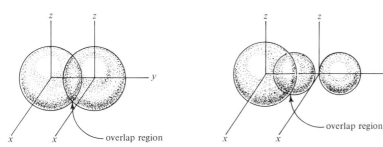

Figure 8-1 *Some s-s and s-p overlaps.*

197

Ch. 8 The Shapes of Molecules

The valence bond method was devised by Pauling in an attempt to resolve the difficulties mentioned above. The method involves the construction of a new set of orbitals from the atomic orbitals in such a way that the angular overlap of orbitals (radial distribution neglected) is a maximum (for maximum bond strength). The process of combination is called hybridization and the new orbitals obtained are called hybrids. The procedure followed in hybridizing atomic orbitals is not important to us at this point.

Bond formation takes place when the appropriate orbitals from other atoms overlap with these hybrids. In the case of carbon the hybrids that provide a maximum overlap with the tetrahedrally arranged hydrogen atoms are a combination of the $2s$ and three $2p$ orbitals. The results of this combination, as well as several other possible combinations, are shown in Fig. 8-2.

In sp^3 hybrid formation, four orbitals are hybridized and four hybrids are obtained. Each of these orbitals is directed toward one of the corners of a tetrahedron. Figure 8-2 lists other hybrids which provide a maximum overlap for some other commonly encountered structures.

Several examples will be used to illustrate the use of the valence bond method. The electronic structure of beryllium is

$$\text{Be} \;\; \overset{..}{\underset{1s}{}},\; \overset{..}{\underset{2s}{}},\; \underset{2p}{-\;-\;-}$$

There are no unpaired electrons; however, by employing the concept of promotion the structure

$$\text{Be} \;\; \overset{..}{\underset{1s}{}},\; \overset{.}{\underset{2s}{}},\; \overset{.}{\underset{2p}{-}}\;-\;-$$

can be obtained. Two chlorine atoms with the electronic structure

$$\text{Cl} \;\; \overset{..}{\underset{3s}{}},\; \overset{..}{}\; \overset{..}{\underset{3p}{}}\; \overset{.}{}$$

can be accommodated by an overlap of the singly occupied beryllium orbitals with the singly occupied p orbital of chlorine. The hybrids on the beryllium are sp hybrids and the structure is linear (from Fig. 8-2). This example illustrates an important concept in the valence bond method. That is, the energy released in bond formation must be sufficient to justify the expenditure of the energy needed for promotion. The promotion of the $1s$ electrons of beryllium, in addition to the $2s$ (to obtain molecules such as BeX_4), is not expected because the energy of the $1s$ electrons is very low and sufficient energy for such a promo-

Sec. I The Valence Bond Method

Orbitals used	Hybrid obtained, # and type	Structure	
$s + p$	3 sp	linear	
$s + 2p$	3 sp^2	planar	
$s + 3p$	4 sp^3	tetrahedral	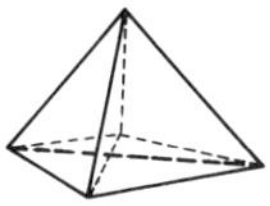
$s + 2p + d$	4 dsp^2	square planar	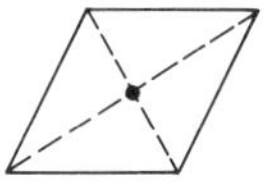
$s + 3p + d$	5 dsp^3 or sp^3d	trigonal bipyramid	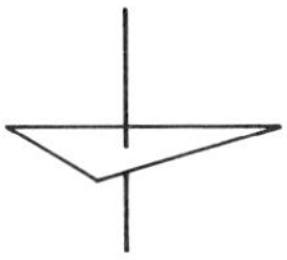
$s + 3p + 2d$	6 d^2sp^3 or sp^3d^2	octahedral	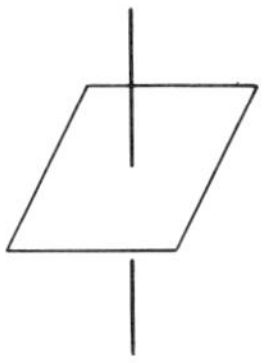
$s + 3p + 3d$	d^3sp^3 or sp^3d^3	pentagonal bipyramid	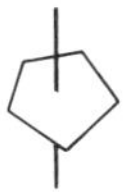

Figure 8-2 *Common hybrid orbitals and the resulting structures.*

tion is not released by the formation of the two extra bonds. We might add that although there might be some disadvantage in having a stoichiometry BeX_4, because of increased repulsion of the X^- groups, it is not so large that it cannot be overcome. Some complex ions of beryllium have this stoichiometry.

The next element, boron, has the electronic structure

$$B \frac{\cdot\cdot}{1s}, \frac{\cdot\cdot}{2s}, \frac{\cdot}{} \frac{}{2p} \frac{}{}$$

With promotion three bonds are possible. The bonds should show sp^2 hybridization, resulting in a flat triangular molecule.

The octet rule is followed most closely by the elements in the second row because these elements are limited to a maximum coordination number of four. Coordination numbers larger than four are observed for elements of the third row, and elements such as phosphorus, sulfur, and iodine can form compounds with coordination numbers of 5, 6, and 7. Since the number of s and p orbitals is limited to 4, something must clearly be done to accommodate these higher coordination numbers. The valence bond formalism requires additional orbitals, and the next available orbitals are the nd. If the $3d$ orbitals are only slightly higher in energy than the $3p$, promotion to them is possible. Table 8-1 shows the results for some selected compounds of phosphorus, sulfur, and iodine.

So far every example treated has involved the promotion concept. It should be remembered, however, that promotion is only justified if sufficient energy is available from bond formation. Promotion need not, and does not always take place. Some examples where promotion does not take place may be found in the second row of the periodic table. The electronic structure of nitrogen and oxygen may be written

$$N\underset{2s}{\stackrel{\cdot\cdot}{=\!=}},\ \underset{2p}{\stackrel{\cdot}{-}\ \stackrel{\cdot}{-}\ \stackrel{\cdot}{-}}$$

$$O\underset{2s}{\stackrel{\cdot\cdot}{=\!=}},\ \underset{2p}{\stackrel{\cdot\cdot}{-}\ \stackrel{\cdot}{-}\ \stackrel{\cdot}{-}}$$

The $3s$ orbital is the lowest energy unoccupied orbital (there are no

Table 8-1 *A Representation of the Bonding in Some Compounds of Phosphorus, Sulfur, and Iodine*

Compound	Electronic Structure	Hybridization and Structure
PCl_5	$P\underset{3s}{\stackrel{\cdot\cdot}{=\!=}},\ \underset{3p}{\stackrel{\cdot\cdot}{-}\ \stackrel{\cdot}{-}\ \stackrel{\cdot}{-}},\ \underset{3d}{-\ -\ -\ -\ -}$ $P\underset{Cl}{\stackrel{\cdot\times}{=\!=}}\ \underset{Cl}{\stackrel{\cdot\times}{-}}\ \underset{Cl}{\stackrel{\cdot\times}{-}}\ \underset{Cl}{\stackrel{\cdot\times}{-}}\ \underset{Cl}{\stackrel{\cdot\times}{-}}\ -\ -\ -\ -$	sp^3d trigonal bipyramid
SF_6	$S\underset{3s}{\stackrel{\cdot\cdot}{=\!=}},\ \underset{3p}{\stackrel{\cdot\cdot}{-}\ \stackrel{\cdot}{-}\ \stackrel{\cdot}{-}},\ \underset{3d}{-\ -\ -\ -\ -}$ $S\underset{F}{\stackrel{\cdot\times}{=\!=}}\ \underset{F}{\stackrel{\cdot\times}{-}}\ \underset{F}{\stackrel{\cdot\times}{-}}\ \underset{F}{\stackrel{\cdot\times}{-}}\ \underset{F}{\stackrel{\cdot\times}{=\!=}}\ \underset{F}{\stackrel{\cdot\times}{-}}\ -\ -\ -$	sp^3d^2 octahedral
IF_7	$I\underset{5s}{\stackrel{\cdot\cdot}{=\!=}},\ \underset{5p}{\stackrel{\cdot\cdot}{=\!=}\ \stackrel{\cdot\cdot}{=\!=}\ \stackrel{\cdot}{-}},\ \underset{5d}{-\ -\ -\ -\ -}$ $I\underset{F}{\stackrel{\cdot\times}{=\!=}}\ \underset{F}{\stackrel{\cdot\times}{-}}\ \underset{F}{\stackrel{\cdot\times}{-}}\ \underset{F}{\stackrel{\cdot\times}{-}}\ \underset{F}{\stackrel{\cdot\times}{-}}\ \underset{F}{\stackrel{\cdot\times}{-}}\ \underset{F}{\stackrel{\cdot\times}{-}}\ -\ -$	sp^3d^3 pentagonal bipyramid

Sec. I The Valence Bond Method

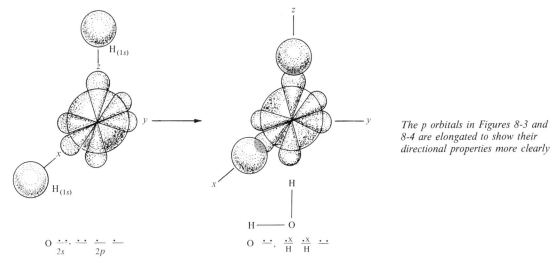

The p orbitals in Figures 8-3 and 8-4 are elongated to show their directional properties more clearly

Figure 8-3 *Bond formation in H_2O using pure p orbitals.*

$2d$ orbitals), and it is much too high in energy to be used. The bond energy is not sufficient to justify promotion and it does not take place. In general promotion to the next higher quantum level is not favorable. The maximum number of bonds formed by the second row elements is limited to the number of unpaired electrons present without promotion to the 3s or 3p orbital. This is the significance of the octet rule.

Another important aspect of structural considerations is illustrated by the compounds of oxygen and nitrogen. Bond formation without promotion or hybridization gives

$$O \underset{}{\overset{..}{\text{—}}}, \underset{}{\overset{..}{\text{—}}} \underset{\text{·×}}{\overset{\text{H}}{\text{—}}} \underset{\text{·×}}{\overset{\text{H}}{\text{—}}}$$

Figure 8-3 gives a diagram of the water molecule using p orbitals for bonding. The remaining two electron pairs are represented by the orbitals that they occupy (an s orbital and the p_y orbital). A little thought will reveal that this is not the most stable structure possible. The structure could be stabilized if the electrons could move further apart. The bonding electrons are localized between the oxygen and the hydrogen and cannot move from this space; however, the lone pairs are not localized and nothing is holding them as they are drawn. It is reasonable to expect that the lone pairs will distort from the original atomic orbitals to new ones with as much spread as possible. The most symmetrical arrangement of four electron pairs is tetrahedral. The repulsion effects are at a minimum in this structure; however, a symmetrical tetrahedron for the four electron pairs is not to be expected for water because two of the electron pairs (the lone pairs) are different from the other two (bond pairs). The bond pairs are localized (held by two nuclei); hence they will not occupy as much space as the lone pairs. Repulsion will be minimized further if the lone pairs spread apart and the bond pairs come together. As a result, the angle between bonded hydrogen atoms is expected to be less than 109° (the tetrahedral angle) but larger than 90°. The angle is found to be 105°.

Another example with which this effect may be illustrated is ammonia (NH_3). The electronic structure of nitrogen is

$$N \frac{\cdot\cdot}{2s}, \frac{\cdot}{} \frac{\cdot}{2p} \frac{\cdot}{}$$

leaving the three p orbitals available for bonding. Figure 8-4 represents ammonia, showing three p bonds at right angles to each other along with the s electrons. This again is not the most stable arrangement. The $2s$ electrons on the nitrogen are expected to be distorted from the original atomic orbital because of repulsions from the bond pairs. The position of minimum repulsion is again between 90° and 109°. The observed bond angle in ammonia (107°) is slightly less than the tetrahedral angle.

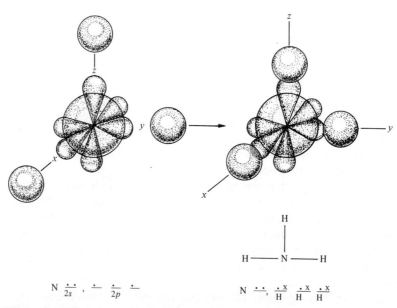

Figure 8-4 *Bond formation in NH_3 using pure p orbitals.*

An alternative approach to the structure of water and ammonia is to start with sp^3 hybridization in both cases, with the lone pairs occupying one or two of the hybrids, as the case may be. The deviation from the tetrahedral angle is then justified in the same way as above. Either method is appropriate and both are used, so it is necessary to be familiar with both.

The promotion concept is a very useful one in many cases; however, if the objective is simply the prediction of structure, the method is cumbersome in some cases. To understand this try to predict the structures of sulfur dioxide (SO_2), $Ag(NH_3)_2^+$, and BF_4^-.

The method may be simplified considerably if we recall the oxidation number method discussed in Chapter 3. The oxidation number method arbitrarily divides the electrons in a bond so that the element with the largest electronegativity gets two, and the element with the smallest electronegativity gets none. In determining the structure of BF_4^- by this method each fluorine has an oxidation number of -1 and the boron an oxidation number of $+3$. The electronic structure of boron(III) is

$$B(III) \frac{\cdot\cdot}{1s}, \frac{}{2s}, \frac{}{} \frac{}{2p} \frac{}{}$$

The four fluoride ions are placed into the four empty valence orbitals

$$B(III) \frac{\cdot\cdot}{}, \frac{xx}{F}, \frac{xx}{F} \frac{xx}{F} \frac{xx}{F}$$

The molecule involves sp^3 hybrids, and it is tetrahedral. $Ag(NH_3)_2^+$ is linear (sp hybridization), and SO_2 is bent because sp^2 hybrids are involved.

II. The Gillespie-Nyholm Method

Some difficulties are encountered in applying the valence bond method to any molecule that has lone pairs involved. As an example, consider iodine trichloride (ICl_3). The valence bond treatment of this molecule gives sp^3d hybrids or a trigonal bipyramid structure. However, there are three possible ways to arrange the two electron pairs in this trigonal bipyramid structure. These are given in Fig. 8-5. The valence bond method tells nothing about the relative stability of these three possibilities, but the structure will depend upon the position taken by the lone pairs. A method is available that is capable of distinguishing between the possible structures of iodine trichloride. We have already applied this method while discussing the structure of water

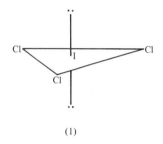

(1)

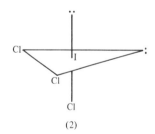

(2)

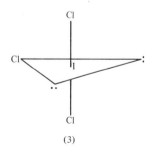

(3)

Figure 8-5 *Possible structures of* ICl_3.

Ch. 8 The Shapes of Molecules

and ammonia. In that discussion it was pointed out that the distribution of four electron pairs should be tetrahedral. In the same way it is reasonable to assign, intuitively, a linear structure to compounds containing two bonding electron pairs and a trigonal structure to those containing three. Five electron pairs are more difficult to see but, of course, it is possible to calculate the most symmetrical distribution of five electron pairs on the surface of a sphere; the result is a trigonal bipyramid. Six electron pairs result in an octahedral structure.

When applying the Gillespie-Nyholm method, the number of electron pairs (bonding and lone pairs) in a molecule are counted and the structure in which the electron repulsions are at a minimum is assigned. Of course, the results obtained are identical to those obtained with the valence bond method.

The real advantage of the Gillespie-Nyholm method lies in the treatment of molecules with lone pairs. When dealing with molecules of this type, the objective is to attempt to minimize electrostatic repulsions. Because of the difference between lone pairs and bond pairs, the various types of possible interactions (lone pair–lone pair, lone pair–bond pair, and bond pair–bond pair) are not given equal weight. Lone pairs occupy more space than bond pairs; therefore, the electrostatic effect of lone pair–lone pair interactions will be greater than that of bond pair–bond pair interactions. Lone pair–bond pair interactions will be intermediate. In applying the Gillespie-Nyholm method, the procedure is to count the number of each type of interaction for each structure. The structure that provides the minimum repulsions will be favored. In counting the interactions, only the closest are considered effective. In the trigonal bipyramid there are six closest interactions (at 90°). The interactions for the three possible structures for iodine trichloride are shown in Fig. 8-6. On this basis, structure (3) is favored since this structure gives the smallest total for the 90° interactions. Experiment shows that this is the actual structure. As a

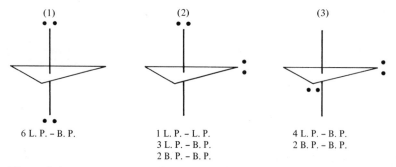

Figure 8-6 *Interactions in* ICl_3.

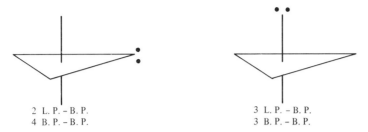

| 2 L. P. - B. P. | 3 L. P. - B. P. |
| 4 B. P. - B. P. | 3 B. P. - B. P. |

Figure 8-7 *Structural possibilities for* SF_4.

general rule the lone pair electrons occupy the position that provides the most space.

Even one lone pair can be important in determining the structure of a molecule. The valence bond or Gillespie-Nyholm method both predict an overall trigonal bipyramid structure for sulfur tetrafluoride (SF_4). This molecule contains five electron pairs including one lone pair. The position occupied by the lone pair may be either equatorial or axial. A count of the interactions involved (Fig. 8-7) shows the equatorial position as the most favorable, and this is the observed structure.

Bonds along the vertical axis in Fig. 8-6 and Fig. 8-7 are called axial; those in the plane perpendicular to this axis are called equatorial.

III. Applications

A. TRANSITION METALS AND COORDINATION COMPOUNDS

The procedure for applying the valence bond method to compounds of the transition elements is only slightly different from that used above. Consider, for example, the complex $Co(NH_3)_6^{3+}$.

The electronic structure of cobalt (0) is

$$Co(0) \underset{3d}{\overset{\cdot\cdot\ \cdot\cdot\ \cdot\ \cdot\ \cdot}{\text{—}}},\ \underset{4s}{\overset{\cdot\cdot}{\text{—}}}\ \underset{4p}{\text{— — —}}$$

Cobalt(III) has the electronic structure

$$Co(III) \underset{3d}{\overset{\cdot\cdot\ \cdot\ \cdot\ \cdot\ \cdot}{\text{—}}},\ \underset{4s}{\text{—}}\ \underset{4p}{\text{— — —}}$$

In $Co(NH_3)_6^{3+}$ the number of groups attached to the cobalt (six) is larger than the number of valence shell orbitals available (four). Instead of a promotion to unpair electrons, what is needed is a pairing up of electrons to give

$$Co(III) \underset{3d}{\overset{\cdot\cdot\ \cdot\cdot\ \cdot\cdot}{\text{—}}}\ \text{— —},\ \underset{4s}{\text{—}}\ \underset{4p}{\text{— — —}}$$

Ch. 8 The Shapes of Molecules

Now six hybridized orbitals (d^2sp^3) on the cobalt(III) are available for overlap with the six orbitals containing the electron pairs on the ammonia. Other examples of this type would be $Fe(CN)_6^{3-}$

$$Fe(III) \; \underset{3d}{\cdot\!\!- \;\; \cdot\!\!- \;\; \cdot \;\; \cdot\!\!- \;\; \cdot\!\!-}, \; \underset{4s}{\overline{}}, \; \underset{4p}{\overline{} \;\; \overline{} \;\; \overline{}}$$

$$Fe(III) \; \underset{3d}{\cdot\!\!\cdot\!\!- \;\; \cdot\!\!\cdot\!\!- \;\; \cdot \;\; +\!\!\!\!+ \;\; +\!\!\!\!+}, \; \underset{4s}{+\!\!\!\!+}, \; \underset{4p}{+\!\!\!\!+ \;\; +\!\!\!\!+ \;\; +\!\!\!\!+} \qquad d^2sp^3 \qquad \text{octahedral}$$

and $Cr(NH_3)_6^{3+}$

$$Cr(III) \; \underset{3d}{\cdot\!\!- \;\; \cdot\!\!- \;\; \cdot \;\; +\!\!\!\!+ \;\; +\!\!\!\!+}, \; \underset{4s}{+\!\!\!\!+}, \; \underset{4p}{+\!\!\!\!+ \;\; +\!\!\!\!+ \;\; +\!\!\!\!+} \qquad d^2sp^3 \qquad \text{octahedral}$$

There are a number of other features that become important in the valence bond treatment of coordination compounds. One is the use of outside (nd) orbitals in the hybridization. For example $Fe(F)_6^{3-}$ is considered an sp^3d^2 hybrid because it is evident from experiment that the molecule contains five unpaired electrons

$$Fe(III) \; \underset{3d}{\cdot\!\!- \;\; \cdot\!\!- \;\; \cdot \;\; \cdot\!\!- \;\; \cdot\!\!-}, \; \underset{4s}{+\!\!\!\!+}, \; \underset{4p}{+\!\!\!\!+ \;\; +\!\!\!\!+ \;\; +\!\!\!\!+}, \; \underset{4d}{+\!\!\!\!+ \;\; +\!\!\!\!+ \;\; \overline{} \;\; \overline{}} \qquad sp^3d^2 \qquad \text{octahedral}$$

These features are developed further in standard inorganic chemistry textbooks, but we will not carry the valence bond approach further here because a much simpler and more successful method is available for coordination compounds. As we shall see the valence bond method cannot account for the magnetic properties of all transition metal compounds.

The usefulness of the valence bond method and the Gillespie-Nyholm method in predicting structures is obvious. However, it would be surprising indeed if either method were applicable without failure to all molecules. Remember that the wave functions used in constructing the hybrids are one-electron functions. They are solutions to the Schrödinger equation for hydrogen, and they are not strictly applicable to any other element. With this in mind, eventual failure is understandable and expected. We shall now examine briefly the structure of some complexes of nickel(II).

The structure of nickel(II) complexes may be obtained using the valence bond method outlined above. Two structures are possible according to this treatment.

$$Ni(II) \; \underset{3d}{\cdot\!\!\cdot\!\!- \;\; \cdot\!\!\cdot\!\!- \;\; \cdot\!\!\cdot\!\!- \;\; \cdot\!\!- \;\; \cdot\!\!-}, \; \underset{4s}{+\!\!\!\!+}, \; \underset{4p}{+\!\!\!\!+ \;\; +\!\!\!\!+ \;\; +\!\!\!\!+} \qquad sp^3$$

$$Ni(II) \; \underset{3d}{\cdot\!\!\cdot\!\!- \;\; \cdot\!\!\cdot\!\!- \;\; \cdot\!\!\cdot\!\!- \;\; +\!\!\!\!+ \;\; \overline{}}, \; \underset{4s}{+\!\!\!\!+}, \; \underset{4p}{+\!\!\!\!+ \;\; +\!\!\!\!+ \;\; \overline{}} \qquad dsp^2$$

A coordination number of five is not favorable to crystal packing, and very few complexes of nickel(II) with all the electrons paired exhibit this coordination number. The valence bond method predicts that nickel(II) complexes will have either two unpaired electrons or none. The magnetic moment will correspond to either two or zero unpaired electrons. It is not possible to have an intermediate number of unpaired electrons because electrons are either paired or unpaired. However, recently a number of compounds that appear to contain one unpaired electron have been prepared and, in addition, it has been found that many of the compounds which were previously thought to be tetrahedral are, in fact, octahedral. Here is a situation that is impossible to explain in valence bond terms.

The Gillespie-Nyholm method also has difficulties with transition metal compounds. In this case the lone pairs cannot be included, or they must be treated in a different manner. The result has been that additional methods have been developed for the transition metal ions. In spite of this, it must be emphasized that the methods presented here are very useful in predicting the structures of compounds of the representative elements.

B. MULTIPLE BONDING

Table 8-2 lists the bond lengths found in a number of molecules. These values are experimental quantities, determined by neutron or X-ray diffraction or one of a number of other methods. Comparison

Table 8-2 *Experimental Bond Lengths for Some Molecules*

Bond	Bond length, Å	Molecule
F-F	1.42	F_2
Cl-Cl	1.99	Cl_2
Br-Br	2.28	Br_2
I-I	2.67	I_2
H-F	0.92	HF
H-Cl	1.27	HCl
H-Br	1.41	HBr
H-I	1.61	HI
O-H	0.96	H_2O
Li-F	2.01	LiF
Li-Cl	2.56	LiCl
Li-Br	2.75	LiBr
Ag-Cl	2.46	AgCl
Hg-F	2.31	Hg_2F_2
Hg-Cl	2.52	Hg_2Cl_2

of these values with the ionic and atomic radii given in Chapter 3 shows that the radii are, in many cases, additive. This generalization is particularly true of elements that are not too different from one another. There are, of course, numerous exceptions to this rule. In most cases the exceptions represent deviation from a bond type. For example, the sum of the ionic radii of silver ion and chloride ion is a great deal larger than the observed distance in silver chloride (AgCl), indicating that a degree of covalency is important in silver chloride. Some covalency is also indicated for the bonds in silver fluoride (AgF).

Another instance where the rule appears to be broken is observed in certain molecules such as ethylene or acetylene (C_2H_2). These examples do not really constitute a failure of the rule but are interpreted in terms of the formation of multiple bonds. All of the bonds in ethane (C_2H_6) are single, electron pair, covalent bonds that may be represented as sp^3 hybrids on each carbon

$$\begin{array}{c} \text{H} \quad \text{H} \\ | \quad | \\ \text{H}-\text{C}-\text{C}-\text{H} \\ | \quad | \\ \text{H} \quad \text{H} \end{array}$$

ethane

In ethylene (C_2H_4) only three atoms are attached to each carbon so the hybridization is sp^2. One orbital on each carbon is not involved in σ bonding, but it may overlap with the orbital of the other carbon, to form a π bond. The π bonds in ethylene are shown in Fig. 8-8.

Multiple bonds are possible in a large number of molecules, particularly in molecules involving carbon, nitrogen, oxygen, and to a lesser extent, fluorine. In these instances double bond formation resulting from $p\pi$-$p\pi$ overlap serves to *delocalize* electrons in the molecule containing them and, hence, to diminish the electron densities on these atoms. This in turn decreases the electron repulsions among the unshared electrons. Some molecules in which multiple bonding is believed to be important are listed in Table 8-3. The tendency toward multiple bond formation is less for larger atoms. In this case the $p\pi$-$p\pi$ overlap is less effective because the orbitals are larger and more diffuse. Also the exceptional size difference between the K and L shells permits better overlap of $2p$ orbitals before there is serious repulsion from the inner shell. In addition, the larger size of the atoms in later rows favors larger coordination numbers and more electrons are used in σ bond formation.

The difference in π bonding capabilities of the second and third row elements is reflected in the structures of the compounds formed. Many compounds of the second row elements form discrete molecules be-

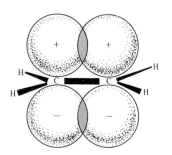

Figure 8-8 *The π bonds in ethylene.*

Table 8-3 *Molecules Involving Multiple Bonds*

CO_2	O=C=O	HCN	H—C≡N
N_2	N≡N	SO_2	S(=O)(=O)
ClO_2	Cl(=O)(O)	SO_4^{2-}	S(=O)(=O⁻)(O⁻)(O)
$IO_2F_2^-$	F,F,I,O,O⁻	C_2H_4	H₂C=CH₂
		C_2H_2	H—C≡C—H

cause of the tendency to form $p\pi$-$p\pi$ bonds (for example, BCl_3, HNO_3, CS_2). The corresponding compounds of the third period elements are polymeric because of the reduced tendency to form $p\pi$-$p\pi$ bonds [that is, $(AlCl_3)_2$, $(HPO_3)_x$, $(SiS_2)_x$].

Some compounds formed between second and third period elements probably do involve π bonds. For example, ions such as SO_4^{2-}, ClO_4^-, have bond lengths that are shorter than the expected single bond distances, and multiple bonding is indicated. The multiple bonds cannot involve $p\pi$-$p\pi$ overlaps in these cases because the hybridization on the central atom is sp^3. In these ions π bonding results from the overlap of the filled p orbitals on the oxygen with the vacant d orbitals on the central atom. $d\pi$-$p\pi$ bonding is primarily important whenever the central atom would otherwise have too large a positive charge. Actually there is good evidence for $d\pi$-$p\pi$ bonding in all compounds of elements in oxidation states of $+6$ and higher.

C. BOND ANGLES

The number of valence electrons, the type of hybrid orbitals, and the magnitude of the observed bond angles for a number of triatomic and polyatomic molecules are given in Table 8-4. Deviations from the predicted valence bond angles are numerous, and they can be rationalized in terms of the Gillespie-Nyholm method. The observed angles for NO_2, O_3, ClO_2, OF_2 can be explained using the Gillespie-Nyholm and valence bond methods.

There are no lone pairs of electrons on the carbon atom in carbon dioxide (CO_2). Each oxygen atom is double-bonded to the carbon. With no lone pairs the molecule should be linear. This is also true

Ch. 8 The Shapes of Molecules

of a great many other compounds that contain 16 valence electrons. Some examples are $SiS_2(g)$, NO_2^+, $AgCl_2^-$, $HgCl_2$, $BeCl_2(g)$, and ZnI_2.

In ozone (O_3) there is a lone pair of electrons on the central oxygen so the molecule must be bent. The expected bond angle is close to 120°. The observed angle is actually 117°.

Nitrogen dioxide (NO_2) is even more interesting. In this molecule there is an odd number of valence electrons (17) so there must be an odd electron in the molecule. A bond angle of 120° would be called for by the sp^2 hybridization of the nitrogen; however, with only one "lone pair" electron on the nitrogen the oxygen atoms would be

Table 8-4 *Bond Angles in Triatomic Molecules*

Molecule or ion	Number of valence electrons	Hybrids	Predicted bond angle	Observed bond angle
CO_2	16	sp	180°	180°
BeF_2				180°
NO_2^+				180°
NO_2	17	sp^2	>120°	134°
O_3	18	sp^2	<120°	117°
SO_2		sp^2	120°	120°
ClO_2	19	sp^3	>109°	117°
OF_2	20	sp^3	<109°	103°
Cl_2O				111°
SCl_2				102°
$TeBr_2$				98°
XeF_2	22		180°	180°
BeH_2	4	sp^2	180°	180°
H_2O	8	sp^3	<109°	105°
H_2S				92°
H_2Se				91°
H_2Te				90°
BH_3	6	sp^3	120°	120°
NH_3	8	sp^3	<109°	107°
PH_3				94°
AsH_3				92°
CH_4	8	sp^3	109°	109°
SiH_4				109°
TiH_4				109°
SnH_4				109°

Table 8-5 *Bond Lengths for N-O Bonds*

Compound	Bond order			Bond length, Å
	σ	π	Total	
NH_2OH	1	0	1	1.46
NH_3OH^+	1	0	1	1.45
NO_3^-	1	$\frac{1}{3}$	$\frac{4}{3}$	1.243
NO_2^-	1	$\frac{1}{2}$	$\frac{3}{2}$	1.236
NO_2	1	$\frac{3}{4}$	$\frac{7}{4}$	1.188
NO_2^+	1	1	2	1.154
NO	1	$\frac{3}{2}$	$\frac{5}{2}$	1.151
NO^+	1	2	3	1.019

expected to spread apart. The bond angle in NO_2 should be greater than 120° but less than 180°. The experimental value is 134°.

The other predicted bond angles in Table 8-4 are similarly obtained. For example, there are three "lone pair" electrons on the chlorine atom of ClO_2. As a result the O-Cl-O bond angle is expected to be larger than the tetrahedral angle of 109° but less than 120°.

One example that might cause difficulty is xenon difluoride (XeF_2) with 22 valence electrons. In the octet rule structure there are three lone pairs of electrons on the xenon atom. The predicted structure for this molecule is a trigonal bipyramid in which the lone pairs occupy the equatorial position and the fluorine atoms occupy the axial positions. The F-Xe-F bond angle will thus be 180°.

The octet rule structure for nitrogen dioxide (NO_2) shows that the molecule contains a total of seven bonding electrons distributed in two bonds. Four of these bonding electrons are involved in σ bonding and three are used for π bonding.

The term bond order is often used to designate the number of bonds between two atoms in a molecule. The bond order in nitrogen dioxide is $\frac{7}{4}$ or $1\frac{3}{4}$. One of the bonds is a σ bond and there is $\frac{3}{4}$ of a π bond. The observed bond length in nitrogen dioxide (given in Table 8-5) is in excellent agreement with this bond order.

D. EXCEPTIONS TO THE GILLESPIE-NYHOLM METHOD

In the discussion above the fact that electron-electron repulsions are extremely important in predicting the structures of molecules and bond angles has been emphasized. These simple considerations can be used to predict or rationalize a great deal of information; however, in some cases other energy terms must be considered. These examples constitute exceptions to the Gillespie-Nyholm method. The first exception that we will consider here occurs in the molecule trisilyl amine [$N(SiH_3)_3$].

Ch. 8 The Shapes of Molecules

In a compound in which one of the hydrogens of methane (CH$_4$) has been replaced by something else, the CH$_3$ group is called a methyl group.

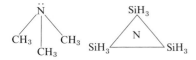

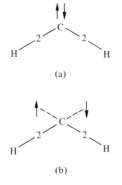

Figure 8-9 *Electron configurations considered for CH$_2$.*

The three silyl (SiH$_3$) groups attached to the nitrogen and the lone pair of electrons in the nitrogen should assume a tetrahedral arrangement as they do in the corresponding carbon compound, trimethyl amine [N(CH$_3$)$_3$]. Actually the silicon atoms are arranged in a plane triangle with the nitrogen in the center. According to the Gillespie-Nyholm method this arrangement is energetically unfavorable so far as electron-electron repulsions of the nitrogen octet are concerned. In the actual molecule this disadvantage is overbalanced by the $d\pi$-$p\pi$ bonding possible in the planar structure. The empty $3d$ orbitals of the three silicon atoms are used for the $d\pi$-$p\pi$ bonding in this molecule.

Another exception to the predictions of the Gillespie-Nyholm method is found in carbene (CH$_2$). Here the carbon atom does not have a completed octet and, as shown in Chapter 4, the lone pair electrons should have parallel spins. For this reason structure (a) of Fig. 8-9 should not be the most stable structure. The observed structure of the carbene molecule is structure (c) of Fig. 8-9. In this structure each of the $2p$ orbitals perpendicular to the molecular axis contains a single electron. Apparently, in this case, the energy gained from the decreased repulsion of the tetrahedral arrangement is not sufficient to justify hybridization. The bent form of carbene, with the electrons paired as they are in Fig. 8-9a, can be prepared, but the molecule transforms to the more stable linear form upon collision with other molecules.

Numerous exceptions to the Gillespie-Nyholm method are found in the solid state, particularly for the large atoms in the last row of the periodic table. The crystal structure of lead oxide (PbO) clearly shows a lone pair of electrons on one side of the lead atom, but lead sulfide (PbS) has the sodium chloride structure and the lone pair is no longer stereochemically active. The effects of the increase in size of the cations in the series germanium (Ge), tin (Sn), lead (Pb), and increasing size of the anions oxide (O^{2-}), sulfide (S^{2-}), selenide (Se^{2-}), and telluride (Te^{2-}) in favoring delocalization of the lone pair electrons is illustrated in Table 8-6.

Anything that can lead to a reduced electron-electron repulsion in the valence shell of atoms can lead to exceptions to the Gillespie-

Table 8-6 *Stereochemical Activity of Lone Pair Electrons in Some Solids*

Cation	Anion			
	O^{2-}	S^{2-}	Se^{2-}	Te^{2-}
Ge^{2+}	?	active	active	?
Sn^{2+}	active	active	inactive	inactive
Pb^{2+}	active	inactive	inactive	inactive

Nyholm predictions. The exceptions are most numerous when large atoms are involved because, on the average, the electrons are further apart. Exceptions also occur when an atom is surrounded by atoms in which the electrons are easily distorted.

We must add here a final word on the Gillespie-Nyholm method. The method has definite merit in spite of the disadvantages that have been given above. Similar disadvantages are expected of any method because of the inherent complexity of the structure of molecules. Methods are being developed that will allow the quantitative prediction of bond angles but, because of the complexity of these methods, there will always be a place for the qualitative methods such as the valence bond and/or Gillespie-Nyholm method.

One real problem with the Gillespie-Nyholm method is that it *appears* to be based upon a particle model of the electron. It is not, nor is any other method. The particle model has not been used successfully in any qualitative or quantitative discussion of the structure of molecules.

QUESTIONS

1. Arrange the following in order of increasing H-X-H angles: CH_4, NH_3, H_2O, SiH_4, H_2S.
2. List the shapes associated with the following hybrid orbitals: sp^2, sp^3, sp^2d, sp^3d^2, d^2sp^3.
3. Describe the shape of SF_4 in terms of hybridization of the sulfur orbitals and again using the Gillespie-Nyholm method.
4. What forces keep the two hydrogens in H_2O from moving apart to a bond angle of 180°?
5. Why is the bond angle in H_2Te smaller than that in H_2S?
6. Which of the following molecules are linear?
   CH_2, H_2Se, HCl, XeF_2, SF_2, $HOCl$, HCN, HNO.
7. Calculate the number of d electrons present in each of the following: Co(III), Fe(II), Mn(IV), Cr(IV), V(II), Cu(II), and Ga(III).
8. How many unpaired d electrons will there be in each part of Question 7 if they unpair to the maximum extent?
9. How many unpaired d electrons will there be in each part of Question 7 if two d orbitals are required for the chemical bonding?
10. Predict the geometry of each of the following molecules: ICl_3, SO_2, CCl_4, PCl_3, PCl_5, SF_4, $BeCl_2(g)$, AsH_3, BeH_2, $HOBr$, $IO_2F_2^-$, BrF_5, and S_2F_2.

PROBLEMS I

1. Estimate the bond length in each of the molecules in Table 8-2 using covalent and metallic radii (Tables 3-6 and 3-7).
2. Estimate the bond length in each of the molecules in Table 8-2 using ionic radii from Table 3-7.

3. Estimate the bond order in each of the following:

ClO_2, CO_3^{2-}, O_3, Cl_2O, NO_2, NO_2^+.

PROBLEMS II
1. What hybridization is appropriate for four bonds at angles of 109°?
2. Why is the bond angle for NH_3 smaller than 109°?
3. Predict the shape of the molecule and bond angle for BiH_3.
4. Predict the bond order and hybridization in $SiO_2(s)$. (Each silicon is bonded to four oxygen atoms.)
5. Predict the bond order and hybridization in $CO_2(g)$.
6. Predict the bond order and shape for NO_2 and for O_3.
7. Which of the following species have sp^2 hybridization?
   NH_3, NO_3^-, CF_4, BF_3, BeF_2, SO_2, SO_3, SO_3^{2-}, SO_4^{2-}.
8. Why do SO_3 and SO_3^{2-} have a different shape?
9. Predict the bond order and bond angle in SO_2.
10. What is the shape of PF_5?
11. Predict the shapes of SF_4 and ClF_3.

SUGGESTED READINGS
ANDER, P., and A. J. SONNESSA, *Principles of Chemistry.* New York: The Macmillan Co., 1965, Chapter 3.

COLE, R. H., and J. S. COLES, *Physical Principles of Chemistry.* San Francisco: W. H. Freeman and Co., 1964, Chapter 6.

MAHAN, B., *College Chemistry.* Reading, Mass.: Addison Wesley, 1966, Chapter 11.

> When I say of motion that it is the genus of which heat is a species I would be understood to mean, not that heat generates motion or that motion generates heat (though both are true in certain cases) but that heat itself, its essence and quiddity, is motion and nothing else ... the body acquires a motion alternative, perpetually quivering, striving, and struggling, and initiated by repercussion, whence springs the fury of fire and heat.
>
> FRANCIS BACON

Chapter 9 The Energy of Molecules in Gases, Liquids, and Solids

I. Gas-Liquid Equilibria: Boiling Points

In Chapter 1 we listed the boiling points of the elements. A far more complete listing of chemical and physical properties could have been made because, as indicated in that chapter, essentially all of the properties of elements and compounds reflect periodicity. For example, the boiling point and enthalpy of formation of oxides might have been listed along with their formulas. There would be no difficulty in assembling most of the data since the boiling point is an easily measured and widely tabulated physical property.

The temperature at which a liquid becomes a gas is easily recognized in an open system by the fact that once bubbles are formed they rapidly grow larger as they rise through the liquid.

A very pure material is required when measuring a boiling point, and corrections for variations in atmospheric pressure must be made. Precautions must be taken to prevent overheating by using a small flame rather than a large one. For the same reason the temperature of the vapor should be measured in preference to that of the boiling liquid. In addition it is usually necessary to protect the temperature-

Ch. 9 The Energy of Molecules in Gases, Liquids, and Solids

Table 9-1 *Accurately Measured Boiling Points*

He	−268.94°C
H_2 (para)	−252.78°C
CO	−191.48°C
O_2	−182.97°C
C_8H_{10}	217.96°C
Hg	356.58°C
S	444.60°C

measuring device from direct heating or cooling by radiation from the surroundings. With these precautions many boiling points can be reproduced within 0.01°C. Several are listed in Table 9-1.

The boiling point of a substance is the temperature at which a pure liquid and its vapor are in equilibrium at 1.00 atm pressure. The equation for the vaporization of water can be written as

$$H_2O(l) = H_2O(g)$$

An equilibrium constant expression can be written for this equilibrium,

$$K = 10^{-\Delta H^0/4.5757T} 10^{\Delta S^0/4.5757} = \frac{a_{vap}}{a_{liq}} = \frac{P_{H_2O}}{X_{H_2O}}$$

The dependence of the boiling point of water on the pressure and on the purity of the water is given by this expression. For a pure substance there is only one temperature at which there is equilibrium between

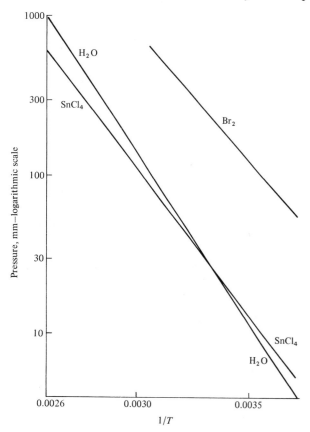

Figure 9-1 *Vapor pressure as a function of temperature.*

liquid and vapor at 1 atm pressure. At this temperature the equilibrium constant is equal to 1.00 and $\Delta H^0/T = \Delta S^0$. The equation indicates the dependence of the equilibrium on the pressure of the vapor and on the purity of the liquid. It implies that a plot of log P against $1/T$ should be a reasonably straight line. Figure 9-1 shows such a plot for water and two other compounds.

Using enthalpy and entropy to express the equilibrium condition has some additional advantages. The slope of the lines in Fig. 9-1 is related to the enthalpy of vaporization (ΔH_v^0) of the substance. The entropy of vaporization (ΔS_v^0) of many substances shows a regularity that is obvious from the values in Table 9-2. The simple generalization that the entropy of vaporization of a substance is approximately 21 eu is known as Trouton's rule. Trouton's rule can be used to estimate the boiling point from the enthalpy of vaporization, or the enthalpy can be estimated from the boiling point. Since ΔH^0 of vaporization in turn is related to the variation of the vapor pressure with temperature (Fig. 9-1), it is possible to estimate the whole vapor pressure against temperature curve from a single measurement of the boiling point at one pressure. With two or more measurements a plot of log P against $1/T$ will permit even more accurate interpolation or extrapolation to other temperatures. Later in this chapter we will be interested in an even more accurate representation of vapor-liquid equilibrium data.

For $X_{H_2O} = 1.00$

$K = P_{H_2O}$
$= 10^{-\Delta H^0/4.5757 T} 10^{\Delta S^0/4.5757}$

$\log P_{H_2O}$
$= (\Delta S^0/4.5757) - (\Delta H^0/4.5757)(1/T)$
Thus the slope is $-(\Delta H^0/4.5757)$.

Table 9-2 *Entropy of Vaporization of Some Substances*

Substance	$\Delta S_v = \Delta H_v/T$, eu
CH_4	19.8
HCl	19.2
NH_3	23.4
HF	24.9
SO_2	23.1
O_2	20.3
CO	20.1

II. Ideal Gases

There are two striking differences between a gas and a liquid in equilibrium at the boiling point. First of all the gas has a much lower density than the liquid. Liquid water at 100°C has a density of 0.846 g cm^{-3} compared to a density of 0.000597 g cm^{-3} for the vapor. The volume of 1 g of vapor is 1420 times the volume of 1 g of liquid. Atoms and molecules are quite closely packed together in solids and liquids, but in a gas the molecules must be much further apart and a gas must be considered mostly empty space.

The second striking difference between a liquid and a gas concerns the variability of the volume of the gas. The volume of a liquid varies only slightly with pressure. The effect is too small, for example, to allow a 1500 ml flask to be filled with 1000 ml of a liquid. However, a gas expands to fill its container no matter how much larger the container is made. Because the molecules are already quite far apart in a gas, there is little to prevent them from moving still further apart. The attractive forces that exist between the molecules of a gas that might prevent them from separating further are quite small. It is easy to

Ch. 9 The Energy of Molecules in Gases, Liquids, and Solids

imagine a hypothetical gas in which these forces are not just small but equal to zero. A hypothetical gas in which the intermolecular forces are equal to zero is called an ideal gas.

The behavior of ideal gases is quite simple and easy to predict. There are no forces between the molecules, so there is no change in the energy of the gas as it expands. The energy of an ideal gas is independent of its volume. For an increase in volume, $\Delta E = 0$. There is, however, a change in the entropy of an ideal gas when the volume is increased. It is quite obvious that the number of possible orientations of the molecules of a gas will be increased if the volume is increased. This is obvious because a state in which all the molecules are concentrated in just the right half of the total volume is clearly more ordered and less probable than a more even (more random) distribution. The expansion of a gas to fill its container completely is a spontaneous process, and as such it corresponds to an increase in entropy.

The term energy is used in a variety of ways, but when it is used with the symbol E it has a precise thermodynamic meaning. Energy and enthalpy are related by the equation $H = E + PV$. The significance of this equation will be clearer later in this chapter.

The dependence of entropy on volume is expressed by a relationship similar to equation 4.3. The relationship is

$$\Delta S = k \ln \frac{(V + \Delta V)}{V}$$

The pressure, volume, temperature, and the amount of an ideal gas are interrelated by the equation $PV = nRT$, where n is the number of moles of gas, P is the pressure, V is the volume, and R is a constant. This relationship can be obtained experimentally from observations on real gases at low pressures, as well as derived in many different ways. Of course, the validity of the ideal gas law does not depend upon the accuracy of the assumptions made in the derivations or upon your ability to follow each step in a derivation, but the derivations do give some additional insight into the nature of gases. It is possible to look on the ideal gas law simply as a definition of the absolute temperature (T) and of the concept of ideality for gases.

The thermodynamic derivation begins with a consideration of the work required to compress a gas contained in a cylinder with a piston. Work is required to compress a gas because expansion is a spontaneous process.

The work function A gives the minimum amount of work that will be required to compress the gas ($\Delta A = W_{\min}$). The work function A is defined as

The definition of the work function A in terms of energy, $A = E - TS$, is precisely analogous to the definition of free energy in terms of enthalpy, $G = H - TS$. Chemists use enthalpy and free energy much more often than energy and the work function. For a full understanding of thermodynamics all four functions are needed.

$$A = E - TS$$

where E is the energy of the system, T is the temperature, and S is the entropy. Since we are considering a compression in which T does not change, we have

$$W_{min} = \Delta A = \Delta E - T\Delta S = 0 - T\left(k \ln \frac{V + \Delta V}{V}\right)$$

$$W_{min} = -kT \ln\left(1 + \frac{\Delta V}{V}\right)$$

When ΔV is small, the logarithmic term may be approximated as

$$\ln\left(1 + \frac{\Delta V}{V}\right) \approx \frac{\Delta V}{V}$$

$$W_{min} = -kT\frac{\Delta V}{V}$$

W_{min} is positive for a compression since ΔV is negative.

Work is defined as force times distance, and pressure is force per unit area. The minimum force required to move the piston is, therefore, $P \times a$ where a is the area of the piston and P is the pressure. The distance the piston moves is $-\Delta V/a$, so

$$W_{min} = (\text{force})(\text{distance}) = (P \times a)(-\Delta V/a) = -P\Delta V$$

Comparing these two expressions, we have

$$P = \frac{kT}{V} \quad \text{or} \quad PV = kT$$

So far the units for the constant k have not been specified. Actually when the symbol k was chosen in equation 4.3 we had the Boltzmann constant, 3.30×10^{-24} cal molecule^{-1} deg^{-1}, in mind. This value of k refers to a single molecule of gas. For N molecules, the equation should be $PV = NkT$. Thus the gas constant R is very closely related to the Boltzmann constant: $Nk = nR$. For 1 mole of gas, N is Avogadro's number (N^0) which is equal to 6.023×10^{23} and

$$R = N^0 k = (6.023 \times 10^{23})(3.30 \times 10^{-24}) =$$
$$1.99 \text{ cal mole}^{-1} \text{ deg}^{-1} = 1.99 \text{ eu}$$

This is the same constant R that was used in connection with equilibrium constants in Chapter 4.

In the gas law calculation that is used most often, calories are not a convenient unit for PV. A more appropriate unit is $PV = 22.4$ liter atm at 273 K or $R = 0.0821$ liter atm mole^{-1} deg^{-1}. This value of R can be used in the equation $n = PV/RT$ to calculate the number of moles of a gas present from measurements of temperature, pressure (in atmospheres), and volume (in liters). It is easier to calculate the volume at standard conditions of pressure and temperature (1 atm and 273 K) and divide by 22.4 liters mole^{-1}, the volume occupied by a mole of gas at standard temperature and pressure (STP). The equation for converting a volume to standard temperature and pressure is quite simple. It is

$$V_{STP} = V_i \left(\frac{P_i}{P_s}\right)\left(\frac{T_s}{T_i}\right)$$

where the subscript i refers to the initial conditions and P_s and T_s are the standard pressure and temperature. It is obvious that the units for P_i and P_s and T_i and T_s must be the same. Notice that the result can always be checked by knowing that an increase in pressure will compress the gas and that the volume increases (at constant pressure) as the temperature increases. It is not, therefore, really necessary to memorize that P_i goes in the numerator and P_s in the denominator, and so forth. Thus the volume at standard temperature and pressure of 20 ml of gas at 2.5 atm and 350 K can be calculated from

$$V_{STP} = 20 \text{ ml} \left(\frac{2.5}{1}\right)\left(\frac{273}{350}\right) = 39 \text{ ml}$$

If the pressure decreases the volume must increase, so the factor (2.5/1) must be greater than 1. Similarly the smaller temperature must be in the numerator to allow for contraction on cooling to 273 K.

The reverse calculation is equally simple. What volume will be occupied by 8 g of O_2 at 500 K and 610 mm pressure? Convert 8 g of O_2 to 0.25 moles of O_2, which will occupy 5.60 liters at standard temperature and pressure. The volume desired is given by

$$V = 5.60 \text{ liters} \left(\frac{P}{P}\right)\left(\frac{T}{T}\right) = 5.60 \left(\frac{760}{610}\right)\left(\frac{500}{273}\right) = 12.8 \text{ liters}$$

Either the pressure or the temperature can be unknown, resulting in a tremendous variety of gas law problems. This is why it is important to recognize that the same general procedure is applicable to all gas law problems instead of memorizing a separate procedure for each new problem.

Since the interaction between ideal gas molecules is negligible, the pressure exerted by a mixture of ideal gases will be the sum of the pressure exerted by each gas. If we have n moles of one gas and m moles of a second, the pressure is given by

$$P = P_1 + P_2$$

$$P_1 = n\frac{RT}{V}$$

$$P_2 = m\frac{RT}{V}$$

or by

$$P = (n + m)\frac{RT}{V}$$

where P_1 and P_2 are called the partial pressures of the two gases. Since one fifth of the molecules in air are oxygen, the partial pressure of oxygen in a sample of air is $\frac{1}{5}$ of the total pressure

$$P_1 = \frac{n}{m + n}P$$

An example of the use of partial pressures occurs when a gas is collected by the displacement of a liquid. Nearly pure oxygen can be collected by allowing bubbles of the gas to rise into an inverted bottle filled with water. In a case like this the gas will contain the vapor of the liquid at a partial pressure equal to its vapor pressure. The de-

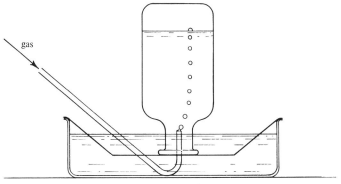

Apparatus for the collection of a gas over water

Table 9-3 *Vapor Pressure of Water*

Temperature, °C	P_{H_2O}, mm	Temperature, °C	P_{H_2O}, mm
0	4.6	20	17.5
10	9.2	21	18.6
20	17.5	22	19.8
30	31.8	23	21.1
40	55.3	24	22.4
50	92.5	25	23.8
60	149	26	25.2
70	234	27	26.7
80	355	28	28.3
90	526	29	30.0
100	760	30	31.8
110	1075		

pendence of the vapor pressure of water on temperature is shown in Table 9-3. To obtain the partial pressure of the gas collected, it is necessary to subtract the partial pressure of the vapor from the measured total pressure. For example, the number of moles of oxygen present in a 29 ml sample collected over water at 25°C and 720 mm pressure can be calculated. The vapor pressure of water at 25°C is 24 mm, so the partial pressure of oxygen in the gas collected is 696 mm. This amount of oxygen would occupy

$$V = 29\left(\frac{696}{760}\right)\left(\frac{273}{298}\right) = 24.4 \text{ ml at STP}$$

and thus, we have $24.4/22400 = 0.00109$ moles of oxygen.

III. Kinetic Theory

The fact that a gas exerts a pressure is completely explained by the increase in entropy when a gas expands. The entropy increase means that the expansion can do work, as evidenced by a piston being raised against an opposing force such as a weight.

It is useful to have a mechanical picture as well as a thermodynamic explanation of a phenomenon as basic as pressure. Pressure is exerted because the gaseous molecules are in motion continually bombarding the walls of the container.

The pressure is proportional to the number of collisions per unit area per second. It also depends on the average force on the wall per collision. By Newton's second law this force is proportional to the momentum of the molecule, mv, where m is the mass and v is the velocity of the molecule. The number of collisions per second is proportional to the concentration (N/V) and velocity (v) of the molecules.

$$P = \frac{CNmv^2}{V}$$

The constant C is equal to $\frac{1}{3}$. The only accurate and convenient way to obtain the value of C uses calculus to add together the effect of molecules moving in all directions. The expression can be written in terms of the kinetic energy of a mole of gas as

$$PV = \frac{2}{3} \frac{Nmv^2}{2}$$

where the quantity $Nmv^2/2$ is the average kinetic energy of a mole of gas. Since

$$PV = RT$$

we have

$$\frac{Nmv^2}{2} = \frac{3}{2} RT$$

The average kinetic energy of a mole of gas is $\frac{3}{2} RT$, and the average energy of a single molecule is $\frac{3}{2} kT$.

The kinetic energy of a gaseous molecule increases as the temperature is raised because a raise in the temperature of a gas requires an energy input. The direct proportionality between the average kinetic energy and T can be considered as a definition of the absolute temperature. We will look at the proportionality constant, $\frac{3}{2}k$, in more detail later in this chapter, but first let us look for further experimental evidence that the molecules are actually in motion.

If pressure is caused by the collision of individual molecules with the walls of the container, it cannot really be uniform and the piston mentioned should be in constant motion. For a piston weighing 10 g the movement is too small to observe; however, a particle like a piece of a pollen grain (which weighs about 10^{-12} g) that is suspended in a gas might have an observable motion. Particles such as these can be seen in a specially lighted microscope, and the predicted motion is readily apparent in both gases and liquids. The movement exhibited is called Brownian motion because it was discovered in 1827 by the

Ch. 9 The Energy of Molecules in Gases, Liquids, and Solids

Table 9-4 *Root Mean Square Velocities Calculated from Molecular Mass for Some Molecules*

Molecule	Molecular weight	Mass, g × 10²³	Root mean square velocity, v (calc.), 273 K	Measured velocity of sound at 0°C, cm sec⁻¹
H_2	2.016	0.334	18.4×10^4	12.84×10^4
He	4.003	0.664	13.0×10^4	9.65×10^4
CH_4	16.04	2.66	6.5×10^4	4.30×10^4
Ne	20.18	3.35	5.8×10^4	4.35×10^4
N_2	28.01	4.65	4.9×10^4	3.34×10^4
HCl	36.46	6.05	4.3×10^4	2.96×10^4
SO_2	64.06	10.6	3.3×10^4	2.13×10^4

Particles such as the pieces of pollen grain observed by Brown weigh around 10^{-12} g. They are known as colloidal particles. From Boltzmann's constant ($k = 1.38 \times 10^{-16}$ erg molecule⁻¹ deg⁻¹) we can calculate that every particle should have an average kinetic energy due to thermal motion of $\frac{3}{2}kT = 6.17 \times 10^{-14}$ erg at 298 K. For a particle of 10^{-12} g this corresponds to a velocity of about 0.35 cm sec⁻¹.

botanist Robert Brown. The measurement of the energy of colloidal particles led to the first experimental evaluation of Boltzmann's constant, and hence of Avogadro's number ($N^0 = R/k$).

The average velocity of the molecules (the root mean square velocity, $\sqrt{v_x^2}$) in a gas at a specified temperature can be calculated using the molecular mass. The velocity obtained is of the same order of magnitude as the velocity of sound in the gas. (See Table 9-4.) Of course, sound is not carried by the actual net movement of molecules. The disturbance is passed from one molecule to another through collision. At 273 K and 1.00 atm pressure a hydrogen molecule can move only about 1.1×10^{-5} cm before colliding with another hydrogen molecule. Molecules of nitrogen can move only about 6×10^{-6} cm at 273 K and 1.00 atm pressure without colliding.

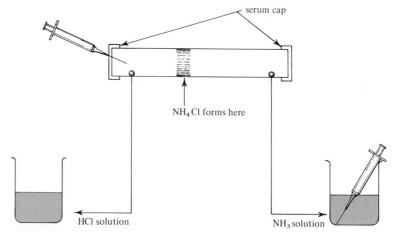

Figure 9-2 *Diffusion of NH_3 and HCl through air.*

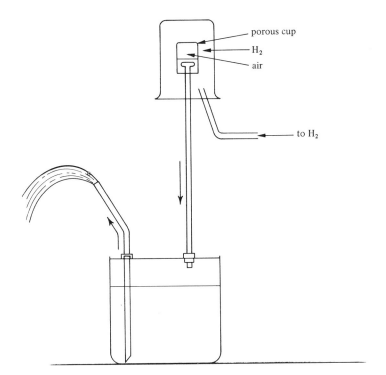

Figure 9-3 *Demonstration of the diffusion of gases.*

Hydrogen is lighter than air. The open beaker is easily kept full of hydrogen.

Since the direction of motion of each molecule is continually altered by collisions with other molecules, the net movement of molecules by diffusion is many orders of magnitude smaller than the average velocity of the molecules or the speed of sound. However, lighter, faster-moving molecules do diffuse faster than heavier molecules. If samples of NH_3 (molecular weight, MW = 17) and HCl (MW = 36.5) are placed at opposite ends of a glass tube filled with air, a coating of solid NH_4Cl will begin to form at some point near the middle of the tube after a certain length of time. The position of the coating indicates the distance each substance has traveled. Figure 9-2 shows a drawing of the apparatus that can be used to demonstrate the faster rate of diffusion of the lighter NH_3 molecules.

An even more striking demonstration of the relative velocity of the molecules of different gases is provided by effusion, the passage of molecules through small holes in a container. If a porous cup has air on one side and hydrogen on the other, the hydrogen will pass through the porous cup so much faster than the air that a large pressure differential will develop. This phenomenon is the heart of a number of demonstrations using hydrogen or helium, one of which is pictured in Fig. 9-3.

225

Table 9-5 *The Relative Rate of Effusion for Some Gases Compared With Nitrogen*

Gas	Molecular weight	Relative rate of effusion*	$\sqrt{\dfrac{\text{Mol wt } N_2}{\text{Mol wt of gas}}}$
H_2	2.016	3.56	3.73
N_2	28.01	1.000	1.000
O_2	32.00	0.937	0.936
CO_2	44.01	0.823	0.798

*Thomas Graham, *Phil. Trans. Roy. Soc.* (*Lond.*), **136**:573 (1846).

When careful measurements are made of the relative rate of effusion of several gases at the same temperature (see Table 9-5), it is found that they are nearly inversely proportional to the square root of the molecular weights. This is just what is predicted from the kinetic theory. Since

$$\tfrac{1}{2}m_1 v_1^2 = \tfrac{1}{2}m_2 v_2^2 = \tfrac{3}{2}kT$$

$$v_1^2/v_2^2 = m_2/m_1$$

$$v_1/v_2 = (m_2/m_1)^{1/2}$$

The relative rate of effusion of the gases shown in Table 9-5 is obtained by comparison with nitrogen.

IV. Boltzmann Constant

One of the mathematical difficulties in kinetic theory lies in the fact that any and all molecular velocities are allowed. There will be a distribution of molecular velocities with some molecules moving slowly and a few moving very fast. On each collision there is a change in the magnitude and direction of the velocity of the molecule. There should be an equilibrium velocity distribution (with a maximum entropy for a given energy), but the classical derivation of this distribution is a real exercise in calculus. On the other hand once the quantum mechanical idea of a restricted number of allowed energy levels is accepted, the individual energy levels may be considered with relatively simple mathematics.

For example, consider a system of just two energy levels, a ground state of energy 0, and a state that is ϵ_1 higher in energy. If we represent the number of molecules in each energy state as N_0 and N_1 respectively, the condition for equilibrium is

$$K = \frac{N_1}{N_0} = e^{-\Delta H^0/RT} e^{\Delta S^0/R}$$

Note that this is an equilibrium expression using the form introduced in Chapter 4 for the treatment of any equilibrium. This expression can be rewritten in terms of the energy of the individual molecules, ϵ_i, in place of the enthalpy per mole, ΔH^0.

The enthalpy per mole, ΔH^0, is equal to the number of molecules in a mole (Avogadro's number, N^0) times the energy of an individual molecule, ϵ_i.

$$\Delta H^0 = N^0 \epsilon_i$$

The term $\Delta H^0/R$ in the equilibrium expression is then

$$\frac{\Delta H^0}{R} = \frac{N^0 \epsilon_i}{R}$$

or

$$\frac{\Delta H^0}{R} = \frac{\epsilon_i}{k}$$

since

$$\frac{N}{R} = \frac{1}{k}$$

With this substitution the equilibrium expression is

$$\frac{N_1}{N_0} = e^{-\epsilon_i/kT}$$

or

$$N_1 = N_0 e^{-\epsilon_i/kT}$$

When the condition for equilibrium is written in terms of the energy of a single quantum level the entropy term drops out because the entropy of such a state is zero.

The quantity $e^{-\epsilon/kT}$ is called the Boltzmann factor. It gives, in

The entropy for a single quantum level is zero.

$S^0 = k \ln W = k \ln 1.0 = 0.0$

$\Delta S^0 = 0.0$

$e^{\Delta S^0/2.303} = e^0 = 1.0$

Ch. 9 The Energy of Molecules in Gases, Liquids, and Solids

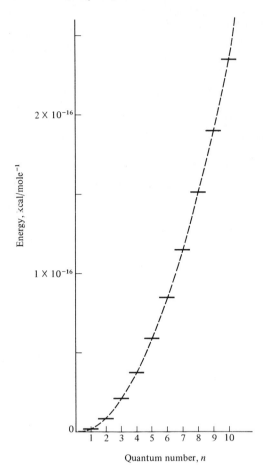

Figure 9-4 *One-dimensional translational energy levels for H_2 in a 1 cm^3 container.*

general, the correct weighting of quantum levels of energy ϵ above the ground state.

If there are ten energy levels that have an energy of ϵ, each would have an equilibrium population of $N_0 e^{-\epsilon/kT}$ for a total of $N_0 10 e^{-\epsilon/kT}$. The equilibrium population of the entire group is given by the Boltzmann factor multiplied by the number of quantum states of energy ϵ (or with energies negligibly different from ϵ).

The quantum mechanical treatment of translational motion of a molecule is beyond the scope of this book. We will present only a few of the results here. As always there are a series of allowed energy levels described by a quantum number. The quantum number n is assigned integral values from 1 to infinity. The dependence of the energy level on the quantum number comes from the Schrödinger equation. In this case the energy of a quantum level is given by

$$\epsilon_n = (h^2/8ml^2)n^2$$

where l is the distance between the walls of the container and m is the mass of the molecules. There is an even distribution of momentum or velocity [$mv_n = (h/2l)n$ and $v_n = (h/2ml)n$], but the energy levels become further apart as n increases. A plot of ϵ_n against n is shown in Fig. 9-4.

The number of molecules in each quantum level

$$N_0 e^{-\epsilon/kT}$$

corresponds to the bell shaped velocity distribution shown in Fig. 9-5.

It takes three quantum numbers to describe three-dimensional motion, and the energy of the levels in a cubic container are given by

$$\epsilon = (h^2/8ml^2)(n_x^2 + n_y^2 + n_z^2)$$

In this case the number of quantum states of a particular energy increases with energy. The number is in fact proportional to v^2 so that the relative number of molecules with the velocity v is given by

$$N(v) = Cv^2 e^{-1/2mv^2/kT}$$

This expression is called the Maxwell-Boltzmann velocity distribution. With the constant evaluated, the expression becomes

$$N(v) = 4\pi(m/2\pi kT)^{3/2} v^2 e^{-1/2mv^2/kT}$$

Plots of the Maxwell-Boltzmann distribution are shown in Fig. 9-6. The exponential term (the Boltzmann factor) ensures that the probability that molecules will have very high velocities is very small. Very low velocities are also improbable, since there are relatively few sets of three quantum numbers for a particularly low energy. There is only

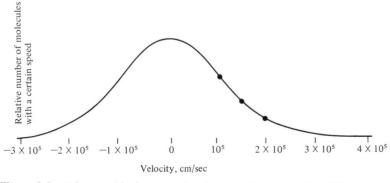

Figure 9-5 *Velocity of hydrogen molecules (one-dimensional) at 273K, in cm sec^{-1}*.

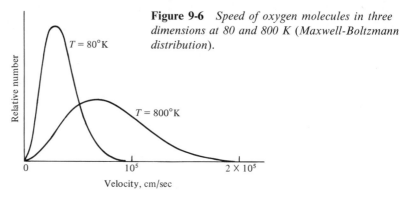

Figure 9-6 *Speed of oxygen molecules in three dimensions at 80 and 800 K (Maxwell-Boltzmann distribution).*

one quantum state of zero energy compared to about 12,600 for $n_x^2 + n_y^2 + n_z^2$ between 10^4 and 1.02×10^4. The maximum in the curve (most probable velocity) at

$$v = (2kT/m)^{1/2}$$

Because of the tail on the curve at high energies, the average energy is somewhat higher than kT.

corresponds to a kinetic energy of kT. The average must be three times the average energy for one-dimensional motion ($3 \times \frac{1}{2}kT$), because the three dimensions are independent.

So far we have only used the Boltzmann factor in discussing the kinetic energy of gaseous molecules. The derivation from equilibrium considerations suggests that this same exponential form should apply to all types of energy. The distribution of molecules in a uniform gravitational field is another particularly simple application. The number of molecules (N_h) per unit volume, at an altitude h, should decrease exponentially with h since the potential energy is mgh, and

$$N_h = N_0 e^{-mgh/kT}$$

The density of the atmosphere does decrease exponentially with altitude. The Boltzmann factor gives the observed equilibrium distribution in this situation also.

V. Heat Capacities

We have found that the average translational energy of the molecules of an ideal gas is equal to $\frac{3}{2}kT$, and the total translational energy of a mole of gas is $\frac{3}{2}RT$. For every degree rise in temperature the translational energy of an ideal gas increases by $\frac{3}{2}R$ or 2.98 cal. The energy needed to raise the temperature of a substance 1.00°C is the heat capacity of the substance. If only translational energy is involved the energy of a gas should increase by $\frac{3}{2}R$ for every degree rise in temperature.

The measured heat capacity of the inert gases provides a simple experimental test of this prediction. The absorption of heat by an *inert gas* goes almost exclusively into increasing the translational energy of the atoms. The heat capacity of helium is 3.01 cal mole^{-1} deg^{-1} or eu. The heat capacity of argon is 3.00 eu. These are heat capacities at constant volume, C_v. The heat capacity at constant pressure (C_p) is larger than the heat capacity at constant volume. When a gas is heated at constant pressure, it expands and does work in the process. The work of expansion is given by $P \Delta V$. From the gas law

$$V_1 = RT_1/P \quad \text{and} \quad V_2 = RT_2/P$$

so

$$V_1 = \frac{RT_1}{P}$$

$$V_2 = \frac{RT_2}{P}$$

$$\Delta V = \left(\frac{R}{P}\right) \Delta T$$

$$P \Delta V = R \Delta T$$

Thus the work of expansion is $R \Delta T$. For a single degree rise in temperature, the heat capacity at constant pressure (C_p) of a nearly ideal gas should be given by $C_p = C_v + R$. This relationship is illustrated in Table 9-6.

It is desirable to have a simple way to include the work of expansion in constant pressure processes. This is the reason enthalpy is used in place of energy in many cases. The definition of enthalpy is $H = E + PV$. In a constant pressure change $\Delta H = \Delta E + P \Delta V$. An energy change ($\Delta E$) can be due to both heat and work. The $P \Delta V$ term corrects for the work of expansion so ΔH is equal to only the heat part of the energy change. This is the reason the heat of a reaction at constant pressure is related to enthalpy instead of energy.

To illustrate the use of enthalpy further let us look again at C_p. Since the heat involved in a constant pressure process is given by the enthalpy (ΔH) we have

$$C_p \Delta T = \Delta H \quad \text{or} \quad C_p = \frac{\Delta H}{\Delta T}$$

Ch. 9 The Energy of Molecules in Gases, Liquids, and Solids

For an ideal gas $H = E + PV = E + RT$. Therefore

$$\Delta H = \Delta E + R\,\Delta T \quad \text{and} \quad C_p = \frac{\Delta E}{\Delta T} + R$$

This is a particularly simple derivation of the relationship $C_p = C_v + R$.

The distinction between G (the free energy) and A (the work function) is just the same as that between H and E. G is defined as $H - TS$ and $A = E - TS$. Thus

$$G = A + PV$$

The free energy (G) is used in place of the work function (A) in constant temperature and constant pressure processes because it includes the work of expansion.

Table 9-6 shows a number of examples of gases where the heat capacity of the gas is substantially larger than the value calculated using the translational energy alone.

For molecules the addition of energy does more than increase the translational energy. Molecules also have rotational and vibrational energy. The addition of a specific amount of energy not only increases the translational motion but the rotational and vibrational motion as well. The division of the added energy between the three types of motion is dependent upon the number of atoms present and the struc-

Table 9-6 *Heat Capacities for Gases*

Gas	Molecular weight	Temperature, °C	C_p, cal mole^{-1} deg^{-1}	C_v,	$C_p - C_v$, or eu
Monatomic gases					
He	4.003	−180	4.99	3.01	1.98
Ar	39.95	−180	5.00	3.01	1.99
		15	5.00	3.00	2.00
Kr	83.80	19	5.01	2.99	2.02
Diatomic gases					
H_2	2.016	−181	5.32	3.34	1.98
		15	6.84	4.84	2.00
O_2	32.00	15	6.96	4.97	1.99
I_2	253.81	25	8.82	6.8	2.0
Polyatomic gases					
CO_2	44.01	15	8.69	6.63	2.06
CH_4	16.04	15	8.47	6.47	2.00
H_2O	18.02	100	8.70	6.58	2.12
		2000	14.93	12.95	1.98

ture of a molecule. To illustrate this we must define what is meant by the term degrees of freedom.

The degrees of freedom of a molecule is the number of coordinates needed to locate all of the mass points in a molecule. In a molecule containing N atoms there are $3N$ degrees of freedom. These will be divided unequally according to each type of motion. For example, for any molecule or atom there are three degrees of translational freedom. In an atom ($3N = 3$) there are zero degrees of rotational and vibrational freedom and there is no contribution of rotational and vibrational energy to the heat capacity. The only contribution to C_v is $(3)(\frac{1}{2}R)$.

$C_v = \frac{3}{2}R = 2.98$ *eu and*
$C_p = \frac{5}{2}R = 4.97$ *eu.*

In a diatomic molecule the total number of degrees of freedom is

$$3N = 6$$

Of these, three represent translational degrees of freedom so there will be three degrees of freedom to be split up into the rotational and vibrational degrees of freedom. The spacing of rotational energy levels is similar to that of translational energy levels, and each rotational degree of freedom makes a contribution of $\frac{1}{2}R$ to the heat capacity.

In a diatomic molecule (or any linear molecule) there are two degrees of rotational freedom. The rotational contribution to the heat capacity of a diatomic molecule is given by

$$E_{\text{rot}} = \tfrac{1}{2}RT \times 2 = RT$$

Five of the total of six degrees of freedom for a diatomic molecule are assigned as translational or rotational degrees of freedom. The one remaining must be a vibrational degree of freedom. The contribution of a vibrational degree of freedom to the heat capacity is somewhat more complex than a simple value of $\frac{1}{2}R$.

The total number of degrees of freedom in a linear triatomic molecule is

$$3N = 9$$

There are three degrees of translational freedom, two degrees of rotational freedom, and four degrees of vibrational freedom.

The total average energy of 1 mole of a linear triatomic gas is

$$E_{\text{tot}} = E_{\text{trans}} + E_{\text{rot}} + E_{\text{vib}}$$

$$E_{\text{tot}} = \tfrac{3}{2}RT + RT + E_{\text{vib}}$$

No simple substitution for E_{vib} will fit the heat capacity for CO_2 given in Table 9-6.

All nonlinear molecules have three rotational degrees of freedom so that nonlinear triatomic molecules have only three vibrational degrees of freedom.

The equality of translational and rotational energies per degree of freedom is a consequence of the law of equipartition of energy which states that each degree of freedom will have, on the average, an equal fraction of the total energy. This "law" can be derived from Newton's laws of motion with certain assumptions, but its success on comparison with experimental data has been somewhat limited as we have seen.

In contrast to gases there can only be a vibrational contribution to the heat capacity in a solid. The atoms in a solid are only free to vibrate about a fixed position. The heat capacity of a number of solids was studied by Dulong and Petit in 1819. They found that the heat capacity of a number of solids was close to $5.96n$ where n is the number of atoms per molecule of the solid. This value can be obtained if RT is used for the average energy per degree of vibrational freedom.

$$3Rn = 5.96n$$

We have seen above how the law of equipartition of energy fails for the vibrational energy of gases. It can be made useful in the case of many solids if RT is used instead of $\frac{1}{2}RT$ for the energy of each vibrational motion. This substitution is justifiable if it is realized that a vibration involves both potential energy and kinetic energy, and a value of $\frac{1}{2}RT$ is assigned for each.

Table 9-7 *Heat Capacities of Some Solids*

Solid	Molecular weight	Heat capacity, C_p, eu		
		Exptl.	Calc. by Dulong-Petit	Calc. by Kopp's rule
Al	26.98	5.9	6	6.2
B	10.81	3.32	6	3.0
Ca	40.08	5.81	6	6.2
CaF_2	78.08	16.4	18	16.2
C graphite	12.01	1.92	6	2.0
C diamond	12.01	1.36	6	2.0
$CuSO_4$	159.63	26.5	36	28.4
$CuSO_4 \cdot 5H_2O$	249.71	68	126	73
Fe	55.85	6.32	6	6.2
K	39.10	7.5	6	6.2
K_2SO_4	174.25	30.7	42	34.6
Sn	118.69	6.43	6	6.2
$ZnCl_2$	136.28	17.05	18	18.6

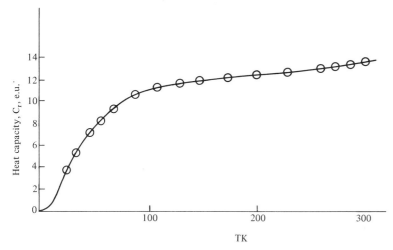

Figure 9-7 *Experimental heat capacity curve to low temperatures.*

Experimental data on the heat capacity of a number of solids is given in Table 9-7 and in Appendix IV. There are some values such as those for the two crystalline forms of carbon which are inconsistent with equipartition of energy or the relationship of Dulong and Petit. In fact all solids have heat capacities much lower than these predictions if the measurements are made at a low temperature. The experimental heat capacity of a typical solid is shown as a function of temperature in Fig. 9-7. Such curves represent one of the simplest experimental demonstrations that classical mechanics is inadequate to describe the motion of atoms, and quantum mechanics must be used instead.

Einstein showed in 1906 that a set of equally spaced vibrational energy levels would give a heat capacity curve (Fig. 9-8) very similar in shape to the experimental curve shown in Fig. 9-7. The calculation uses the Boltzmann factor to determine the relative number of atoms in each vibrational energy level. The slight discrepancy between the experimental curve and Einstein's is entirely explained by the fact that

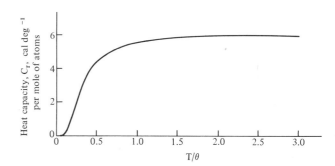

Figure 9-8 *Heat capacity curve calculated from Einstein function.*

Table 9-8 *Approximate Contribution of Various Elements to the Heat Capacity Per Mole of Solid (near room temperature)*

Element	H	Li	Be	B	C	N	O	F
Contribution to C_p, eu	2.5	5.0	3.5	3.0	2.0	3.0	4.0	5.0
Number of bonds		1	2	3	4	3	2	1

the actual vibrational energy levels are not exactly evenly spaced. Heat capacities slightly above $5.96n$ can be due to a slightly closer spacing of the higher energy vibrational levels, a situation often found spectroscopically. Heat capacities much lower than the limiting value are predicted whenever the spacing of the vibrational levels is appreciable compared to kT. When ϵ/kT is large, the Boltzmann factor ($e^{-\epsilon/kT}$) is very small and there are very few molecules in even the first allowed vibrational energy level. In this limit the vibrational contribution to the heat capacity becomes negligible.

The spacing of vibrational energy levels is inversely proportional to the square root of the mass that moves during the vibration. It is generally the lighter atoms that show an exceptionally low contribution to the vibrational heat capacity at room temperature.

This is evident for diatomic molecules, and it is also reflected in Kopp's rule for estimating the heat capacity of a solid at room temperature. The contribution to the heat capacity of the light elements is shown in Table 9-8. The value 6.2 eu is used for all atoms heavier than fluorine. The assigned values depend on the number of bonds an atom normally forms as well as its atomic weight.

To illustrate the use of Kopp's rule we can calculate a value for the heat capacity of K_2SO_4. The two potassium atoms and the sulfur contribute 12.4 and 6.2 eu respectively. The oxygen contribution is $4 \times 4.0 = 16.0$. The calculated heat capacity for K_2SO_4 is 34.6 eu. This can be compared to experimental values of 30.7 at 0°C and 33.3 at 100°C. A number of other Kopp-rule estimates are included in Table 9-7 along with the experimental values.

VI. Real Gases

Real gases do not meet any of the criteria of an ideal gas except in the limit of low pressure. In a real gas there is an attractive force between the molecules and as a result the energy will be lowered as the molecules approach each other. Because of the attractive force there will be fewer collisions with the walls and the pressure will be less than nRT/V. In this situation the activity of the gas molecules will be less than the ideal pressure.

On the other hand the repulsive forces between molecules will lead to an increase in energy as the pressure increases. This will result in a pressure greater than nRT/V. In this case the activity of the real gas is greater than the ideal pressure.

The two possibilities result in positive and negative deviations from ideality. Both types of deviations from ideality are observed because all molecules attract one another (all gases can be condensed to a liquid at low enough temperatures), and all molecules have some size and repel each other at short distances.

The deviations from ideality can be represented by the ratio of the activity to the pressure called the activity coefficient (γ). The ratio of the actual pressure of the gas to that calculated for an ideal gas $P/(RT/V)$, called the compressibility factor (κ) may also be used to represent deviation from ideality. The activity coefficient and the compressibility factor are shown in Figs. 9-9 and 9-10 as a function of temperature and pressure for oxygen gas. If the deviation from the ideal behavior is small, the approximation $\gamma = \kappa$ holds.

A. THE CRITICAL POINT

When a tube containing a liquid in equilibrium with its gas is heated the pressure increases according to the relationship

$$P = \kappa = 10^{-\Delta H/2.3RT}10^{\Delta S/2.3R}$$

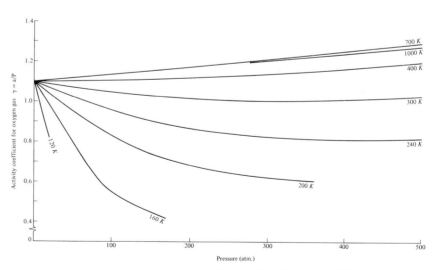

Figure 9-9 *Activity coefficient of oxygen.*

Ch. 9 The Energy of Molecules in Gases, Liquids, and Solids

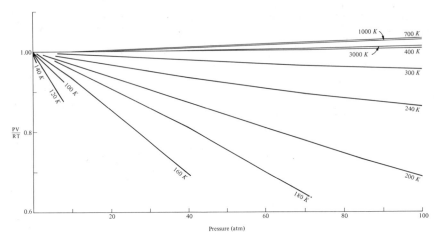

Figure 9-10 Compressibility factor, $\kappa = PV/RT$, for oxygen.

Some typical values of the critical constants of a number of gases are given in Table 9-9.

As the temperature increases the liquid expands and the density of the liquid decreases. The density of the gas increases because the increase in pressure overbalances the increase in temperature. As the density of the phases approaches one another, the difference between the two phases becomes more difficult to see. Finally at the critical temperature (T_c) the difference disappears. The pressure in the tube at this point is called the critical pressure. Figure 9-11 shows some photographs of carbon dioxide (CO_2) sealed in glass tubes at several temperatures near its critical temperature (31.1°C). The tubes also contain three glass bulbs which give some indication of the density of the phases.

The critical temperature and critical pressure of a gas are characteristic of that particular gas. We may use the symbols T_c and P_c for the critical temperature and critical pressure respectively. The values of the critical temperature and pressure of a gas are dependent on the intermolecular force present. They provide a convenient way to summarize data which is similarly dependent upon the forces between the molecules of gas. The equation

$$\left(P + \frac{27R^2T_c^2}{64P_cV^2}\right)\left(V - \frac{RT_c}{8P_c}\right) = RT$$

is useful for calculations when the ideal gas law is not quite accurate enough. This is one form of the equation known as van der Waal's equation of state for gases. It is customary to plot the compressibility factor and activity coefficient of gases against P/P_c where P_c is the critical pressure. The graphs shown in Figs. 9-12 and 9-13 provide a reasonably accurate representation of data for a variety of different gases over a wide range of temperatures.

B. THE EXACT TREATMENT OF EQUILIBRIA

The value of the vapor pressure of water from 0°C up to the critical point is shown in Fig. 9-14. The equation

$$K = 10^{-\Delta H^0/2.3RT} 10^{\Delta S^0/2.3R} = a_{H_2O(g)} \approx P_{H_2O}$$

gives the activity of the gas in equilibrium with the liquid at 1 atm pressure. This is entirely adequate at low pressures where the vapor is nearly ideal and the pressure above the liquid is not much different from 1 atm, but it works very poorly near the critical point. It can

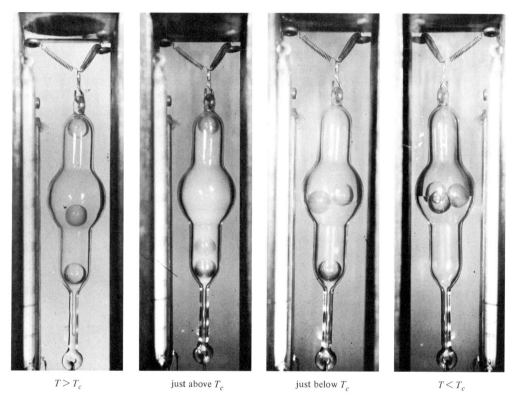

$T > T_c$ just above T_c just below T_c $T < T_c$

T close to the critical temperature

Figure 9-11 *Carbon dioxide in a sealed tube at about 70 atm pressure with three glass floats of different densities at a series of temperatures, T. (After Ray Rakow, Chemical and Engineering News.)*

Ch. 9 The Energy of Molecules in Gases, Liquids, and Solids

Table 9-9 *Critical Constants of Some Gases*

Gas	Formula	T_c, °C	P_c, atm	Density at critical point, g cm^{-3}
Ammonia	NH_3	132	112	0.235
Argon	Ar	−122	48	0.531
Carbon dioxide	CO_2	31	73	0.460
Carbon monoxide	CO	−139	35	0.311
Chlorine	Cl_2	144	76	0.573
Helium	He	−268	2.3	0.0693
Hydrogen	H_2	−240	13	0.0310
Neon	Ne	−229	26	0.484
Nitric oxide	NO	−94	65	0.52
Nitrogen	N_2	−147	34	0.311
Oxygen	O_2	−119	56	0.430
Water	H_2O	374	220	0.307

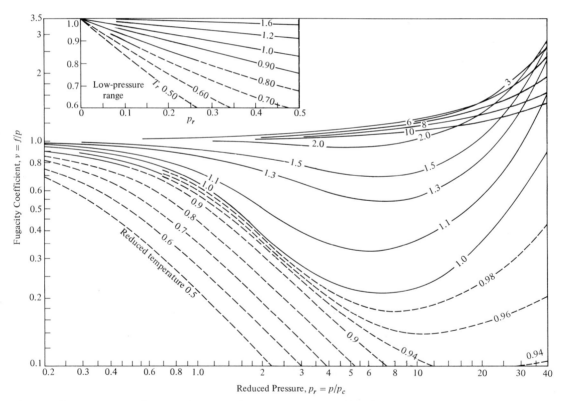

Figure 9-12 *Activity coefficients of gases. (After O. A. Hougen and K. M. Watson, Chemical Process Principles, New York: John Wiley, 1949.)*

Sec. VI Real Gases

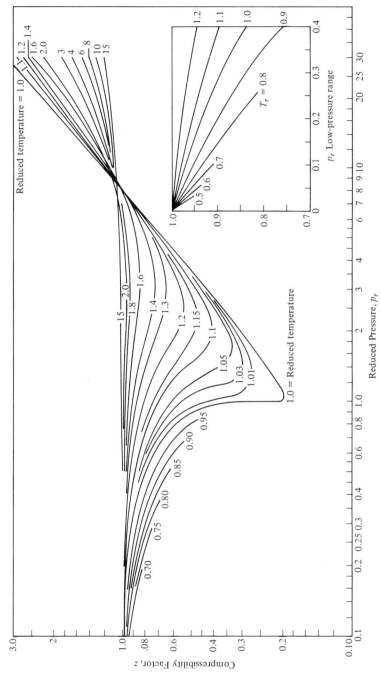

Figure 9-13 *Compressibility factor of gases. (After O. A. Hougen and K. M. Watson, Chemical Process Principles, New York: John Wiley, 1949.)*

Ch. 9 The Energy of Molecules in Gases, Liquids, and Solids

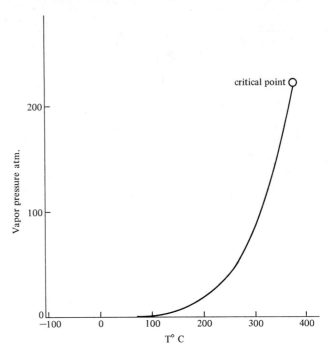

Figure 9-14 *Vapor pressure of water.*

be shown that this equation predicts that the slope of the line in a pressure against temperature plot is given by

$$\text{Slope} = (\Delta H^0/T)/(P/RT) = C\,\Delta H^0/TV$$

where C is a constant (0.0413 l atm cal^{-1}). An equation very similar to this, proposed by Clapeyron in 1834, was later shown by Clausius to be exact. The Clausius-Clapeyron equation is

$$\text{Slope} = C\frac{\Delta H}{T\,\Delta V}$$

where ΔH is the enthalpy of vaporization, and ΔV is the volume change on vaporization

$$\Delta V = V_g - V_l$$

The inclusion of the volume of the liquid corrects for the fact that the liquid is at pressure P instead of 1 atm. The corrections for the deviations from ideality of the vapor are included in the substitution of ΔH at the pressure P in place of the standard enthalpy of vaporization at 1 atm. The Clausius-Clapeyron equation applies to any change of state, including melting points and equilibria between different crystal forms.

The enthalpy of fusion of ice is 1440 cal mole^{-1} and the molar volumes of the solid and liquid at 0°C are 19.6 and 18.0 ml respectively. The line for the melting equilibrium has a negative slope.

This behavior is unusual. Of the common substances only antimony and bismuth behave like water.

$$\text{Slope} = C1440/273(-.0016) = -136 \text{ atm deg}^{-1}$$

This line is one of those shown in the phase diagram for water (Fig. 9-15). Note that the negative slope of this line expresses the fact that the melting point of ice is below 0°C at high pressures. Each area of the phase diagram shows the region of temperatures and pressures over which a particular phase can exist. Each line expresses the conditions required for equilibrium between the two phases on opposite sides of the line.

At the intersection of two lines there is only one temperature and only one pressure at which the three pure phases can be in equilibrium. An example of such a triple point is that for solid, liquid and gaseous water at 273.160 K and 0.00603 atm. The triple point of carbon dioxide is shown in the phase diagram for carbon dioxide in Fig. 9-16. There is an increase in volume when solid carbon dioxide melts, so the solid-liquid equilibrium line in Fig. 9-16 has a positive slope.

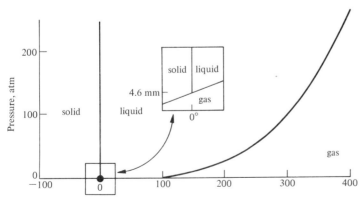

Figure 9-15 *Phase diagram for* H_2O.

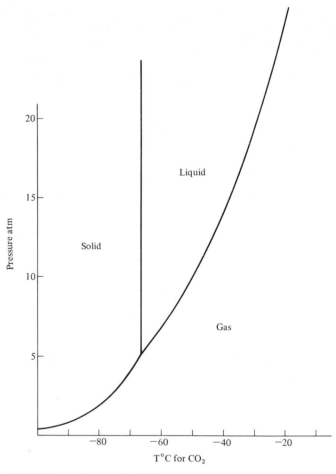

Figure 9-16 *Phase diagram for* CO_2.

QUESTIONS

1. How does the energy of a water molecule vary with the bond angle?
2. Why do attractive forces between molecules reduce the pressure?
3. What effect will attractive forces between the molecules of a gas have on its heat capacity?
4. Is the Clausius-Clapeyron equation applicable to an equilibrium like $CaCO_3 \rightleftharpoons CaO + CO_2$?
5. What is the effect of increasing pressure on any equilibrium?
6. Show how increasing temperature favors high-energy quantum states.

Sec. VI   Real Gases

PROBLEMS I
1. Calculate the total energy of 1 mole of N_2 gas at 0°C, 25°C, and 1000°C.
2. Estimate the heat capacity of linear N_2O and Br_2. Calculate the heat capacity of $CaCO_3$ and $K_2S_2O_3$.
3. Calculate the density of air ($\frac{4}{5}N_2$ and $\frac{1}{5}O_2$) at standard temperature and pressure. How much will the density be reduced if the temperature is raised to 300°C at a pressure of 1.00 atm?
4. Cool 7.00 liters of steam at 2.00 atm pressure and 163°C to standard temperature and pressure. What is the actual volume of the liquid obtained? What volume would it occupy as an ideal gas at standard temperature and pressure?
5. Gaseous PH_3 was collected from the reaction of Ca_3P_2 with water by holding an inverted graduated cylinder filled with water over the reacting solid. At a temperature of 30°C and 640 mm pressure there were 24.3 ml of gas in the cylinder. How many moles of PH_3 were collected?
6. What pressure is necessary to melt ice at -5°C?
7. What is the average energy of mercury atoms in the gas phase at 450°C?
8. The potential energy of N_2 molecules at the surface of the moon is -19.5 kcal mole^{-1}. Calculate the relative equilibrium concentrations of N_2 molecules at the moon's surface and in outer space.
9. From the thermodynamic data in Appendix IV calculate the vapor pressure of Br_2 at 25°C.
10. Estimate the boiling point of $SiCl_4$ from the data in Appendix IV.
11. Stannic chloride, $SnCl_4$, boils at 114°C to give a vapor with a density of 0.0085 g cm^{-3}. Calculate the compressibility factor for $SnCl_4$ gas at the boiling point.
12. Estimate the heat of vaporization of $SnCl_4$ from the data in Problem 11.
13. The critical point of $SnCl_4$ is 319°C and 36.95 atm pressure. Calculate the density of $SnCl_4$ gas at the boiling point using van der Waal's equation.
14. Calculate the root mean square velocity and average kinetic energy of $SnCl_4$ molecules at the critical point.
15. Estimate the heat capacity of $BaCO_3$ from Kopp's rule.
16. List the degrees of freedom of the $SnCl_4$ molecule.

PROBLEMS II
1. A 10 liter container of CO_2 gas is heated. What will happen to the pressure?
2. If the pressure of the gas in Problem 1 is 1 atm at -15°C, what will the pressure be at 100°C?
3. To what volume must the CO_2 expand at 100°C to bring the pressure back down to 1 atm?
4. What will the volume of the CO_2 gas from Problem 3 be at 273 K and 1.00 atm pressure?
5. How many moles of CO_2 are present?
6. What weight of CO_2 is present?
7. Calculate the volume of 100 g N_2 at standard temperature and pressure.
8. Calculate the pressure of 100 g N_2 gas at 200°C in a 1000 ml container.

245

9. The vapor pressure of mercury at 200°C is 17.3 mm. Calculate the total pressure of nitrogen and mercury if a drop of liquid mercury is added to the 1000 ml container in Problem 8 at 200°C.
10. Calculate the weight of mercury present as gas in Problem 9.
11. The heat capacity of cobalt is 0.104 cal g^{-1} deg^{-1}. Calculate the heat capacity per mole.
12. The heat capacity of another metal is 0.059 cal g^{-1} deg^{-1}. Estimate the atomic weight of this metal.
13. The metal in Problem 12 forms an oxide which is 23.1% oxygen. Calculate the accurate atomic weight of the metal.
14. Estimate the heat capacity of PdO_2 using Kopp's rule.
15. Classify (as translational, rotational and vibrational) the 9 degrees of freedom of the O_3 molecule.
16. The heat capacity, C_p, of O_3 is 9.37 cal mole^{-1} deg^{-1}. Calculate C_v for O_3.
17. What is the vibrational contribution to the heat capacity of ozone?

SUGGESTED READING

HILDEBRAND, J. H., *Introduction to Molecular Kinetic Theory.* New York: Reinhold Publishing Corporation, 1963.

Size is not grandeur...

THOMAS HENRY HUXLEY

Chapter 10 Hydrogen

I. Isotopes of Hydrogen

The hydrogen atom is the simplest of all atoms. The nucleus consists of one proton and there is only one electron. Three isotopes exist, ^{1}H, ^{2}H (deuterium), and ^{3}H (tritium). Tritium may be produced in nuclear reactors by the reaction of thermal neutrons with lithium. It exists only in very small amounts naturally (about 1 part in 10^{17} of normal hydrogen), and even this much is found only because it is continually being produced in the upper atmosphere by cosmic ray induced nuclear reactions. It is radioactive with a half-life of 12.4 years. Deuterium occurs naturally to the extent of about 0.0156%. It may be separated from water as D_2O by fractional distillation or electrolysis. Deuterium oxide is made in ton quantities and is used in nuclear reactors.

Although the isotopes do not differ in the type of reactions that they undergo, the rate of reaction will depend upon the isotope because there is such a great difference in mass. The isotopes of hydrogen will behave quite differently in any situation where the mass number is important. The effect is much more pronounced for the isotopes of hydrogen than for the isotopes of any other element. The difference in reaction rate is a powerful tool in the study of reaction mechanisms. For example, the reaction of methyl radicals (CH_3) with deuterium molecules (D_2) is about three times slower than the corresponding reaction with hydrogen. The difference results from the slightly stronger bonding in deuterium. The dissociation energies of hydrogen and

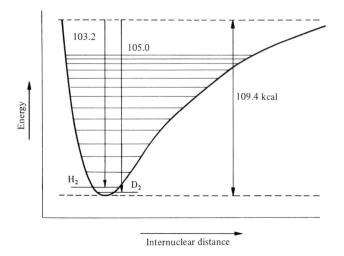

Figure 10-1 *Dissociation energy of H_2 and D_2. (After W. J. Moore, Physical Chemistry, Englewood Cliffs, N.J.: Prentice-Hall, 1950.)*

deuterium at absolute zero (0 K) are shown in Fig. 10-1. There is a very small difference in the free energy of formation of hydrogen and deuterium compounds. The equilibrium constant for the reaction

$$H_2O + HDS \rightleftharpoons HDO + H_2S$$

may be calculated from the thermodynamic data given in Table 10-1.

II. Types of Hydrogen Compounds

Hydrogen is an element of intermediate electronegativity. It reacts with the elements of low electronegativity to form saltlike compounds (saline hydrides) in which it may be considered to be the negative ion. With elements of high electronegativity it reacts to form volatile compounds in which it may be considered to be at least partially positive. (You may recall from Chapter 7 that hydrogen fluoride is 45% ionic.) The reaction with the elements at the extreme ends of the periodic table is not highly exothermic, except with oxygen, fluorine, and chlorine. These three elements may react explosively with hydrogen. With the elements oxygen and fluorine the reaction is sufficiently exothermic to be useful as a chemical rocket propellant. The formulas for the hydrides of the elements are shown in Table 10-2. The enthalpies of formation of hydrogen compounds are given in Table 10-3. The compounds formed with the transition elements are often called hydrides, but they do not contain hydride ions and they are most often nonstoichiometric.

A. SALTLIKE HYDRIDES OR SALINE HYDRIDES

Hydrogen can form saltlike hydrides with only the most electropositive metals because of the very low electron affinity of hydrogen. These hydrides have typical ionic lattices in which the metals are arranged in a cubic close packed array with the hydride ion occupying the octahedral holes. In general the hydrides occupy a smaller volume than the metal itself. The ionic radius of the hydride ion ranges from 1.44 Å in LiH to 1.54 Å in RbH. When compared to the ionic radius of fluoride (F^-) and oxide (O^{2-}) this value is unexpectedly large. The relative stability of the hydrides is apparent from the data in Table 10-3 and from the bond energy data given in Table 10-4. Of the saltlike hydrides, those of lithium and calcium are the most stable. The stability of all the hydrides of the remaining elements in groups IA and IIA is very nearly the same.

The white, highly reactive solid hydrides are best prepared by direct reaction of the elements with hydrogen gas at temperatures up to 700°C. Rapid reactions take place with lithium at 600°C and sodium at 300–400°C. Temperatures between 150 and 300°C are required for the remaining alkali and alkaline earth metals (except for beryllium and magnesium). The hydrides react with extreme vigor with any substance capable of supplying a trace of H_3O^+ ion. The standard potential of the $H_2 \mid H^-$ couple is estimated at -2.25 volts, so H^- is one of the most powerful reducing agents known. From the oxidation state maps in Fig. 10-2 we see that the saline hydrides are well out

BeH_2 and MgH_2 are prepared indirectly through organometallic intermediates

$$2Be + 2CH_3Br \longrightarrow$$
$$Be(CH_3)_2 + BeBr_2$$
$$Be(CH_3)_2 + LiAlH_4 \longrightarrow$$
$$BeH_2 + LiAlH_2(CH_3)_2$$

and

$$2Mg + 2C_2H_5Br \longrightarrow$$
$$Mg(C_2H_5)_2 + MgBr_2$$
$$Mg(C_2H_5)_2 \xrightarrow{heat}$$
$$MgH_2 + 2C_2H_4$$

Table 10-1 *Thermodynamic Properties of Some Compounds of Hydrogen and Deuterium*

		$H°$, kcal mole^{-1}	$G°$, kcal mole^{-1}	$S°$, eu
H_2	(g)	0.000	0.000	31.208
HD	(g)	0.076	-0.350	34.343
D_2	(g)	0.000	0.000	34.620
H_2O	(g)	-57.796	-54.634	45.103
HDO	(g)	-58.628	-55.719	47.658
D_2O	(g)	-59.560	-56.059	47.378
H_2S	(g)	-4.93	-8.02	49.2
HDS	(g)	-5.24	-8.64	51.9
D_2S	(g)	-5.71	-8.48	51.5
S	(s)	0	0	7.60
H	(g)	52.095	48.581	27.391
D	(g)	52.981	49.360	29.455

Ch. 10 Hydrogen

Table 10-2 *Hydrides*

IA	IIA	IIIB	IVB	VB	VIB	VIIB	VIII			IB	IIB	IIIA	IVA	VA	VIA	VIIA	Inert gases
LiH	BeH$_2$											B$_2$H$_6$	CH$_4$	NH$_3$	H$_2$O	HF	
NaH	MgH$_2$											AlH$_3$	SiH$_4$	PH$_3$	H$_2$S	HCl	
KH	CaH$_2$											Ga$_2$H$_6$	GeH$_4$	AsH$_3$	H$_2$Se	HBr	
RbH	SrH$_2$											InH$_3$	SnH$_4$	SbH$_3$	H$_2$Te	HI	
CsH	BaH$_2$											TlH$_3$	PbH$_4$	BiH$_3$			

Sec. II Types of Hydrogen Compounds

Table 10-3 *Standard Enthalpies of Formation, H^0, of Hydrides (kcal mole^{-1})*

IA	IIA	IIIB	IVB	VB	VIB	VIIB	VIII	VIII	VIII	IB	IIB	IIIA	IVA	VA	VIA	VIIA	Inert gases
−21.7												B$_2$H$_6$ 8.5	−17.88	−11.0	−57.8(g) −68.3(l)	−64.8	
−13.5	−18.2											AlH$_3$(c) −11	8.2	1.3	−4.93	−22.06	
−13.8	−45.2				Cr$_7$H$_2$ −3.8					CuH 5.1		—	21.7	15.88	7.1	−8.70	
−12	−42.3							Pd$_2$H −4.7				—	38.9	34.68	23.8	6.33	
−20	−40.9											—	—	—	—	—	

Table 10-4 *Bond Energies for M-H bonds in Gaseous Molecules,** ($kcal\ mole^{-1}$)

IA	IIA	IIIB	IVB	VB	VIB	VIIB	VIII			IB	IIB	IIIA	IVA	VA	VIA	VIIA	Inert gases
59	54											89	103	109	119	135	
48	47											68	79	85	92	103	
44	57									63	20	65	71	77	78	87	
40	39										16	58	61	65	65	71	
43	43									69	10	47		59			

*Most of the data is from the references in Appendix IV or from J. A. Kerr and A. F. Trotman-Dickinson, *Handbook of Chemistry and Physics*, (Cleveland, Ohio: The Chemical Rubber Co., 1964), p. F94.

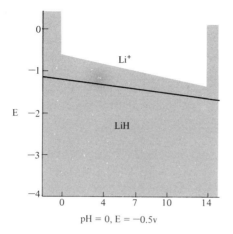
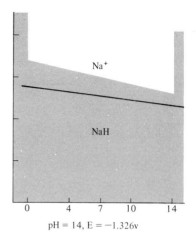

Figure 10-2 E_h-pH diagrams.

of the water range in the lower left hand corner. The reaction with water may be written

$$MH + H_3O^+ \longrightarrow M^+ + H_2 + H_2O$$

or

$$MH + H_2O \longrightarrow M^+ + H_2 + OH^-$$

Many of the saline hydrides ignite spontaneously in dry air.

The enthalpies of formation (Table 10-3) and the bond dissociation energies (Table 10-4) show irregularities in both groups IA and IIA. LiH and CaH_2 show similar enthalpies of formation and those of NaH, KH, and RbH are nearly similar. The bond energy of LiH is the largest of the group while the RbH bond energy is the lowest. In group IIA a maximum bond energy and a minimum enthalpy of formation occurs at CaH_2. The bond energy of MgH_2 is particularly low. Comparison of the enthalpy of formation of the hydrides with those of the halogen compounds (in Chapter 6) shows the relative stability of the compounds. The difference is due to the difference in bond energy of the diatomic molecules (X_2) and the difference in electron affinity. The electron affinity of hydrogen is 16.5 kcal mole^{-1} whereas that of iodine is 71 kcal mole^{-1}. The bond energy of H_2 is 104 kcal mole^{-1} whereas that of I_2 is only 52 kcal mole^{-1}. Both of these effects indicate that an alkali metal should react more vigorously with I_2 than with H_2.

The free energy of formation of the saline hydrides may be estimated if the entropies are known. The entropy of formation of the saline hydrides is largely due to the entropy of the hydrogen gas used in the reaction

$$2\text{Li} + \text{H}_2 \longrightarrow 2\text{LiH}$$
$$S^0 = 13.4 \quad 31.2 \quad\quad 9.6$$
$$\Delta S^0 = -35.0 \text{ eu at } 25°\text{C}$$

This is a significant entropy decrease. The free energies of formation estimated using this value are given in Table 10-5. The exothermicity of the compounds formed just barely balances the unfavorable entropy changes in all but a few cases. Only these cases show any degree of thermal stability.

Technically the term hydride refers to the H⁻ ion and ionic solids containing H⁻ ions. However, the term has been extended to include all binary hydrogen compounds. Note, however, that metallic hydrides and volatile hydrides do not contain hydride ions.

B. THE METALLIC HYDRIDES

These materials are prepared by heating the metal in the presence of hydrogen. Of this type, only the hydrides of titanium (Ti), uranium (U), zirconium (Zr), lanthanum (La), niobium (Nb), manganese (Mn), and palladium (Pd) are formed exothermically. The reaction with manganese and palladium is just barely exothermic. These hydrides have a negative free energy of formation only because the entropy change is not as unfavorable as it is in the case of the saline hydrides. The larger entropy of the metallic hydrides results from the random distribution of hydrogen atoms over only a fraction of the available sites. In many cases hydrogen dissolves in the metal up to a certain point and then a new solid phase with excess hydrogen is formed. Therefore, well defined stoichiometries are not usually isolated.

C. VOLATILE HYDRIDES

In the volatile hydrogen compounds the entropy effect is much smaller because the products of the reaction are gases. The formation of all of the hydrogen halides results in a positive entropy change, therefore, the reaction

$$\text{H}_2 + \text{X}_2 \longrightarrow 2\text{HX}$$

is favored. When coupled with a favorable enthalpy term, this entropy change indicates that the reaction will be spontaneous and complete. Alternatively it means that decomposition temperatures will be quite high; as a result these compounds (HF, HCl, HBr) will be dissociated only slightly at fairly high temperatures.

From the signs of the H^0 and the $T\Delta S^0$ terms shown in Table 10-7, we expect all of the halogens except iodine to react completely with hydrogen. The formation of hydrogen iodide (HI) is an equilibrium reaction. The formation of hydrogen iodide is possible in spite of an

Table 10-5 G^0 of Formation of Hydrides*

IA	IIA	IIIB	IVB	VB	VIB	VIIB	VIII	IB	IIB	IIIA	IVA	VA	VIA	VIIA	Inert gases
−16										20.7	−12.13	−3.98	−54.63	−65.3	
−8	−8										13.6	3.2	−8.02	−22.8	
−9	−35										21.7	16.47	3.8	−12.77	
−7	−32										45.0	35.31	—	0.41	
−15	−30														

*Estimate of $\Delta S^0 = -35$ at 298 K used for groups I and IIA. Measured $T\,\Delta S^0$ values for other groups are in Table 10-6.

Ch. 10   Hydrogen

Table 10-6 $T \Delta S^0$ *Values for Volatile Hydrides at 25°C (kcal mole^{-1})*

IA	IIA	IIIB	IVB	VB	VIB	VIIB	VIII			IB	IIB	IIIA	IVA	VA	VIA	VIIA	Inert gases
												−12.2	−5.75	−7.12	−3.15	+0.5	
													−5.4	−1.9	+3.09	+0.8	
													0	−0.59	+3.3	+4.07	
													−6.1	−0.62		+5.99	

256

Table 10-7 Signs of H^0 and ΔS^0, $H^0/\Delta S^0$ for the reaction of hydrogen with nonmetals

	IVA	VA	VIA	VIIA	Inert gases
	−/−	−/−	−/−	−/−	
	+/−	+/−	−/+	−/+	
	+/−	+/−	+/−	−/+	
	+/−	+/−	+/+	+/+	

Ch. 10 Hydrogen

unfavorable enthalpy term because the entropy term is favorable. The decomposition of hydrogen iodide is accompanied by a favorable enthalpy term and an unfavorable entropy term. As a result equilibrium is achieved at the temperature required to promote the formation of hydrogen iodide.

The relationship between the binding energy of the volatile hydrides and the signs of the enthalpy and the entropy terms is interesting. Notice the separation of carbon, nitrogen, and oxygen and of fluorine, sulfur, chlorine and bromine from the remaining elements in both Table 10-7 and Table 10-3. In Table 10-7 especially notice that the enthalpy of formation of methane (CH_4), ammonia (NH_3) and water is favorable whereas the entropy term is unfavorable. From the G^0 values we see that the reaction of these elements with hydrogen is thermodynamically favored at room temperature. Ammonia could be prepared in good yield at room temperature, but the reaction is too slow to be practical. The commercial procedures for the preparation of ammonia use moderately high temperatures to increase the rate, but at too high a temperature the equilibrium is unfavorable. Similarly the reaction of carbon or oxygen with hydrogen at room temperature is too slow to be of any significance. Methane and water can be prepared at elevated temperatures, but again decomposition occurs at still higher temperatures. Here are three examples where a reaction is thermodynamically favorable and yet does not proceed to any appreciable degree at room temperature. All three compounds, water, methane, and ammonia are stable at room temperature.

From the data given in Tables 10-4 and 10-7 we see that hydrogen sulfide (H_2S) belongs in the same category as hydrogen fluoride and hydrogen chloride. The formation of hydrogen sulfide represents the transition between the reaction of hydrogen and oxygen where enthalpy favors the reaction but entropy does not and the reaction between hydrogen and selenium (Se) where entropy favors the reaction but enthalpy does not. Like the formation of hydrogen fluoride and hydrogen chloride the formation of hydrogen sulfide from the elements should be spontaneous and complete.

For the hydrides that form with unfavorable enthalpy changes and favorable entropy changes, equilibrium mixtures should be the rule. This is true, as indicated earlier with hydrogen iodide. Hydrogen selenide (H_2Se) can be prepared from the elements but the decomposition temperature should be fairly low. Hydrogen telluride (H_2Te) should behave similarly; however, with tellurium the enthalpy term is so unfavorable that little hydrogen telluride is expected—that is, the equilibrium constant for the formation of hydrogen telluride is small over the whole range of temperatures up to the boiling point of tellurium. However, hydrogen telluride can easily be prepared because the reaction

$$Na_2Te + 2H_3O^+ \longrightarrow 2Na^+ + H_2Te + 2H_2O$$

is faster than the reaction

$$Na_2Te + 2H_3O^+ \longrightarrow 2Na^+ + Te + H_2 + 2H_2O$$

The production of hydrides from the elements should be very difficult for the block of elements where the enthalpy change is positive and the entropy change is negative. Only trace amounts of phosphine (PH_3) and silane (SiH_4) can be prepared in this way. Heating these substances can only promote the decomposition reaction because of the unfavorable entropy of formation.

1. GROUP IVA HYDRIDES

Table 10-8 gives the Born-Haber terms involved in the formation of the group IVA hydrides. The bond energies of the group IVA hydrides decrease as expected down the group. These data adequately illustrate the differences between the elements in each row. Carbon forms a stable hydride, whereas the remainder are all formed endothermically. The stability of the group IVA hydrides decreases rapidly from methane (CH_4) to silane (SiH_4) while germane (GeH_4) and stannane (SnH_4) are highly unstable. Decomposition temperatures are 800°, 450°, 285°, and 25°C respectively down the group. Plumbane (PbH_4) can be prepared, but it is so unstable that no reliable thermodynamic data are available for it.

The physical properties of group IVA hydrides are given in Table 10-9. Although the boiling points change in the expected manner, the melting points do not show a regular trend. A corresponding regularity in ΔH_v (the enthalpy of vaporization) and irregularity in ΔH_f (the enthalpy of fusion) is apparent. While methane and silane melt at very nearly the same temperature ($-183°C$) the ΔH_f of methane is larger than that of silane.

The compound CH_4 has the common name methane, and the ending -ane is used for the compounds of hydrogen with the other members of group IVA (silane, germane, stannane, and so forth).

Table 10-8 *Born-Haber Cycle Data (kcal mole^{-1})*

Term	CH_4	SiH_4	GeH_4	SnH_4
H^0	−17.9	8.2	21.7	38.9
H^* atoms†	187	117	96	76
Bond energy	103	79	71	61

†H^0 of hydrogen atoms = 52 kcal mole^{-1}.

Table 10-9 *Physical Properties of Group IVA Hydrides*

Compound	Melting point, °C	Boiling point, °C	Density, g cm^{-3}	ΔH_f, kcal mole^{-1}	ΔH_v, kcal mole^{-1}
CH_4	−182.48	−161.49	0.415$^{-164°C}$	0.225	1.955
SiH_4	−184.7	−111.4	0.68$^{-185°C}$	0.159	2.9
GeH_4	−165.9	−88.36	1.523$^{-142°C}$	0.200	3.361
SnH_4	−150	−41.8			4.4

Chemically, methane is relatively inert, although it burns in air. Silane and other silicon-hydrogen compounds such as Si_2H_6, Si_3H_8, and Si_4H_{10} are spontaneously inflammable and may react explosively with air or any other oxidizing agent. Hydrolysis takes place in water. Analogous derivatives of carbon and silicon hydrides can be prepared, for example, CH_3Cl and SiH_3Cl, $(CH_3)_3N$ and $(SiH_3)_3N$, $(CH_3)_2O$ and $(SiH_3)_2O$ are just a few. It should be emphasized that these examples are selected to indicate similarities, but in fact the chemistry of silicon is extremely limited compared to that of carbon. The germanium hydrides are not spontaneously inflammable and they are not hydrolyzed by H_2O or OH^-. Several germanes are known (Ge_2H_6 and Ge_3H_8) and also several stannanes (SnH_4 and Sn_2H_6). The reactivity of stannane (SnH_4) appears to parallel that of germane (GeH_4) rather than silane (SiH_4).

2. GROUP VA HYDRIDES

The physical properties of group VA hydrides in Table 10-10 illustrate an important feature of hydride chemistry. Ammonia (NH_3) would boil and melt at a considerably lower temperature if the melting and boiling points were dependent only upon molecular weight. In fact, ammonia freezes at higher temperatures than all other hydrides

Table 10-10 *Physical Properties Group VA Hydrides*

Compound	Melting point, °C	Boiling point, °C	Density, g cm^{-1}	ΔH_f, kcal mole^{-1}	ΔH_v, kcal mole^{-1}
NH_3	−77.74	−33.40	0.817$^{-79°}$	1.35	5.581
PH_3	−133.75	−87.72	0.746$^{-90°}$	0.27	3.490
AsH_3	−116.3	−62.5	—	0.56	4.18
SbH_3	−88	−17	2.204^{bp}	—	
BiH_3		22			

of the group except BiH$_3$, and it boils at a higher temperature than phosphine (PH$_3$) and arsine (AsH$_3$). Both the crystalline phase and liquid phase of ammonia are stabilized by intermolecular interactions that originate in and involve the hydrogen atoms of one molecule and the electron pair of another molecule. This type of bonding is called hydrogen bonding.

Chemically the hydrides of group VA elements are again quite different. One of the main reasons is the ability of ammonia to behave as a base. The molecule has an appreciable dipole moment because of the structure of the molecule and the polarity of the N-H bonds. The permanent dipole of ammonia is not as large as that of water, but it is a better electron donor because the electronegativity of nitrogen is lower than that of oxygen. As a result NH$_3$ will combine with H$^+$ and certain other acids more readily than H$_2$O does. Cu^{++}, Co^{++}, and Ni^{++} are cations that may be used to illustrate this point because they form complex ammines (M(NH$_3$)$_6^{2+}$) in water. Phosphine does not react with H$_3$O$^+$ in water, and it forms complexes with only a few transition metal cations. As a general rule the basicity of hydrides decreases down the family.

What is the percent ionic character of the N-H bonds in NH$_3$?

The permanent dipole moment of ammonia is 1.5 debye, that of water is 1.85.

The polar nature of ammonia makes it a good solvent for polar substances, although it is not as good as water, which has a dielectric constant of 81 at 25°C. Some interesting chemical consequences are apparent when reactions in water are compared with reactions in ammonia. Here we simply point out that ammonia undergoes a self dissociation reaction that is similar to that of water

$$2NH_3 \rightleftharpoons NH_4^+ + NH_2^-$$

for which the equilibrium constant at −50°C is

$$K = [NH_4^+][NH_2^-] = 10^{-30}$$

This may be compared with that of water

$$2H_2O \rightleftharpoons H_3O^+ + OH^-$$

for which the equilibrium constant at 25°C is

$$K = [H_3O^+][OH^-] = 10^{-14}$$

In Table 10-11 a further comparison of the ammonia and water systems is made.

Some reactions peculiar to liquid ammonia may be in order here, especially those involving the most active metals. Since ammonia is

Table 10-11 *Analogous Compounds of H$_2$O and NH$_3$*

H$_2$O	NH$_3$
(H$_3$O$^+$)(Cl$^-$)	(NH$_4^+$)(Cl$^-$)
(Na$^+$)(OH$^-$)	(Na$^+$)(NH$_2^-$)
Li$_2$O	Li$_2$NH
MgO	Mg$_3$N$_2$
Cu(H$_2$O)$_6^{2+}$	Cu(NH$_3$)$_6^{2+}$
Hg(OH)Cl	Hg(NH$_2$)Cl
Na$_2$Zn(OH)$_4$	Na$_2$Zn(NH$_2$)$_4$
CO(OH)$_2$	C(NH)(NH$_2$)$_2$
HClO$_4 \cdot$ H$_2$O	NH$_4$ClO$_4$

a fair reducing agent (it can exist in water only because of hydrogen overvoltage), the oxidation-reduction range in liquid ammonia is considerably altered from that in water; therefore, much stronger reducing agents can be used in liquid ammonia (but the useful oxidizing agents are limited). The resistance of ammonia to reduction allows solution of some metals without the reduction of the solvent. The resulting blue solutions contain metal ions and solvated electrons. All of the alkali metals, calcium (Ca), strontium (Sr), and barium (Ba), as well as some of the lanthanides can be dissolved in liquid ammonia. The solutions are stable for long periods of time and may be used as powerful reducing agents. As a result some novel preparations are possible in liquid ammonia.

For example it is possible to reduce some transition metal compounds to unusually low oxidation states in liquid ammonia

$$Na \xrightarrow{NH_3} Na^+ + e^-$$

$$Ni(CN)_4^{2-} + e^- \xrightarrow{NH_3} Ni(CN)_4^{3-}$$

$$Ni(CN)_4^{3-} + e^- \xrightarrow{NH_3} Ni(CN)_4^{4-}$$

$$4Na^+ + Ni(CN)_4^{4-} \xrightarrow{NH_3} Na_4Ni(CN)_4(s)$$

The blue solutions containing solvated electrons generally react slowly to produce molecular hydrogen.

$$e^- + NH_3 \longrightarrow NH_2^- + \tfrac{1}{2}H_2$$

$$Na + NH_3 \longrightarrow NaNH_2 + \tfrac{1}{2}H_2$$

The presence of a transition metal salt such as $FeCl_2$ catalyzes the reaction.

The normal combustion reaction of ammonia in oxygen may be represented by the equation

$$4NH_3(g) + 3O_2(g) \longrightarrow 2N_2(g) + 6H_2O(g) \qquad (10.1)$$

This is an extremely exothermic reaction for which ΔH^0 is -302.72 kcal mole^{-1}. The standard entropy change ($+31.26$ eu) is also highly favorable. The standard free energy change that may be obtained from these data corresponds to an equilibrium constant ($K_{25°C}$) of 10^{228}. Ammonia and oxygen represent a highly unstable mixture from a thermodynamic point of view. The reverse reaction never takes place, and water vapor in the atmosphere is not likely to react with nitrogen in this fashion. Another reaction that is also highly favored is

$$4NH_3(g) + 5O_2(g) \longrightarrow 4NO(g) + 6H_2O(g)$$

This reaction is less exothermic than the first, but the enthalpy change is still very favorable ($\Delta H^0 = -195.3$ kcal mole^{-1}). The entropy change (+43.2) is even more favorable than for the oxidation of ammonia to nitrogen. The second reaction is also essentially complete because the equilibrium constant ($K_{25°C}$) is 10^{168}.

A temperature of 750–900°C and a catalyst are usually employed in the industrial preparation of nitric oxide (NO) by this method. The exothermicity of the reaction is sufficient to make the wire catalyst glow. The nitric oxide produced is allowed to react with oxygen

$$2NO + O_2 \longrightarrow 2NO_2$$

and the nitrogen dioxide (NO$_2$) produced is bubbled into water to give nitric acid (HNO$_3$)

$$3NO_2 + H_2O \longrightarrow 2HNO_3 + NO$$

Although direct reaction of nitrogen, oxygen, and water is thermodynamically unfavorable, the total process using ammonia as an intermediate is very highly favorable from a thermodynamic point of view. In this way nitric acid is produced from atmospheric nitrogen using the low oxidation state compound, ammonia, as an intermediate. The equilibrium constant expression for the direct reaction of nitrogen, oxygen, and water is

$$K = \frac{(a_{H^+})^2(a_{NO_3^-})^2}{(a_{H_2O})(P_{O_2})^{5/2}(P_{N_2})} = \frac{(H^+)^2(NO_3^-)^2}{P_{O_2}^{5/2} P_{N_2}} = 3.5 \times 10^{-3}$$

From the magnitude of this equilibrium constant we might expect a substantial concentration of NO$_3^-$ in the oceans if equilibrium were established. The reaction of nitrogen, oxygen, and water does occur naturally through the reaction sequence

$$N_2 + O_2 \longrightarrow 2NO$$

$$2NO + O_2 \longrightarrow 2NO_2$$

$$3NO_2 + H_2O \longrightarrow 2H^+ + 2NO_3^- + NO$$

Electrical energy in the form of lightning permits the first step to proceed at a measurable rate. However, all the nitrate ions produced in this way are used by living organisms before they have a chance to accumulate in the oceans. Ammonia is the most important hydride of the group VA elements. The chemistry of the remaining hydrides is much more limited.

Table 10-12 *Physical Properties of Group VIA Hydrides*

Compound	Melting point, °C	Boiling point, °C	Density at boiling point, g cm^{-3}	ΔH_f, kcal mole^{-1}	ΔH_v, at boiling point, kcal mole^{-1}
H_2O	0.00	100.00	0.958	1.436	9.717
H_2S	−85.53	−60.31	0.993	0.5682	4.463
H_2Se	−65.73	−41.3	2.004	0.601	4.62
H_2Te	−51	2.3	2.650		5.55

Phosphine (PH_3) is a colorless, poisonous gas. It is a stronger reducing agent than ammonia but solutions in water are still possible. It is less soluble in water than ammonia because the electron pair on the phosphorus atom is not basic. The decomposition temperatures of salts such as PH_4X, where X is a halide, are several hundred degrees below those of NH_4X. The hydrides of arsenic (As), antimony (Sb), and bismuth (Bi) are less stable than ammonia and phosphine. They decompose leaving a metallic mirror when exposed to a hot glass surface.

3. Properties of Group VIA Hydrides

Some of the physical properties of the group VIA hydrides are given in Table 10-12. Here we see the importance of hydrogen bonding in its most important case. The liquid range of water is long because the liquid phase is stabilized by hydrogen bonding. ΔH_v is the highest of all the hydrides of the nonmetals. Water boils at an unusually high temperature. The type of structure that can result from intermolecular hydrogen bonding is shown in Fig. 10-3.

Although the difference in boiling point between ammonia and phosphine is only 54°C, water boils 160°C higher than hydrogen sulfide. The properties of hydrogen sulfide are only slightly affected by hydrogen bonding. The chemistry of water must, of course, be a prominent feature of all parts of this book.

Hydrogen sulfide is most noticeable for the number of insoluble compounds that it forms. As a result many metals are found in nature as sulfides. The acidity of hydrogen sulfide, hydrogen selenide, and hydrogen telluride is considered in the next chapter.

4. Properties of the Group VIIA Hydrides

The properties of the group VIIA hydrides are given in Table 10-13. The liquid range of hydrogen fluoride (HF) is as great as that of water because of the effects of hydrogen bonding. The hydrogen halides complete the scheme of changing acidity across a row from the very

Sec. II Types of Hydrogen Compounds

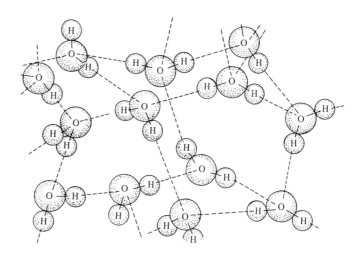

Figure 10-3 *Hydrogen bonding in water. (Each oxygen is in the center of a tetrahedron of hydrogen atoms.)*

basic hydrides of groups IA and IIA through the amphoteric hydrides of VA and VIA. The hydrogen halides are the strongest acids of the binary hydrides in each row. The free anions are reducing agents and weak bases. Liquid hydrogen fluoride is another nonaqueous solvent which has shown preparative promise.

Liquid HF is particularly useful for fluorination reactions, since it is one of a very small number of solvents that does not react with F_2. An example is the preparation of thallium trifluoride.

$$TlF + F_2 \xrightarrow{HF} TlF_3$$

Table 10-13 *Properties of Group VIIA Hydrides*

Compound	Melting point, °C	Boiling point, °C	Density at boiling point, g cm^{-3}	ΔH_f, kcal mole^{-1}	ΔH_v, kcal mole^{-1}
HF*	−83.07	19.9	0.991	1.094	1.85
HCl	−114.19	−85.03	1.187	0.4760	3.86
HBr	−86.86	−66.72	2.160	0.5751	4.210
HI	−50.79	−35.35	2.799	0.6863	4.724

*Hydrogen fluoride (HF) boils at 19.9°C with a heat of vaporization of 1.85 kcal mole^{-1}. Most of the hydrogen bonds are not broken, as the vapor contains such species as $(HF)_6$, and so forth. To break all the hydrogen bonds much more energy is required.

$$HF(l) \longrightarrow HF(g) \quad \Delta H^0 = 6.8 \text{ kcal mole}^{-1}$$

At both ends of the periodic table reaction of the elements with hydrogen is exothermic. Compounds in the center become progressively less stable until the minimum is reached in groups IVA and VA. The only elements in these groups that form stable hydrides are those of the second row.

III. Hydrogen Bonding

Hydrogen bonding is primarily an electrostatic effect that involves an interaction between molecules that have large dipole moments. The partial charges on the hydrogen and fluorine atoms in the hydrogen fluoride molecule may be calculated from the dipole moment. In this molecule the charges are very high (± 0.45); as a result the interaction between the fluorine of one molecule and the hydrogen of another molecule will be exceptionally strong. This example represents one of the strongest hydrogen bonding systems that involve neutral molecules. The strongest hydrogen bond known is formed between HF and F^-.

The electrostatic energy of attraction between two HF molecules arranged as shown in Fig. 10-4 may be calculated from the interatomic distances shown in this figure. The value of 2.50 Å represents the F–F distance in solid HF; 0.92 Å is the H–F distance in gaseous HF. In the actual molecule (H_2F_2) the charges as well as the distances are probably somewhat different; it should therefore be emphasized that the calculation that follows is for purposes of illustration only.

In the dimer the interaction between adjacent positive and negative charges are partially offset by two repulsions between like atoms at 2.50 Å. In addition there is one long-range attraction between a hydrogen that is 3.42 Å from a fluorine. The energy of each of these interactions is given by $q_1 q_2/r$. The attractions contribute toward a lower (more negative) energy so the sign of the attraction terms is negative. The repulsions tend to make the energy more positive;

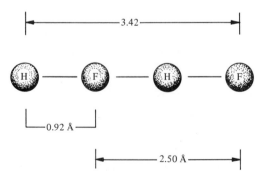

Figure 10-4 *Hydrogen bonding in* HF.

therefore, the sign of these terms is positive. The total interaction energy is given by

$$E_{tot} = -\text{attraction} + \text{repulsion}$$

$$E_{tot} = -\frac{q^2}{r_1} - \frac{q^2}{r_2} + \frac{q^2}{r_3} + \frac{q^2}{r_4}$$

To obtain the energy in ergs, the charges must be in statcoulombs and the distance must be in centimeters. It is thus necessary to convert electron charge to statcoulombs by multiplying the electron charge by 4.80×10^{-10} statcoulombs and the distance in angstroms by 10^{-8} cm. Substituting into the above expression we obtain

$$E_{tot} = -\frac{(0.45)^2(4.8 \times 10^{-10})^2}{1.58 \times 10^{-8}} - \frac{(0.45)^2(4.8 \times 10^{-10})^2}{3.42 \times 10^{-8}}$$

$$+ \frac{(0.45)^2(4.8 \times 10^{-10})^2}{2.50 \times 10^{-8}} + \frac{(0.45)^2(4.8 \times 10^{-10})^2}{2.50 \times 10^{-8}}$$

$$E_{tot} = -(4.31 \times 10^{-12}) + (3.73 \times 10^{-12}) = 5.8 \times 10^{-13} \text{ erg}$$

For 1 mole the energy is

$$(-5.8 \times 10^{-13})(6.022 \times 10^{23}) = -35 \times 10^{10} \text{ ergs mole}^{-1}$$

Conversion to kilocalories per mole will allow a better comparison of this energy with that of chemical bonds. This conversion may be made using 1 kcal $= 4.18 \times 10^{10}$ ergs.

$$\frac{-35 \times 10^{+10} \text{ ergs mole}^{-1}}{4.18 \times 10^{+10} \text{ ergs kcal}^{-1}} = -8.4 \text{ kcal mole}^{-1}$$

The agreement between this value and the observed hydrogen bond energy in HF (6.8 kcal mole^{-1}) supports the argument that hydrogen bonding is essentially an electrostatic effect.

The presence of hydrogen bonding influences nearly all properties of those materials where this type of bonding is possible. We have already seen that the melting and boiling points of NH_3, H_2O, and HF are much larger than the molecular weight would indicate. The heat capacity of water is almost twice (18 eu) the value predicted from Kopp's rule (9 eu). The enthalpy of vaporization of water is 10.519 kcal mole^{-1} at 25°C. This is considerably higher than the value expected for the interaction between molecules where the force of attraction is due simply to van der Waal attractions (see ΔH_v for methane,

Table 10-9). The effects of hydrogen bonding are particularly pronounced for water since each water molecule can be involved in four hydrogen bonds. (See ΔH_v for NH_3 and HF, Table 10-10 and 10-13.) The crystal structure of ice shows each oxygen surrounded tetrahedrally by four other oxygens to which it is hydrogen bonded. The structure of NH_4F is essentially the same as that of ice. Not as much hydrogen bonding is possible in either pure NH_3 or pure HF. Ammonia suffers from a deficiency of electron pairs so that all of the hydrogens in the molecule cannot engage in hydrogen bonding. HF has a sufficient number of electrons, but in this molecule there is a shortage of protons. As a result, water boils and freezes higher than either ammonia or hydrogen fluoride. The reason for the long liquid range of water and other hydrogen bonded materials is that only about 14% of the hydrogen bonds are broken when a sufficient amount of energy is added to melt the solid. (The enthalpy of fusion of ice is large, 1.44 kcal mole^{-1}.) More hydrogen bonds are broken as the temperature is increased (that is, the heat capacity is large) but over 75% of the hydrogen bonds are still present in the liquid at 100°C. The ordering present in liquid water is reflected in the entropy of vaporization which is greater than the usual 20–21 eu. In addition the solution of a substance in water may result in a considerable entropy change. The consequences of this effect will be an important aspect of the remainder of this book.

The fact that hydrogen bonds are intermediate in energy between chemical bonds and van der Waal attractions has some very important consequences. It means that only a small amount of energy is required to shift an equilibrium between entirely different structures if the structures are maintained by hydrogen bonds. Such changes are involved in equilibria between different structures of protein molecules. In a similar way the two strands of deoxyribonucleic acid (DNA, the basic genetic material of cells) are held together by hydrogen bonds.

QUESTIONS

1. Define or state briefly the meaning of the following terms: isotope, saline hydride, metallic hydride, volatile hydride, half-life, radioactive.
2. Why do isotopes undergo the same type of reactions?
3. Suggest a reason for the differences in rate observed for reactions of the isotopes of hydrogen.
4. Would the difference in rate be as great for the isotopes of iodine?
5. Classify the hydrides of the following elements according to the terms saltlike, metallic, or volatile: Cs, Pb, Po, Fe, Nb, Se, I.
6. Suggest a reason for the difference in reaction temperature needed to prepare the hydrides of the groups IA and IIA metals.
7. How does the electron affinity of an atom affect the stability of hydrides and fluorides?

8. Discuss the relative stability of the volatile hydrides in terms of Table 10-7.
9. Discuss the preparation of the volatile hydrides in terms of the data in Table 10-7.
10. What is the meaning of the term dielectric constant?
11. From the following list of hydrides select two which fit the specification given in each part below: RbH, BaH$_2$, CuH, Cr$_7$H$_2$, HCl, CH$_4$, NH$_3$, B$_2$H$_6$, H$_2$Se
    (a) saline hydrides
    (b) volatile
    (c) H^0 is positive
    (d) metallic hydrides
    (e) have low ΔH_v
    (f) hydrogen bonded in the liquid
    (g) react vigorously with water
12. (a) Explain why water has an entropy of vaporization larger than the value (21 eu) given by Trouton's rule. (b) Name another hydride that will have an entropy of vaporization larger than 21eu.

PROBLEMS I

1. From the data in Table 10-1 calculate the equilibrium constant for the reaction H$_2$O + HDS $\longrightarrow$ HDO + H$_2$S.
2. Calculate the potential of the H$_2$ | H$^-$ couple at pH 5, 3, and 1.
3. From the bond energy data in Table 10-4 and the H^* of atoms from Chapter 4, calculate the enthalpy of formation of the hydrides of the group VA elements.
4. Using the appropriate data, calculate ΔS_v and ΔS_f of the hydrides of group VIIA. Compare these values with those calculated for group VIA and IVA elements.
5. From the data in Table 10-3 calculate the H-S bond energy. How does this compare to the value predicted by Pauling's rule? $\alpha_{S-S} = 72$ kcal mole^{-1}.
6. From the data in Table 10-4 calculate H^0 for SbH$_3$(g) and for NaH(g). What is the heat of vaporization for NaH? How does this compare to ΔH_v for water?

PROBLEMS II

1. Explain briefly the following sequence of heats of vaporization.

Compound	ΔH_v, kcal mole^{-1}
SiH$_4$	2.9
PH$_3$	3.49
H$_2$S	4.46
HCl	3.86

2. For the reaction SiH$_4$(l) $\longrightarrow$ SiH$_4$(g), $\Delta S = +17.9$ eu. Calculate K for this reaction.

3. Calculate K for the vaporization of SiH_4 at $-111°C$ (162 K).
4. H_2S boils at $-60°C$. Calculate ΔG^0 for the reaction $H_2S(l) \longrightarrow H_2S(g)$ at $-60°C$.
5. Calculate ΔS_v for H_2S.
6. Why is ΔS_v for H_2S larger than ΔS_v for SiH_4?
7. At what temperature will the vapor pressure of $H_2S(l)$ be 0.10 atm?
8. The bond dissociation energy of $KH(g)$ is 44 kcal mole. Calculate H^0 for gaseous potassium hydride.
9. H^0 for solid potassium hydride (KH) is -14 kcal mole^{-1}. Calculate ΔH for the reaction $KH(s) \longrightarrow KH(g)$.
10. Assuming that ΔS^0 for the reaction in Problem 9 is $+25$ eu, at what temperature would KH boil?
11. Why is KH so much less volatile than the materials listed in Problem 1?
12. ΔS^0 for the reaction $2KH(s) \longrightarrow 2K(s) + H_2(g)$ is approximately $+38$ eu. At what temperature will solid KH decompose to give H_2 at 1 atm pressure?
13. Why can't liquid KH be prepared?

SUGGESTED READINGS

HOLIDAY, A. K., and A. G. MASSEY, *Inorganic Chemistry in Non-aqueous Solvents*. Oxford; Pergamon Press, 1965.

HURD, D. T., *Chemistry of the Hydrides*. New York: J. Wiley & Sons, Inc. 1952.

MORTIMER, C. E. *Chemistry, A Conceptual Approach*. New York: Reinhold Publishers, 1967, Chapters 10 and 13.

SISLER, H., C. A. VAN DER WERF, and A. W. DAVIDSON, *College Chemistry*. New York: Macmillan, 1967, Chapter 13.

... and this diversity must be attributed to the keener or blunter edges of the different sorts of acids; and so likewise this difference of the points in subtilty is the cause that one acid can penetrate and dissolve well one sort of mixt, that another can't rarifie at all: Thus Vinegar dissolves Lead, which aqua fortis can't: Aqua fortis dissolves Quick-silver, which Vinegar will not touch; Aqua Regalis dissolves Gold, whenas Aqua fortis cannot meddle with it; on the contrary Aqua fortis dissolves Silver, but can do nothing with Gold, and so of the rest.

As for Alkali's, they are soon known by pouring an acid upon them, for presently or soon after, there rises a violent Ebullition, which remains until the acid finds no more bodies to rarifie.*

NICOLAS LEMERY
A Course of Chymistry London, 1686

Chapter 11 Acids and Bases

According to Brønsted and Lowry an acid is a proton donor and a base is a proton acceptor. Therefore, any chemical substance that contains a proton can behave as an acid under the appropriate conditions, and any chemical substance that contains an unshared pair of electrons can function as a base. Substances that contain both an electron pair and a proton can theoretically exhibit either behavior. The potential acid and base behavior of water and ammonia may be illustrated by the equations

$$H_2O + H_2O = H_3O^+ + OH^-$$

$$NH_3 + NH_3 = NH_4^+ + NH_2^-$$

An even more versatile definition of an acid may be obtained if we simply change the emphasis and state that acids are electron-pair acceptors and bases are electron-pair donors. Thus H^+ and BF_3 are acids because they can accept a share in a pair of electrons. The acid behavior in BF_3 can be represented by the equation

$$BF_3 + F^- \longrightarrow BF_4^-$$

$$B\overset{..}{\underset{F}{\vphantom{|}}}, \overset{..}{\underset{F}{\vphantom{|}}}\overset{..}{\underset{F}{\vphantom{|}}} \quad \overset{..}{\underset{}{\vphantom{|}}}^- \qquad B\overset{..}{\underset{F}{\vphantom{|}}}, \overset{..}{\underset{F}{\vphantom{|}}}\overset{..}{\underset{F}{\vphantom{|}}}\overset{..}{\underset{F}{\vphantom{|}}}^-$$

*By permission from H. M. Leicester and H. S. Klickstein, *A Source Book in Chemistry 1400-1900*, (Cambridge, Mass.: Harvard University Press, 1952).

Ch. 11 Acids and Bases

This more general definition was proposed by G. N. Lewis, the originator of the electron-pair bond concept. Although the two definitions overlap to a large degree, each has its own area in which the other definition is inappropriate or cumbersome. In some instances it is necessary to consider both definitions for even a single molecule.

Because almost all substances are inherently acidic or basic, acidity ranks with oxidation-reduction capacity in importance, and a great many chemical reactions can be considered to be acid-base reactions.

I. The Relative Strength of Acids

A. CONJUGATE ACID-BASE PAIRS

When water behaves as a base the hydronium ion (H_3O^+) formed is called the conjugate acid of water.

$$H_2O + HCl = H_3O^+ + Cl^-$$

When water behaves as an acid

$$H_2O + NH_2^- = NH_3 + OH^-$$

the hydroxide ion formed is the conjugate base of water. In the reactions cited above the other conjugate acid-base pairs are hydrochloric acid (HCl) with the chloride ion (Cl^-), and the amide ion (NH_2^-) with ammonia (NH_3). Hydrochloric acid is the conjugate acid of chloride ion and ammonia is the conjugate acid of amide ion.

The conjugate acid-base concept is useful in rationalizing acid or base strength. For example, compare the two substances water and ammonia. Ammonia is a stronger base than water; therefore, the conjugate acid of ammonia (NH_4^+, ammonium ion) must be a weaker acid than the conjugate acid of water (H_3O^+). Also ammonia is a weaker acid than water; therefore, the conjugate base of ammonia (NH_2^-) must be a stronger base than OH^-. In another example acetic acid (CH_3COOH) is a stronger acid than water; therefore, the acetate ion (CH_3COO^-) must be a weaker base than the hydroxide ion (OH^-). The value of this concept lies in the fact that once some knowledge of the relative strength of acids and bases is obtained, it is possible to extend that knowledge through the concept of conjugate acids and bases.

It is impossible to measure the inherent strength of a Brønsted acid because the proton cannot exist free in solution. Therefore, the strength of an acid is a relative concept in that one acid is compared with another by reacting each with the same base. The relative strength

The structural formula of acetic acid is

```
      H   O
      |   ||
   H—C—C—O—H
      |
      H
```

of many acids may be easily determined in water solution. This subject will be discussed in detail in Section II, Acid-Base Equilibria.

In this section we wish to examine the strength of acids from the viewpoint of electronic structure. To do this it is convenient to divide the acids into two groups, the binary hydrogen compounds (for example H_2O, HF, HCl, and so forth) and the oxyacids ($HClO_4$, and so forth). All of the metal ions (Al^{3+} or $Al(H_2O)_6^{3+}$) are included in the latter category. Each of these groups will be discussed in turn, and the correlation between acid strength and position in the periodic table will be emphasized.

B. SIMPLE HYDRIDES

From the Lewis definition of a base, we expect to find a correlation between acid-base strength and electronegativity. The anion of a very electronegative atom, such as fluorine, should be a poor base, and the conjugate acid (HF) should be a stronger acid than water or ammonia. The correlation between acid strength and position in the periodic table is a reflection of the relationship between acid strength and electronegativity. The electronegativity of the elements of the first row of the periodic table increases toward the right, and the acidity of the hydrides of these elements also increases toward the right. The statement, the greater the electronegativity of the element the greater the acidity of the hydride of that element, holds in every period. This holds true because the sizes of the elements in a period are not widely different.

The trend down a group, such as the hydrogen halides, represents quite a different situation. Although iodine is less electronegative than fluorine and hydriodic acid (HI) is expected to be a weaker acid than hydrofluoric acid (HF), the exact reverse is found to be the case. This trend may be explained by examining the ionization process

$$HX(aq) = H^+(aq) + X^-(aq)$$

in terms of the Born-Haber cycle data given in Table 11-1. From these

Table 11-1 *Free Energy Changes for the Dissociation of Hydrogen Halides in Aqueous Solution (kcal mole^{-1})*

	HF	HCl	HBr	HI
$HX(aq) = HX(g)$	5.7	-1	-1	-1
$HX(g) = H(g) + X(g)$	127.8	96.5	81.0	65.0
$H(g) = H^+(g) + e^-$	315.3	315.3	315.3	315.3
$X(g) + e^- = X^-(g)$	-83.0	-87.6	-82.5	-75.3
$H^+(g) + X^-(g) = H^+(aq) + X^-(aq)$	-361.5	-332.8	-325.7	-317.7
$HX(aq) = H^+(aq) + X^-(aq)$	4.3	-10	-13	-14
pK_a	3.3	-7	-9.5	-10

data we see that all of the free energy changes are nearly identical except for the bond dissociation and hydration reaction. The dominant factor in determining the strength of a hydrogen halide acid is the bond dissociation energy. This is the only term that is substantially more positive for hydrofluoric acid than for hydriodic acid. It is different for two reasons. First, hydrofluoric acid is more ionic than hydriodic acid; secondly, fluoride ion is much smaller than iodide ion. Therefore, the charge concentration is much larger and as a result the electrostatic attraction for the proton is much greater. The net result is that hydriodic acid is a stronger acid than hydrofluoric because the bond in hydrofluoric acid is nearly twice as strong as the bond in hydriodic acid.

Electronegativity may be used to predict the relative acidity of binary hydrogen compounds if the sizes of the anions are similar, but reverse trends are observed whenever size becomes a factor. When this is the case the acidity varies with size. This same trend is noted in all compounds of hydrogen with nonmetals.

C. THE OXYACIDS

The largest group of related acids is made up of the hydrates of metal ions and the oxyacids of the elements at the extreme right of the periodic table. The ionization of a hydrated cation may be represented by the equation

$$M(H_2O)_n^{x+} + H_2O = M(H_2O)_{(n-1)}(OH)^{(x-1)+} + H_3O^+$$

Equilibrium constants cover a wide range of values and it is sometimes desirable to list the negative logarithm of the value, pK_a, instead of K_a.

From a consideration of size and charge only, the acidity of cations should decrease downward in the main groups and increase across a period. Table 11-2 lists the pK_a values for some hydrated cations. The observed trends are generally as expected; however, some exceptions are obvious. The pK_a for $Be(H_2O)_4^{2+}$ seems too low, and Fe(III) has a larger pK_a than Al(III). In this connection it should be noted that the pK_a values listed in Table 11-2 refer to the monomolecular species, and this species only exists at high dilution. The main hydrolysis product for Fe^{3+} is $Fe_2(OH)_2^{4+}$, which is diamagnetic rather than paramagnetic. Both Fe^{3+} and $Fe(OH)^{2+}$ are paramagnetic. The tendency toward the formation of polymeric species and the spin states of the ion will affect the pK_a data.

Al^{3+} is smaller than Fe^{3+}, yet Fe^{3+} is the stronger acid.

This distinction is experimental. In a diamagnetic substance there are no unpaired electrons. Paramagnetic substances have unpaired electrons. The electronic structure of Fe(III) may be either

Fe(III) ⟂ ⟂ ⟂ ⟂ ⟂, —, — — —

or

Fe(III) ⟂⟂ ⟂⟂ ⟂ — —, —, — — —

The spin state refers to the electronic structure of the metal, that is, diamagnetic or paramagnetic. In order for a substance containing Fe(III) to be diamagnetic there must be an interaction between the iron atoms. A single Fe(III) atom will have either five or one unpaired electrons.

A measure of the acidity of a cation is also available from the enthalpy of hydration or the free energy of hydration. The enthalpy of hydration data for some cations of interest are given in Table 11-3. The free energy of hydration of a number of cations is plotted against ionic radius in Fig. 11-1. Examination of these data shows that the free energy and enthalpy changes follow the expected trends in most

Table 11-2 pK_a for Cations

Ch. 11 Acids and Bases

Table 11-3 *Enthalpy of Hydration of Ions*

$-\Delta H$ hydration, kcal mole^{-1}

Li$^+$	229.24	Be^{2+}	806		
Na$^+$	201.56	Mg^{2+}	668.57	Al^{3+}	1333
K$^+$	181.68	Ca^{2+}	590.41	Sc^{3+}	1259.0
Rb$^+$	175.7	Sr^{2+}	555.12	Y^{3+}	1170
Cs$^+$	167.8	Ba^{2+}	521.43	La^{3+}	1139.6
Ag$^+$	217.43	Sn^{2+}	580.8	Ga^{3+}	1433.7
Tl$^+$	183.1	Pb^{2+}	564.5	In^{3+}	1294
		Cr^{2+}	665.3	Tl^{3+}	1293.2
		Mn^{2+}	651.90	Cr^{3+}	1370
		Fe^{2+}	676.1	Fe^{3+}	1369.4
		Co^{2+}	689.8		
		Ni^{2+}	710.2		
		Cu^{2+}	713.73		
		Zn^{2+}	698.70		
		Cd^{2+}	642.08		
		Hg^{2+}	646.8		

cases. The ions with a single charge have the smallest hydration enthalpies and free energies. The values for the 2+ ions are larger, and 3+ ions have the largest values of the cations shown. In Fig. 11-1 notice that the 2+ and 3+ transition metal ions have values that are larger than their radius would indicate. Some of the same irregularities noted above are apparent in these data.

In spite of the irregularities noted, there are a number of properties that correlate well with these data. For example, there is a good correlation with amphoteric character of the oxides. It will be noted that, for the most part, the amphoteric cations exhibit the largest values of the free energies of hydration given in the figure. There are several exceptions to this, the most obvious one being Pb(II). In addition, those cations that form complex ammines are grouped together in the center region section A of Fig. 11-1.

A substance that can act as both an acid and a base is called amphoteric.

PbO is amphoteric but the free energy of hydration of Pb^{2+} is much smaller than that of other 2+ ions with amphoteric oxides.

With regard to the exceptions that have been noted above, it is obvious that factors such as coordination number (Be and Zn coordinate four water molecules, while most of the other ions coordinate six) and structure (again Be and Zn form tetrahedral hydrates, coordination number of four, whereas most of the others are octahedral) are likely to be important in addition to the tendency toward polymerization. We have considered only the combination of size and charge to correlate the data, and it does a fair job.

Sec. I The Relative Strength of Acids

There is a striking regularity to the acidity of the remaining oxyacids that may be correlated nicely with position in the periodic table. The relationship may be illustrated by considering either the oxidation state of the central ion or the stoichiometry of the compound. In effect, these two points of view mean the same thing, as we saw in Chapter 3, because stoichiometry is often determined by oxidation state. In terms of oxidation state, an acid such as perchloric ($HClO_4$) may be consid-

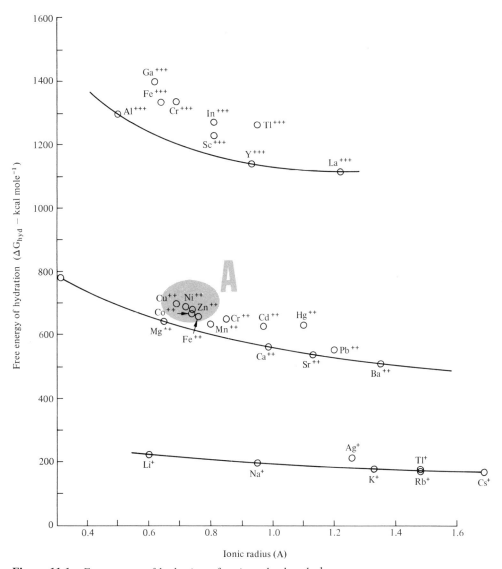

Figure 11-1 *Free energy of hydration of cations, kcal mole^{-1}.*

ered to contain Cl^{7+}. This cation would be such an extremely strong acid that the hydrate $Cl(H_2O)_4^{7+}$ does not exist because it loses a sufficient number of protons to produce the perchlorate ion (ClO_4^-) or perchloric acid ($HClO_4$). Taking this approach to arrive at a sequence of acid strength, we simply compare the oxidation state of the hypothetical cations of which these acids are composed. Thus sulfuric acid (H_2SO_4) is a stronger acid than sulfurous acid (H_2SO_3). In the sequence phosphoric acid (H_3PO_4), sulfuric acid (H_2SO_4), and perchloric acid ($HClO_4$), we compare P^{5+}, S^{6+}, and Cl^{7+} to arrive at the relative acid strength of $H_3PO_4 < H_2SO_4 < HClO_4$. In the same way perchloric acid is stronger than chloric acid ($HClO_3$); chloric acid is stronger than hypochloric acid ($HClO_2$); and finally, hypochloric acid is stronger than hypochlorous acid ($HClO$). From these considerations we expect the strongest acid to contain the cation with the highest oxidation state and smallest size. For compounds with the same stoichiometry, acid strength will then decrease toward the left of the periodic table and down a group.

An alternative approach to the strength of these acids involves a consideration of the electron density on the anion. The electron density on the oxygen atoms in perchlorate ion (ClO_4^-) must be lower than the electron density on the oxygen atoms in sulfate ion (SO_4^{2-}) because the total charge on the ion is smaller, and in both cases it is spread over four oxygen atoms. Therefore,

$$\begin{array}{c} \text{O} \\ | \\ \text{O}-\text{S}-\text{O}^{2-} \\ | \\ \text{O} \end{array} \text{ is more basic than } \begin{array}{c} \text{O} \\ | \\ \text{O}-\text{Cl}-\text{O}^- \\ | \\ \text{O} \end{array}$$

In the same way sulfuric acid is a stronger acid than phosphoric acid, and sulfuric acid is stronger than sulfurous acid (H_2SO_3). pK_a values for a number of oxyacids are given in Table 11-4. Those that have formulas H_nXO_4 all follow the predicted trends as do those that have the formula H_nXO_3.

This simple introduction to acids and bases is meant to give some feel for the concept of acid-base strength. The acid-base reaction involves sharing an electron pair. From this point of view a vast number of reactions may be generalized in the acid-base terminology. It is, therefore, necessary to inquire much more fully into the nature of acids and bases and expand the simple ideas presented here.

In terms of the molecular orbital terminology used in Chapter 7, an acid is a substance that contains low lying empty molecular orbitals. The lower the energy of these empty orbitals, the stronger the acid. Any factor that tends to lower the energy of the molecular orbital promotes acid strength. On the other hand, any substance that has

Table 11-4 pK_a's for Some Inorganic Oxyacids in Aqueous Solution

Acid	pK_a	Acid	pK_a	Acid	pK_a	Acid	pK_a
HClO	7.2	HNO_2	3.3	HNO_3	−1.4	$HClO_4$	(−10)
HBrO	8.7	$HClO_2$	2.0	$HClO_3$	−1		
HIO	10.0	H_2CO_3	3.9				
H_3BO_3	9.2	H_2SO_3	1.9	H_2SO_4	(−3)		
H_3AsO_3	9.2	H_2SeO_3	2.6				
H_3SbO_3	11.0	H_2TeO_3	2.7				
H_4SiO_4	10.0	H_3PO_4	2.1				
H_4GeO_4	8.6	H_3AsO_4	2.3				
H_6TeO_6	8.8	H_5IO_6	1.6				

electrons in a high-energy nonbonding level or antibonding level will be a base. The factors that lead to higher orbital energies will increase the basic nature of a substance.

We have continually repeated the idea that electronegativity is one approach to orbital energies. We have also seen that the usefulness of electronegativity has its limits in certain cases because of the electrostatic factors that may become important. The bond energies of covalent molecules can be represented with good accuracy in terms of electronegativity differences; however, as the degree of ionic character of a bond increases the deviations from the covalent model are the result of the increasing importance of the electrostatic factors. No simple and completely satisfactory method that allows a simple picture of real orbital energies in those cases where the deviations become large has been devised. There are obviously two starting points for the treatment of problems where neither the covalent nor ionic approach alone is satisfactory. One method might begin with the covalent model and add a term that includes electrostatic effects (this was the method used in the discussion of bond energies), or a method that begins with the ionic model and adds a term that includes increasing covalency might be used.

The second method has received considerable attention in the last few years. In this approach the ions and molecules are discussed under the general headings of hard and soft acids and bases. The concept is based upon the experimental observations that certain cations (the hard acids) form the most stable compounds with the lightest members of a group (that is, a hard base such as fluoride ion rather than iodide ion), whereas the soft acids form the most stable compounds with the heavier elements of a group (that is, a soft base such as iodide ion rather than fluoride ion).

The concept of hard and soft acids and bases has been discussed from a number of different theoretical approaches, none of which has proved to be entirely satisfactory. The objective of such studies has been to obtain an ordering of cations in terms of increasing softness.

This approach and its application to acidity, solubility, and coordination tendencies will be examined in detail in Chapters 13, 14, and 15. In the next section we shall expand the earlier treatment of estimating orbital energies using electronegativities.

D. MOLECULAR ORBITALS AND THE ACIDITY OF SIMPLE HYDROGEN COMPOUNDS

The striking trend in the acidity of the hydrides of the first row elements is easily understood from simple electronegativity considerations. In previous chapters we used the electronegativity of an atom to estimate the energy of an orbital from a qualitative standpoint. A quantitative estimate of the energy of an orbital is prerequisite to a more complete discussion. In nearly all covalent molecules electronegativities provide an easy means of estimating these energies. Ionization energies alone overestimate the attractive power of atoms for electrons to a considerable degree because the electrostatic effects of adding electrons to an atom are, of course, reflected in the magnitude of the potential. Thus the ionization energy of chlorine is 300 kcal mole^{-1}, but the electron affinity (see Table 11-5) is only -86 kcal mole^{-1}.

A measure of the energy of an orbital was suggested by Mulliken as an alternative measure of electronegativity. Mulliken's electronega-

Table 11-5 *Electron Affinities*

Atom	Electron affinity	
	ev	kcal mole^{-1}*
H	-0.747	-17.2
F	-3.56	-82.1
Cl	-3.75	-86.4
Br	-3.52	-81.2
I	-3.08	-71.0
O	-1.42	-32.8
N	-0.1	-2
S	-2.07	-47.7
Li	-0.54	-12
Na	-0.74	-17
Be	~ -0.6	-14
Mg	~ -0.3	-7

*The values given to tenths of a kcal mole^{-1} in Table 11-5 are all from "Technical Note" 270-3.

Sec. I The Relative Strength of Acids

Table 11-6 *Electronegativity of the Halogens*

Atom	Mulliken values		Pauling value, Pauling scale
	Kcal mole^{-1}	Pauling scale	
F	242.0	3.78	4.0
Cl	193.3	3.03	3.0
Br	177.2	2.77	2.8
I	156.1	2.44	2.5

If a SCSE correction is made to the ionization potentials, even closer agreement to the Pauling values can be obtained. This gives the values F, 3.92; Cl, 3.08; Br, 2.82; I, 2.47.

tivity scale is based on the average of the negative of the ionization energy and the electron affinity.

$$\text{Orbital energy} = \frac{-\text{IP} + \text{EA}}{2}$$

For chlorine this amounts to -193 kcal mole^{-1}. Ionization energies were given in Table 3-2. In Table 11-6 the Mulliken electronegativities converted to the Pauling scale are given.

The energy of the atomic orbitals of hydrogen and chlorine obtained from Mulliken electronegativities are shown in Fig. 11-2 as levels A and B. These values may also be obtained by multiplying the Pauling values by a factor (-64) which converts the values of the Pauling scale to the Mulliken scale in kilocalories per mole. The bonding molecular orbital in hydrogen chloride (HCl) is shown as level C. The difference between C and B is given as Y. The difference between A and B is given as X.

The total bond energy measured for hydrogen chloride is 103 kcal mole^{-1}. This does not represent the total attractive energy between hydrogen and chlorine, however, because the attractive force is partially balanced by a repulsive force between the atoms. In general, at the equilibrium internuclear distance, the net repulsive energy is about two thirds the total attractive energy. With a bond energy of 103 kcal, the total attractive energy for hydrogen chloride is $3 \times 103 = 309$ kcal. This is the energy required for heterolytic cleavage of HCl

$$\text{HCl(g)} = \text{H(g)} + \text{Cl(g)}$$

To affect heterolytic cleavage, one electron must be raised from the bonding level C to A and one must be raised to B. The energy required to cleave the hydrogen chloride bond heterolytically is thus $2Y + X = 309$ kcal, where X is the difference between the energy of the chlorine atomic orbital and the atomic orbital of hydrogen. Since $X = 59$, Y may be obtained as 125; so C then lies at -318 kcal,

[(−193) + (−125)]. The molecular orbital of hydrogen chloride lies 155 kcal below the average energy of the hydrogen and chlorine atomic orbitals. The energy of the antibonding level (C*) is 155 kcal above this energy center at −8 kcal. The relative energies of the levels in other HX molecules of interest are given in Fig. 11-3.

From the diagram in Fig. 11-2 it would appear that it might take less energy to promote the homolytic dissociation reaction

$$HCl(g) = H^+(g) + Cl^-(g)$$

than the heterolytic dissociation reaction

$$HCl(g) = H(g) + Cl(g)$$

This is not the case in the gas phase where the heterolytic dissociation reaction is favored.

The diagram is, of course, not applicable to the creation of ions. The orbital energies are those of the atomic orbitals, not those of the

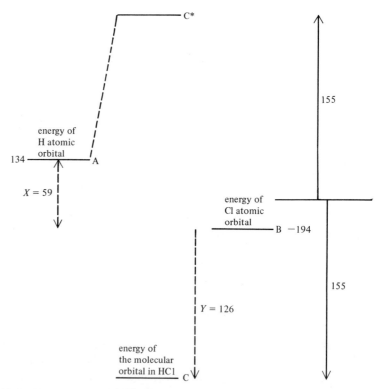

Figure 11-2 *Molecular orbital diagram for HCl.*

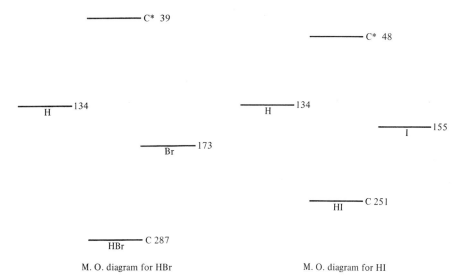

Figure 11.3 *Molecular orbital diagrams for* HBr *and* HI.

ions. Considerably more energy is required to make Cl⁻ ions and H⁺ ions. This energy can be estimated from the processes

$$HCl(g) = H(g) + Cl(g) \qquad \Delta H = 103$$
$$H(g) = H^+(g) + e^- \qquad \Delta H = 313$$
$$\underline{Cl(g) + e^- = Cl^-(g) \qquad \Delta H = -86}$$
$$HCl(g) = H^+(g) + Cl^-(g) \qquad \Delta H = 330 \text{ kcal}$$

Only at the expense of large amounts of energy could such a process take place. The energy supplied when the ions are hydrated is sufficiently large so that the homolytic dissociation (ionization) is favored in water solution.

II. Acid-Base Equilibria

The reaction of an acid to produce its conjugate base and a proton is called dissociation or ionization, and the degree to which dissociation

$$BH = B^- + H^+$$

or ionization takes place depends upon the strength of the acid.

Another way of thinking of this is to consider the strength of the conjugate base B⁻. The stronger the conjugate base (as a base) the weaker the acid. An acid will not undergo dissociation or will dissociate only very slightly unless a base is available that is stronger than its conjugate base. For example, hydrochloric acid is a fairly strong acid that dissociates completely into H⁺ and Cl⁻ in water. In this instance the solvent water acts as a base

$$HCl + H_2O = H_3O^+ + Cl^-$$

The reaction takes place to a large extent because H_2O is a stronger base than Cl^-. As a result of the reaction a weaker acid than HCl (that is, H_3O^+) is produced. Reactions of this type are called neutralization reactions. In a neutralization reaction the stronger acid and base are on the same side of the equation.

In contrast to its behavior in water, hydrochloric acid is only very slightly dissociated in the solvent glacial acetic acid; probably it is not dissociated at all in sulfuric acid (H_2SO_4). Neither acetic acid nor sulfuric acid is a very strong base; in fact, these substances are weaker bases than Cl^-. The dissociation of hydrochloric acid in glacial acetic acid (HAc) may be represented by the equation

Glacial acetic acid is

$$CH_3-\overset{\overset{\displaystyle O}{\|}}{C}-OH \text{ or HAc.}$$

$$HCl + HAc \rightleftharpoons H_2Ac^+ + Cl^-$$

In this reaction the lengths of the arrows indicate roughly the extent to which the reaction takes place. The neutralization reaction in this case is the backward reaction. H_2Ac^+ is a stronger acid than HCl and Cl^- is a stronger base than HAc.

As indicated above, although a material may be inherently an acid, it can function as an acid only when in contact with a base. Acid-base reactions are fundamentally competition reactions for a proton. The base that wins the competition is the one that produces the weakest acid (has the weakest conjugate acid) and the more neutral solution. In all cases this is the reaction that occurs to the largest degree.

The strong acids are those whose conjugate base is very weak compared to water. Some examples are $HClO_4$, H_2SO_4, HNO_3, and HCl. All of these substances react with water to produce H_3O^+ according to the equation

$$HA + H_2O \rightleftharpoons H_3O^+ + A^-$$

The weak acids have a conjugate base that is a stronger base than water. These substances react with water according to the equation

$$HA + H_2O \rightleftharpoons H_3O^+ + A^-$$

In the dissociation of a strong acid a weaker acid is produced, but in the dissociation of a weak acid a stronger acid and stronger base are produced. Thus the first reaction proceeds to completion while the second takes place to only a slight degree. Another way of stating the difference between strong and weak acids is by noting the direction in which the reaction tends to proceed. The backward reaction for the ionization of hydrochloric acid in water may be neglected, but the backward reaction for acetic acid in water may not.

As pointed out above, acidity and basicity are inherent properties which are manifested only under the proper conditions. For the most part the conditions used in previous examples have been water solutions. A change in the conditions can lead to some very interesting consequences for some materials. HCl could behave as either an acid or a base if the conditions were right. In order to observe basic behavior in HCl, an acid must be present that has a conjugate base that is weaker than HCl; however, it is doubtful that a chemical substance exists other than gaseous H^+ that can put another proton on HCl. $HClO_4$ is the strongest uncharged protonic acid known, and it is not strong enough to form H_2Cl^+ from HCl. In $HClO_4$ solution, however, HCl is a very weak acid and is undissociated.

For weaker acids both acidic and basic properties are readily observable. In pure acetic acid (acetic acid is an acid in water) the ion equivalent to the hydronium ion is H_2Ac^+, and the very strong acids such as $HClO_4$ will dissociate according to the equation

$$HClO_4 + HAc \rightleftharpoons H_2Ac^+ + ClO_4^-$$

In this reaction acetic acid is behaving as a base. Basicity of certain other acids is also observable from self-dissociation reactions. For example, H_2SO_4 undergoes self-dissociation to a considerable degree.

$$H_2SO_4 + H_2SO_4 \rightleftharpoons H_3SO_4^+ + HSO_4^-$$

It is important to realize that water is not the only substance that exhibits both acidic and basic properties; many other substances have the potential to do so and will if the conditions are appropriate. Amphoterism then, a term often applied to the acid-base behavior of water and hydroxides, must be extended to include many other substances.

In this section we are particularly concerned with acid-base equilibria. Other types of equilibria will be discussed in subsequent chapters.

Equilibrium constant expressions may be easily written for any acid-base reaction. For example, the reaction between acetic acid and the base water may be represented by

$$HAc + H_2O \rightleftharpoons H_3O^+ + Ac^-$$

and the equilibrium constant expression may be written

$$K_a = \frac{(a_{H_3O^+})(a_{Ac^-})}{(a_{HAc})(a_{H_2O})}$$

Since we are using mole fraction to approximate the activity of the solvent for dilute solutions, a_{H_2O} is essentially constant. As a result the equilibrium constant expression may be written

$$K_a = \frac{[H_3O^+][Ac^-]}{[HAc]}$$

The reaction between an acid such as hydrochloric and the base water

$$HCl + H_2O \longrightarrow H_3O^+ + Cl^-$$

goes essentially to completion. In this case the minimum free energy of the system corresponds to very nearly 100% products and the equilibrium constant

$$K_a = \frac{[H_3O^+][Cl^-]}{[HCl]}$$

is very large.

A. THE RELATIVE STRENGTH OF ACIDS REVISITED

From the above discussion it is obvious that the value of pK_a represents the strength of acids. The pK_a values listed in Table 11-4 are relative strengths in that they are measured relative to the base water. Of course, the values would be different in a different solvent. Using pK_a data, a scale of acidity such as that shown in Table 11-7 can be established. Substances that are low on this scale will be weak acids while those that are high will be strong acids. By using the conjugate acid-base pair concept a scale of the relative strength of bases can be established. This is also given in Table 11-7.

B. EQUILIBRIUM CALCULATIONS

Some acid-base equilibrium calculations can become quite complicated, and it is often necessary to simplify the calculations involved.

Table 11-7 *A Scale of Acid and Base Strength*

Acid formula	Base formula	Acid formula	Base formula
$HClO_4$	ClO_4^-	HNO_3	NO_2^-
H_2SO_4	HSO_4^-	H_3BO_3	$H_2BO_3^-$
↑HCl	Cl$^-$	↑HClO	ClO$^-$
HNO_3	NO_3^-	H_2S	HS^-
$HClO_3$	ClO_3^-	NH_4^+	NH_3
H_3O^+	H_2O	HCN	CN^-
HIO_3	IO_3^-	HCO_3^-	CO_3^{2-}
H_5IO_6	↓$H_4IO_6^-$	C_6H_5OH	↓$C_6H_5O^-$
H_2SO_3	HSO_3^-	H_2O	OH^-
$HClO_2$	ClO_2^-	C_2H_5OH	$C_2H_5O^-$
H_3PO_4	$H_2PO_4^-$	NH_3	NH_2^-
H_3AsO_4	$H_2AsO_4^-$	CH_3NH_2	CH_3NH^-
H_2SeO_3	$HSeO_3^-$	H_2	H^-
H_2TeO_3	$HTeO_3^-$	CH_4	CH_3^-

(Left pair: increasing acidity / increasing basicity. Right pair: increasing acidity / increasing basicity.)

This is usually not a serious problem because the ionization constants of the weak monoprotic acids (an acid containing only one ionizable hydrogen) generally vary in accuracy between ±0.5 and ±10%. The values available for polyprotic acids, such as hydrogen sulfide (H_2S) and phosphoric acid (H_3PO_4), may have much larger limits of error.

One of the most precisely determined equilibrium constants is the dissociation constant of water

$$2H_2O \longrightarrow H_3O^+ + OH^-$$

$$K_w = (a_{H_3O^+})(a_{OH^-})$$

The value of this constant at 25°C is $(1.008 \pm 0.001) \times 10^{-14}$.

There are a large number of types of calculations that may be performed using acid-base equilibrium data. If the concentration of all of the species in solution is specified, the equilibrium constant may be calculated. If the equilibrium constant is known, the concentration of one or more of the species in solution may be calculated. In the case of a weak acid where ionization is slight, the usual practice is to assume that the concentration of weak acid in solution is not altered by ionization. The validity of such an assumption can be readily checked. Another possibility in equilibrium calculations is the presence of a substance that produces one of the ions that the acid produces (the common ion effect). This can be either a strong acid that supplies hydrogen ions or a salt that supplies the anion. If a substance contains more than one ionizable hydrogen, it is called a polyprotic acid, and equilibrium constant expressions may be written for the ionization of each hydrogen. All of the above types of calculations are applicable in these cases, but some care must be exercised to use the proper equilibrium constant.

Some sample problems that illustrate the types of calculations that become important are in order here.

Example 1. Calculate the concentration of all species in a 0.010 molar (M) solution of acetic acid (HAc). The ionization constant (K_a) of acetic acid is 1.75×10^{-5}.

To begin, we write the equation that represents ionization and the equilibrium expression

$$HAc + H_2O \longrightarrow H_3O^+ + Ac^-$$

$$K_a = \frac{[H_3O^+][Ac^-]}{[HAc]}$$

For every mole of H_3O^+ formed 1 mole of Ac^- will be formed and 1 mole of HAc will be used up. Thus $[H_3O^+] = [Ac^-]$ and $[HAc] = 0.010 - [H_3O^+]$. The total concentration $[HAc] + [H_3O^+]$ must equal 0.010 mole liter^{-1}. Because of the uncertainty in the equilibrium constant data and because HAc is a weak acid, it is reasonable to assume that the amount of HAc that dissociates is negligible. We then have $[HAc] = 0.010$. (As the acid becomes stronger the error that results from the assumption may be larger.)

Substituting into the K_a expression

$$K_a = \frac{[H_3O^+]^2}{[0.01]} = 1.75 \times 10^{-5}$$

we obtain

$$[H_3O^+] = 4.18 \times 10^{-4} \; M$$

$$[Ac^-] = 4.18 \times 10^{-4} \; M$$

The validity of the assumption may be easily checked. If we use 5% as the allowable limit, the sum of the HAc and H_3O^+ concentration (or the sum of the HAc and Ac^- concentration)

$$[HAc] + [H_3O^+] = [HAc] + [Ac^-] = (1.00 \times 10^{-2}) + (4.18 \times 10^{-4})$$

Sec. II Acid-Base Equilibria

is within this limit. The [OH$^-$] may be obtained from K_w and [H$_3$O$^+$].

$$K_w = [H_3O^+][OH^-]$$

$$[OH^-] = \frac{K_w}{[H_3O^+]} = \frac{10^{-14}}{4.18 \times 10^{-4}} = 2.4 \times 10^{-11}$$

All problems involving weak acids may be treated in the same general way. In fact the procedure is formally the same for weak bases.

Example 2. Calculate the [H$_3$O$^+$] in a solution prepared by diluting 0.10 mole of HCN to 1.00 liter with water. The K_a of HCN is 7.2×10^{-10}.

Answer: [H$_3$O$^+$] = 8.5×10^{-6} M

The addition of a strong acid to a solution of a weak acid decreases the degree of dissociation of the weak acid. If the concentration of strong acid is large compared to the concentration of weak acid, the pH is controlled by the concentration of strong acid.

Example 3. Calculate the pH and the concentration of Ac$^-$ in a solution containing 0.1 M HCl and 0.01 M HAc.

In Example 1 the concentration of Ac$^-$ was negligible compared to the concentration of HAc. In this problem the Ac$^-$ concentration is even less than it was in Example 1 because the presence of HCl decreases the ionization of HAc. Therefore, the assumption that the concentration of HAc = 0.01 M is even more justified in this problem. The H$_3$O$^+$ concentration is 0.1 M from the HCl plus the amount produced from the ionization of HAc. In Example 1 we found that the H$_3$O$^+$ concentration in a solution of the same concentration of HAc was 4.18×10^{-4} M. With strong acid present this will be reduced even further and may certainly be neglected compared to the concentration of HCl.

From the equilibrium constant expression

$$K_a = \frac{[H^+][Ac^-]}{[HAc]} = 1.75 \times 10^{-5}$$

$$K_a = \frac{1.0 \times 10^{-1}[Ac^-]}{1.0 \times 10^{-2}} = 1.75 \times 10^{-5}$$

$$[Ac^-] = \frac{(1.75 \times 10^{-5})(1 \times 10^{-2})}{1.1 \times 10^{-1}}$$

$$[Ac^-] = 1.75 \times 10^{-6}\ M$$

The pH is given by

$$pH = -\log [H_3O^+]$$
$$pH = -\log (1 \times 10^{-1})$$
$$pH = 1$$

From K_w the OH⁻ concentration is 1×10^{-13}.

Example 4. Calculate the concentration of Ac⁻ and the pH of a solution containing 0.1 M HCl and 0.015 M HAc.

The same general procedure is used when the conjugate base of the acid is added in the form of a salt.

Example 5. What is the pH of a solution prepared by adding 0.10 moles of HAc and 0.15 moles of sodium acetate (NaAc) in 1.00 liter of solution?

The Ac⁻ from NaAc will prevent the HAc from ionizing to any appreciable extent. Thus [HAc] = 0.10 M and [Ac⁻] = 0.15 M are good approximations so that

$$\frac{[H_3O^+][Ac^-]}{[HAc]} = \frac{0.15[H_3O^+]}{0.10} = 1.75 \times 10^{-5}$$

$$[H_3O^+] = \frac{1.75 \times 10^{-5}(0.10)}{0.15} = 1.17 \times 10^{-5}\ M$$

$$pH = -\log (1.17 \times 10^{-5}) = 5 - \log 1.17$$
$$= 5 - 0.07 = 4.93$$

Equilibrium constant calculations involving polyprotic acids are solved by the same general methods as those involving monoprotic acids but they are more complex because of the additional equilibrium involved. The ionization of polyprotic acids occurs stepwise; each ionization step involving an equilibrium constant expression.

Example 6. Calculate the [H_3O^+], [HS⁻], and [S²⁻] in a solution containing 0.1 M H₂S.
The equilibria involved are

$$K_{a_1} = \frac{[H_3O^+][HS^-]}{[H_2S]}$$

$$K_{a_2} = \frac{[H_3O^+][S^{2-}]}{[HS^-]}$$

The problem is simplified considerably if the following assumptions are made. Since K_{a_2} is small (1×10^{-14}), only a small amount of HS^- is ionized to produce S^{2-}. K_{a_1} is also quite small (1.1×10^{-7}); therefore, $[H_2S] \gg [HS^-]$ so that $[H_2S] = 0.10\ M$. With these approximations we obtain from K_{a_1}

$$K_{a_1} = 1.1 \times 10^{-7} = \frac{[H_3O^+]^2}{[0.10]}$$

$$[H_3O^+] = [HS^-] = 1.0 \times 10^{-4}\ M$$

The S^{2-} concentration may now be determined from the expression for K_{a_2}

$$K_{a_2} = 1 \times 10^{-14} = \frac{[H_3O^+][S^{2-}]}{[HS^-]}$$

$$K_{a_2} = \frac{1 \times 10^{-4}[S^{2-}]}{1 \times 10^{-4}} = 1 \times 10^{-14}\ M$$

$$[S^{2-}] = 1 \times 10^{-14}\ M$$

The approximations may be checked from the equations

$$[H_3O^+] = [HS^-] + 2[S^{2-}]$$

$$1 \times 10^{-4} = 1 \times 10^{-4} + 2(1 \times 10^{-14})$$

and

$$[H_2S] + [HS^-] + [S^{2-}] = 0.10\ M$$

$$0.10 + (1 \times 10^{-4}) + (1 \times 10^{-14}) = 0.100\ M$$

The results are within the 5% allowable limit.

Example 7. Calculate the $[H_3O^+]$, $[HS^-]$ and $[S^{2-}]$ in a solution containing 0.10 M H_2S and 0.10 M HCl.

C. BUFFERS

A buffer solution is a solution which shows little change in acidity when small amounts of acid or base are added. In order to react in the manner described a buffer must contain an acid to react with the added base and a base to react with an added acid

$$H_3O^+ + A^- = HA + H_2O$$

$$OH^- + HA = H_2O + A^-$$

When an acid is added to a buffer solution the relative concentrations of A^- and HA are changed. Thus the pH of a buffer cannot remain constant when an acid or base is added; however, if both HA (a weak acid) and A^- (a weak base) are present in reasonably high concentration, the pH change will be small. For high buffering capacity the concentration of the buffering substances should be high. The choice of the buffering components is also important to buffer capacity. It is best to choose buffering components so that the pK_a of the weak acid is as close as possible to the pH desired. Under this condition, $[HA]/[A^-]$ is close to unity, and the buffering capacity is a maximum.

Most body fluids are buffers, an excellent example being blood. The pH of blood is constant at 7.3–7.5 even though large amounts of acids and bases are continually being introduced by absorption and digestion of foods, muscular activity, and respiration. The buffering components of blood are phosphate, carbonate, and proteins.

Most biological processes are critically dependent upon pH. A difference of a few tenths of a pH unit from the normal may result in the death of the organism. The yield of an industrial fermentation process or of a process for antibiotic production may be influenced considerably if the pH is altered appreciably from the optimum. Rates of food spoilage, production of bacteria, and digestion are all highly dependent upon pH. In analytical applications the pH must often be controlled during separations. (See Chapters 14 and 19.)

Example 8. Calculate the increase in $[H_3O^+]$ when 0.010 mole of HCl are added to 1.00 liter of an acetate buffer with initial concentration of 0.10 M HAc and 0.15 M NaAc.

Answer: $[H_3O^+]$ changes from 1.2×10^{-5} to 1.4×10^{-5}

QUESTIONS

1. Define the terms: acid, base, conjugate acid, conjugate base.
2. Are the following substances likely to be acids or bases: SO_2, $HMnO_4$, $OAs(OH)_3$, $ClOH$, SO_3?

Sec. II Acid-Base Equilibria

3. Explain the trends in acidity of the simple hydrides across a row and down a group.
4. Write the reactions for the successive ionization of $Al(H_2O)_6^{3+}$.
5. Describe the trends in acidity of the oxyacids.
6. Order the following anions in terms of increasing basicity:

   CO_3^{2-} and NO_3^-
   ClO_4^-, SO_4^{2-} and PO_4^{3-}
   ClO^-, ClO_4^- and ClO_3^-
   SO_3^{2-}, SeO_3^{2-} and TeO_3^{2-}

7. Look up the solubilities of several salts of each anion in a handbook. Is there any correlation between basicity and solubility?
8. What factors are most important in determining the acidity of a hydrated cation or an oxyacid?
9. Order the following in terms of increasing acidity:
   (a) Mg^{2+}, Ca^{2+}, Sr^{2+}
   (b) $HClO_3$, H_2SO_3, H_3PO_3
   (c) Na^+, Mg^{2+}, Al^{3+}
   (d) H_2SO_3, H_2SO_4, HSO_4^-
   (e) Mn^{2+}, Fe^{2+}, Co^{2+}, Ni^{2+}
10. Write balanced chemical equations for the following acid-base reactions:
    (a) __HBr + __H_2O ⟶
    (b) __KNH_2 + __H_2O ⟶
    (c) __Na_2O + HCl ⟶
    (d) __ClO^- + __H_2O ⟶
    (e) __NH_3 + __HF ⟶
    (f) __CaO + __CO_2 ⟶
11. Label each material given in Question 10 as a Brønsted or Lewis acid or base.
12. From the following list pick as many conjugate acid-base pairs as you can. Each species may be used more than once. H_2S, H_2O, OH^-, Cl^-, H_3O^+, H^-, NH_4^+, S^{2-}, H_2, NH_3, S, HCl, O_2, CO_3^{2-}
13. Order the following ions in order of increasing hydration enthalpy.
    (a) Ca^{2+}, Co^{3+}, Cd^{2+}
    (b) Cr^{2+}, Fe^{2+}, Sn^{2+}
    (c) Mg^{2+}, Sr^{2+}, Ca^{2+}
    (d) Na^+, Mg^{2+}, Al^{3+}

PROBLEMS I

1. Calculate the K_a of a number of the acids listed in Table 11-4.
2. A 0.1 M solution of NH_4OH is found to have an OH^- concentration of 0.25×10^{-5} M. What is the concentration of NH_4^+? $K_{bNH_4OH} = 1.8 \times 10^{-5}$.
3. It is necessary to make the H_3O^+ concentration in a 0.05 M solution of HCN equal to 3.5×10^{-8}. Can this be accomplished by adding KCN? What must the concentration of CN^- be in such a solution?
4. If solid NaOH (0.1 mole) is added to 1 liter of 0.125 M HAc, what is the final H^+ concentration?

5. What concentration of H$^+$ is required to obtain a solution that has a S^{2-} concentration of 10^{-5}, 10^{-10}, 10^{-15}?
6. HClO is an acid with an ionization constant of 3.2×10^{-8}. Calculate the pH of a 0.20 M solution of HClO.
7. Calculate the pH of a 0.20 M solution of NaClO.
8. H^0 values for Sr^{2+}(g), Zn^{2+}(g), and Pb^{2+}(g) are 427.75, 665.09, and 567.25 kcal mole^{-1} respectively. Calculate the enthalpy of hydration for these three ions.

PROBLEMS II

1. What is the pH of a 0.020 M solution of HI?
2. The ionization constant of HAc is 1.8×10^{-5}. Calculate the pH of a 0.020 M solution of HAc.
3. The ionization constant of Sc^{3+} is 10^{-5}. Write the equilibrium constant expression for the reaction Sc^{3+} + H$_2$O $\longrightarrow$ ScOH^{2+} + H$^+$.
4. What is the pH of a 0.020 M solution of ScCl$_3$?
5. Which is more acid, Sc^{3+} or Fe^{3+}?
6. What is the conjugate base of Fe^{3+}?
7. A 0.020 M solution of FeCl$_3$ has (FeOH^{2+}) = 0.010 M and a pH of 2.0. Calculate the ionization constant of Fe^{3+}.
8. Calculate the concentration of FeOH^{2+} in a solution containing 0.020 mole of Fe(NO$_3$)$_3$ and 1.00 mole of HNO$_3$ in 1 liter.
9. Calculate the pH of a buffer solution containing 0.03 M HAc and 0.02 M NaAc.
10. Calculate the pH of 1.00 liter of the buffer solution in Problem 9 after the addition of 0.005 mole of NaOH.
11. What is the pH of a solution of 0.005 mole of NaOH in water?

SUGGESTED READINGS

BANKS, J. E., *Chemical Equilibrium and Solutions.* New York: McGraw-Hill, 1967, Chapter 3.

MORTIMER, C. E., *Chemistry, A Conceptual Approach.* New York: Reinhold Publishers, 1967, Chapters 10 and 13.

RICH, R., *Periodic Correlations.* New York: W. A. Benjamin, Inc., 1965.

VAN DER WERF, C. A., *Acids, Bases and the Chemistry of the Covalent Bond.* New York: Reinhold Publishers, 1961.

The atoms which towered around him were much larger than any he had seen before, and he could count as many as twenty-nine electrons in each of them. If he had known his physics better he would have recognized them as atoms of copper, but at these close quarters the group as a whole did not look like copper at all. Also they were spaced rather close to one another forming a regular pattern which extended as far as he could see. But what surprised Mr. Tompkins most was the fact that these atoms did not seem to be very particular about holding on to their quota of electrons, particularly their outer electrons. In fact the outer orbits were mostly empty, and crowds of unattached electrons were drifting lazily through space, stopping from time to time but never for very long, on the outskirts of one atom or another. Rather tired after his breakneck flight through space, Mr. Tompkins tried at first to get a little rest on a steady orbit of one of the copper atoms. However he was soon infected with the prevailing vagabondish feeling of the crowd, and he joined the rest of the electrons in their nowhere-in-particular motion.*

G. GAMOW

Chapter 12 Molecular Orbitals

I. Introduction to Metals

From the discussion of chemical bonding in Chapter 7 we expect that there should be a relationship between properties of materials (such as those given in Table 3-1) and the number of bonding and antibonding electrons. However, to gain a fuller understanding of this relationship, the bonding involved must be considered in more detail. To do this we will consider again the simplest molecules of all, the gaseous diatomic homonuclear molecules.

The molecular orbital diagram for homonuclear diatomic molecules developed in Chapter 7 (Fig. 7-11) is repeated here as Fig. 12-1. Similar molecular orbital diagrams are applicable to heteronuclear diatomic molecules, such as CN, if the two atoms are not too different. These diagrams may be used to predict many facts about diatomic molecules, including the exceptional stability of molecules with ten valence electrons. (See Table 12-1 and Fig. 12-2.) With this number of electrons the bonding levels are completely filled. The bonding in molecules with less than ten electrons is weaker because fewer bonds are involved, while antibonding orbitals must be occupied in molecules with more than ten electrons.

*Mr. Tompkins Explores the Atom, (New York: The Macmillan Company, 1944).

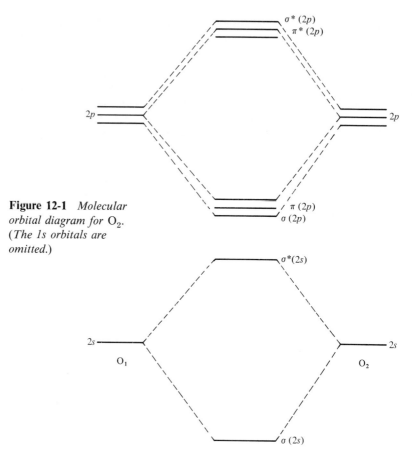

Figure 12-1 *Molecular orbital diagram for O_2. (The 1s orbitals are omitted.)*

From the data in Table 12-1 we see that the diatomic molecules of the elements of the second row generally have the largest bond energies. The decrease in bond energy that is noted down every group (except VIIA) is exceptionally large for the molecules in groups IIIA to VA. The data are not available for group IIA diatomic molecules, but some recent evidence indicates that the bond dissociation energy of Mg_2 is much greater than the van der Waal's attraction in He_2. The bond energy in Mg_2 is estimated as 6 kcal mole^{-1}.

This fact, and the stability of the diatomic molecules of the elements of group IIB are difficult to explain on the basis of Fig. 12-1. From this molecular orbital diagram we expect that there should be no net bonding for any of the group IIA or IIB diatomics because the bonding and antibonding levels should both be occupied by two electrons. It is obvious from the limited amount of experimental data that is currently available that the molecular orbital diagram given in Fig. 12-1 is not appropriate for the group IIA and group IIB diatomic

Table 12-1 *Bond Dissociation Energies of Gaseous Homonuclear Diatomic Molecules (not corrected for spin correlation stabilization energy).*

IA	IIA	IIIB	IVB	VB	VIB	VIIB	VIII	VIII	VIII	IB	IIB	IIIA	IVA	VA	VIA	VIIA	Inert gases
104.2																	
26												70.5	151	226	119.2	38	
18.2												39.6	75.8	116.7	84	58	
13										48		27.6	66.9	91.5	66	46	
11										40	3	25.3	47	69.1	53.8	36.1	
11										53	4		24	40			

Sec. I Introduction to Metals

Ch. 12 Molecular Orbitals

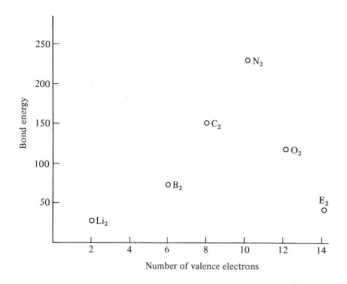

Figure 12-2 *Bond dissociation energies of gaseous homonuclear diatomic molecules.*

molecules. In fact, we have already seen that the diagram must be altered to that shown in Fig. 7-13 in order to account for the experimentally observed ground state of B_2 in which there are two unpaired electrons. As noted in Chapter 7 it is necessary to modify the simple picture to include hybridization or mixing. This will be the subject of the first part of this chapter.

As we proceed it should become clear that the introduction of mixing results in a much clearer picture of the bonding involved in diatomic molecules and solids. In addition, along the way we shall be led to an unexpected prediction. That is, although the bond dissociation energies should decrease down the group there should be at least two exceptions. One of these has already been noted (see data for group VIIA). Another exception is in group IIA. In this case (where only Mg_2 has been observed) we predict that the bond dissociation energy should increase down the group. This might also be true for the elements in group IIB.

To obtain an idea of what mixing is, we return to the method used to construct the molecular orbital diagram in Fig. 12-1. In Chapter 7 the molecular orbitals were obtained by visualizing the $s_1 - s_2$, $p_{z_1} - p_{z_2}$ and so forth overlaps. However, the $2s$ orbital of one atom does not only overlap with $2s$ orbitals of other atoms. There may be a significant interaction between the $2s$ orbital of one atom and *any other orbital* from another atom that has a lobe directed along the molecular axis. This type of interaction is shown in Fig. 12-3. The interaction may be neglected when the difference in energy between the bonding $\sigma(s)$ and the antibonding $\sigma^*(s)$ functions is small compared to the separation between the atomic s and p orbitals. The $\sigma(s)$-$\sigma^*(s)$

The term lobe is used to describe a region of reasonably high electron density for an electron. It is the area between nodes.

Sec. I   Introduction to Metals

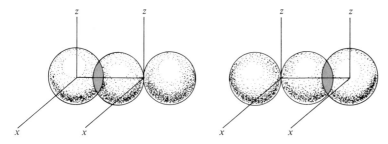

Figure 12-3 *The s-p overlaps in diatomic molecules.*

energy difference is strongly dependent upon the interatomic distance. The $s-p$ interaction will not be very important when the interatomic distance is large, as it is in O_2 and F_2; therefore, Fig. 12-1 is applicable in these cases. For elements to the right of beryllium and to the left of oxygen the bonding interaction is strong and the interatomic distance is small so that the $s-p$ interaction may not be neglected. When mixing is very important, as it is in B_2 and N_2, the result is the order shown in Figs. 7-13 and 12-4. The four σ functions shown in these

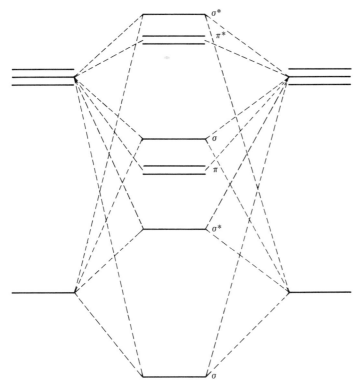

Figure 12-4 *Molecular orbital diagram for B_2, C_2, N_2.*

299

Ch. 12 Molecular Orbitals

figures result from the combination of four atomic orbitals. The four orbitals involved are one s and one p from each atom.

Another way of stating this quantum mechanical fact deals with the symmetry of the atoms involved. As the atoms approach each other they cannot remain spherical. All atomic orbitals are distorted, so it is no longer correct to designate the orbitals as s and p. Hybrid orbitals are formed. There are still four orbitals formed (four orbitals from four). The lowest energy molecular orbital is described as a bonding orbital made up of overlapping hybrid orbitals.

The criterion given above for determining whether mixing should be important depends upon the variation in orbital energies with interatomic distance. This variation is conveniently shown by a correlation diagram such as the one given in Fig. 12-5. The levels on the extreme right of the diagram correspond to the separate atoms, whereas the levels on the left correspond to the energy levels in the united atom (if the two nuclei practically overlap). Of course diatomic molecules will never reach this extreme or even approach it very closely. The

When mixing is substantial it is not possible to classify the two σ orbitals of intermediate energy as either bonding or antibonding. From an energetic point of view, both are antibonding with respect to an electron from the s orbital but they are bonding for an electron from a 2p orbital.

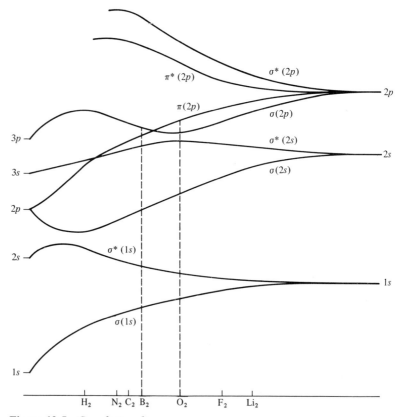

Figure 12-5 *Correlation diagram.*

300

really important part of the diagram is to the right of H_2, but the left-hand side does give important insight into how the energy of the orbitals changes in the intermediate region.

At the extreme right of the diagram we visualize the molecule as two separate atoms. As the atoms are brought closer together, the orbitals begin to overlap. The resulting molecular orbitals are labeled in terms of the atomic orbitals involved and the bonding or antibonding nature of the interaction. The relative ordering of the energy levels for molecules with large internuclear distance (when mixing is not important) will be found near the right-hand side of the diagram. In this section of the diagram, since the overlap is small, the effect of mixing is slight and the ordering in a molecule, such as Li_2, is "normal." The correct ordering of levels for Be_2 is just to the right of Li_2. The molecule is not stable because the effect of mixing has not yet altered the energy of the $\sigma^*(2s)$ to a significant degree. It is not reasonable to expect, however, that the bond energy in Be_2 will be zero, since even at this distance there may be some mixing of a p orbital into the $\sigma(2s)$ bonding state. Be_2 is not expected to be very stable, but it is not incapable of existence. Passing further to the left, we find B_2 and C_2 in the region of strong interactions between the $2s$ and $2p$ orbitals. The $\pi(2p)$ state is now lower in energy than the $\sigma(2p)$. The crossing of these two levels accounts for the ground state of B_2 (two unpaired electrons). The shortest bond distance of all the second row diatomics is found for N_2, and it may be placed just to the left of C_2. Because O_2 and F_2 contain antibonding electrons, the bonding is less effective and the internuclear distance is increased in these molecules. To place O_2 and F_2 on the diagram, we must move back toward the right. Therefore, the ordering of the levels inverts back to the original order $[\sigma(2p) < \pi(2p)]$ in these molecules.

This diagram is not strictly applicable to the elements in other rows of the periodic table because the energies of the atomic orbitals of the heavier elements are originally closer together. Because of this, mixing should become important earlier in the row and continue to be important later in the row. This should counteract the effect of the increased size of heavier elements on the bond energy for the group IIA diatomic molecules. The bond energy ordinarily decreases with increased size. From this we expect that Li_2 should have the largest bond energy of the diatomic molecules of the alkali metals and the difference in successive bond dissociation energies should become very small down the group. A glance at Table 12-1 will confirm this expectation.

The diatomic molecules of the alkaline earth metals should be unstable unless mixing or hybridization of the s and p orbitals is important. As we have seen, mixing does become more important down a group; therefore we expect that Mg_2 will be more stable than Be_2. The reasons for expecting an increased contribution to bonding, due to mixing, as one goes down a group in the periodic table appear to be quite good. We predict that Ca_2 and Ba_2 should be somewhat more stable than the recently discovered Mg_2. A similar interaction of the s, p, and d orbitals is apparent in the bonding of Zn_2, Cd_2, and Hg_2.

To completely interpret the data from Table 12-1 for the elements at the right hand side of the periodic table, the exceptional π bonding possible in the diatomic molecules of elements of the second row must also be considered. The large difference in bond energy between N_2 and P_2 may be attributed to the π bonding capabilities of nitrogen.

This difference is indicative of the fact that π bonding is exceptionally important in the chemistry of the elements from boron to fluorine.

It is the exceptional $p\pi$-$p\pi$ overlap in the second row diatomic molecules that is partially responsible for the irregularity observed in the bond energies of group VIIA diatomics. In this group the stability provided by the $p\pi$-$p\pi$ overlaps is cancelled because the π^* level is filled so that in this case the π orbital overlap affects the bond energy in a different way. The overlap is so great between atoms as small as fluorine that the electron repulsions become excessive and, as a result the atoms move apart, thereby weakening the σ bond interaction. (The interatomic distance in F_2 is larger than twice the covalent radius of fluorine.) In addition, mixing becomes more important in the heavier atoms and this leads to increased bond energies. The net result is that the dissociation energy of Cl_2 is greater than that of F_2. From the third through the sixth rows the same general successive decrease in dissociation energy is observed in all families of the periodic table except IIA and IIB.

Much of the above discussion is applicable to heteronuclear diatomic molecules, except that the crossing of the $\sigma(ns)$ and $\pi(np)$ levels should take place at larger values of r. The correlation diagram for heteronuclear diatomic molecules is not given here because in this case the usefulness is essentially limited to single molecules. When the atomic orbitals are very different in energy, as they are in LiF and BeO, Fig. 12-6 is the appropriate molecular orbital diagram. In this figure there are two nonbonding functions [$2s$ and $\sigma(n)$]. The lowest lying function is mostly nonmetal $2s$, whereas the higher energy nonbonding level is mostly metal $2s$. One $\sigma(p)$ and two $\pi(p)$ or a total of three bonding functions are shown.

The major difference between the diatomic fluorides and homonuclear diatomic molecules is the larger number of electrons on the fluorine. Fluorides with fewer than ten electrons will maintain as many bonding electrons as possible by decreasing the number of nonbonding electrons. The compound LiF has a dissociation energy of 124 kcal mole^{-1} compared to the maximum of 195 kcal for BF and the mini-

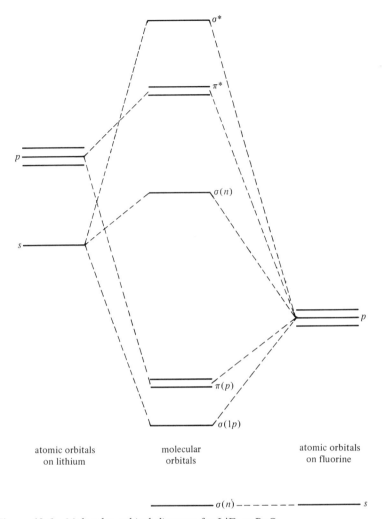

Figure 12-6 *Molecular orbital diagram for* LiF *or* BeO.

mum of 38 kcal in F_2. Since oxygen has one less valence electron than fluorine, oxygen compounds will have a lower dissociation energy than the corresponding fluorides with elements early in a row, but with fewer electrons the maximum bonding occurs at C and Si rather than B and Al. For the same reason the bonding in nitric oxide (NO) and O_2 is considerably stronger than that observed in the corresponding fluorides (NF and OF). As stated earlier the maximum bond energy is observed in all cases at ten valence electrons irrespective of the electronegativity difference. The maximum bonding for nitrides occurs at N_2.

In this section we have seen the importance of mixing in determining the relative energy of molecular orbitals. We have also seen, once again,

The bond dissociation energies of heteronuclear diatomic molecules are given in Table 12-2.

Ch. 12 Molecular Orbitals

Table 12-2 *Bond Dissociation Energies for Gaseous Diatomic Molecules (uncorrected for spin correlation stabilization energy, values in kcal mole^{-1})*

	Number of valence electrons							
	7	8	9	10	11	12	13	14
Fluorides 2nd row		LiF 138	BeF 161	BF 195	CF 107	NF 63	OF 45	F$_2$ 38
3rd row		NaF 115	MgF 121	AlF 158	SiF 126	PF 106	SF 66	ClF 60
Oxides 2nd row	LiO 79	BeO 106	BO 171	CO 257	NO 153	O$_2$ 119	FO 45	
3rd row	NaO	MgO 101	AlO 116	SiO 193	PO 142	SO 125	ClO 64	
Nitrides 2nd row	BeN	BN 152	CN 195	N$_2$ 226	NO 153	NF 61		
3rd row	MgN	AlN (71)	SiN 105	PN 165	SN 105	NCl —		
Chlorides 2nd row		LiCl 113	BeCl 110	BCl 119	CCl 81	NCl —	OCl 64	ClF 60
3rd row		NaCl 98.5	MgCl 91	AlCl 118	SiCl 93	PCl 69	SCl —	Cl$_2$ 58

that the number of orbitals is conserved when atomic orbitals are combined to form molecular orbitals. Before proceeding to the bonding in solids we shall examine three more examples in order to illustrate the principle of the conservation of orbitals.

The structural formula of the molecule we shall consider (butadiene, C_4H_6) may be written

$$\begin{array}{c} \text{H} \quad \text{H} \quad \text{H} \quad \text{H} \\ | \quad | \quad | \quad | \\ \text{H}-\text{C}=\text{C}-\text{C}=\text{C}-\text{H} \end{array}$$

The σ bonding carbon skeleton of this molecule may be represented as shown in Fig. 12-7. The four p_z orbitals available for π bonding (one on each carbon atom) are also shown. In this figure the carbon atoms are numbered for later reference. The molecular orbitals that may be written for the π system in this molecule are easily visualized. Of course, the total number of possible combinations is four since the π system is composed of one $2p$ orbital from each carbon atom. The low energy bonding state is visualized if each p orbital overlaps with the *p* orbitals on the neighboring carbon atoms. This may be represented as

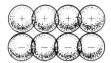

or

$$\psi_1 + \psi_2 + \psi_3 + \psi_4$$

Another possible combination is

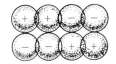

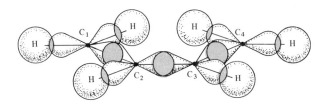

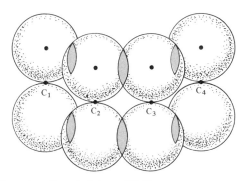

Figure 12-7 *The σ bonding skeleton of* C_4H_6.

This combination places a node between carbon atoms 2 and 3; however, there is bonding between atoms 1 and 2 and 3 and 4. The combination is represented by the function

$$\psi_1 + \psi_2 - \psi_3 - \psi_4$$

In a similar manner, additional combinations may be written as

$$\psi_1 - \psi_2 + \psi_3 - \psi_4$$

and

$$\psi_1 - \psi_2 - \psi_3 + \psi_4$$

These are shown in the drawings

and

In the combination $\psi_1 - \psi_2 + \psi_3 - \psi_4$ there is a node between each pair of neighboring carbon atoms. This definitely corresponds to an antibonding function. The combination $\psi_1 - \psi_2 - \psi_3 + \psi_4$ is not fully antibonding because there is bonding between atoms 2 and 3. There are two nodes in this combination, one between atoms 1 and 2 and the other between atoms 3 and 4. A summary of the possible combinations and the nodes in these combinations is given in Table 12-3.

Table 12-3

Combination	Number of nodes perpendicular to the bonds
$\psi_1 + \psi_2 + \psi_3 + \psi_4$	zero
$\psi_1 + \psi_2 - \psi_3 - \psi_4$	one
$\psi_1 - \psi_2 - \psi_3 + \psi_4$	two
$\psi_1 - \psi_2 + \psi_3 - \psi_4$	three

$$\underline{} \quad \psi_1 - \psi_2 + \psi_3 - \psi_4 \, \pi^*$$

$$\underline{} \quad \psi_1 - \psi_2 - \psi_3 + \psi_4 \, \pi^*$$

$$\underline{XX} \quad \psi_1 + \psi_2 - \psi_3 - \psi_4 \, \pi$$

Figure 12-8 *Molecular orbitals in butadiene, C_4H_6.*

$$\underline{XX} \quad \psi_1 + \psi_2 + \psi_3 + \psi_4 \, \pi$$

We have, of course, encountered a similar situation previously. In Chapter 2 the relative energy of atomic orbitals was given as $s < p < d$. The number of angular nodes in these functions are none in the s function, one in the p function, and 2 in the d function. Similarly the relative energies of the combinations given in Table 12-3 may be judged from the number of nodes that are perpendicular to the bond axis. Once the relative energies are known, the molecular orbital diagram for the π orbitals of butadiene may be written as shown in Fig. 12-8. The four electrons that constitute the π bonds in this molecule are placed in the two lowest levels.

The same principles are applicable to any molecule. The butadiene molecule was chosen because it is so easy to visualize the overlaps, and all of the possible combinations are easily written. For molecules that are more symmetrical, the overlaps are more difficult to visualize, and the combinations become more difficult to write. In a molecule such as benzene (C_6H_6), where six p orbitals are involved in the π bonding system, there are six possible combinations or six molecular orbitals. These are shown in Fig. 12-9, where the energy of an orbital

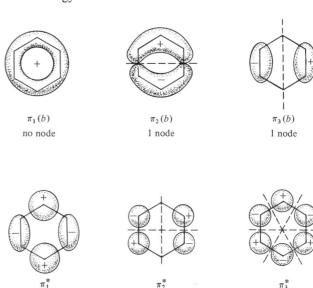

Figure 12-9 *Molecular orbitals for benzene, C_6H_6.*

307

is indicated by designating the number of nodes perpendicular to the bonds. Four different energy levels are obtained from the six orbitals. They have anywhere from zero to three nodes. Only four different energy levels are obtained because some of the combinations have the same number of nodes and the same energy. When this happens the orbitals are said to be degenerate. In a molecule with eight interacting orbitals there will be eight different possible combinations. Similarly in a metallic crystal with $4N$ interacting atomic orbitals there will be $4N$ possible combinations and a very large number of possible energy levels.

II. The Band Theory of Bonding in Solids

The molecular orbital, linear combination of atomic orbitals (LCAO), method may be used to account for the properties of solids. The overlap of atomic orbitals results in molecular orbitals in much the same way as in diatomic molecules. However, in all solids it is definite that both the s and p levels are involved, and that mixing takes place for even the group IIA elements. Otherwise, stable solids would not be expected for the elements in this group. There is another very important restriction placed on the molecular orbitals of a solid. In a solid, such as lithium metal, each lithium atom is surrounded by eight other lithium atoms. If only the s orbitals are considered (as a first approximation) we can visualize a molecular orbital that extends over the entire crystal by allowing the $2s$ orbital of each lithium atom to overlap with the $2s$ orbitals of all eight neighbors. The number of electrons that can occupy such an orbital is, of course, restricted to two by the Pauli exclusion principle. No two electrons can have the same four quantum numbers. If the spins are parallel, then the space coordinates for the electrons must be different. There are N atoms in a crystal; therefore there are N possible combinations of $2s$ orbitals, each with a different energy. Instead of obtaining one molecular orbital a large number of molecular orbitals are obtained that differ from each other in energy. In practice the energy difference between adjacent orbitals is found to be very small; so small in fact, that in many cases it is not detectable. The result is that the electrons in a solid such as lithium exist in an energy band that consists of a large number of closely spaced molecular orbitals. For a one-electron atom such as lithium, $N/2$ of the orbitals will be occupied. At low temperatures this corresponds to the bottom half of the band or the bonding molecular orbitals.

If only the s orbitals were involved in magnesium metal (Mg) the band would be completely occupied and there would be no net bond-

At higher temperatures the upper levels (the antibonding levels) will also be partially occupied up to energies where $e^{-\epsilon/kt}$ becomes negligible, and there will be corresponding vacancies in some lower energy orbitals.

Sec. II The Band Theory of Bonding in Solids

ing. However, unlike the inert gases in which the solids are held together only by van der Waal's forces, magnesium has empty *p* orbitals only slightly higher in energy than the *s* levels. With magnesium a sufficiently small Mg-Mg distance will result in the interaction of the *s* and *p* orbitals similar to that shown in Fig. 12-4.

If we think of only two magnesium atoms interacting at small distances, the result would be given by Fig. 12-10, which is similar to the diagram for N_2 and CO. If a large number of atoms interact in this way, a large number of molecular orbitals will result and bands

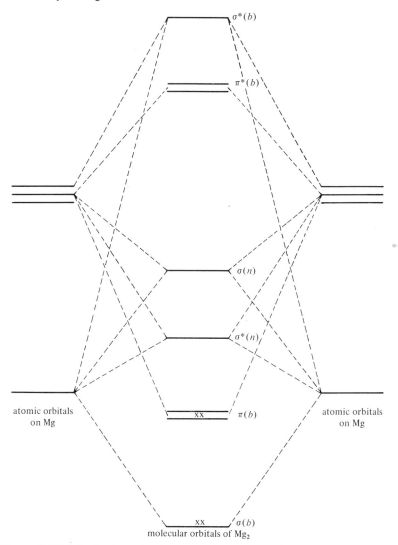

Figure 12-10 *Molecular orbital diagram for two very close Mg atoms.*

309

Ch. 12 Molecular Orbitals

The metallic radius of Mg is 1.60 Å.

will be generated. Simply add a large number of bonding levels to those shown. The actual situation in magnesium will naturally be different from that shown in this figure because each magnesium atom is surrounded by a large number of atoms. However, this simple picture can be used to understand the origin of the bonding and of the bands. Figure 12-11 shows only one set of bonding molecular orbitals for the interaction of a large number (Mg_n) of magnesium atoms.

The bands involved in sodium and magnesium are shown diagramatically in Fig. 12-12. In this figure $N(E)$ represents the number of energy levels and E represents the energy. The filled levels are shown as hatched. The dip in the magnesium diagram is related to the mixing of the s and p bands. Figure 12-13 shows the relationship between

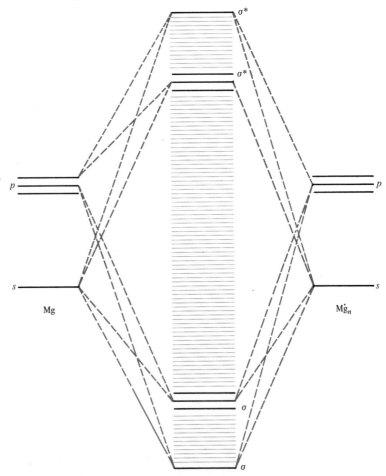

Figure 12-11 *One set of molecular orbitals for* Mg *metal.*

310

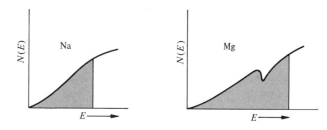

Figure 12-12 *Energy bands in Na and Mg.*

band spread and internuclear distance. As the internuclear separation decreases, the overlap between orbitals increases; mixing becomes more important and the *s* and *p* bands spread. The *s* and *p* bands begin to overlap in the region of about 7 Å. At the internuclear distance r_0 observed in the metal, the overlap of the bands and the mixing is considerable.

In solids the effect of mixing produces exactly the same trends in bond energy as that observed in the bond energies of diatomic molecules. Going down a group, such as group IA, there are two opposing effects, increased size and increased mixing. In the diatomic molecules the effect of increasing size predominates in most groups. This same effect should be noted here; therefore, the M-M bond energies, the H^0 of atoms, and the melting point of the solids should decrease down the group. Again the successive differences should become smaller going down the group. This same general trend might be expected

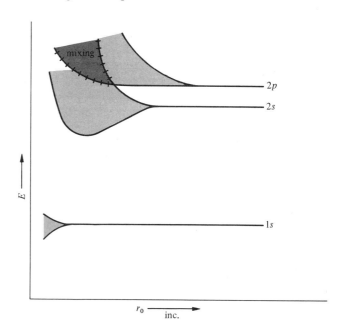

Figure 12-13 *Band spread versus internuclear distance.*

Ch. 12 Molecular Orbitals

in all groups of metals. The melting points and H^0 of atoms have been given in previous chapters. In Table 12-4 the M-M bond energies in solids are given. The general trends are apparent from these data with one important exception. An irregularity is noted in group IIA. The bond energy, melting point, and H^0 of the atoms is greater for beryllium than for magnesium, but the values for magnesium are smaller than those for calcium. This irregularity is caused by the final complication in our consideration of the bonding in solids. The increased values observed for calcium are due to the availability of the $3d$ orbitals for bonding. The effects of introducing a contribution from d orbitals is much the same as introducing a contribution from the p orbitals. The bonds are strengthened in both cases.

III. Metals and Nonmetals

Any continuous structure may be considered in terms of the band model. In the second row, lithium and beryllium are metals, whereas boron is on the borderline between a metal and a nonmetal. Carbon is a nonmetal. If we look at carbon (diamond) using Fig. 12-14 (where only one set of bonding molecular orbitals is shown for clarity), the nature of the bonding is apparent. In a diamond crystal the coordination number of each carbon atom is four, so each carbon atom contributes one electron to four bonds. As a result the bonding levels are completely filled. In diamond the s-p interaction is so strong that the bonding molecular orbitals are completely separate from the antibonding levels. That is, there is a gap between the allowed energy levels. This is shown in Fig. 12-15.

The presence of a band gap is typical of the nonmetals that form continuous solids. The band gap in carbon, the difference between the filled s-p band and the empty s-p antibonding band, is 6.0 ev. No other element approaches the band gap observed for carbon. This is true because as the atomic weight increases the atomic energy levels become closer together, and the possibility of overlapping bands becomes greater.

If we try to apply Fig. 12-14 or 12-15 to nitrogen, with five valence electrons, some of the antibonding levels would be occupied. Diatomic molecules containing five electrons per atom are the most stable of all diatomics. In a continuous solid the most stable situation results when there are four electrons per atom. As a result none of the nonmetals in row 2, groups VA, VIA or VIIA, form continuous solids. This is also reflected in the enthalpy of formation data for atoms discussed in Chapter 4. In a continuous solid any number of electrons beyond four per atom must occupy antibonding orbitals. From this

Table 12-4 M-M Bond Energies (for metals or single covalent bonds, spin correlation stabilization energy is included, values in kcal mole^{-1})

IA	IIA	IIIB	IVB	VB	VIB	VIIB	VIII	IB	IIB	IIIA	IVA	VA	VIA	VIIA	Inert gases
104															
77	78									90	91	70	50	38	
52	35									52	58	67	75	58	
42	46	62	60							44	48	60	60	46	
41	39	68	75							39	38	50	51	36	
38	42	67								29					

Ch. 12 Molecular Orbitals

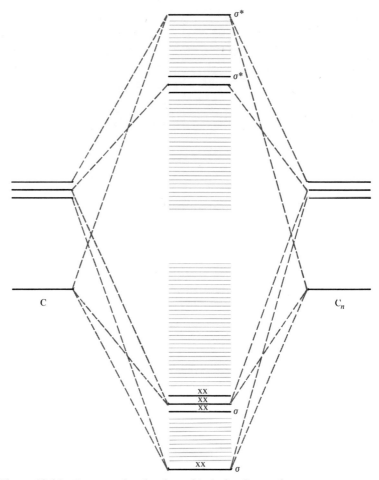

Figure 12-14 *One set of molecular orbitals for diamond.*

it is expected that the elements in group IV of the representative elements should have an optimum number of bonding electrons. Elements to the left of group IV have less than the maximum number of bonding electrons, whereas antibonding levels must be used in elements to the right of group IV. In Table 12-5 the observed energy required to form atoms from the elements is given for the second and third rows of the periodic table. The full data for the whole periodic table is shown in Tables 4-5 and 4-6 in Chapter 4. For the transition elements the bonding, nonbonding, or antibonding character of the successive electrons (see Table 4-6) is much clearer after the correction is made for the spin correlation stabilization energy in the free atom.

It is interesting to note that the remaining elements in group IVA would remain nonmetallic if it were not for the possibility of including

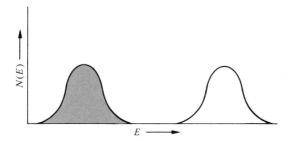

Figure 12-15 *Energy bands in diamond.*

d orbitals in the bonding. Germanium (Ge), tin (Sn), and lead (Pb) all have *nd* orbitals. The result is a continuous increase in metallic character with increasing *d* contribution.

A clear indication of the magnitude of the band gap in the group IVA elements is revealed by the electrical conductance of the substance. For an appreciable current to flow, it must be possible to put electrons into orbitals with a momentum in a particular direction. This will only be possible if there are a large number of empty orbitals only slightly higher in energy than those that are filled. In an insulator (such as diamond) the band gap is too large. A semiconductor (silicon, Si) is an insulator at low temperatures and a conductor at high temperatures. There is a band gap in silicon, but at higher temperatures some electrons will occupy the high-energy antibonding orbitals. This leaves a deficit of electrons in the bonding band. In a conductor the band is only partially filled.

The energy gaps between filled and empty orbitals in solid silicon and germanium are such that they are useful as semiconductors.

The energy gap between bonding and antibonding orbitals depends very strongly on the precise packing in the crystal. Tin (Sn) crystallizes in two crystal structures. One is a semiconductor whereas the other is metallic. In a similar way graphite is a fair electrical conductor whereas diamond is a very good insulator.

Table 12-5 *Bonding Electrons and H^0 of Atoms*

Valence electrons per atom	1	2	3	4	5	6	7	8
Net bonding electrons per atom	1	2	3	4	3	2	1	0
Element	Li	Be	B	C	N	O	F	Ne
H^0 for gaseous atoms, kcal mole^{-1}	37	78	141	172	113	60	19	0
H^*	37	78	141	187	161	76	19	0
Element	Na	Mg	Al	Si	P	S	Cl	Ar
H^0 for gaseous atoms, kcal mole^{-1}	26	35	78	109	75	67	29	0
H^*	26	35	78	117	99	75	29	0

Ch. 12 Molecular Orbitals

QUESTIONS

1. Which of the following has the largest bond energy: Al_2, Si_2, P_2, S_2, Cl_2?
2. Which of the molecules listed in Question 1 are stable?
3. Why is phosphorus found as a solid instead of as $P_2(g)$?
4. Arrange the following molecules in order of increasing bond energy: O_2, S_2, Se_2, C_2.
5. Find the data for Si_2 in Table 12-1 and place it into the list in Question 4.
6. Why can't an s orbital overlap with a p orbital to form a π bond?
7. Why do the two π bonding orbitals in C_2 have the same energy?
8. What mixing of orbitals is evident in C_4H_6?
9. Estimate the amount of π bond character present in each of the three C-C bonds in C_4H_6. Would there be more or less in $C_4H_6^+$?
10. Interpret the boiling points of the elements from sodium through sulfur in terms of band theory and the number of electrons present.
11. Which is stronger in a solid, a Na-Na bond or a Mg-Mg bond?
12. What are the requirements for an element to be a semiconductor?
13. Predict the electrical properties for solids of the composition: GaAs, GaGe, and GeAs.
14. Which element in row 3 of the periodic table has the highest boiling point? Why?
15. Why does magnesium have the lowest boiling point of the group IIA metals?
16. Describe the electron distribution and bonding for O_2, CN, Na_2, Na(s) and Si(s) including a qualitative molecular orbital diagram.
17. In each set of four or five species select the one specified at the left.

    (a) Shortest bond length B_2 N_2 O_2 F_2
    (b) Weakest bond Be_2 B_2 N_2 O_2 F_2
    (c) Greatest mixing of orbitals Mg_2 Al_2 P_2 S_2 Cl_2
    (d) Best electrical conductor   Al   Si   P S_8 Cl_2
    (e) Strongest bond O_2 S_2 Se_2 Te_2 Po_2
    (f) Strongest bond F_2 Cl_2 Br_2 I_2
    (g) Strongest bond   LiF(g)   BeF(g)   BF(g)   CF(g)   NF(g)
    (h) Weakest bond B_2 Al_2 Ga_2 In_2 Tl_2
    (i) Weakest bond   BO   CO   NO O_2 OF
    (j) Greatest mixing of orbitals   K   Ca   Sc   Ti
    (k) Best electrical conductor   InSn(s)   Sn   SnSb(s)   Sb   InSb
    (l) Most antibonding electrons   BeN(g)   BN(g)   CN(g) N_2(g)   NO(g)

PROBLEMS I

1. Calculate ΔH for the following reactions using data from Table 12-1 and Table 12-2:

   $Li + ClF \longrightarrow LiF + Cl$
   $CO + F \longrightarrow CF + O$
   $B + F_2 \longrightarrow BF + F$
   $Li + CO \longrightarrow LiO + C$

$$N_2 + O_2 \longrightarrow 2NO$$
$$Si_2 + O_2 \longrightarrow 2SiO$$

2. ΔH^0 for the reaction $N_2 \longrightarrow 2N$ is $+226$ kcal mole^{-1}. Calculate ΔH^0 for $N_2 \longrightarrow 2N^*$.
3. Correct the values for the IVA, VA and VIA diatomic molecules in Table 12-1 for spin correlation stabilization energy. Are the same trends shown?
4. If the reaction $2Mg(g) \longrightarrow Mg_2(g)$ has $\Delta H^0 = -6$ kcal mole^{-1} and $\Delta S^0 = -24$ eu, calculate the pressure of Mg_2 in magnesium vapor at the boiling point, $1110\,°C$.

PROBLEMS II

1. H^0 of oxygen atoms is $+59.6$ kcal mole^{-1}, and SCSE $= 16$ kcal mole^{-1}. Calculate H^* for oxygen atoms.
2. H^0 for boron atoms is $+134.5$ kcal mole^{-1}. Calculate H^* for boron atoms.
3. H^0 for $BO(g)$ is $+6$ kcal mole^{-1}. Calculate ΔH for the reaction $BO(g) \longrightarrow B(g) + O^*(g)$.
4. H^0 for carbon atoms is 171 kcal mole^{-1}. Calculate H^* for carbon atoms.
5. H^0 for CO is -26.4 kcal mole^{-1}. Calculate the bonding present in CO correcting for spin correlation stabilization energy in the atoms.
6. Why is the bonding in $CO(g)$ stronger than in $BO(g)$?
7. H^* for nitrogen atoms is 161 kcal mole^{-1} and H^0 for NO is $+21.6$ kcal mole^{-1}. Calculate the bond dissociation energy of NO.
8. Why is the bonding stronger in CO than in NO?
9. Quantitatively how much is the bonding weakened by still another electron (compare NO with O_2)?
10. How many bonding molecular orbitals are filled in solid calcium?
11. How many bonds must be broken to remove a calcium atom from solid calcium?
12. H^0 for the reaction $Ca(s) \longrightarrow Ca(g)$ is $+46$ kcal mole^{-1}. What is the average energy of Ca-Ca bonds in solid calcium?
13. $H^0 = +35$ kcal mole^{-1} for $Mg(s) \longrightarrow Mg(g)$. Calculate the Mg-Mg bond energy in the metal.
14. Why is the Ca-Ca bond stronger than the Mg-Mg bond?
15. For the reaction $Na(s) \longrightarrow Na(g)$, $\Delta H^0 = +25.8$ kcal mole^{-1} and $\Delta S^0 = +24.5$ eu. Calculate the Na-Na bond energy in solid sodium.
16. Calculate the vapor pressure of solid sodium at $25\,°C$.
17. Compare the bond energies in Table 12-4. Which of the metals in families IA and IIA will have the lowest boiling points?
18. Examine the electrical conductivities of the metals in Table 1-4. What effect does mixing d orbitals with s orbitals have on electrical conductivity?

... There does not exist in all nature a single phaenomenon but what is so connected with certain conditions, that when they are absent, the phaenomenon shall either not appear, or be varied occasionally. It is of consequence to science, that the changes and the combination of causes in every operation should be accurately known, as far as a knowledge of them is attainable; and the utility of a strict inquiry into attractions will, I hope, clearly appear from many instances in the following pages.

But if, on the contrary, a fixed order does really take place, will it not, when once ascertained by experience, serve as a key to unlock the innermost sanctuaries of nature, and to solve the most difficult problems, whether analytical or synthetical? I maintain, therefore, not only that the doctrine deserves to be cultivated, but that the whole of chemistry rests upon it, as upon a solid foundation, at least if we wish to have the science in a rational form, and that each circumstance of its operations should be clearly and justly explained. Let him who doubts of this consider the following observations without prejudice, and bring them to the test of experiment.*

T. BERGMAN
De Attractionibus Electivis (1775)

Chapter 13 Polarizing Ability and Solubility

I. The Alkaline Earth Metals

In the earth's crust there are more magnesium atoms than calcium atoms, but the weight of calcium is larger. In the deeper layers of the earth (mantle), magnesium is much more abundant than calcium.

Calcium (Ca) is the fifth most abundant element found in the earth's crust, and as such it is the most abundant alkaline earth metal. The principal ore containing calcium is calcium carbonate ($CaCO_3$), but calcium sulfate ($CaSO_4$) is also quite abundant. Magnesium (Mg), the second most abundant metallic element in the sea, is obtained commercially by taking advantage of its insoluble hydroxide. Strontium (Sr) is frequently found accompanying calcium, and there are significant deposits of barium sulfate ($BaSO_4$).

The electrode potentials of the alkaline earth metals are similar to those of the group IA metals. Consequently these metals are not found free in nature.

Although the electrode potentials of calcium, strontium and barium are similar, magnesium has a slightly lower potential; beryllium (Be) has the lowest electrode potential of the group. Beryllium and magnesium are stable in air because of the formation of a protective oxide

*Used by permission from H. M. Leicester and H. S. Klickstein, *A Source Book in Chemistry 1400–1900*, (Cambridge, Mass.: Harvard University Press, 1952).

coating. The application of magnesium as a structural metal is only possible because of the resistant protective coating. Calcium, strontium, and barium react slowly with water because of the oxide coating.

Magnesium alloys are widely used in aircraft construction.

The similarity between calcium (Ca^{2+}) and strontium (Sr^{2+}) ions results in the retention of strontium in biological systems and makes the removal of radioactive strontium ions very difficult. There is no known function for strontium ions in biological systems, although it is used in medicine as an anion carrier (for bromide and salicylate). The chloride has been suggested as a toothpaste ingredient because of reports that it reduces the sensitivity of the teeth to hot and cold.

Barium (Ba^{2+}) salts are extremely toxic. The fatal dose of barium chloride ($BaCl_2$) is 0.8–0.9 g and it has been used as a rodenticide. Beryllium compounds are also extremely toxic. The sweet taste of many beryllium salts has probably resulted in a number of accidental poisonings. The element is not abundant; it occurs to the extent of 0.0006%. Only one ore, beryl ($Be_3Al_2Si_6O_{18}$), is found in sufficient quantities to be of commercial importance. Emerald is crystalline beryl containing chromium (Cr^{3+}) ions.

The name glucinium (for sweet) was suggested by the editors of Annales de Chemie *which published the paper of N. L. Vauguelin who first isolated the oxide.*

Magnesium ions are essential to life. They are involved in several enzymes and are essential to the functioning of the neuromuscular system. Magnesium sulfate is often used as an antidote for barium poisoning and as a laxative. The mild antacid behavior of magnesium hydroxide [$Mg(OH)_2$] corresponds to the general acid-base trend where the strongest bases and weakest acids occur at the bottom of a family. In accordance with this barium hydroxide [$Ba(OH)_2$] is a strong base, and beryllium hydroxide [$Be(OH)_2$] is amphoteric.

Calcium ion is essential to human life because of the calcium present in the bones, and in addition, calcium is essential for proper heart action. A delicate balance between the calcium and potassium ion concentration must be maintained for a proper beat. An excess of calcium ion results in contraction (systolic arrest) while an excess of potassium ion results in relaxation (diastolic arrest). Calcium ion is also required for coagulation of blood.

The alkaline earth metals react with air at elevated temperatures to produce nitrides and oxides. The reaction is used in the production of vacuum tubes where an alloy of barium is used as a "getter" to remove the last traces of air. Barium is produced for this purpose by the reduction of the oxide with aluminum according to the equation

$$3BaO + 2Al \longrightarrow Al_2O_3 + 3Ba$$

The barium is mixed with additional aluminum and heated to a high temperature to form a compound close to the stoichiometry of $BaAl_4$, which is used as the getter.

Ch. 13 Polarizing Ability and Solubility

From this discussion it is apparent that the chemistry of the alkaline earth metals is much more interesting and varied than that of the alkali metals. One of the obviously important and interesting features of the chemistry of the group is associated with the solubility relations.

II. The Solubility of Inorganic Salts

In the introduction to Chapter 4 the relationship between thermodynamic data and the nature of atoms and ions was discussed. An understanding of chemical behavior depends upon a knowledge of thermodynamic data and its interpretation. This is one objective of the remainder of this chapter. We shall examine the solubility of salts from the thermodynamic point of view and attempt to understand solubility on the basis of the nature of the ions involved.

A covalent model based on bond energies is discussed in Chapters 23 and 24.

Once again a choice of the beginning model must be made. Is it more appropriate to begin with the ionic model and add a term to account for deviations from this model, or should the covalent model be used as a basis with appropriate corrections made for ionic character? Both approaches have merit and both have been widely used. It is perhaps most important to see that both approaches can lead to similar predictions.

Z is used to represent the charge of an ion of radius r.

We shall begin here with a discussion of the ionic model because this is undoubtedly the more appropriate model for the alkaline earth and alkali metal ions. From the ionic point of view we are concerned with the size and charge of the cations and anions. These can be combined into the term Z/r^2 which is called the polarizing ability of the cation. A certain degree of covalent character can be allowed in the bonding by referring to the ease of polarization of the anion.

A. POLARIZING ABILITY OF CATIONS AND EASE OF POLARIZATION OF ANIONS

Ionization potentials and electronegativities are extremely useful in understanding the chemical behavior of atoms. Low ionization potential and low electronegativity lead to metallic character, whereas high ionization potential and high electronegativity result in nonmetallic character. Once ions or molecules are formed the concept of electronegativity loses some of its significance. For example, what is the electronegativity of a fluoride ion? In this section we will introduce two concepts that are important when dealing with ions. They are the polarizing ability of cations and the ease of polarization of anions.

The polarizing ability of a cation is defined as the ability of the cation to attract electrons to itself. The ease of polarization of an anion is defined as the ease with which an anion can be polarized by a cation.

Polarization of an anion amounts to a deformation of the spherical electron distribution of the anion toward the center of positive charge. The greater the polarizing ability of the cation, the greater the distortion. As the polarizing ability increases in a series of cations, the bonding with a specific anion becomes more covalent. This results in a trend from more ionic properties for compounds of the least polarizing cation to more covalent properties for compounds of the more polarizing cation. For example, in the series Na^+, Li^+, Ag^+, the polarizing ability increases and the solubility of the halides in water decreases.

In order to understand the trends in polarizing ability for ions of the representative elements only two factors, the size of the ion and the charge on the ion, must be considered. The polarizing ability of these cations will vary inversely with size and directly with charge. Thus the polarizing ability of cations decreases downward in each main group and increases across each row.

The concept has been put on a quantitative basis by using the formula polarizing ability (PA) $= Z/(r + 0.50)^2$. Table 13-1 lists some polarizing abilities calculated in this way. The formula given above seems to give adequate results for some of the representative elements. The trends are reasonable, at least, even if there might be some question about the absolute magnitude; however, this simple formula is not adequate for the transition metal ions and the ions immediately following the transition series (groups IB and IIB). The cations in these groups (the transition metal ions and the ions following them) behave differently than the calculated polarizing ability would indicate. For example, from the formula the polarizing ability of Zn(II) should be approximately the same as that of Mg(II), and the value for Ag^+ should also be very similar to that of K^+. Zn(II) and Mg(II) are similar in some ways (they compete with each other in certain biological systems) but they are very different in other ways. Zn(II) is amphoteric, Mg(II) is not. Zn(II) forms an insoluble sulfide, Mg(II) does not. Similar differences are noted between Ag^+ and K^+.

From this behavior it is expected that the polarizing ability of a transition element and a post transition element is greater than that calculated using the formula PA $= Z/r^2$. The difference in electron configuration of these two sets of ions suggests that this might be an important factor in influencing the chemical behavior of these two sets of ions. The post transition metal ions contain filled d shells just slightly lower in energy than the valence orbitals, whereas in the transition metals the d level is partly filled. The excess polarizing ability of these cations can be explained if these facts are taken into account.

For ions that contain d electrons the effective nuclear charge must be considered rather than the apparent nuclear charge. It will be

The constant 0.50 Å is added to the ionic radius of the cation to allow for the fact that the electrons of the anion are, on the average, some distance beyond the ionic radius of the cation.

Table 13-1 *Polarizing Abilities of Some Cations,* $Z/(r + 0.50)^2$

IA	IIA	IIIB	IVB	VB	VIB	VIIB	VIII			IB	IIB	IIIA	IVA	VA	VIA	VIIA	Inert gases
0.83	3.05																
0.48	1.51											+3 6.12					
0.30	0.90	+3 1.75	+4 2.87	+5 4.21								+3 3.00	+4 4.83	+5 7.09			
0.26	0.75	+3 1.47															
0.21	0.58	+3 1.10															

recalled that d electrons do not provide complete screening of the nucleus. Each d electron has a definite probability of being further from the nucleus than the electrons of the anion. In order to include this factor the formula can be modified to

Screening and penetration are considered in Chapters 2 and 3.

$$PA = \frac{Z + S}{(r + 0.50)^2}$$

where S is the increased charge due to ineffective screening of the nuclear charge by the d electrons. One of Slater's rules for determining the effective nuclear charge may be used to calculate S. Each d electron only shields 85% of a single nuclear charge, therefore, $S = 0.15 \times$ (the number of d electrons). The polarizing ability of an ion such as Cr(III) is then found to be

Slater's rule is designed to estimate the effective nuclear charge of an atom.

$$PA = \frac{Z + S}{(r + 0.50)^2} = \frac{3 + (0.15 \times 3)}{(0.69 + 0.50)^2} = \frac{3.45}{(1.19)^2} = 2.45$$

The polarizing ability of Cu(II) is

$$PA = \frac{Z + S}{(r + 0.50)^2} = \frac{2 + (0.15 \times 9)}{(0.69 + 0.50)^2} = 2.37$$

The results of this calculation are given in Table 13-2 for a number of transition metal ions.

This correction helps considerably. However, the values still do not include one important factor. More realistic values for these ions are probably higher, because a partly filled d shell can be easily distorted from the completely spherical arrangement. An ion such as Al(III) has an electron configuration

$$\text{Al(III)} \; \overset{..}{\underset{2s}{\cdot\cdot}}, \; \overset{..}{\underset{2p}{\cdot\cdot}} \;\overset{..}{\cdot\cdot}\; \overset{..}{\cdot\cdot}, \; -, \; -\;-\;-$$

with a filled $2s$ and $2p$ level. Al(III) is spherical because it is difficult to make any appreciable change in a filled inner $2p$ subshell in order to alter the electron distribution present in the ion. This is not true for an ion such as Ni(II). In this case only the free gaseous ion is spherical (a definite pattern is obtained in the Stern-Gerlach experiment). If an external force is applied to this ion, the ion cannot remain spherical. For example, in the complex Ni(H$_2$O)$_6^{2+}$, the d electrons of the nickel will orient themselves so that electron-electron repulsions (that is repulsions between nickel electrons and water electrons) are at a minimum. The water molecules assume an octahedral structure

Ch. 13 Polarizing Abilities and Solubility

Table 13-2 Polarizing Abilities of Cations with d Electrons in the $n-1$ Shell, $(Z+S)/(r+0.50)^2$

IA	IIA	IIIB	IVB	VB	VIB	VIIB	VIII	VIII	VIII	IB	IIB	IIIA	IVA	VA	VIA	VIIA	Inert gases
					+3 2.45	+2 1.63	+2 1.83 / +3 2.89	+2 1.98	+2 2.15	+1 1.17 / +2 2.37	+2 2.28	+3 3.59	+4 5.18	+5 6.91			
										+1 0.81	+2 1.62	+3 2.62	+4 3.76	+5 5.18			
											+2 1.37	+3 2.14	+4 3.06	+5 4.22			

about the nickel as shown in Fig. 13-1. We may tentatively assume that the molecule is held together by electrostatic forces due to the positively charged nickel ion and an electron pair of the water. From Fig. 13-2 it can be seen that the $d_{x^2-y^2}$ and d_{z^2} orbitals of the nickel are oriented toward the vertices of an octahedron. The electrostatic interaction between ligand electron pairs and metal electrons will be greater for electrons in these two orbitals (the $d_{x^2-y^2}$ and the d_{z^2}) than for the electrons in the other three d orbitals (d_{xz}, d_{xy}, d_{yz}). The d electrons in Ni(II) arrange themselves so that the $d_{x^2-y^2}$ and d_{z^2} orbitals contain only one electron each

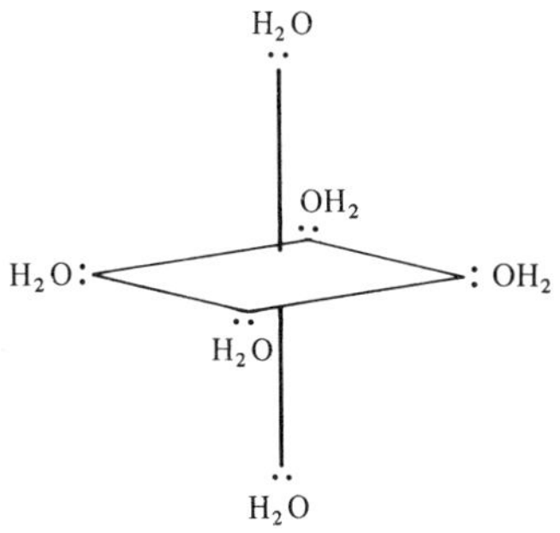

$$\text{Ni(II)} \quad \underset{d_{xz}}{\overset{\cdot\cdot}{\rule{1em}{0.4pt}}} \; \underset{d_{yz}}{\overset{\cdot\cdot}{\rule{1em}{0.4pt}}} \; \underset{d_{xy}}{\overset{\cdot\cdot}{\rule{1em}{0.4pt}}} \; \underset{d_{x^2-y^2}}{\overset{\cdot}{\rule{1em}{0.4pt}}} \; \underset{d_z^2}{\overset{\cdot}{\rule{1em}{0.4pt}}}, \quad \text{—}, \quad \text{— — —}$$

while the remaining three orbitals each contain two. As a result the distribution of d electrons in Ni(II) is not spherical. If we look at the electron distribution over the surface of a sphere, we see that there

Figure 13-1 An octahedral complex.

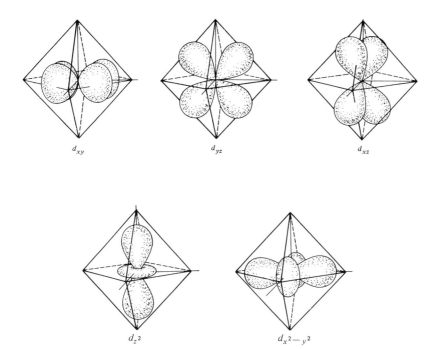

Figure 13-2 The d orbitals in an octahedron.

is less d electron density in the region directed toward the corners of the octahedron than that directed toward the edges and faces. As a result the polarizing ability of Ni(II) is dependent upon direction because the shielding is not uniform over the surface of a sphere. The polarizing ability is greatest along the lines pointing toward the corners of the octahedron (which is why the structure is octahedral). While it is apparent that the calculated polarizing abilities are too low, it is difficult to estimate what the actual values should be.

Another case where the polarizing ability must be modified involves the post transition elements such as Ag(I) and Cd(II). These ions may be more highly polarizing than the size and charge indicate because the d electrons are relatively loosely held. Just as the electron distribution can distort from the spherical in Ni(II), the d electrons in Ag(I) or Cd(II) can be distorted by certain ligands to give a nonspherical ion that is more highly polarizing because the effective radius of the ion is reduced. In the case of Ag(I) and Cd(II), only the anions that are easiest to polarize can cause distortion, thus the most stable salts of these cations are formed with the most easily polarized anions. This is the behavior of all typical soft cations.

From the above discussion it is obvious that any question of relative stability of a given cation-anion pair depends upon the anion as well as the cation. The ease of polarization of a simple anion depends upon size, nuclear charge and the magnitude of the negative charge. For the simple anions the ease of polarization increases down the periodic table as the size increases. For anions that are about the same size (O^{2-} and F^-), the ion with the largest charge is easiest to polarize. The ease of polarization of some of the simple anions is given in Table 13-3.

The trends in the ease of polarization of complex anions can be predicted from the oxidation state of the central atom and the charge on the anion. The larger the charge the more easily polarized the anion, and the smaller the oxidation state of the central atom the more easily polarized the anion. Thus in the sequence ClO_4^-, SO_4^{2-}, PO_4^{3-}, the hardest anion to polarize would be ClO_4^-, SO_4^{2-} would be easier and PO_4^{3-} would be easier still. Similarly the ease of polarization should increase in the series XO_4^{n-}, XO_3^{n-}, XO_2^{n-}, XO^{n-}. This trend follows the trends in acid strength encountered in Chapter 11. The sequence

Table 13-3 *Order of Increasing Ease of Polarization of Some Anions*

$$F^- < SO_4^{2-} < NO_3^- < H_2O < NH_3 < O^{2-} < OH^- < Cl^- < Br^- < I^- < S^{2-}$$

in Table 13-3 can also be used to predict changes in the ease of polarization when an element in the outer sphere of a complex anion is changed. Thus $S_2O_3^{2-}$ is much more easily polarized than SO_4^{2-}.

Once the concept of polarizing ability is introduced, several points become immediately obvious. For example, the enthalpy and free energy of hydration data presented in Chapter 11 become clearer. The cations that deviate from the lines shown in Fig. 11-1 (page 277) are those for which softness or distortion from spherical symmetry is expected. The cations in this figure are arranged approximately in order of increasing acidity. Thus we see that the cations that have a low polarizing ability have basic hydroxides; those of intermediate polarizing ability have amphoteric hydroxides, and those of very high polarizing ability (for example, Cl^{7+} not shown in the figure) form very strong acids. On the scale used above the cations may be arranged roughly as follows:

$$PA = 0\text{-}2.9 \text{ basic hydroxide}$$

$$PA = 3.0\text{-}4.4 \text{ amphoteric hydroxide}$$

$$PA > 4.5 \text{ acidic hydroxide or oxyacid}$$

Another feature that correlates well with polarizing ability is complex ammine formation. Those cations with low polarizing ability (with small ΔG_{hyd}) have no particular preference for the ammonia molecule over water; therefore complex ammines are not formed in water solution. The cations in the center are better polarizers [Ni(II), Cu(II), and so forth]. These cations form complex ammines. The cations with high polarizing ability are too acidic to form complex ammines. The reaction

$$M(H_2O)_6^{n+} + NH_3 = M(H_2O)_5(OH)^{(n-1)+} + NH_4^+$$

takes place instead, with the eventual precipitation of the metal as the hydroxide.

The pattern in acid-base behavior that becomes clear from a consideration of the polarizing ability of a cation is simple. We can represent it as shown in Fig. 13-3. The same type of diagram (shown in Fig. 13-4) is applicable to complex ammine formation. We shall see in the next chapter that this type of pattern is applicable to ligands other than water and ammonia.

In the next section the concept of polarizing ability will be applied to solubility considerations after a brief discussion of the measurement of solubility.

Ch. 13 Polarizing Ability and Solubility

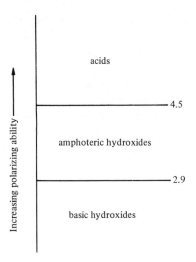

Figure 13-3 *Polarizing ability and acidity of cations.*

B. MEASUREMENT OF SOLUBILITY

The solubility of a salt may be determined by measuring the concentration of either the cation or anion in a saturated solution. If activities are taken as concentrations, the equilibrium constant for the reaction may be calculated from

$$K = \frac{[M^+][X^-]}{a_{MX}}$$

or

The activity of a pure solid is 1.

$$K_{sp} = [M^+][X^-]$$

where the K_{sp} is called the solubility product. The limitations of this method for calculating solubility products are obvious. If the salt is slightly soluble, the method can be used to approximate K_{sp}'s; however, if the salt is very insoluble, the concentration of ions in solution will be so small that alternative methods must be used. In one alternative using

$$\Delta G^0 = -RT \ln K$$

the solubility product may be calculated if the thermodynamic data are available. The thermodynamic data that are needed to calculate the equilibrium constants for a number of substances are listed in Table 13-4 and Appendix IV.

The data in this table require some explanation, especially the method that is used to obtain the H^0 and G^0 values for aqueous ions.

The enthalpy of formation of a single ion cannot be directly measured. To obtain data on ions it is usually assumed that H^0 for $H^+(aq) = 0$. When this assumption is made the enthalpy of formation of a number of other ions can then be calculated. For example, from

$$\tfrac{1}{2}H_2 + \tfrac{1}{2}Cl_2 = H^+(aq) + Cl^-(aq)$$
$$\Delta H^0 = -39.95$$

we obtain

$$\Delta H^0 = [(H^0_{Cl^-(aq)} + H^0_{H^+(aq)}) - (\tfrac{1}{2}H^0_{Cl_2} + \tfrac{1}{2}H^0_{H_2})] \qquad \text{H^0 of an element is 0.}$$
$$H^0_{Cl^-} = -39.95$$

Similarly if the enthalpy change for the reaction

$$Na(s) + \tfrac{1}{2}Cl_2(g) + H_2O(l) = Na^+(aq) + Cl^-(aq)$$

is

$$\Delta H^0 = -97.34 \text{ kcal}$$

then

$$\Delta H^0 = [(H^0_{Na^+(aq)} + H^0_{Cl^-(aq)}) - (H^0_{Na(s)} + \tfrac{1}{2}H^0_{Cl_2(g)})]$$

and

$$H^0_{Na^+} = \Delta H^0 - H^0_{Cl^-(aq)} = -97.34 - (-39.95)$$
$$= -57.39 \text{ kcal mole}^{-1}$$

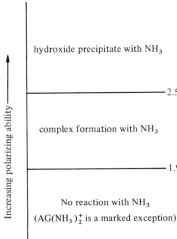

Figure 13-4 *Relationship between polarizing ability and reaction with NH_3.*

Ch. 13 Polarizing Ability and Solubility

The medicinal value of substances such as Pepto Bismol is highly dependent upon the fact that many bismuth salts are insoluble. Bismuth and other heavy metals are highly toxic, and the success of their application depends upon the insolubility of their salts.

$\Delta G^0 = -n\mathcal{F}E^0$, *from Chapter 5.*

The enthalpies of formation of $Cl^-(aq)$ and $Na^+(aq)$ are -39.95 kcal mole^{-1} and -57.39 kcal mole^{-1} respectively.

Additional calculations of this type lead to a whole series of values for the enthalpy of formation of ions. The solubility product of Bi_2S_3 obtained from the data in Table 13-4 is 2×10^{-99}.

Of course, solubility products can be obtained from redox potential data because of the relationship between ΔG^0 and E^0. This represents a method of obtaining the equilibrium constants of many extremely

Table 13-4 *Thermodynamic Data for Ions in Aqueous Solution and Some Solids (for additional data see Appendix IV)*

Ion or solid	H^0, kcal mole^{-1}	G^0, kcal mole^{-1}	S^0, eu
H^+	0.00	0.00	0.00
F^-	-79.50	-66.64	-3.3
Cl^-	-39.95	-31.37	13.5
Br^-	-29.05	-24.85	19.7
I^-	-13.19	-12.33	26.6
OH^-	-54.97	-37.59	-2.57
S^{2-}	7.9	20.5	-3.5
SO_4^{2-}	-217.32	-177.97	4.8
Na^+	-57.28	-62.59	14.4
Cs^+	-59.2	-67.41	31.8
Ag^+	25.23	18.43	17.37
Mg^{2+}	-110.41	-108.99	-28.2
Ca^{2+}	-129.77	-132.18	-13.2
Sr^{2+}	-130.38	-133.2	-9.4
Ba^{2+}	-128.67	-134.0	5
Fe^{2+}	-21.3	-18.85	-32.9
Zn^{2+}	-36.78	-35.14	-26.8
Bi^{3+}		19.8	
Al^{3+}	-127	-116	-76.9
CaF_2	-290.3	-277.7	16.46
$NaCl$	-98.23	-91.79	17.33
$ZnCl_2$	-99.20	-88.30	26.64
$AgCl$	-30.37	-26.24	23.0
$Mg(OH)_2$	-221.00	-199.25	15.10
$Fe(OH)_2$	-136.0	-116.3	21
$Al(OH)_3$	-306.2	-273.3	16.76
CaS	-115.3	-114.1	13.5
FeS	-23.9	-24.0	14.41
Bi_2S_3	-34.2	-33.6	47.9
ZnS	-49.23	-48.11	13.8
$MgSO_4$	-305.5	-278.2	21.9
$BaSO_4$	-350.2	-323.4	31.6

insoluble substances. One example will be used to illustrate the method.

The standard cell potentials for the half-reactions

$$AgCl + e^- = Ag + Cl^-$$

and

$$Ag^+ + e^- = Ag$$

are 0.222 and 0.800 volt respectively. Therefore, ΔE^0 for the overall reaction

$$AgCl = Ag^+ + Cl^-$$

is -0.578. Using

$$\Delta G^0 = -RT \ln K = -n\mathcal{F} E^0$$

$$RT \ln K_{sp} = \mathcal{F}(-0.578)$$

$$\ln K_{sp} = -\frac{\mathcal{F}}{RT}(0.578)$$

$$\log K_{sp} = -\frac{0.578}{0.059} = -9.8$$

we obtain the solubility product as

$$K_{sp} = 10^{-9.8} = 1.5 \times 10^{-10}$$

The solubility products of a number of inorganic salts are listed in Table 13-5. From this data we see that the most insoluble substances are generally sulfides. Hydroxides of the same metals are usually more soluble. Although the data in this table is necessarily not complete (it does not contain all of the insoluble salts), it does contain nearly all of the chlorides that are insoluble. Nearly all other chlorides and many fluorides are soluble. Of the salts containing complex anions, such as PO_4^{3-} and CrO_4^{2-}, almost all are insoluble except perchlorates, nitrates, and sulfates. However, SO_4^{2-} forms insoluble salts with Pb^{2+} and with the larger group IIA cations (all except Be^{2+} and Mg^{2+}).

The solubility of a number of salts is apparent from the preceding discussion and the data presented in Tables 13-5 and 13-6. Thermodynamic data may be used to obtain this information, but a deeper understanding of solubility in terms of the relationship between the

Ch. 13 Polarizing Ability and Solubility

Table 13-5 *Solubility Products of Salts*

Substance	K_{sp}	Substance	K_{sp}
Fluorides		**Sulfides**	
BaF_2	5.3×10^{-6}	MnS	4×10^{-14}
PbF_2	6.9×10^{-9}	FeS	1.6×10^{-19}
CoF_2	1.2×10^{-6}	NiS	4×10^{-22}
SrF_2	5.2×10^{-9}	ZnS	2.8×10^{-25}
CaF_2	1.2×10^{-9}	SnS	3.5×10^{-28}
MgF_2	9.3×10^{-11}	PbS	9×10^{-29}
Chlorides		CdS	1.4×10^{-29}
$PbCl_2$	1.7×10^{-5}	Bi_2S_3	2×10^{-99}
TlCl	1.9×10^{-4}	SnS_2	10^{-70}
CuCl	1.8×10^{-7}	CuS	1.2×10^{-36}
Hg_2Cl_2	1.5×10^{-18}	Cu_2S	2×10^{-48}
AgCl	1.7×10^{-10}	Ag_2S	6×10^{-50}
		HgS	2×10^{-53}
Hydroxides		**Phosphates**	
TlOH	8.0×10^{-2}	$AlPO_4$	5.8×10^{-19}
$Ca(OH)_2$	7.8×10^{-6}	Ag_3PO_4	1×10^{-21}
$Mg(OH)_2$	9.3×10^{-12}	$FePO_4$	1.3×10^{-22}
$Mn(OH)_2$	2.1×10^{-13}	$Ca_3(PO_4)_2$	2×10^{-29}
$Cd(OH)_2$	5.1×10^{-15}	$Sr_3(PO_4)_2$	1×10^{-31}
$Fe(OH)_2$	4.7×10^{-16}	$Ba_3(PO_4)_2$	6.0×10^{-39}
$Co(OH)_2$	2.5×10^{-16}	**Chromates**	
$Zn(OH)_2$	4.3×10^{-17}	$Cu(CrO_4)$	3.6×10^{-6}
$Cr(OH)_2$	10^{-17}	$Sr(CrO_4)$	5×10^{-6}
$Cu(OH)_2$	2.2×10^{-21}	$Hg_2(CrO_4)$	2×10^{-9}
$Al(OH)_3$	6×10^{-33}	$BaCrO_4$	1.2×10^{-10}
$Sn(OH)_2$	6×10^{-27}	Tl_2CrO_4	9.8×10^{-13}
$Mn(OH)_3$	10^{-36}		
$Fe(OH)_3$	2.5×10^{-39}		
$Co(OH)_3$	10^{-43}		
$Sn(OH)_4$	4×10^{-65}		

thermodynamic data and the nature of the ions involved requires that the solution process be examined in detail.

C. THE SOLUTION PROCESS

The solubility of a salt depends upon a number of factors, and it is not possible to specify any single one as being the most important. The relative magnitude of each factor must be considered. The reaction involved in the solution of a simple salt may be represented as

$$MX(s) = M^+(aq) + X^-(aq)$$

Sec. II The Solubility of Inorganic Salts

Table 13-6 ΔG_{soln} *for some Salts*

Substance	ΔG_{soln}, kcal mole^{-1}	Substance	ΔG_{soln}, kcal mole^{-1}
ZnBr$_2$	−10.24	AgCl	+13.30
ZnI$_2$	−9.86	AgBr	+16.74
ZnSO$_4$	−4.1	AgI	+21.92
Ni(OH)$_2$	+20.8	NiF$_2$	+0.2
MnS	+18.2	CoF$_2$	+8.4
FeS	+25.7	MnCl$_2$	−12.0

The solution process depends upon the solvent approaching the crystal and pulling the ions apart. In the simplest possible terms a crystal will dissolve if the attraction between the solvent and the ions is larger than the attraction between the ions in the crystal. The enthalpy terms of interest are the energy required to form free gaseous ions

$$MX(s) \xrightarrow{-\Delta H_u} M^+(g) + X^-(g)$$

and the hydration enthalpy of the cation and the anion

$$M^+(g) + nH_2O \xrightarrow{\Delta H_c} M(H_2O)_n^+$$

$$X^-(g) + nH_2O \xrightarrow{\Delta H_a} X(H_2O)_n^-$$

As usual the solution process may be represented by a Born-Haber cycle. In the cycle shown in Fig. 13-5, ΔH_u is the lattice enthalpy, ΔH_{hyd}

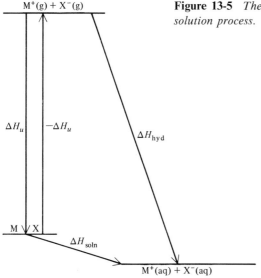

Figure 13-5 *The Born-Haber cycle for the solution process.*

Table 13-7 *Thermodynamic Changes Corresponding to the Process MX (c) $\longrightarrow$ M⁺(aq) + X⁻(aq) at 25°C (kcal mole⁻¹)*

Compound	$-\Delta H_u$	$-\Delta H_{hyd}$	ΔH_{soln}
LiF	245.8	245.5	+0.25
NaF	218.2	217.9	+0.2
KF	193.8	198.0	−4.2
RbF	184.9	192.0	−7.1
CsF	172.3	184.1	−11.8
LiCl	203.0	211.9	−8.93
NaCl	185.3	184.3	+0.92
KCl	168.5	164.4	+4.1
RbCl	162.4	158.4	+4.06
CsCl	154.8	150.5	+4.3
LiI	181.8	196.9	−15.1
NaI	167.6	169.3	−1.8
KI	155.2	149.4	+4.8
RbI	149.8	143.4	+6.4
CsI	143.6	135.5	+8.1

is the total enthalpy of hydration of anion and cation, and ΔH_{soln} represents the enthalpy of solution. From the Born-Haber cycle

$$\Delta H_{soln} = -\Delta H_u + \Delta H_{hyd}$$

As a first approximation, if ΔH_{soln} is negative, the substance will be soluble. (Of course this neglects entropy changes, a very dangerous thing to do. They will be included later.) The sign of ΔH_{soln} thus depends upon the relative magnitude of ΔH_u and ΔH_{hyd}. If ΔH_{hyd} is larger than ΔH_u, the substance will be soluble. ΔH_{soln} will be negative. If ΔH_u is larger, ΔH_{soln} will be positive and the substance will be insoluble.

In Chapter 6 the effects of size and charge on the lattice energy were emphasized and the radius-ratio effect was also noted. At this time it is necessary to point out two additional factors that become important in some cases. These factors are the polarizing ability of the cation and the ease of polarization of the anion. If a polarizing cation is combined with an anion that is easily polarized, the lattice energy will be more negative than that expected on the basis of size and charge alone. The effect is especially noticeable for transition metal ions where the lattice energy often increases faster than the size decreases. It is also apparent from the lattice energies listed (Table 6-6) for pairs of cations of similar size, such as Ca^{2+}-Cd^{2+} and Sr^{2+}-Hg^{2+}.

The polarizing ability of a cation also influences the enthalpy or free energy of hydration, as was pointed out earlier in this chapter. Both of these effects, the influence of polarizing ability on lattice enthalpies and hydration enthalpies, represent deviations from the purely ionic model. The greater the effect, the larger the deviation and the greater the covalency of the bonding. In the remainder of this chapter these considerations will be used to interpret relative solubilities.

1. THE SOLUBILITY OF GROUP IA SALTS

The elements of group IA form the largest cations with the lowest charge. As a result the lattice enthalpies of salts and hydration enthalpies of the ions are small. The size and charge of these cations would seem to indicate that polarization is negligible.

In Table 13-7 the lattice enthalpies and total hydration enthalpies of the alkali metal halides are compared. Lithium salts show the largest lattice enthalpies as well as the largest hydration enthalpies. The ΔH_{soln} values calculated from these quantities are given in Table 13-7. When these data are plotted as they are in Fig. 13-6 some interesting features are apparent.

The enthalpy of solution for combinations containing large anions and large cations (chlorides, bromides, and iodides of K^+, Rb^+, and Cs^+) is positive whereas the enthalpy of solution for combinations

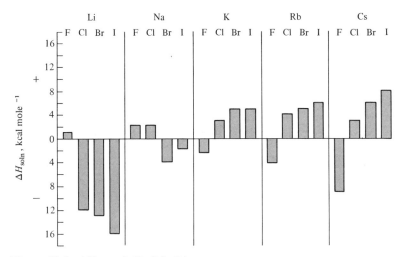

Figure 13-6 ΔH_{soln} of alkali halides.

Table 13-8 Solubilities of Group 1A Salts (g per 100 ml)

	Li^+	Na^+	K^+	Rb^+	Cs^+
F^-	0.27	4.22	92.3	130.6	367
Cl^-	63.7	35.7	34.7	77	162.2
Br^-	145	116	53.5	98	124
I^-	165	184	127	152	44
NO_3^-	90	92	13.3	44.3	9.2
ClO_4^-	60		1.5	1.0	2.0
SO_4^{2-}	25	44	12	42	180

containing smaller anions and the same cations (fluorides of K^+, Rb^+, and Cs^+) is negative. On the other hand, the enthalpy of solution for the fluorides of large cations (Cs^+, Rb^+, K^+) is negative, whereas the fluorides of smaller cations (Na^+, Li^+) have positive enthalpies of solution. The data demonstrate very nicely the importance of size in the determination of the lattice energy and the effect of size on ΔH_{soln}. Although the lattice enthalpies of cesium (Cs) salts decrease in the order, $CsF > CsCl > CsBr > CsI$, the decrease in this series is not as large as it is in the series LiF, LiCl, LiBr, and LiI. The total difference in lattice energy for the lithium salts is 64.0 kcal mole^{-1} whereas the difference for the cesium series is 28.7 kcal mole^{-1}. In other words, the lithium salts become less stable as the size of the anion increases whereas the cesium salts become relatively more stable. This is the same effect of size that was noted in Chapter 6.

From these data alone we might conclude that the solubility trends shown by group IA fluorides will be reversed for the chlorides, bromides, and iodides. That is, LiF should be less soluble than CsF, but CsI should be less soluble than LiI. The solubility data given in Table 13-8 show that these trends are indeed observed.

Table 13-9 gives the lattice enthalpies of the alkali metal nitrates. The difference between the lattice enthalpies of lithium and cesium nitrates is about one half of the difference between the fluorides of these metals or about the same as the difference between the bromides. Thus we might expect $LiNO_3$ to be more soluble than $CsNO_3$. From the solubility data on the salts of complex anions given in Table 13-8 we see that this is the case and, as a matter of fact, the same trend is observed for a number of salts that involve large complex anions. In many cases the solubilities of the salts of large anions decrease in the order $Li > Na > K \approx Rb \approx Cs$. The decrease is large enough for the perchlorates so that the K^+, Rb^+, and Cs^+ salts of this anion are only moderately soluble.

Table 13-9 Lattice Enthalpies of the Group 1A Nitrates

Salt	Lattice enthalpy, ΔH_u, kcal mole^{-1}
$LiNO_3$	−195
$NaNO_3$	−178
KNO_3	−160
$RbNO_3$	−155
$CsNO_3$	−148

From the preceding discussion it is fairly obvious that, although considerable insight into solubility *trends* and the factors affecting solubility may be gained by examining the calculated enthalpy of solution, there are some difficulties when an attempt is made to decide if a particular substance is soluble on the basis of the sign of ΔH_{soln}. It is well known, for example, that all of the halides of the group IA metal ions are soluble or slightly soluble, and yet the enthalpies of solution are positive for many of the halides. Of course, the sign of ΔH may not be used as the criterion for solubility. The sign of the free energy change is the criterion for spontaneity so the entropy changes must also be considered.

The magnitude of the entropy term depends upon the nature of the substance, and the relative magnitude of the term is difficult to estimate on a simple basis. It is clear, however, that the net entropy change for the solution process will be more negative or more unfavorable the smaller the ion and the larger the charge of the ion. This is true because the smaller the size and the greater the charge of the ion the greater the hydration of the ion, and the greater the hydration the more ordered the structure of the solution around the ion. The data given in Table 13-10 support this argument. The positive values given for large cations and anions with low charges may be rationalized because the introduction of these cations into water does not result in a more ordered system. For these ions there is an increase in disorder because the ordered structure of water is partially destroyed.

With the entropy data available it is possible to interpret the solubilities of the fluorides and chlorides discussed in a qualitative manner above. The free energy change that corresponds to the solution of a number of salts is given in Table 13-11. Negative values of the free energy change correspond to a soluble salt, whereas positive values indicate an insoluble substance. The data confirm the solubility trends predicted above. For both LiF and NaF the enthalpy term and the entropy term are unfavorable. The corresponding ΔG values are positive but not highly so, and the equilibrium constants are less than 1.00 but they are not extremely small. LiF ($K_{sp} = 10^{-2}$) is the least soluble of the group IA fluorides. With the larger cations (CsF and RbF) both the enthalpy term and the entropy term are favorable, and the salts are soluble. The solubility trend is reversed for the chlorides, bromides, and iodides. For the chlorides the only salt with a favorable enthalpy term is LiCl. All of the chlorides and iodides are soluble, however, because the entropy term is sufficiently positive in each case to compensate for the unfavorable enthalpy term. As predicted, CsI is the least soluble iodide.

Table 13-10 $T \Delta S^0$ Values for Individual Ions crystal $\longrightarrow$ aqueous solution at 25°C (kcal mole^{-1})

	IA	IIA	IIIB	IVB	VB	VIB	VIIB	VIII			IB	IIB	IIIA	IVA	VA	VIA	VIIA	Inert gases
	1.7															OH⁻ −0.8	−1 −0.8	
	0.3	−13.4					+2 −11.8	+2 −14.6	+2 −14.5	+2 −16.2	+2 −12.7	+2 −13.4					−1 +2.8	
	2.8	−9.2									+1 −0.5	+2 −10.9	+1 +2.0	+2 −6.0			−1 +3.8	
	3.4	−8.5															−1 +5.8	
	3.7	−5.9																

NO_3^- +6.2 CO_3^{2-} −5.0
MnO_4^- −7.2 SO_4^{2-} −1.6
ClO_3^- +5.9

2. Solubility of Group IIA Halides

In general the lattice enthalpies of the group IIA salts should be larger than those of group IA because of the increased charge. At the same time, however, hydration enthalpies should also be larger. From the data given in Table 13-12, we see that the two effects almost balance, and the ΔH_{soln} for these salts is again small except for MgI_2. For the fluorides the entropy terms are fairly large and unfavorable and, as a result, these fluorides are insoluble. The chlorides, bromides and iodides of all the group IIA elements are soluble.

One way the value of ΔH_{soln} (that is, ΔH_{soln} for MgF_2 is negative instead of being more positive than the value for CaF_2) can be explained is to postulate that, although polarizing ability is not important for Ca^{2+}, Sr^{2+}, and Ba^{2+}, it might be for Mg^{2+}. The enthalpy of hydration of Mg^{2+} is high because of this and a favorable enthalpy term results, but the entropy term is still predominant so MgF_2 is insoluble. The polarizing ability of Mg^{2+} would not be expected to contribute much to the lattice enthalpy of MgF_2 because of the low ease of polarization of F^-.

The effect of size on the enthalpy of solution of group IIA salts is nicely illustrated when the thermodynamic data are plotted as in Fig. 13-7. For every cation, ΔH_{soln} of the iodide is negative, but it becomes more positive as the size of the cation increases. However, in this group, ΔH_{soln} does not become positive even with Ba^{2+}. This inversion did take place in the group IA halides and as a result CsI is less soluble than BaI_2.

The solubility of the group IA and IIA fluorides sets the pattern for further solubility considerations. In the following discussion it will be assumed that F^- is nonpolarizable and continued reference will be made to the group IA and IIA fluorides. It should be emphasized that the discussion deals with relative solubilities. The procedure involves comparison of the effects of size, charge, polarizing ability, and ease

Table 13-11 *Thermodynamic Changes Corresponding to the Process* $MX(c) \longrightarrow M^+(aq) + X^-(aq)$ *at* $25°C$

Compound	ΔG, kcal mole^{-1}
LiF	+2.7
NaF	+0.7
KF	−5.8
RbF	−9
CsF	−17
LiCl	−9.80
NaCl	−2.17
KCl	−1.36
RbCl	−4
CsCl	−4
LiI	−18.4
NaI	−6.92
KI	−2.57
RbI	−2.0
CsI	−0.0

Table 13-12 *Thermodynamic Changes Corresponding to the Process* $MX_2(s) \longrightarrow M^{2+}(aq) + 2X^-(aq)$ *at* $25°C$ *(kcal mole^{-1})*

Compound	$-\Delta H_u$	$-\Delta H_{\text{hyd}}$	ΔH_{soln}	$T\Delta S_{\text{soln}}$	ΔG_{soln}
MgF_2	700.5	701.2	−0.7	−14.5	+13.7
CaF_2	624.5	623.0	+1.5	−10.7	+12.2
SrF_2	588.7	587.7	+0.9	−10.5	+11.3
BaF_2	553.2	554.0	−0.8	−7.9	+7.2
MgI_2	553.2	604.0	−50.8	+1.6	−49.2
CaI_2	497.4	525.8	−28.4	+1.8	−30.2
SrI_2	469.3	490.5	−21.3	+1.9	−23.2
BaI_2	445.7	456.8	−11.1	+4.4	−15.5

Ch. 13 Polarizing Ability and Solubility

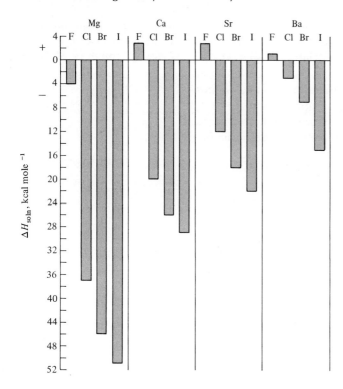

Figure 13-7 ΔH_{soln} of alkaline earth salts.

of polarization on lattice enthalpy and hydration enthalpy. Some of the available thermodynamic data are given in Table 13-13. Since some of the H^0 values for transition metal fluorides are not accurately known, this table includes calculated lattice enthalpies and ΔH_{soln} values based on lattice enthalpy as well as those calculated from the H^0 values.

3. The Solubility of Transition Metal Halides

The approach to solubilities used here requires a comparison of the ions involved and an evaluation of the effects of differences on the magnitude of lattice enthalpies and hydration enthalpies. The starting point for such a comparison is Mg^{2+}. Mn(II) will be the first transition metal ion considered.

The Mn^{2+} ion is larger than Mg^{2+} (0.80 Å compared with 0.65 Å). Because of the difference in size, the lattice enthalpy and hydration enthalpy of MnF_2 should be less than those for Mg^{2+}. However, because of the presence of d electrons the effective charge of Mn^{2+} is greater than 2 and the ion has a higher polarizing ability than Mg^{2+}. Moreover, the d electron cloud of Mn^{2+} may be distorted by an easily polarized anion; therefore Mn^{2+} can exhibit an even greater polarizing ability. For example, the lattice enthalpy of MnS is greater than the lattice enthalpy of MgS. The larger polarizing ability of Mn^{2+} is also apparent from the values of ΔH_{hyd} of the cations.

Although ΔH_{hyd} of Mn(II) is less than that of Mg^{2+}, the difference between the values (-652 and -669 kcal mole^{-1}) is less than the difference between ΔH_u of the fluorides of the two metal ions ($\Delta H_u MnF_2 = -662$ kcal mole^{-1}, $\Delta H_u MgF_2 = -700$ kcal mole^{-1}). As a consequence, the ΔH_{soln} of MnF_2 is fairly negative (-23 kcal mole^{-1}). In addition, because of the larger size of Mn(II), the entropy term for the dissolution of MnF_2 is a not as unfavorable as the entropy term for MgF_2. For both these reasons MnF_2 should be more soluble than MgF_2.

This argument may seem to be inconsistent because the effect of polarizing ability on the entropy term has not been considered. It has not been considered and it is not necessary to do so because the $T\Delta S$ term amounts to 10–20 kcal compared to an enthalpy term of the order of 600 kcal.

Going from Mn^{2+} to Fe^{2+} a fairly large increase in lattice enthalpy is observed whereas the corresponding increase in ΔH_{hyd} is only moderate (34 compared with 24 kcal mole^{-1}). The result is a small negative enthalpy term leading to an insoluble salt because of a highly negative entropy term that is comparable in magnitude to that of Mg^{2+}. A similar increase in lattice enthalpy compared with hydration enthalpy results in a more insoluble CoF_2 salt. From CoF_2 to the end of the first transition series the general trend is toward increasing solubility because of the fairly regular increase in lattice enthalpy with decreasing

Table 13-13 *Thermodynamic Changes Corresponding to the Process* $MX_2(c) \longrightarrow M^{2+}(aq) + 2X^-(aq)$ *at* $25°C$ *(kcal mole^{-1})*

Compound	$-\Delta H_u$	$-\Delta H_{hyd}$	ΔH^*_{soln}	$\Delta H^{\dagger}_{soln}$	$T\Delta S$	ΔG
MnF_2	662	685	-23	-23	-13.7	-9
FeF_2	696	709	-13	-12 ± 5	-17.9	$+4$
CoF_2	(713)	722	(-7)	-7.5	-15.6	$+8$
NiF_2	728	743	-15	-16.2	-16.5	$+0.2$
CuF_2	727	746	-17	-13.8	-16 ± 1	$+2$
ZnF_2	710	732	-22	-13.1	-15.2	$+2.1$
$MnCl_2$	598	617	-19	-17.63	-5.63	-11.95
$FeCl_2$	624	641	-17	-19.5	-10.1	-9.33
$CoCl_2$	642	655	-13	-19.1	-7.9	-11.2
$NiCl_2$	658	675	-17	-19.8	-8.1	-11.7
$CuCl_2$	666	679	-13	-11.8	-6.6	-5.2
$ZnCl_2$	648	664	-16	-17.48	-7.87	-9.58

*Calculated from ΔH_u and ΔH_{hyd}.
†Calculated from H^0 value of the solid.

size (that is, an increase of 14 kcal from Co^{2+} to Cu^{2+}) and an irregular increase (23 kcal) in the hydration enthalpy with decreasing size. From Fe(II) to Cu(II) the hydration enthalpy increases faster than the lattice enthalpy because of the increasing polarizing ability of the cations. Water is more easily polarized than fluoride. The result is an increasing solubility across the series. NiF_2, CuF_2, and ZnF_2 are slightly soluble with FeF_2 and CoF_2 being less soluble.

The solubility trends observed for the chlorides, bromides, and iodides of the transition metals are the reverse of those observed for the fluorides. Although all of the +2 chlorides are soluble, $CuCl_2$ is less soluble than $FeCl_2$. In the case of the chlorides the change in lattice enthalpy between $FeCl_2$ and $CuCl_2$ is 42 kcal (compared to 31 kcal for the fluorides) whereas the change in hydration enthalpy is the same as that of the fluorides (37 kcal). In this case the lattice enthalpy is increasing faster than the hydration enthalpy as we move from Fe(II) to Cu(II).

This subtle difference between the chemistry of fluoride and chloride arises because chloride is easier to polarize than water whereas fluoride is harder to polarize than water. The more highly polarizing cations will combine preferentially with the substance that is most easily polarized. Cu(II) combines with H_2O in preference to F^- but with Cl^- in preference to H_2O. The transition metal chlorides are soluble because of the large size of Cl^- in spite of the effect of the polarizing cation.

4. SOLUBILITY OF HYDROXIDES

Hydroxide (OH^-) is like F^- in that the size is similar and the charge is the same. It is more easily polarized than H_2O or F^-, but it is harder to polarize than Cl^-.

Because of the similarities between F^- and OH^- the entropy term for hydroxides should be very nearly the same as that for fluorides, and the solubilities of hydroxides are expected to follow the same trends as the fluorides with a few exceptions. In general on the basis of the enthalpy term, hydroxides of group IA and IIA cations should be more soluble than the corresponding fluorides, whereas the hydroxides of the fairly good polarizing cations should be less soluble than the fluorides. Because of the small size of OH^-, hydroxides will have a fairly high lattice enthalpy that will be even larger for highly polarizing cations.

More specifically, the group IA cations all form soluble hydroxides. Of these, LiOH is expected to be the least soluble, but it is not expected to be as insoluble as LiF because of the greater hydration energy of OH^-. With the exceptions of Be^{2+} and Mg^{2+} the group IIA cations are incapable of polarizing OH^-. The remaining hydroxides of group

IIA are more soluble than the fluorides because of the greater hydration energy of OH^-.

The transition metal (+2) fluorides are just on the verge of solubility. The hydroxides are insoluble because OH^- is easier to polarize than water, and solubility decreases across the period because of the increasing polarizing ability of the cations. The K_{sp}'s of the transition metal hydroxides demonstrate this trend. $Mn(OH)_2$ hydroxide has a solubility product of 2.1×10^{-13} compared to the value 2.2×10^{-21} for $Cu(OH)_2$. For the IB group cations the entropy term is more favorable because of the large size of these cations; however, on the basis of enthalpy the hydroxides of these cations should be less soluble than the fluorides. These trends are observed. The group IB hydroxides are insoluble, but the solubility products are not exceedingly small. The values of the solubility products indicate that hydroxide is not a particularly good anion for inducing softness in these cations. On the other hand, the K_{sp} values for the group IVA cations are very small, and these hydroxides are quite insoluble.

5. Solubility of Sulfides

The sulfide ion is very large and easily polarized. Only the sulfides that contain a low polarizing cation or cation with little or no induced polarizing ability will be soluble. This includes the group IA and IIA cations. A trend similar to the chlorides and hydroxides is noted for the sulfides of the +2 ions from Mn^{2+} to Cu^{2+}. The change in lattice enthalpy across this series is 92 kcal mole^{-1} compared to 52 kcal mole^{-1} for fluorides. The K_{sp}'s of these salts decrease much more rapidly across the series than the K_{sp}'s of the hydroxides. Many of the elements toward the right and bottom of the periodic table form highly insoluble substances with sulfide ions.

6. The Solubility of Salts of Complex Anions

The complex anions can generally be classified into three categories. Typical members of these categories are ClO_4^-, NO_3^-; SO_4^{2-}; PO_4^{3-}, $C_2O_4^{2-}$, CO_3^{2-}. The obvious difference between these examples is an increasing charge, ClO_4^-, SO_4^{2-}, and PO_4^{3-}. There is also an increase in the ease of polarization and a decrease in the ΔS_{soln} from perchlorate to phosphate. There is, of course, a distinct relationship between each of these factors and the solubility of the salts of these anions.

These ions are perchlorate, nitrate, sulfate, phosphate, oxalate, and carbonate.

A set of solubility rules may be formulated for salts of these anions from the considerations that have been set down in this chapter. All perchlorates will be soluble because of a favorable entropy and enthalpy term (perchlorate cannot be polarized). Sulfate is slightly easier to polarize, and the salts of this anion might be expected to be less soluble than perchlorates. Sulfate is not as easy to polarize as water however; so in spite of an unfavorable entropy term the sulfate of any good polarizing cation should be soluble. As a result all transition metal sulfates are soluble but the group IIA sulfates (except for Mg^{2+} and Be^{2+}) as well as $PbSO_4$ are insoluble. The solubilities of the group IIA sulfates also demonstrate the effect of relative size. The K_{sp} data show that the larger cations form the least soluble sulfates. Magnesium sulfate is easily dissolved in water.

The hydrate of magnesium sulfate ($MgSO_4 \cdot 7H_2O$) that crystallizes from a concentrated solution is Epsom salt. The hydrate of calcium sulfate ($CaSO_4 \cdot 2H_2O$) is obtained when a lower hydrate ($CaSO_4 \cdot \frac{1}{2}H_2O$), plaster of Paris, is mixed with a specified quantity of water

$$2CaSO_4 \cdot \tfrac{1}{2}H_2O + 3H_2O \longrightarrow 2(CaSO_4 \cdot 2H_2O)$$

In the final category all factors lead to insolubility so that all phosphates and carbonates are insoluble except those of the cations of group IA.

D. SOLUBILITY EQUILIBRIUM

In the preceding discussion K_{sp}'s were used as a measure of solubility. The solubility product may also be viewed from another point of view. It expresses the condition necessary for precipitation of an insoluble salt. If the product of the concentration of ions exceeds the K_{sp}, precipitation will occur and will continue until the concentration of ions in solution just satisfies the K_{sp} expression.

For example, if the sulfate ion concentration is 10^{-2}, precipitation of $BaSO_4$ will not occur until the concentration of Ba^{2+} is 1.6×10^{-7}. On the other hand, when the Ba^{2+} concentration is 10^{-2}, the concentration of SO_4^{2-} remaining in solution will be 1.6×10^{-7}. From this it is apparent that the complete removal of an ion from solution is not possible; however, if the salt is fairly insoluble, the concentration can be reduced to a negligible amount.

One type of problem that must be illustrated here involves substances such as CaF_2. This salt is soluble to the extent of 2.15×10^{-4} mole liter^{-1}. From

$$CaF_2 = Ca^{2+} + 2F^-$$

we see that for every mole of CaF_2 that dissolves, 2 moles of F^- are produced. To obtain the fluoride ion concentration the solubility must be multiplied by 2. The K_{sp} is then calculated from

$$K_{sp} = (2.15 \times 10^{-4})(2 \times 2.15 \times 10^{-4})^2 = 4.0 \times 10^{-11}$$

In order to dissolve a precipitate the product of the ion concentrations must be less than the K_{sp}. The process of redissolving a precipitate then requires that a method be found to decrease the concentration of one of the ions in solution. Salts containing the anion of a weak acid may be more soluble in acid solution because of the formation of undissociated weak acid,

$$MX = M^+ + X^-$$

$$H^+ + X^- = HX$$

The insoluble salts of Ac^-, CO_3^{2-}, S^{2-}, etc. are possible examples of salts that may dissolve in acids.

Equilibrium calculations may be used to decide whether the salt will be affected by an acid. In order to effect solution, the concentration of one or both of the ions must be reduced. Whether the anion concentration is reduced sufficiently to cause the salt to dissolve depends upon the strength of the acid (K_a) and the solubility of the salt (K_{sp}). Thus, some salts containing the anion of a weak acid will dissolve whereas others will not. If the acid is not weak enough to cause sufficient reduction of the anion concentration, the salt will not dissolve. MnS and ZnS dissolve easily in acids, but acid alone will not dissolve some other sulfides (HgS, PbS, and so forth). In these cases other methods must be used.

Another method of dissolving a salt involves producing a complex ion. AgCl may be dissolved in a water solution containing ammonia.

$$AgCl = Ag^+ + Cl^-$$

$$Ag^+ + 2NH_3 = Ag(NH_3)_2^+$$

A change in oxidation state may also be used to dissolve a sparingly soluble salt. This is also involved in dissolving PbS. $S^=$ is converted to free S with nitric acid (HNO_3). On the other hand, $BaSO_4$ which is not soluble in acid may be dissolved if SO_4^{2-} is reduced to S^{2-}. Carbon is a satisfactory reducing agent.

$$BaSO_4 + 4C \longrightarrow BaS + 4CO$$

BaS is soluble in water.

$$BaS + 2H_2O \longrightarrow H_2S + Ba^{2+} + 2OH^-$$

There are a few special and interesting examples where the relationship between basicity of the anion and the solubility of the salt is very important. Those discussed here involve the carbonate ion. The solubility of carbonates is strongly dependent upon acidity because the carbonate ion is a base.

All of the solution processes that have been used thus far have used the Brønsted acid (H^+). A Lewis acid (CO_2) may also be used to dissolve $CaCO_3$ if the pressure of CO_2 is large enough. The equation that represents the reaction is

$$CO_2(g) + H_2O + CaCO_3 \longrightarrow Ca^{2+} + 2HCO_3^-$$

The action of CO_2 and H_2O on limestone is partly responsible for the formation of extensive caves in limestone deposits. The equilibrium expression for the reaction is

$$K = \frac{(a_{Ca^{2+}})(a_{HCO_3^-})^2}{P_{CO_2}} = 6 \times 10^{+5} = 10^{9,310/2.303RT} 10^{-57.8/2.303R}$$

This particular equilibrium is responsible also for much of the chemistry of ground waters and of limestone deposits. The respiration of microscopic animals and the roots of plants tend to make soils acid.

$$C_6H_{12}O_6 + 6O_2 \longrightarrow 6CO_2 + 6H_2O$$
glucose

$$CO_2 + H_2O \rightleftharpoons H_2CO_3$$
carbonic acid

The moisture in the soil is in equilibrium with a pressure of CO_2 that is substantially higher than that in the atmosphere. The equilibrium expression for the reaction

$$CO_2 + H_2O + CaCO_3(s) \rightleftharpoons Ca^{2+} + 2HCO_3^-$$

shows that under such conditions $CaCO_3$ will be quite soluble. If there is an extensive layer of limestone beneath the soil, this dissolving will enlarge any cracks into extensive, but nearly featureless, caves. Where large amounts of rock have dissolved near the surface, the ground may sag into the cavities, forming sinkholes, directing more and more of the water falling on the ground into the cave system.

The equilibrium is, however, shifted drastically at some point when erosion or other action opens the cave to the atmosphere. The droplets of water on the inside of the cave will lose CO_2 to become supersaturated with respect to $CaCO_3$. Thus a cave open to the atmosphere will tend to fill up with stalactites, stalagmites, and other deposits of $CaCO_3$ on the walls of the cave.

The $CaCO_3$ solubility equilibrium is also shifted in favor of the solid by raising the temperature; therefore $CaCO_3$ and $MgCO_3$ are deposited as scale in water heaters and boilers. Such waters are said to possess temporary hardness since Ca^{2+} and Mg^{2+} are removed by precipitation when the water boils.

$$Ca^{2+} + 2HCO_3^- \longrightarrow CO_2 + H_2O + CaCO_3(s)$$

Hard water can also be softened using calcium oxide.

$$CaO(s) + H_2O \longrightarrow Ca(OH)_2(s)$$

$$Ca^{2+} + 2HCO_3^- + Ca(OH)_2(s) \longrightarrow 2CaCO_3(s) + 2H_2O$$

The use of limestone to lower the acidity of a soil according to the equation

$$CaCO_3 + 2H^+ \longrightarrow Ca^{2+} + CO_2 + H_2O$$

is another example of the relationship between solubility and the basicity of the anion. An even better base, CaO, can be prepared by simply heating limestone

$$CaCO_3 \xrightarrow{\Delta} CaO + CO_2(g)$$

Sea water contains $0.054\ M\ Mg^{2+}$ and $0.010\ M\ Ca^{2+}$. It also contains carbonate, primarily as HCO_3^- at a concentration of $0.0024\ M$. The total ion concentration of sea water is fairly high so that the activity coefficients of ions are much less than 1.00. However, the total ion composition of ocean water can be determined, and the individual activity coefficients can be calculated or estimated within 10–20%. The results are shown in Table 13-14. The activity of Ca^{2+} in sea water is

$$\gamma[Ca^{2+}] = 0.25\ (0.010) = 2.5 \times 10^{-3}$$

Since the solubility product of calcium carbonate is 4.7×10^{-9}, precipitation of $CaCO_3$ will not begin until the activity of CO_3^{2-} is 1.9×10^{-6}.

$$K_{sp} = [Ca^{2+}][CO_3^{2-}] = 4.7 \times 10^{-9} = (2.5 \times 10^{-3})[CO_3^{2-}]$$

$$[CO_3^{2-}] = 1.9 \times 10^{-6}\ M$$

Table 13-14 *Activity Coefficients for the Ions in Sea Water**

Ion	γ	Concentration	Activity
Na^+	0.75	0.48	0.36
K^+	0.63	0.010	0.0063
Mg^{2+}	0.31	0.054	0.017
Ca^{2+}	0.25	0.010	0.0025
Cl^-	0.64	0.56	0.36
HCO_3^-	0.47	0.0024	0.0011
SO_4^{2-}	0.065	0.028	0.0018

*Adapted from R. M. Garrels and M. E. Thompson, *Am. J. Sci.,* **260**:57 (1962).

The pH at which sea water will be saturated with calcium carbonate may be calculated from the second ionization constant of carbonic acid.

$$K_{a_2} = 4.7 \times 10^{-11} = \frac{(a_{H^+})(a_{CO_3^{2-}})}{a_{HCO_3^-}}$$

$$a_{H^+} = 4.7 \times 10^{-11} \frac{(a_{HCO_3^-})}{(a_{CO_3^{2-}})} = 4.7 \times 10^{-11} \frac{(1.1 \times 10^{-3})}{(1.9 \times 10^{-6})} = 2.7 \times 10^{-8}$$

$$a_{H^+} = 10^{0.43} \times 10^{-8} = 10^{-7.57}$$

$$pH = 7.57$$

At a pH lower than 7.6, $CaCO_3$ will dissolve. At a pH greater than 7.6, $CaCO_3$ will precipitate. The amount of Mg^{2+} in sea water is similarly determined by the solubility of $MgCO_3$ and of dolomite, $CaMg(CO_3)_2$.

In the surface layers of the ocean the process of photosynthesis

$$6HCO_3^- + 6H_2O \rightleftharpoons C_6H_{12}O_6 + 6OH^- + 6O_2$$

raises the pH to around 8. Under these conditions large reefs of limestone and dolomite may be built up. Many deposits are now part of the continents so that limestone (which is $CaCO_3$) is the second most abundant rock.

Some of the oldest limestone deposits (Precambrian) show patterns indicating that the material was precipitated in the presence of algae. As organisms evolved in this environment some of them adapted to make use of the supersaturation of $CaCO_3$ for the formation of protective shells of $CaCO_3$. In some cases, as the corals, the protection of an external $CaCO_3$ skeleton is gained at the expense of mobility.

Recent limestone deposits and the analysis of current reefs indicate that the shells and skeletons of marine animals constitute most of the calcium carbonate being precipitated from the oceans. The fact that limestone will deposit is determined by the way the equilibrium is shifted as described above; the particular way it deposits is largely controlled by the organisms present.

QUESTIONS

1. Write equations for the reaction of the alkaline earth metals with oxygen, nitrogen, and carbon.
2. What metal might be used to reduce BaO?
3. Define or state briefly the meaning of the terms: softness, polarizing ability, ease of polarization, high spin and low spin complex.
4. What is the electronic structure of high spin Co^{2+}, of low-spin Co^{2+}?
5. List the factors that affect the lattice energy of a salt.
6. Explain the effect of each of the factors listed in Question 5.
7. What methods would be appropriate for determining the solubility of LiCl, MgF_2?
8. Design a method that could be used to determine quantitatively the solubility of a chloride.
9. Discuss the solubility trends observed for the group IA halides.
10. What are the solubilities of group IIA halides?
11. How do the solubilities of group IIA sulfates, chromates, and oxalates vary?
12. How do these trends compare with the trends observed for the fluorides and hydroxides? Suggest a reason for the difference if any.
13. What ions may be used to separate the group IIA ions from the transition metal ions Fe^{2+}, Co^{2+}, and Ni^{2+}?
14. Order each of the following sets according to the ease of polarization (hardest to polarize at the left, easiest to polarize at the right).
    (a) Cl^-, Br^-, F^-, S^{2-}
    (b) F^-, H_2O, OH^-
    (c) ClO_3^-, PO_4^{3-}, SO_3^{2-}
15. Which of the following cations has the lowest polarizing ability? Li^+, Be^{2+}, K^+, Sr^{2+}
16. On the basis of qualitative arguments relating to expected lattice energies, hydration energies, and entropy effects, which are expected to be the least soluble in each of the following sets?
    (a) ZnF_2, $Zn(ClO_4)_2$, ZnS
    (b) $Ca(ClO_4)_2$, $CaSO_4$, $Ca_3(PO_4)_2$
    (c) NaCl, AgCl, $MgCl_2$
    (d) NaOH, $Ca(OH)_2$, $Ba(OH)_2$

PROBLEMS I

1. Calculate the concentration of H^+ ion required to dissolve AgAc.
2. Will CaF_2 dissolve in acid solution?

Ch. 13 Polarizing Ability and Solubility

3. The ideas discussed in Problem 2 are also very useful in separations. If a substance is soluble in acid, precipitation will be prevented if carried out in acid solution.
   Calculate the concentration of H^+ required to effect a separation of BaF_2 and CaF_2 by precipitating CaF_2 with $0.01\ M\ F^-$. Is such a separation feasible?
4. Can the transition metal ions be separated from the cations of group IIA by precipitation with F^-?
5. Using the data in Appendix IV calculate ΔH_{soln} for the iodides of the cations of group IA and IIA.
6. Using the appropriate thermodynamic data verify your answer to Question 13.
7. If S^{2-} is used as a precipitating agent calculate the H^+ concentration required to separate Cu^{2+}, As^{3+}, Sb^{3+}, and Hg^{2+} from Ni^{2+}, Co^{2+}, Zn^{2+}.
8. Calculate the solubility of FeF_2, $Mg(OH)_2$, LiF, H_2S, and CdS.
9. Calculate the K_{sp} of each of the salts in Problem 8.
10. The partial pressure of CO_2 in the atmosphere is $10^{-3.5}$ atm. How much $CaCO_3$ will dissolve in air-saturated water?
11. The solubility of $BaCrO_4$ (which dissociates into Ba^{2+} and CrO_4^{2-}), is 1.1×10^{-5} moles liter^{-1}. Calculate the K_{sp}.
12. The solubility product of CaF_2 is 1.6×10^{-10}. Calculate the concentration of F^- ions in a $0.40\ M$ solution of $CaCl_2$ saturated with CaF_2.
13. Given the following thermodynamic data at $25°C$ for the solution process of $AgBr$, calculate the enthalpy change on dissolving $AgBr$. $\Delta H_u = -215$ kcal mole^{-1}; $\Delta H_{hyd} = -194$ kcal mole^{-1}.
14. $T\Delta S° = 3.4$ kcal mole^{-1} for the reaction $AgBr \longrightarrow Ag^+ + Br^-$. Calculate K_{sp} for $AgBr$.

PROBLEMS II

1. Write the equilibrium constant expression for the chemical reaction $CaF_2 \longrightarrow Ca^{2+} + 2F^-$.
2. What is the activity of pure solid CaF_2?
3. The solubility product of CaF_2 is 1.0×10^{-10}. Calculate the concentration of Ca^{2+} in a $0.10\ M$ NaF solution saturated with CaF_2.
4. What concentration of F^- is present in a solution saturated with CaF_2 if $[Ca^{2+}] = 0.10\ M$?
5. The ionization constant of HF is 7×10^{-4}. At what pH will $[F^-] = 3.16 \times 10^{-5}$ if $[HF] = 0.20\ M$?
6. The ionic radius of Sc^{3+} is 0.81 Å. Calculate the polarizing ability of Sc^{3+}.
7. How many $3d$ electrons does $Fe(III)$ have?
8. Calculate S for $Fe(III)$.
9. Calculate the polarizing ability of Fe^{3+} using Table 3-7 for the ionic radius.
10. Calculate the polarizing ability of Ag^+.
11. Which of the three ions Sc^{3+}, Fe^{3+}, and Ag^+ forms the least soluble fluoride?

Coordination chemistry is the meeting place of all chemistry . . .
J. CHATT

Chapter 14 Coordination Chemistry

Aside from metallurgical aspects to be treated in Chapter 17, the most important feature of the transition metal ions is the enhanced ability to form a group of compounds generally known as complexes or coordination compounds. The term enhanced is used because *all* metal ions and some nonmetals form coordination compounds, but the transition metal ions are capable of forming many, many more than the ions of elements such as the alkali metals. The alkali metals are, for the most part, restricted to coordination compounds involving oxygen; whereas those of the transition metals may involve the halogens, nitrogen, sulfur, and carbon.

As the term implies, coordination compounds are compounds in which the bonding may be considered to be of the coordinate covalent type. Thus, a coordination compound is made up of an electron acceptor (central metal ion) and an electron donor. The electron acceptor is usually a metal ion such as iron(II) or nickel(II) whereas the donor (called a ligand) may be any group or ion that has a pair of electrons for coordination and is basic in character. Ligands may include neutral molecules such as ammonia and water or negative ions such as cyanide (CN^-) and chloride. From one point of view coordination compounds are the neutralization products of an acid (electron acceptor) and a base (electron donor).

A coordinate covalent bond is one in which a pair of electrons is shared between two atoms. Both electrons came from the same atom.

Although this is a convenient way of describing or defining a coordination compound (as containing a coordinate covalent bond) it must not be regarded as a restriction of the type of bonding responsible for holding a complex together. As we shall see, the bonding may be ionic or completely covalent.

It is found that more than one ligand is usually required for a single acceptor. The number of ligands required is called the coordination number. The coordination number of the larger representative elements is generally found to be six while that of the transition elements is in many cases variable, being four or six for the ions of the first transition series. A few examples that will serve to illustrate this are given in Table 14-1. From these examples it should be noted that one of the factors involved in determining the coordination number is the oxidation number of the metal. A low coordination number is often found for the lower oxidation states, but there is also a dependence upon the ligand. With chloride, iron(II) can have a coordination number of four, whereas cyanide (CN^-) gives a coordination number of six; with nickel(II), both chloride and cyanide (in spite of the great difference in these ions) favor the coordination number four whereas ammonia favors six. The variable coordination number and variable structure of the complexes of the transition elements are all explained by the theories of bonding which will be applied in this chapter and the next.

Coordination compounds have been found to be important in

Table 14-1 *Typical Coordination Complexes*

Ion	Coordination number	Compound or ion formed
Fe(II)	4 or 6	$Fe(Cl)_4^{2-}$
		$Fe(CN)_6^{4-}$
Fe(III)	6	$Fe(CN)_6^{3-}$
Ni(II)	4 or 6	$Ni(Cl)_4^{2-}$
		$Ni(CN)_4^{2-}$
		$Ni(NH_3)_6^{2+}$
Co(II)	4 or 6	$Co(Cl)_4^{2-}$
		$Co(CN)_6^{4-}$
Co(III)	6	$Co(CN)_6^{3-}$
Pd or Pt(II)	4	$Pt(Cl)_4^{2-}$
		$Pt(CN)_4^{2-}$
Pd or Pt(IV)	6	$Pt(Cl)_6^{2-}$
		$Pt(CN)_6^{2-}$

electrodeposition. Research and development have produced coordination compounds that are useful in water softening, blueprint making, catalysis of organic reactions, and as antiknock agents [Fe(CO)$_5$].

Every element in the first transition series except nickel and scandium has been found to be an essential element that is vital to some metabolic function in plants or animals. For example, many enzymes are coordination compounds of iron(II) or iron(III). Vitamin B$_{12}$ is a compound containing cobalt(III), and hemoglobin contains iron. The results of the deficiencies of these elements are well known, and other deficiencies are being discovered. For example, a copper deficiency, or inability to manufacture a copper-containing protein, is associated with leukemia, neoplastic diseases, and schizophrenia. Manganese deficiencies result in impaired growth, bone deformation, and abnormalities of the central nervous system.

Another interesting area of research is the toxicity of these ions. Although all are essential, all are toxic in larger quantities. Iron, cobalt, and nickel have been shown to be carcinogenic agents.

For these reasons the study of coordination chemistry is important. One ultimate objective would be the determination of the role of metal ions in biological functions. Some research is presently being carried out on enzymes in which the focus of attention is the metal ion. In addition, enzyme models and synthetic oxygen carriers are being studied. These are compounds, prepared in the laboratory, that have the same type of activity (but the activity is always decreased) as the natural oxygen carrier or enzyme.

Another interesting and promising area of research is the application of the principles of coordination chemistry to the removal of toxic metal ions from the body. Lead and mercury poisoning have been successfully treated by administering complexing agents. There are ligands that have been designed for their ability to coordinate with mercury or lead, rendering these ions incapable of lodging in various parts of the body, thus facilitating their elimination.

This approach is also reasonably promising for the removal of radioactive metal ions from the body. In addition to the practical aspects, coordination compounds have proven to be excellent materials with which to test and build bonding theories.

I. Types of Coordination Compounds

There are many possible compounds of the type shown in Table 14-1. A distinctive feature of coordination chemistry is its scope. Any of the common anions as well as some that are not so common, in addition to almost any neutral species that contains an electron pair,

Ch. 14 Coordination Chemistry

might be incorporated into a coordination compound. The names and formulas of only a fraction of the possible ligand groups are listed in Table 14-2. Many of the compounds listed are usually classified as organic compounds; that is, compounds that contain a skeleton of carbon chains or rings with the functional groups (that is, O, N, S, and so forth) acting as the donor atom. However, in some instances there is a direct bond between the metal and carbon. Compounds of the latter type are usually called organometallic compounds.

Some of the compounds that result upon combination with a metal are given in Fig. 14-1. This figure shows that, in general, ligand groups are not limited to attachment to a metal ion in a single coordination position. Chelates are formed when the ligand coordinates in more

Two structures are shown for the complex molecules in this table. One shows the position of each atom. The other is a shorthand notation of the same molecule. In the shorthand system a corner represents a carbon atom and a sufficient number of hydrogens so that there are four bonds to each carbon. A circle within a ring represents a π bonding system with one bond for each carbon atom.

Table 14-2 Some Typical Ligands

Name	Formula	and structure
Aquo	H_2O	
Ammine	NH_3	
Fluoro	F^-	
Thio	S^{2-}	
Carbonato	CO_3^{2-}	
Oxalato	$C_2O_4^{2-}$	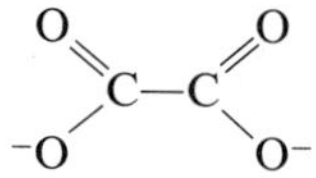 $\begin{array}{c} O \quad\quad O \\ \diagdown\;\;\diagup \\ C - C \\ \diagup\;\;\diagdown \\ {}^-O \quad\quad O^- \end{array}$
Ethylenediamine	$C_2N_2H_8$	$NH_2-CH_2-CH_2-NH_2$
Benzene	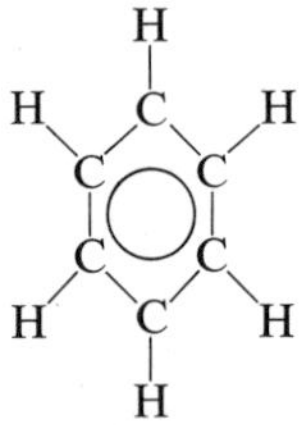 (hexagonal C/H ring structure)	 (benzene ring with circle)
8-Hydroxyquinolato	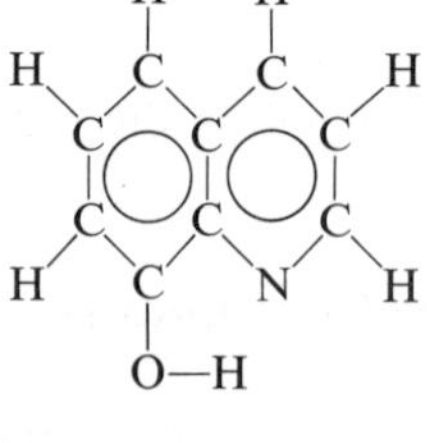 (fused ring structure with N and O–H)	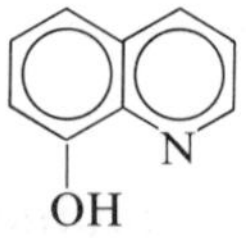 (shorthand fused ring with N, OH)
Dithiooxalato	$C_2S_2O_2^{2-}$	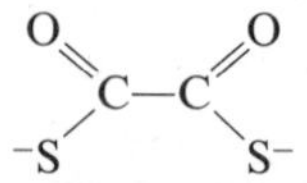 $\begin{array}{c} O \quad\quad O \\ \diagdown\;\;\diagup \\ C - C \\ \diagup\;\;\diagdown \\ {}^-S \quad\quad S^- \end{array}$

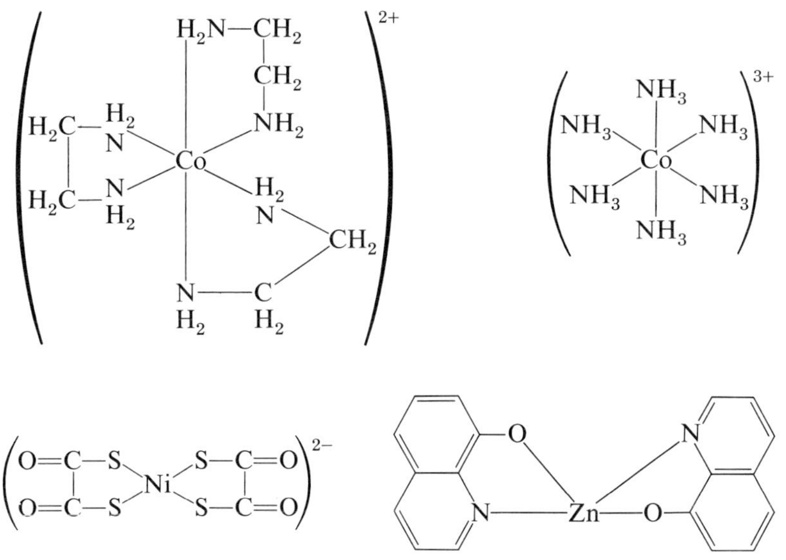

Figure 14-1 *Typical coordination compounds.*

than one position. Unidentate, bidentate, tridentate, quadridentate and sexadentate ligands are known, depending upon the number of donor atoms (capable of coordination) in the ligand.

The ligands listed in Table 14-2 are often classified in terms of the type of bonding involved in the complex. The simple ligands necessarily form complexes through coordination of a single electron pair to form a σ bond. Ammonia, as shown in the complex given in Fig. 14-1, is an example of this type of donor.

Other substances such as water, oxide ion, and sulfide ion can, and probably do, complex in some instances as both σ and π donors. There is also a group of substances that form σ bonds by donating a pair of electrons which also behave as π acceptors. Examples of compounds in this category include carbon monoxide and phenanthroline. The bonding involved with both of these types is shown in Fig. 14-2. Organometallic compounds represent an additional class of compounds, some of which may be classified as π donors. The compounds that involve π acceptors are unique; therefore some space will be devoted to them.

The carbon monoxide in carbonyls is a σ donor and π acceptor ligand. The formulas and some thermodynamic data for some simple metal carbonyls are given in Table 14-3. The structures of some of these compounds are listed in the table as well. Three basic structures, octahedral, trigonal bipyramid, and tetrahedral are observed most often. In addition to the simple carbonyls a number of complex

Ch. 14 Coordination Chemistry

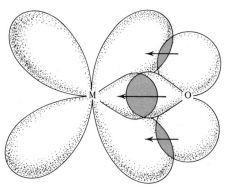

Oxide is σ donor and a π donor

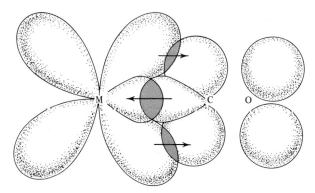

Figure 14-2 $d\pi$-$p\pi$ bonding in complexes.

Co is σ donor and a π acceptor

Table 14-3 *Simple Metal Carbonyls*

	Melting point, °C	H^0, kcal mole^{-1} at 25°C	Structures
$Ni(CO)_4$		−151.6	tetrahedral
$Fe(CO)_5$	−20	−187.8	trigonal bipyramid
$Co(CO)_6$	sublimes in vacuum	−257.6	octahedral
$V(CO)_6$	decomposes at 70		octahedral
$Mo(CO)_6$	sublimes in vacuum	−234.8	octahedral
$W(CO)_6$	sublimes in vacuum	−227.3	octahedral
$Ru(CO)_5$	−22		trigonal bipyramid
$Os(CO)_5$	−15		trigonal bipyramid

carbonyls are known. The structures of the complex carbonyls given in Fig. 14-3 involve bridging carbon monoxide groups between metal atoms.

The valence bond method is applicable to the prediction of the structure of simple and even complex carbonyls if care is exercised.

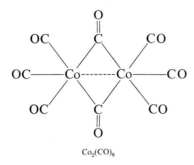

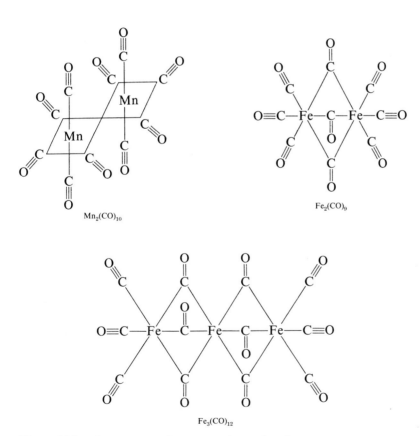

Figure 14-3 *The structure of some complex carbonyls.*

The structure of Fe(CO)$_5$ is trigonal bipyramid with dsp^3 hybridization.

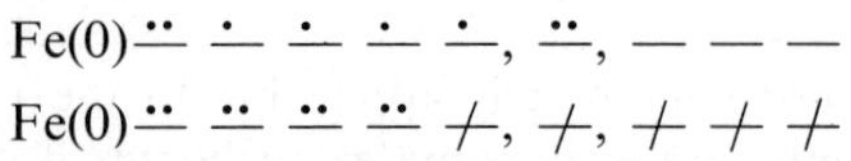

Fe(0) ·· ·· ·· ·· , ·· , — — — —
Fe(0) ·· ·· ·· ·· +, +, + + +

For a molecule that involves bridges it is first necessary to determine the most likely number of bridges consistent with the stoichiometry. For Fe$_2$CO$_9$ there are two possibilities. The structure involving one bridge results in two trigonal bipyramids.

The second possibility involves three bridging CO groups and results in an overall octahedral structure about each iron.

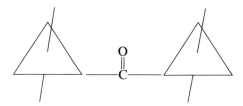

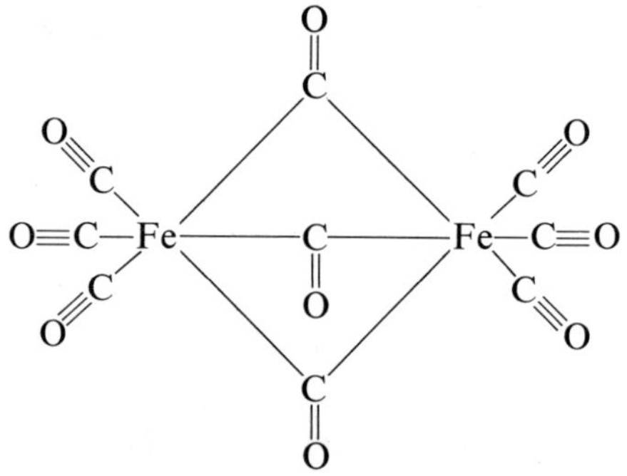

Both of these structures would require a compound with two unpaired electrons as shown in Fig. 14-4. Experiment shows that the molecule is diamagnetic so that a metal-metal bond, shown in the figure as a dotted line, must be postulated. This possibility seems unlikely for the trigonal bipyramid structure because of the excessive distance between the metal atoms. From infrared studies which allow a distinction between $>$C=O and —C≡O the correct structure has been found to be the triply bridged structure.

In general the carbonyls are all diamagnetic complexes in which the metal is in a low oxidation state. In any situation in which there

The use of infrared spectra to determine structure is considered in Chapter 16.

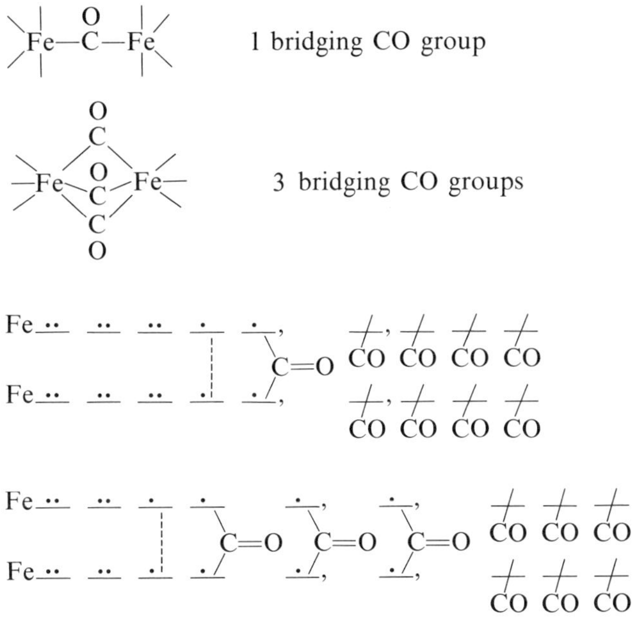

Figure 14-4 *Possible structures of* $Fe_2(CO)_9$.

is an odd number of electrons per metal atom the compounds will involve metal-metal bonds. $Fe_2(CO)_9$ and $Mn_2(CO)_{10}$ are two examples. In $Mn_2(CO)_{10}$, two $Mn(CO)_5$ monomers are joined at one corner of an octahedron by a metal-metal bond.

The carbonyls are low melting and boiling, inflammable liquids or solids. $Fe(CO)_5$ and $Ni(CO)_4$ require very specialized equipment for study since they may form explosive mixtures in air. All carbonyls are highly toxic substances. If the vapors are inhaled, decomposition results in the deposition of finely divided metal in the lungs and the liberation of CO.

The carbonyls undergo a tremendous variety of reactions, and as such they are important beginning materials in the syntheses of many derivatives such as halides and nitrosyls. A few examples are given in Table 14-4.

Table 14-4 *Halide and Nitrosyl Derivatives of Some Carbonyls*

$Mn(NO)(CO)_4$	$Mn(CO)_5X$	$Mn(CO)_4X_2$
$Fe(NO)_2(CO)_2$	$Fe(CO)_2X_1$	$Fe(CO)_2X_2$
$Co(NO)(CO)_3$	$Co(CO)X_2$	

II. Isomerism in Coordination Compounds

Most of the examples of simple complexes used early in this chapter have contained only one kind of ligand. Of course, this need not always be the case and it is quite possible to have any number of mixed complexes as indicated in the last section. If we consider first the possibility that only two different ligands are involved in a six-coordinate complex, an interesting situation results. There are two different ways in which these ligands may be arranged. For example, two chlorides and four ammonia molecules can be arranged around a cobalt ion in two different octahedral structures

(a)   (b)

Cis-trans isomers are not possible in tetrahedral complexes because in this structure all ligands are cis to all others.

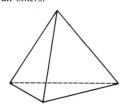

In (a) the chlorines are arranged across from each other (trans) and in (b) the chlorines are next to each other (cis). Both of these compounds can be prepared. The name geometrical isomer is given to compounds of this type because the two forms have identical molecular weights and formulas. They differ only in the arrangement of the ligands around the central metal. Geometrical isomerism is not limited to octahedral complexes, but may also be found in square planar complexes.

When two geometrical isomers exist they are usually not equally stable. The trans isomer is usually the more stable for reasons that relate back to the Gillespie-Nyholm idea. It may be possible to convert the less stable isomer to the more stable one by the application of heat.

The difference in molecular symmetry is most important in determining the properties of cis-trans isomers. Any physical or chemical property that is dependent upon molecular symmetry will be different for the two isomers. The solubility of the isomers in various substances might well be different because the difference in symmetry would result in a different packing in the crystal lattice and, thus, a different lattice energy. This difference in solubility can sometimes be used to advantage in separating mixtures of cis-trans isomers.

Another property that must be different for cis-trans isomers is the dipole moment. The vibrational-rotational spectrum must also differ and, although it is not obvious at this point, the visible spectra must also be different.

Optical isomerism is a special type of geometrical isomerism that will be exhibited by any molecule that does not contain a plane of symmetry or a center of inversion. The cis isomer of $Co(en)_2Cl_2^+$ is an example. This ion can be resolved into the separated optical isomers, called diastereoisomers, which rotate the plane of polarized light passing through a solution containing the isomer. The only difference in properties of optical isomers is the direction of rotation of polarized light (right or left). This type of isomerism is possible for octahedral and tetrahedral complexes but not for square planar.

III. Nomenclature

The following are the recommendations of the International Union of Pure and Applied Chemistry on nomenclature.

1. The oxidation number of the central atom is indicated by a Roman numeral in parentheses after the name of the complex.
2. The suffix -ate is added to the name of the central atom when naming anionic complexes. In neutral or cationic complexes, the name of the central atom is that of the element.
3. In both formulas and names, the ligands are given in the order:
   (a) simple anions (that is, fluoride, sulfide)
   (b) other inorganic anions
   (c) organic anions
   (d) H_2O, NH_3
   (e) other inorganic neutral ligands
   (f) neutral organic ligands
4. The suffix -o is added to the name of the simple anionic ligands. The name of the molecule is used for neutral ligands. The only exceptions to this are water and ammonia, which are given the names aquo and ammine, respectively. .
5. The prefixes, di, tri, tetra, penta, hexa, hepta, octa, ennea, deca (two to ten respectively), are used to show the number of individual ligands of one kind. The name of the more complicated ligands

is set off by parentheses and the prefixes bis, tris, tetrakis (two to four) are added to designate the number.
6. The symbol μ (mu) is used to indicate a bridging ligand group. It precedes the name of the ligand.

These rules are applied to some specific compounds in Table 14-5.

IV. Coordinating Tendencies

The coordination compounds discussed here exhibit a wide range of stability. The stability of a complex and the relative tendency of a metal ion to combine with a particular ligand group are dependent upon many factors. To simplify the discussion of coordinating tenden-

Table 14-5 *Names and Formulas of Some Complexes*

Diamidotetraamminecobalt(III) ethanolate
$[Co(NH_2)_2(NH_3)_4]OC_2H_5$

Sodium bis(thiosulfato)argentate(I)
$Na_3[Ag(S_2O_3)_2]$

Ammonium tetrathiocyanatodiammineferrate(III)
$NH_4[Fe(SCN)_4(NH_3)_2]$

Ammonium tetrathiocyanatodiamminechromate(III)
$NH_4[Cr(SCN)_4(NH_3)_2]$

Potassium oxotetrafluorochromate(V)
$K[CrOF_4]$

Bis(8-quinolinolato)silver(II)
$[Ag(oxinate)_2]$

Tetrapyridineplatinum(II) tetrachloroplatinate(II)
$[Pt(py)_4][PtCl_4]$

Tris(ethylenediamine)cobalt(III) sulfate
$[Co(en)_3]_2(SO_4)_3$

Hydroxopentaaquoaluminum(III) ion
$[Al(OH)(H_2O)_5]^{2+}$

Potassium hexacyanoferrate(III)
$K_3[Fe(CN)_6]$

Potassium pentacyanocarbonylferrate(II)
$K_3[Fe(CN)_5CO]$

Bis(cyclopentadienyl)iron(II)
$[Fe(C_5H_5)_2]$

Table 14-6 *Hard and Soft Acids*

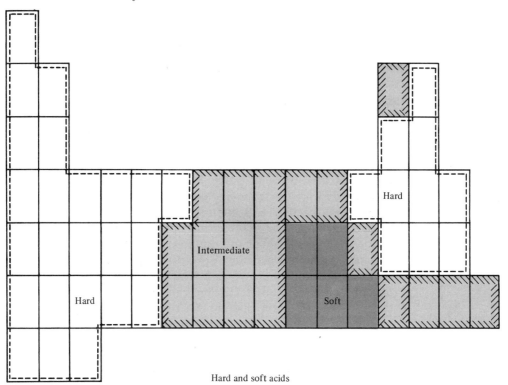

Hard and soft acids

cies, this section will deal with the formation of simple complexes formed by simple ligands such as F^-, Cl^-, OH^-, S^{2-}, NH_3, and H_2O.

The factors that must be considered in a discussion of the tendency for metals to form complexes with the simple anions are closely related to those involved in Chapter 13, where solubility was discussed. More complicated ligands are best treated by other methods; in fact, compounds such as the carbonyls can only be treated in terms of the molecular orbital method. This will be done in Chapter 15.

The most general classification of the coordinating tendencies of metal ions was suggested in 1958 by Arland, Chatt, and Davies. In this classification the ions are divided into two groups which are currently called hard and soft, respectively. The hard metal ions form the most stable complexes with the first element in a family (that is, fluoride ion rather than chloride ion), whereas the soft metal ions form the most stable complexes with the heavier elements of a family (that is, chloride ion rather than fluoride ion). The concept of hard and soft acids and bases introduced in Chapter 11 follows this definition. The metals of the periodic table are shown in Table 14-6 according to this scheme.

Ch. 14 Coordination Chemistry

The concept of hard and soft acids and bases is an extremely useful one even though it has not as yet been put on a quantitative basis. We shall examine the consequences of this classification after a brief discussion of the observed coordinating tendencies of some metal ions.

A. COMPLEXES OF VARIOUS CATIONS

A measure of coordinating ability of the metal ions in Group IA may be obtained by considering the stable hydrate salts. Many of the salts of lithium and sodium are stable as hydrates; however, more hydrates of lithium are known. This trend continues down the group until no stable hydrates of cesium salts are known.

In Chapter 13 we saw that the fluorides of group IIA are insoluble in water. Therefore, we see that a marked preference for oxygen and fluorine ligands is characteristic of these representative elements.

The ions of the group IIA elements are expected to be better coordinators than those of group IA. The element beryllium differs significantly from the remaining elements of the group because of its exceptionally small size. Beryllium ion is strongly hydrated in nearly all inorganic salts. The next element in the group, magnesium, forms an ion which is still small and proves also to be a strong coordinator of water. Magnesium perchlorate [$Mg(ClO_4)_2$] is an excellent desiccant. Calcium, strontium, and barium ions show a decreasing tendency toward hydrate formation. Calcium chloride ($CaCl_2$), calcium sulfate ($CaSO_4$) and strontium chloride ($SrCl_2$) are dehydrating agents that form hexahydrates but only the dihydrate of barium chloride ($BaCl_2$) is known.

Beryllium complexes with ethers (that is, in CH_3—CH_2—O—CH_2—CH_3 oxygen is the donor atom) are known, and the halogens are potential donors with Be^{2+}. $BeCl_4^{2-}$ is known, but of the group IIA cations only Be^{2+} coordinates the halogens. The chloro complexes of the remaining elements can be made in the solid, but they decompose immediately in water. Beryllium is not capable of forming a stable ammine complex in water. $Be(NH_3)_4^{2+}$ salts are thermally stable; however, the ammonia molecules are readily replaced by water.

Note that the basic strength is not the dominating factor here (ammonia is a stronger base than water). The increased stability of the hydrate is connected with the larger permanent dipole moment of water, or equivalently the smaller size and greater electronegativity of the oxygen atom as compared to the nitrogen atom.

Of the ions of group IIIA, Al(III) is a fairly powerful coordinator and hexahydrate salts of Al(III) are common. The hydrate even persists when an aqueous HCl solution of $AlCl_3$ is evaporated. As might be expected, Ga(III), In(III), and Tl(III) are weaker coordinators of water and fluoride.

Halogen coordination is important in group IIIA and the tendencies for coordination are illustrative. Al(III) forms the hexafluoride AlF_6^{3-} and the tetrachloride, $AlCl_4^-$, which decomposes to the hydrate in water. Iodide and bromide complexes of Al(III) are unknown. Of the remaining elements only Ga(III) forms a fluoride complex, whereas In(III) and Tl(III) form only chloride, bromide, and iodide complexes. The smaller ions in the group (hard acids) coordinate best with the smaller halides, whereas the larger (soft acids) coordinate best with the larger more polarizable halides. Thus this family illustrates the full range from cations that prefer oxide and fluoride as ligands to those that coordinate best with iodide and sulfide.

In addition it should be noted that ammine complexes of Al(III) decompose to the hydrate in water while Cr(III), which is very nearly the same size as Al(III), forms many ammine complexes. The difference in chemical behavior between Al(III) and Cr(III) is almost as great as the difference between K^+ and Ag^+. K^+ and Ag^+ are very different chemically. K^+ forms an ionic chloride while AgCl is fairly covalent. K^+ coordinates only with water and a few other oxygen donors whereas Ag^+ coordinates with chloride, cyanide (CN^-), ammonia, and sulfur-containing compounds.

Boron (B) is like beryllium in that it is unique in its group because of size considerations. Boron forms many complexes and molecular addition compounds. BF_4^- is stable in water.

The coordinating tendencies of the transition elements are interesting and varied. With vanadium, coordination occurs mostly with oxygen donor ligands. The tendency towards coordination of vanadium with ammonia is only slight. Chromium(II) coordinates with nitrogen donors but the ammines are unstable in water. The ammines of chromium(III) are kinetically stable in water, but they are not truly thermodynamically stable. Iron and manganese do not form complexes with ammonia; these metals precipitate as hydroxides instead. The ions of cobalt, nickel, copper and zinc, all form stable ammines in water. Complex ammine formation is a property that is often used to separate the ions of these metals from those that form insoluble hydroxides with ammonia. In general, the coordinating tendencies of the first transition metal ions with ammonia increase toward the end of the series.

All of the elements from vanadium to zinc form very stable cyanide complexes, but with sulfur donors, the coordinating tendencies increase toward copper. Vanadium, chromium, and manganese exist in nature as the oxides; iron is found as both the oxide and sulfide whereas cobalt, nickel, copper, and zinc are found largely as the sulfides.

B. AN INTERPRETATION OF COORDINATING TENDENCIES WITH SIMPLE LIGANDS

As we stated earlier, the field of coordination chemistry is vast. A complete description of coordinating tendencies would be a formidable job indeed. It cannot be accomplished in this one book. However, some insight into coordinating tendencies is possible, using the ideas that have been developed in previous chapters.

The factors that influence the coordinating tendency of a metal ion are size, charge, and electronic structure. Since these are the factors that determine polarizing ability and softness (or induced polarizing ability), the coordinating tendencies of an ion may be expressed in these terms. In general, highly polarizing cations will be the best coordinators. As indicated in Chapter 13 high polarizing ability can arise in two ways. A cation will be highly polarizing if it is very small and has a high charge [Al(III)]. Cations of this type are hard. In addition a cation may become highly polarizing through distortion. The distortion of a cation can only be accomplished by a ligand group which is easily polarized. This behavior is described by the statement that the cations are soft acids and they show a preference for the easily polarized anions. This kind of polarizing ability must be induced by the anion.

In ligands the factors that lead to good coordinating abilities are either small size or high ease of polarization.

On the basis of electrostatics only, the most stable complex will be formed between cations that are highly polarizing and the ligand group that combines the smallest size and largest charge. These factors will lead to a favorable (negative) enthalpy term. If a larger ligand group is substituted for a smaller one, the enthalpy term will be less favorable. If a ligand group is easily polarized and the cation is soft, induced polarizing ability will make the enthalpy term more favorable than if the cation were hard.

In complex formation the size and charge considerations predominate as long as the metal is small and highly charged and cannot be distorted. When the metal ion is subject to distortion, the size and charge considerations become less important. In this case the stability that is lost due to larger size of the ligand is regained because of the distortion of the metal and ligand. This, of course, is covalent bonding.

This situation was observed in Chapter 13 in the lattice energies. The result there was that fluorides generally become more soluble across the transition series from ferrous ion (Fe^{2+}) to cupric ion (Cu^{2+}) whereas chlorides become less soluble. A very similar situation is also apparent in the lattice energies of the silver halides (Table 14-7). Although silver fluoride has the largest lattice energy, the influence of the ease of polarization of the ligand is readily apparent in the values for silver bromide and silver iodide.

A quantitative ordering of the cations beyond the $(Z + S)/(r + 0.50)^2$ term of polarizing ability is beyond the scope of this book. Quantum mechanical calculations would be necessary and, in any case, one order cannot be satisfactory for all ligands. In addition there should be a gradual transition, and no firm lines can be drawn to indicate where a particular behavior will stop and a different behavior will begin. In Table 14-8 the ions are listed in the order of increasing polarizing ability. This provides a reasonable order for the data on

Table 14-7 *Thermodynamic Changes Corresponding to the Process* $AgX(s) \longrightarrow Ag^+(aq) + X^-(aq)$ *at* $25°C$ *(kcal mole^{-1})*

Compound	U	ΔH_{hyd}	ΔH^*_{soln}	$\Delta H^\dagger_{soln}$	$T\Delta S$	ΔG
AgF	231	233	−2	−4.8	−1.4	−3.5
AgCl	217.5	202	+15.5	+15.6	+2.4	+13.3
AgBr	215	194	+21	+20.2	+3.4	+16.8
AgI	212.5	183	+29.5	+26.8	+4.9	+21.9

*Calculated from U and ΔH_{hyd}.
†Calculated from H^0 of the solid.

hydration energy, acidity, and solubility of hydroxides and fluorides. An entirely different order is needed to interpret the data for more easily polarized ligands, as indicated by the data on the solubility of sulfides listed in the last column of the table.

Note that the elements with strongly basic soluble hydroxides are grouped at the top of the table. As the polarizing ability increases, the hydroxide generally becomes less and less soluble in acids. A continued increase in polarizing ability results in hydroxides which act as both acids and bases. These hydroxides are amphoteric. Aluminum hydroxide is a typical amphoteric hydroxide. The reactions

$$3H_2O + Al(OH)_3 + 2H^+ \longrightarrow Al(OH)(H_2O)_5^{2+}$$

and

$$Al(OH)_3 + OH^- + 2H_2O \longrightarrow Al(OH)_4(H_2O)_2^-$$

limit the region in which aluminum hydroxide can be precipitated to the pH range of 4–11. The hydroxides of cations of still higher polarizing ability behave as acids only, coordinating hydroxide and oxide as ligands in place of water and hydroxide. Thus solutions of arsenic(V) contain the complex ions

$$AsO_4^{3-} \quad As(OH)O_3^{2-} \quad As(OH)_2O_2^- \quad As(OH)_3O$$

or, as they are more commonly written

$$AsO_4^{3-} \quad HAsO_4^{2-} \quad H_2AsO_4^- \quad H_3AsO_4$$

C. POLARIZING ABILITY AND COORDINATION NUMBER

The term coordination number has been used previously. It refers to the number of ligand groups that are attached to a metal ion. The

Table 14-8 *Polarizing Ability and Properties*

Ion	Polarizing ability $(Z + S)/(r + 0.5)^2$	$-\Delta H_{hyd}$, kcal mole^{-1}	pK_a	K_{sp} of fluoride	Solubility of hydroxide K_{sp}	Solubility of hydroxide ppt. over pH range	K_{sp} of sulfide
Cs$^+$	0.21	168		sol	sol		sol
Rb$^+$	0.26	176		sol	sol		sol
K$^+$	0.30	182		sol	sol		sol
Na$^+$	0.48	202	15	sol	sol		sol
Ba^{2+}	0.58	521	13	5×10^{-6}	sol	14–15	sol
Sr^{2+}	0.75	555	13	5×10^{-9}	sol	13–15	sol
Ag$^+$	0.81	217	12	sol	2×10^{-8}	8–15	6×10^{-50}
Li$^+$	0.83	229	14	10^{-2}	sol		sol
Ca^{2+}	0.90	590	12.7	1×10^{-9}	8×10^{-6}	12–15	sol
La^{3+}	1.10	1140	3.1	insol	10^{-20}	8–15	10^{-13}
Cu$^+$	1.17	128		sol		4–15	2×10^{-48}
Hg^{2+}	1.37	647	2.5		2×10^{-20}	2–15	2×10^{-53}
Y^{3+}	1.47	1170		insol	10^{-29}	7–15	
Lu^{3+}	1.47			insol		7–15	
Mg^{2+}	1.51	669	11.4	10^{-8}	9×10^{-12}	9–15	
Cd^{2+}	1.62	642	9.0	sol	5×10^{-15}	8–15	1×10^{-29}
Mn^{2+}	1.63	652		sol	2×10^{-13}	8–15	4×10^{-14}
Sc^{3+}	1.75	1259	4.6	insol	10^{-30}	6–15	
Fe^{2+}	1.83	676	10		5×10^{-16}	8–15	2×10^{-19}
Co^{2+}	1.98	690	10	1×10^{-6}	2×10^{-16}	7–15	
Ni^{2+}	2.15	710	10.6		2×10^{-16}	7–15	4×10^{-22}
Zn^{2+}	2.28	699	9.7	3×10^{-2}	4×10^{-17}	7–14*	3×10^{-25}
Cu^{2+}	2.37	714	7.5	3×10^{-2}	2×10^{-21}	6–15	1×10^{-36}
Cr^{3+}	2.45	1370	3.8		10^{-31}	5–15*	
In^{3+}	2.62		4.6		10^{-33}	4–15	
Ti^{4+}	2.87	2305				1–15	
Fe^{3+}	2.89	1369	2.2		2×10^{-39}	3–15	
Al^{3+}	3.00	1333	5		6×10^{-33}	4–11	
Be^{2+}	3.05	806				5–13	amphoteric
Ga^{3+}	3.59						
Sn^{4+}	3.76	2232			4×10^{-65}	0–10	insol
V^{5+}	4.21	3840				0–4	
Si^{4+}	4.83	2530				−1–13	
Sb^{5+}	5.18	3394				−1–3	insol
Ge^{4+}	5.18	2476				−1–9	
B^{3+}	6.12						
As^{5+}	6.91	3946			sol		insol
P^{5+}	7.09	4204			sol		

*Zinc is amphoteric although it is much larger than many basic +2 ions. Chromium(III) shows amphoterism only because the approach to equilibrium is slow—see Chapter 21.

fact that the coordination number varies from metal to metal and from ligand to ligand was indicated at the beginning of this chapter. It is the purpose of this section to inquire more fully into the coordination numbers observed in complexes and the factors that affect complex stability. In previous discussions the coordination number observed in a number of complexes was noted. We shall repeat briefly some of this information while discussing the data in Table 14-8. The coordination number of many of the complex ammines is six; however, Cd^{2+} forms $Cd(NH_3)_4^{2+}$ whereas Ag^+ forms $Ag(NH_3)_2^+$ and Hg^{2+} precipitates as the hydroxide instead of forming a complex ammine.

In addition many of the first transition metal ions form complex fluorides (CrF_6^{3-}, FeF_3^{3-}, CoF_6^{3-}) where the coordination number is six. These ions do not form hexacoordinate complex chlorides, bromides, or iodides. There is no evidence of an anionic chloride complex of Ni(II), such as $NiCl_4^{2-}$, in 12 M HCl. Co^{2+} and Cu^{2+} do form such complexes, and the chloro, bromo, and iodo complexes of Ag^+, Cd^{2+}, and Hg^{2+} are exceptionally stable. In these complexes the coordination number is four except for the case of Ag^+ where it is two.

An observed coordination number may be understood on the basis of the polarizing ability of the cation and the ease of polarization of the anion. For a particular cation with a certain charge, the number of anions required to neutralize the charge—satisfy the coordination number of the cation—depends upon the ease of polarization of the anion. The greater the ease of polarization of the anion the lower the coordination number. For a particular metal the coordination number exhibited with fluoride should be larger, for example, than that exhibited with chloride.

In the same way the number of ligands required to satisfy different metal ions will depend upon the polarizing ability of the cation. Those cations with a high polarizing ability will require a larger number of ligands than those with a smaller polarizing ability. The trend in coordination number observed for the ammines of Ni^{2+}, Cd^{2+}, and Ag^+ indicates a decreased polarizing ability from Ni^{2+} to Ag^+.

Because of the dependence of coordination number on polarizing ability, the most stable complex formed by a metal-anion pair may involve more or fewer anions than the number of positive charges on the metal. This leads to the possibility of cationic and anionic complexes.

A cationic complex carries a positive charge. In a complex of this type the number of coordinated anionic ligands is less than the number of positive charges on the metal. An anionic complex carries a negative charge. In this case a larger number of ligands are coordinated than the positive charge on the cation.

From the above discussion we expect cations of very low polarizing ability to form cationic complexes with anions that are difficult to polarize

$$M^{2+} + F^- = MF^+$$

if they form complexes at all. Cations with a slightly greater polarizing ability would form neutral species and might precipitate as insoluble fluorides. Finally a cation with a high polarizing ability should form an anionic complex.

$$Fe^{3+} + 6F^- = FeF_6^{3-}$$

and

$$As^{5+} + 6F^- = AsF_6^-$$

The general scheme should be roughly like that presented in Fig. 14-5. The +1 cations will not react with fluoride, except for lithium ion which would lie at the bottom of the precipitation range. A number of +2 cations form precipitates.

The behavior in the precipitate range is complicated. As the cations that form precipitates become more polarizing, the solubility decreases and passes through a minimum. As shown in Chapter 13, induced polarizing ability may play a significant role. Beyond this point fluo-

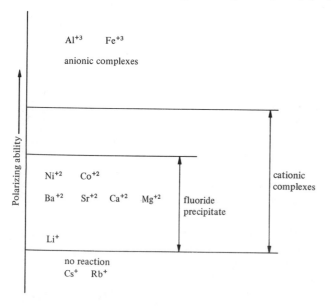

Figure 14-5 *Behavior of cations with fluoride.*

rides become more soluble as the polarizing ability of the cation increases. As we have seen, the fluorides of the group IIA ions are the least soluble. Most of the transition metal fluorides are more soluble as are the group IA fluorides. This has to do with the fact that water is more easily polarized than fluoride.

As usual, to obtain an understanding of the chemical behavior of cations and anions we must examine the enthalpy and entropy data and attempt to interpret it in terms of the nature of the cation and anion.

The enthalpy change for a reaction such as

$$Fe(H_2O)_6^{3+} + F^- = Fe(H_2O)_5F^{2+} + H_2O$$

is expected to be unfavorable because fluoride replaces water, and fluoride is harder to polarize than water. On the other hand the entropy term that accompanies a reaction of this type is expected to be favorable for two reasons. The number of charged species in water is reduced, and the magnitude of the charge is reduced. On the other hand if the cation and anion are very large, the entropy term involved in complex formation will be unfavorable because the decreased number of particles restores the ordered structure of water. Therefore, complex formation such as

$$Cs^+ + I^- = CsI$$

is not likely on the basis of either entropy or enthalpy.

QUESTIONS

1. Write the formulas for the complex ammines and complex cyanides of the first transition series.
2. What is a σ donor?
3. What is a π donor?
4. Write formulas for complexes that involve the ligands in Table 14-2.
5. Determine the structure of the carbonyls in Table 14-3 using the valence bond method.
6. What are geometric isomers? What are optical isomers?
7. How do geometric isomers differ?
8. How do optical isomers differ?
9. Name the compounds given in answer to Questions 1, 4, and 5.
10. Describe the general trends in coordinating tendencies for the elements of groups I, II, and III and the first transition elements.
11. What is polarizing ability and induced polarizing ability?
12. Calculate the polarizing ability of the cations of the first transition series using $Z/(r + 0.50)^2$ and $(Z + S)/(r + 0.50)^2$.
13. Which cations are expected to form complex fluorides? Which should form complex chlorides? complex ammines?

14. How does entropy affect the stability of a complex?
15. How is anhydrous $BaSO_4$ made? How would you prepare anhydrous $BeSO_4$? What hydrates of $CaSO_4$ are known?
16. Give the full systematic names of FeF_6^{3-} and $Co(NH_3)_5Cl^{2+}$.
17. From the cations Al^{3+}, Ni^{2+}, Zn^{2+}, Cd^{2+}, Ag^+, Co^{2+}, Li^+ select two that
    (a) are hard acids
    (b) are soft acids
    (c) have amphoteric hydroxides
    (d) form insoluble fluorides
    (e) form an insoluble sulfide
    (f) form complex ammines
18. From the anions PO_4^{3-}, MnO_4^-, S^{2-}, CO_3^{2-}, F^- select two that
    (a) form anionic coordination complexes
    (b) form few insoluble salts
    (c) are soft bases
    (d) form insoluble salts with the group IIA cations
    (e) form insoluble salts with transition metal ions
    (f) might be used to separate the group IIA cations from the transition metal ions Fe^{2+}, Co^{2+}, Ni^{2+}, Cu^{2+}.

PROBLEMS I

1. The equilibrium constant for the reaction $Zn(NH_3)_4^{2+} \longrightarrow Zn^{2+} + 4NH_3$ is 3.4×10^{-10}. Calculate $[Zn^{2+}]$ in a solution with $[Zn(NH_3)_4^{2+}] = 0.10\ M$ and $[NH_3] = 0.20\ M$.
2. The solubility product of $Zn(OH)_2$ is 5×10^{-17}. What pH would be required to precipitate $Zn(OH)_2$ from the solution in Problem 1?

PROBLEMS II

1. What is the coordination number of copper in $CuCl_4^{2-}$?
2. What is the name for Cl^- as a ligand?
3. Name the coordination compound shown in Problem 1.
4. What orbitals are involved in the overlap when ligands act as donors in π bonding?
5. What is the lowest energy empty orbital in a CO molecule?
6. What orbital on CO can accept electrons in π bonding?
7. What orbitals of Fe(0) are σ acceptors in $Fe(CO)_5$?
8. What is the hybridization of iron orbitals in $Fe(CO)_5$?
9. What is the shape of $Fe(CO)_5$?
10. What filled orbitals of iron in $Fe(CO)_5$ can act as π donors?
11. How many valence electrons are there in NO compared to CO?
12. What atom has one less d electron than Fe(0)?
13. What is the structure of $Mn(NO)(CO)_4$?
14. How many isomers of $Mn(NO)(CO)_4$ do you expect to exist?
15. Draw the structure of the *cis*-dichlorotetramminechromium(III) ion.
16. What is the polarizing ability of In^{3+}?
17. Should In_2O_3 be acidic, basic, or amphoteric?
18. Do you expect In^{3+} to form negatively charged complexes with fluoride?

SUGGESTED READINGS

Mahan, B. H., *College Chemistry.* Reading, Mass.: Addison-Wesley, 1966, Chapter 12.

Rich, R., *Periodic Correlations.* New York: W. A. Benjamin, 1965, Chapters 2, 3, and 4.

Sanderson, R. T., *Chemical Periodicity.* New York: Reinhold, 1968.

> The inspiration came to him like a flash. One morning at two o'clock he awoke with a start, the long sought after solution of this problem had lodged in his brain. He arose from his bed and by five o'clock in the afternoon the essential points of the coordination theory were achieved.*
>
> PAUL PFEIFER

Chapter 15 The Ligand Field

I. Introduction to Ligand Fields

In earlier chapters the valence bond method was applied to the prediction of the structure of coordination compounds. Some of the difficulties encountered when the valence bond method is applied to such compounds were mentioned there. In this chapter we shall describe an alternate method of treating the bonding in coordination compounds. This method is capable of explaining the discrepancies mentioned earlier as well as adding much more. The method is called the ligand field method. We have, in fact, already introduced some of the essential features of the method in previous chapters. In Chapters 7 and 12, the molecular orbital method was discussed. The ligand field method is nothing more than the molecular orbital method applied to complexes. In Chapter 13 we mentioned the excess polarizing ability of cations of the transition metal ions. This, too, is an important feature of the ligand field method.

To apply the molecular orbital method to any molecule we need only set up the molecule and visualize the overlap of orbitals. The treatment given here is entirely qualitative, based upon visual inspection of the molecule. More rigorous methods are, of course, available,

---

*G. B. Kauffman, *Alfred Werner, Founder of Coordination Chemistry*, (Berlin-Heidelberg-New York: Springer Verlag, 1966).

and the use of these methods can add some information to the result we will obtain; but these results are, in fact, only details that are not usually necessary and may be added without explanation to the results we can achieve.

We will refer to the combination of overlapping atomic orbitals as a linear combination of atomic orbitals or LCAO. Since this discussion is qualitative, the mathematical approximation need not concern us. We can regard the term LCAO as a useful designation for molecular orbitals.

When writing LCAOs, there will usually appear to be a large number of possible combinations of atomic orbitals. Of all the possible combinations that may be written, only a certain number may be used as LCAOs because each LCAO must be independent. Two LCAOs which are not independent of each other are simply different forms of the same orbital, and therefore will give an identical probability density. This is easily illustrated using the LCAOs for a diatomic molecule written in Chapter 7 ($\psi_1 + \psi_2$ and $\psi_1 - \psi_2$). If the two combinations are multiplied by -1, we have two more, or a total of four combinations. They are not independent of each other. The probability densities of ($\psi_1 + \psi_2$) and ($-\psi_1 - \psi_2$) are identical, and the probability densities of ($-\psi_1 + \psi_2$) and ($+\psi_1 - \psi_2$) are identical. This is because the probability density is obtained by squaring the wave function. Only two of the four are independent. *The key to the number of independent molecular orbitals is always given by the number of atomic orbitals involved.* In the diatomic molecules discussed in Chapters 7 and 12, there are four LCAOs of σ symmetry and four of π symmetry or a total of eight.

In other molecules such as NO_2, SO_2, BF_3, and so forth, where one atom may be considered at the center of the coordinate system and no bonds are formed between the remaining atoms in the molecule, a simplification is possible. In situations such as this, the LCAOs may be written in two parts. One part is the atomic orbital of the atom at the center of the coordinate system. The other part consists of the various possible combinations of bonded atom orbitals (ligand orbitals). Thus the general form of the bonding and antibonding functions of NO_2 will be

$$\psi_{MO} = \psi_N + (\psi_{O_1} + \psi_{O_2})$$
$$\psi_{MO} = \psi_N - (\psi_{O_1} + \psi_{O_2})$$

In this case there will be two independent combinations of oxygen orbitals involved in σ bonding. These principles will be illustrated for a number of molecules. The first molecule that will be treated is BeH_2.

Ch. 15 The Ligand Field

A. MOLECULAR ORBITALS FOR BERYLLIUM HYDRIDE (BeH$_2$)

The number of independent combinations of hydrogen functions in beryllium hydride is *two*. They may be represented as

$$(H_1 + H_2) \quad \text{and} \quad (H_1 - H_2)$$

The symbol H is used to designate the 1s function of hydrogen. The general form of the bonding and antibonding function will be

$$\psi_{MO} = a_1\psi_{Be} + a_2(H_1 + H_2)$$
$$\psi_{MO} = a_1\psi_{Be} - a_2(H_1 + H_2)$$

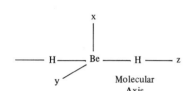

Figure 15-1 *Linear BeH$_2$ in a cartesian coordinate system.*

With molecules containing more than two atoms, it is often desirable and sometimes necessary to specify the direction of the z axis. For a linear molecule like beryllium hydride the z axis is the molecular axis. The appropriate coordinate system for beryllium hydride is shown in Fig. 15-1.

The bonding is visualized by placing the orbitals into the coordinate system. In Fig. 15-2 the s and p orbitals of beryllium are shown along with the s orbitals of hydrogen. Two beryllium orbitals, the 2s and $2p_z$, will be of proper symmetry for bonding.

Examination of Fig. 15-3 shows an effective overlap for the Be$_{2s}$ and H$_{1s}$ functions. The σ bonding function is

$$\psi_{MO} = a_1\psi_{Be_{2s}} + a_2(H_1 + H_2)$$

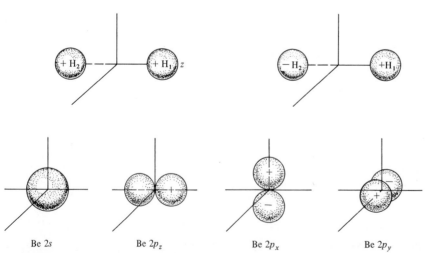

Figure 15-2 *Atomic orbitals in BeH$_2$.*

Sec. I Introduction to Ligand Fields

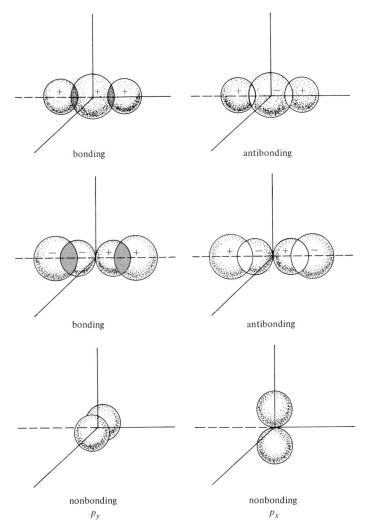

Figure 15-3 *LCAOs for* BeH$_2$.

The σ antibonding function is

$$\psi_{MO} = a_1\psi_{Be_{2s}} - a_2(H_1 + H_2)$$

The p_z function of beryllium will form an effective bond if the sign of the 1s function of hydrogen atom (2) is changed. The combination of hydrogen orbitals that is appropriate for bonding with the p_z orbital of beryllium, is (H$_1$ − H$_2$) which is the second of the two independent combinations possible for two hydrogen atoms. The beryllium p_x and p_y orbitals are not appropriate for bonding as may be readily seen

Ch. 15 The Ligand Field

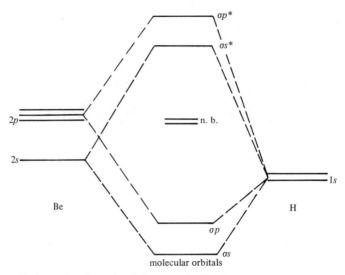

Figure 15-4 *Molecular orbital diagram for* BeH_2.

from Fig. 15-3. These functions are labeled nonbonding. The molecular orbital diagram can now be drawn (Fig. 15-4). In this diagram the atomic orbitals are indicated on the left and right. The molecular orbitals are in the center.

Hydrogen is more electronegative than beryllium, consequently the bonding function will be concentrated on the hydrogen atoms and the antibonding function will be concentrated on the beryllium atoms. This situation will be reflected in the magnitude of the coefficients, a_1 and a_2, in the functions.

The treatment given for beryllium hydride is slightly different from that used for diatomic molecules. In the case of beryllium hydride, mixing or hybridization is not necessary. It would be possible to include it, but in this particular instance the result would not be different. *For molecules of this type, it is only necessary to use the mixing concept if more than one central atom orbital can combine with a single ligand combination.* This situation will be met first in ammonia. We will consider the more symmetrical molecules, boron trifluoride (BF_3) and methane (CH_4) first.

B. MOLECULAR ORBITALS FOR BORON TRIFLUORIDE (BF_3)

The structure of boron trifluoride may be obtained using the valence bond method.

$$B\!:\!\!\cdot, \; \cdot \; - \; - \; -$$
$$BF_3\!:\!\!\cdot\!\!\!\downarrow\!\!\uparrow, \; \cdot\!\!\!\downarrow\!\!\uparrow \; \cdot\!\!\!\downarrow\!\!\uparrow \; -$$

378

Sec. I Introduction to Ligand Fields

The electron involved from the fluorine atom is a p electron.

$$F \stackrel{\cdot\cdot}{\underset{}{}}, \; \frac{\cdot\cdot}{p_x} \; \frac{\cdot\cdot}{p_y} \; \frac{\cdot}{p_z}$$

In a molecule as complicated as boron trifluoride, it is especially important that the coordinates be properly placed and labeled. In addition, it is convenient to think of a framework of ligand orbitals and separately place each of the orbitals of boron into this framework, being careful with the coordinates. The p orbitals of the ligands are placed so that one of them, which we will call p_σ, is directed toward the central atom. The appropriate coordinate system and a framework of fluorine p_σ orbitals is shown in Fig. 15-5.

Since there are three fluorine atoms, there must be three independent

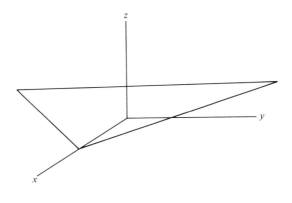

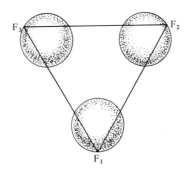

Figure 15-5 *Skeleton of fluorine p_σ orbitals.*

Ch. 15 The Ligand Field

To be strictly accurate, numerical coefficients as in

$$\frac{2}{\sqrt{6}}F_1 - \frac{1}{\sqrt{6}}F_2 - \frac{1}{\sqrt{6}}F_3$$

should be present, but they are an unnecessary complication in a qualitative discussion.

LCAOs involved in σ bonding. These are

$$F_1 + F_2 + F_3$$
$$F_2 - F_3$$
$$F_1 - F_2 - F_3$$

where the symbol F is used to represent the bonding p_σ orbital on the fluorine atom. In this case it is not difficult to write the independent LCAOs down immediately. The most important thing to realize is that there must be three of them.

We now place the orbitals of boron into this framework. The boron $2s$ orbital overlaps effectively with all three fluorine p_σ orbitals. The appropriate combination of ligand orbitals for this interaction is $(F_1 + F_2 + F_3)$. The function describing this bonding state is

$$\psi_{MO} = a_1 \psi_{B_{2s}} + a_2(F_1 + F_2 + F_3)$$

The p_y orbital of boron overlaps effectively with the p_σ orbitals of fluorine atoms 2 and 3. The sign of either F_2 or F_3 must be changed, however. Fluorine atom (1) does not overlap effectively with the p_y of boron. This is shown in Fig. 15-6. The function describing this bonding state is

$$\psi_{MO} = a_1 \psi_{B_{p_y}} + a_2(F_2 - F_3)$$

The p_x orbital of boron overlaps effectively with the orbitals from all three fluorines as shown in Fig. 15-7 where the signs of the orbitals

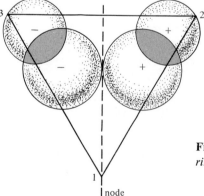

Figure 15-6 *Overlap of p_y with fluorine p_σ.*

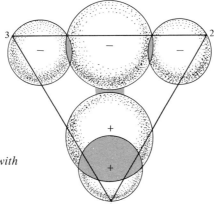

Figure 15-7 *Overlap of boron p_x with fluorine p_σ orbitals.*

are those of the third combination of fluorine p_σ orbitals.

$$\psi_{MO} = a_1 \psi_{Bp_x} + a_2(F_1 - F_2 - F_3)$$

In addition to the σ bonding described, π bonding is possible in this molecule. Figure 15-8 shows the orbital arrangement of the π bonds. In this case there is an overlap of the p_z orbital of boron with the sum of the three p orbitals of fluorine that are perpendicular to the plane.

There are three fluorine p_z orbitals (one from each atom) involved in π bonding so there will be three LCAOs. The effective bonding orbital is

$$\psi_{MO} = a_1 \psi_{Bp_z} + a_2(F_{1\pi} + F_{2\pi} + F_{3\pi})$$

Figure 15-8 *The π bonding in BF_3.*

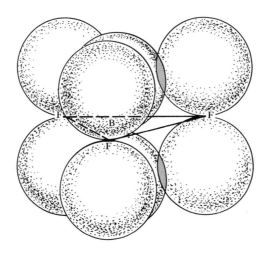

Ch. 15 The Ligand Field

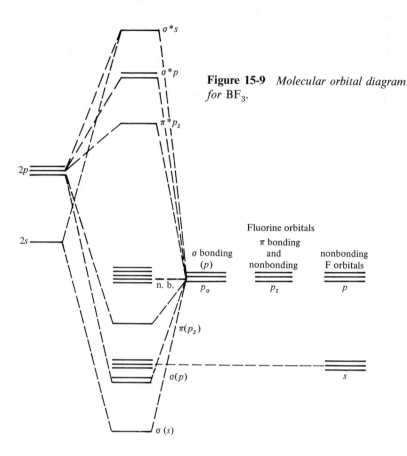

Figure 15-9 *Molecular orbital diagram for* BF_3.

The two remaining LCAOs must be ineffective since they can only be obtained by changing the sign of one of the fluorine functions, and there are no *s* or *p* orbitals on the boron which can match the signs obtained. *In this case, it is not necessary to get the actual form of the additional functions. It is sufficient to know that there are two more LCAOs which are nonbonding.* Each of these LCAOs is, of course, capable of holding an electron pair.

Each of the three fluorine atoms has a third *p* orbital and an *s* orbital making a total of six additional orbitals. Because all the appropriate bonding orbitals for boron have been used, these fluorine orbitals will be nonbonding orbitals.

It is easy to check the number of nonbonding orbitals since the molecular orbital diagram, Fig. 15-9, must show as many molecular orbitals as there were atomic orbitals to start with. In this case there are four atoms with four valence orbitals each, or a total of 16 atomic orbitals. Since four bonding and four antibonding LCAOs have been identified, there must be eight nonbonding orbitals on the fluorine atoms.

Each fluorine atom contributes seven valence electrons to boron trifluoride and the boron contributes three, for a total of 24 valence electrons. As a result, the four bonding orbitals and all eight nonbonding orbitals will be occupied with electrons. The molecular orbital diagram shown in Fig. 15-9 is also applicable to the planar ions NO_3^- and CO_3^{2-} (which also contain four atoms and 24 valence electrons).

C. MOLECULAR ORBITALS IN METHANE (CH_4)

It is very helpful to visualize the tetrahedron (the structure of CH_4), inside of a cube; therefore, the s and p orbitals of carbon are drawn in a cube in Fig. 15-10. The positive lobe of the p_z orbital of carbon will overlap with the orbitals of H_1 and H_2, and the negative lobe will overlap with the orbitals of H_3 and H_4. The bonding LCAO is

$$\psi_{MO} = a_1 \psi_{C_{p_z}} + a_2(H_1 + H_2 - H_3 - H_4)$$

The positive lobe of the p_x orbital of carbon overlaps with the s orbital of H_1 and H_3, and the negative lobe overlaps with the orbitals of H_2 and H_4. The LCAO is

$$\psi_{MO} = a_1 \psi_{C_{p_x}} + a_2(H_1 - H_2 + H_3 - H_4)$$

Similarily the LCAO involving the p_y orbital is

$$\psi_{MO} = a_1 \psi_{C_{p_y}} + a_2(-H_1 + H_2 + H_3 - H_4)$$

The fourth linear combination of hydrogen orbitals

$$H_1 + H_2 + H_3 + H_4$$

has the appropriate symmetry for overlap with the $2s$ orbital of carbon. The LCAO is

$$\psi_{MO} = a_1 \psi_{C_s} + a_2(H_1 + H_2 + H_3 + H_4)$$

The molecular orbital diagram for methane is shown in Fig. 15-11.

D. DEGENERATE STATES IN MOLECULES

In the molecular orbital diagrams for boron trifluoride and methane, some of the bonding functions have the same energy. These states are degenerate. It is generally easy to recognize degeneracies in molecules by considering the relationship between the central atom and the

Ch. 15 The Ligand Field

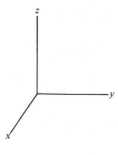

coordinate system

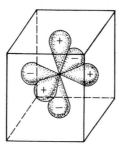

Figure 15-10 *Orbital arrangement in* CH_4.

orbitals of carbon

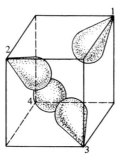

positions of the ligands

surrounding groups. The p orbitals of carbon are originally degenerate. In the molecule methane, each of the p orbitals is arranged in an exactly equivalent way with respect to the hydrogen atoms. In other words, the labeling of the orbital as p_x, p_y, or p_z is arbitrary. The originally degenerate functions are still degenerate. The s orbital is originally different in energy from the p orbitals, and in the molecule the energy is still different. The situation in boron trifluoride is slightly different. The p orbitals of boron are originally degenerate, but in the molecule the degeneracy is split. The p_x and p_y orbitals lie in the same plane as the fluorine atoms, whereas the p_z is out of the plane. The energy of the p_z orbital cannot be the same as the energy of the p_x orbital or the p_y orbital because the relationship between the p_z orbital and ligands is different from the relationship between the p_x, p_y and the ligands. That the energy of the p_z must be different seems obvious. It is not so obvious that the energies of the molecular orbitals that result from the p_x and p_y orbital must be the same, but they are nevertheless equivalent.

E. MOLECULAR ORBITALS FOR AMMONIA (NH$_3$) AND WATER (H$_2$O)

The coordinate system and ligand skeleton appropriate for ammonia are given in Fig. 15-12. The appropriate overlaps are given in Fig. 15-13, and the LCAOs are given in Table 15-1. Here we see an example in which mixing is essential. The table shows four molecular orbitals; however, we know that there can only be three independent combinations for three H ligands. The combination $(H_1 + H_2 + H_3)$ is appropriate for overlap with both $\psi_{N_{2s}}$ and $\psi_{N_{2p_z}}$. Using the mixing concept,

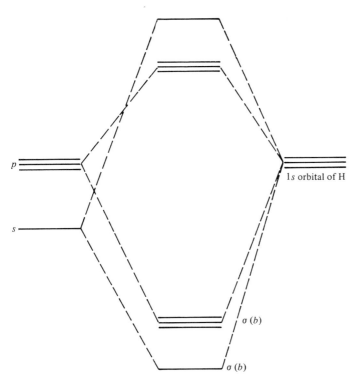

Figure 15-11 *Molecular orbital diagram for* CH$_4$.

Ch. 15 The Ligand Field

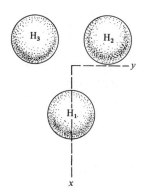

Figure 15-12 *Ligand skeleton in* NH_3.

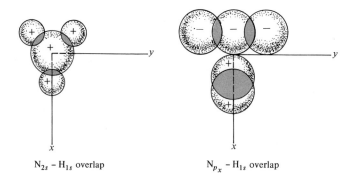

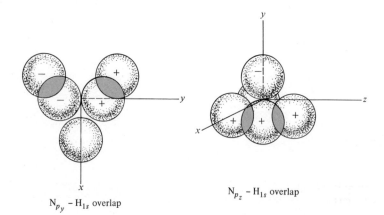

Figure 15-13 *Effective overlaps for* NH_3.

Sec. I   Introduction to Ligand Fields

Table 15-1 *LCAOs for Ammonia Without Mixing*

$\psi_{MO} = \psi_{N_{2s}} + (H_1 + H_2 + H_3)$
$\psi_{N_{2p_z}} + (H_1 + H_2 + H_3)$
$\psi_{N_{2p_x}} + (H_1 - H_2 - H_3)$
$\psi_{N_{2p_y}} + (H_2 - H_3)$

Table 15-2 *LCAOs for Ammonia*

$\psi_{MO}(\sigma) = a_1\psi_{N_{2s}} + a_2\psi_{N_{2p}} + a_3(H_1 + H_2 + H_3)$
$\psi_{MO}(nb) = a_1\psi_{N_{2s}} - a_2\psi_{N_{2p}} + a_3(H_1 + H_2 + H_3)$
$\psi_{MO}(\sigma^*) = a_1\psi_{N_{2s}} + a_2\psi_{N_{2p}} - a_3(H_1 + H_2 + H_3)$

the three functions $\psi_{N_{2s}}$, $\psi_{N_{2p}}$ and $(H_1 + H_2 + H_3)$ are combined and the three molecular orbitals given in Table 15-2 are obtained.

Only one of the three combinations is a bonding orbital. There is also one antibonding LCAO. The third function is basically a nonbonding orbital on the nitrogen. The energy of each of these orbitals is indicated in Fig. 15-14, and the full molecular orbital diagram for ammonia is given in Fig. 15-15.

The results of a similar treatment for water are given in Fig. 15-16 and Table 15-3. Note that mixing is once again necessary.

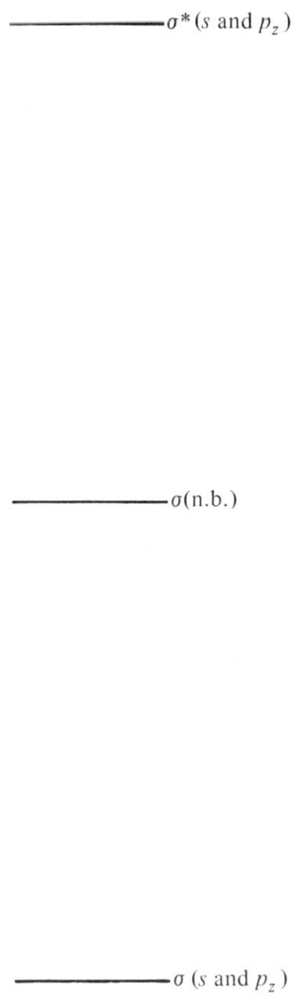

————— $\sigma^*(s$ and $p_z)$

————— $\sigma(n.b.)$

————— σ (s and p_z)

Figure 15-14 *The energies of the bonding, nonbonding, and antibonding orbitals that result from the mixing of N_{2s}, N_{2p}, and $(H_1 + H_2 + H_3)$ functions.*

387

Ch. 15 The Ligand Field

Table 15-3 *LCAOs for Water*

$\psi_{MO}(\sigma_{sp}) = a_1\psi_{O_{2s}} + a_2\psi_{O_{2p_y}} + a_3(H_1 + H_2)$

$\psi_{MO}(nb) = a_1\psi_{O_{2s}} - a_2\psi_{O_{2p_z}} + a_3(H_1 + H_2)$

$\psi_{MO}(\sigma_{sp}^*) = a_1\psi_{O_{2s}} + a_2\psi_{O_{2p_z}} - a_3(H_1 + H_2)$

$\psi_{MO}(\sigma_p) = a_1\psi_{O_{2p_x}} + a_2(H_1 - H_2)$

$\psi_{MO}(\sigma_p^*) = a_1\psi_{O_{2p_x}} - a_2(H_1 - H_2)$

F. COMPLEXES INVOLVING *d* ORBITALS

The application of the method to a coordination compound, such as that shown in Fig. 15-17, follows the same procedure given in the previous discussion. There are six ligand σ bonding functions; therefore, there must be six possible σ bonding LCAOs. When the orbitals from the metal are placed into the skeleton of the ligand σ bonding orbitals, the overlaps shown in Fig. 15-18 may be visualized. The *s* and *p* orbitals of the central atom and two of the five *d* orbitals are

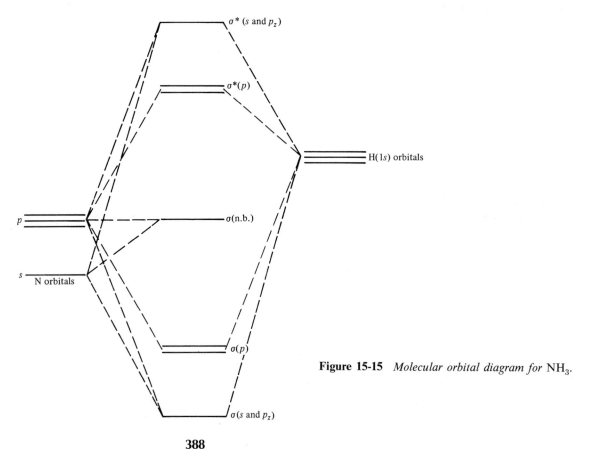

Figure 15-15 *Molecular orbital diagram for* NH_3.

Sec. I Introduction to Ligand Fields

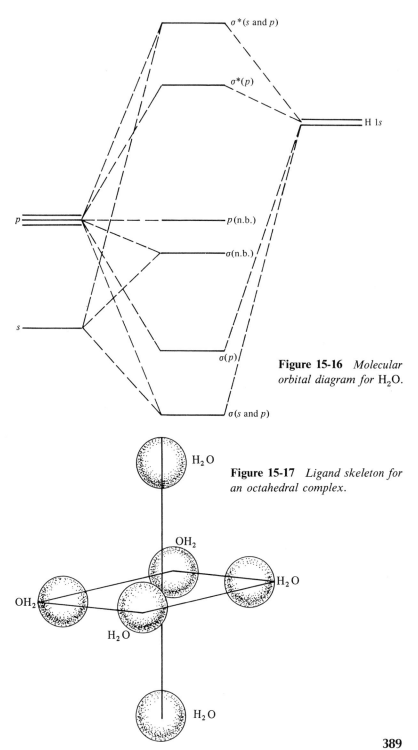

Figure 15-16 Molecular orbital diagram for H_2O.

Figure 15-17 Ligand skeleton for an octahedral complex.

Ch. 15 The Ligand Field

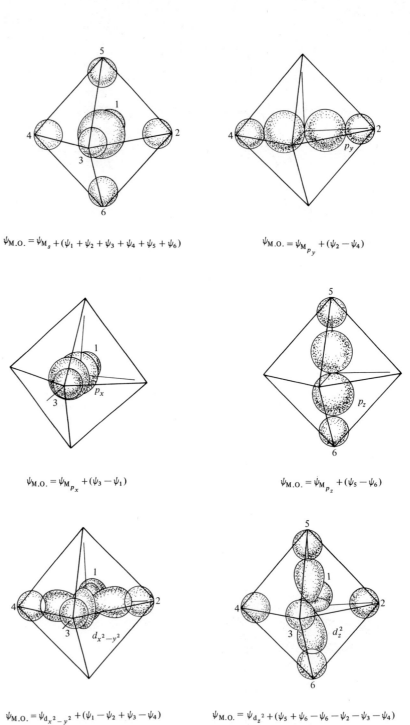

$\psi_{M.O.} = \psi_{M_s} + (\psi_1 + \psi_2 + \psi_3 + \psi_4 + \psi_5 + \psi_6)$

$\psi_{M.O.} = \psi_{M_{p_y}} + (\psi_2 - \psi_4)$

$\psi_{M.O.} = \psi_{M_{p_x}} + (\psi_3 - \psi_1)$

$\psi_{M.O.} = \psi_{M_{p_z}} + (\psi_5 - \psi_6)$

$\psi_{M.O.} = \psi_{d_{x^2-y^2}} + (\psi_1 - \psi_2 + \psi_3 - \psi_4)$

$\psi_{M.O.} = \psi_{d_{z^2}} + (\psi_5 + \psi_6 - \psi_6 - \psi_2 - \psi_3 - \psi_4)$

Figure 15-18 *Linear combinations for an octahedral complex.*

involved in σ bonding. Of the bonding functions, those involving the p_x, p_y, and p_z of the metal are all degenerate, and the $d_{x^2-y^2}$ and d_{z^2} orbitals are also degenerate. It is easy to see that the p orbitals have the same energy. The fact that the $d_{x^2-y^2}$ and d_{z^2} are degenerate is not so obvious.

The overlap of the ligand $p\sigma$ functions with the d_{xy}, d_{xz}, d_{yz}, orbitals is not effective. This is shown in Fig. 15-19. In this figure a node from the d_{xy} cuts the ligand orbitals in half. Electrons that occupy these orbitals will be nonbonding electrons. These three orbitals are also degenerate because the relationship between each orbital and the ligands is identical.

In an octahedral complex the d orbitals of the metal are split into two groups. One group contains the nonbonding d_{xy}, d_{yz} and d_{xz}, while the other contains the bonding $d_{x^2-y^2}$, d_{z^2}. The degeneracy of the d orbitals (five-fold in the free ion) has been split.

Symbols can be assigned to the two degenerate sets of orbitals. For our purposes, the names are purely arbitrary, but we will refer to the t_{2g} orbitals (d_{xy}, d_{xz}, d_{yz}) and use the symbol e_g to represent the degenerate pair ($d_{x^2-y^2}$, d_{z^2}).

The σ bonding molecular orbital diagram is shown in Fig. 15-20. In this diagram the d_{xz}, d_{yz}, and d_{xy} are written as nonbonding functions. *The total number of electrons is determined by adding two electrons for each ligand plus the number of electrons on the metal ion.*

The complete bonding scheme that includes π bonding may be developed in a similar manner. The complete molecular orbital diagrams are complicated and they depend upon the ligand present. The essential features of the diagram involving π bonding are shown in

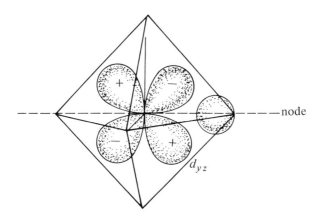

Figure 15-19 *Nonbonding d orbitals.*

Figs. 15-21 and 15-22. Some of the levels have been omitted from both Fig. 15-21 and Fig. 15-22 for clarity.

The bonding scheme shown in Fig. 15-21 is appropriate for ligands which can behave as π acceptors. Ligands of this type include carbon monoxide and cyanide ion. The molecular orbital diagram given in this figure is not different *in its essential features* from the diagram given for σ bonding only (Fig. 15-20). The important point is that the t_{2g} level is bonding and the e_g level is antibonding.

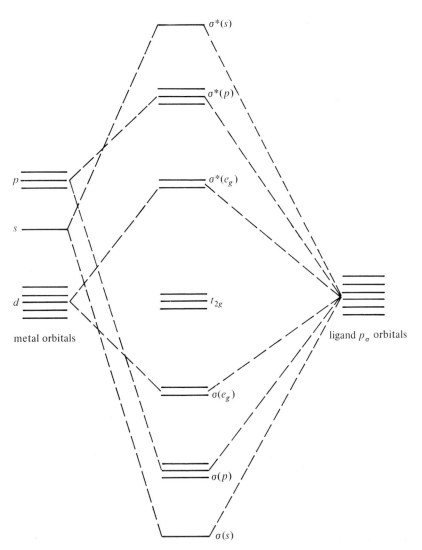

Figure 15-20 *The σ bonding molecular orbital diagram for an octahedral complex.*

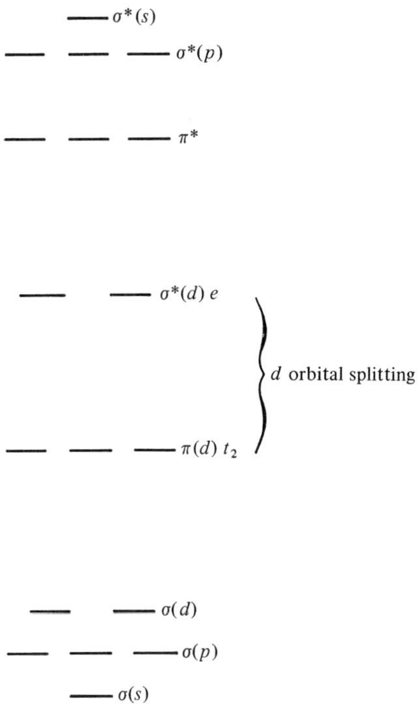

Sec. I Introduction to Ligand Fields

Figure 15-21 *Principal features of the molecular orbital diagram for octahedral complexes involving π bonding groups such as CN⁻.*

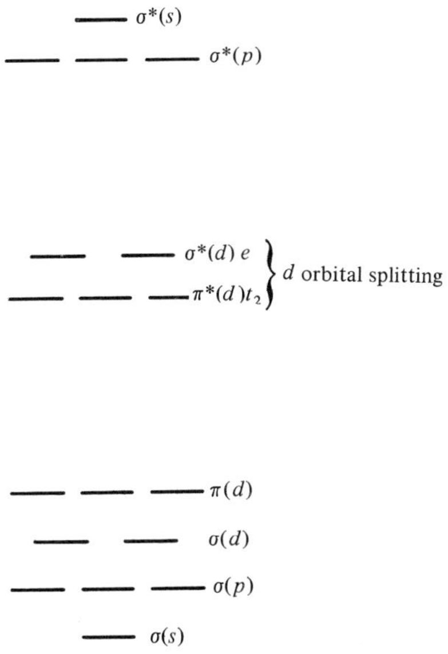

$\sigma^*(s)$
$\sigma^*(p)$
π^*

$\sigma^*(d)\,e$ ⎫
                ⎬ d orbital splitting
$\pi(d)\,t_2$ ⎭

$\sigma(d)$
$\sigma(p)$
$\sigma(s)$

Figure 15-22 *Principal features of the molecular orbital diagram for octahedral complexes involving π donor ligands such as I⁻.*

$\sigma^*(s)$
$\sigma^*(p)$

$\sigma^*(d)\,e$ ⎫
               ⎬ d orbital splitting
$\pi^*(d)t_2$ ⎭

$\pi(d)$
$\sigma(d)$
$\sigma(p)$
$\sigma(s)$

Ch. 15 The Ligand Field

The molecular orbital diagram given in Fig. 15-22 includes π bonding from the ligands. The diagram is appropriate for ligands which are good π donors. This would include chloride, iodide, and sulfide ions.

The difference between Figs. 15-21 and 15-22 that is important involves the t_{2g} level. In Fig. 15-21 the t_{2g} level is bonding. In Fig. 15-22 the t_{2g} level is antibonding (π^*).

Molecular orbital diagrams for other structures may be constructed by visualizing overlaps. However the process quickly becomes cumbersome. The diagrams that are appropriate for some other situations are given below. In a tetrahedral complex, it is the t_2 orbitals which take part in σ bonding, and the relative energy of the t_2 and e levels is reversed (Fig. 15-23). Figure 15-24 shows the molecular orbital diagram for square planar complexes.

The subscript g is dropped in the tetrahedron.

In Figs. 15-21, 15-22, and 15-24 only a portion of the complete diagram is shown, and some of the levels are specially marked. The diagrams are simplified and certain of the levels are emphasized in this way to illustrate the point that these are the levels that are important for our purpose.

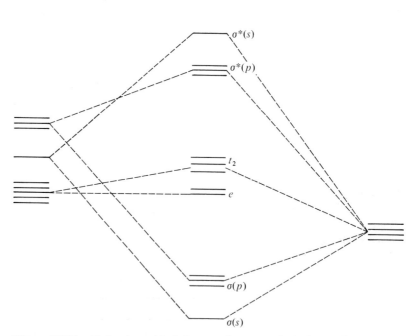

Figure 15-23 *Molecular orbital diagram for a tetrahedral complex (σ bonding only).*

Sec. I Introduction to Ligand Fields

In the diagram (for any structure) the bonding functions and the nonbonding functions on the ligand will be filled by ligand electrons. The metal electrons will occupy the nonbonding d levels (if there are any) and the antibonding σ^* and π^* levels that are lowest in energy. For example the molecular orbital diagrams for the hydrates of some of the ions of the first transition series are given in Fig. 15-25. These may be compared with the molecular orbital diagrams for the cyanides in Fig. 15-26. The ligand electrons occupy the $\sigma(s)$, the $\sigma(p)$, and the $\sigma(e_g)$, in both the hydrate and the cyanides. In these examples π bonding is not being included so the metal electrons are distributed among the nonbonding t_{2g} and the antibonding $\sigma^*(e_g)$. The highest energy antibonding levels remain empty.

Another point may be illustrated using Figs. 15-25 and 15-26. That is, there are two ways of arranging the electrons in any ion that contains more than three d electrons. The diagrams for $Cr(H_2O)_6^{3+}$ and $Cr(CN)_6^{3-}$ are basically the same (the relative energy of the levels would be different however) but there are two electronic structures possible for Fe^{2+} and Co^{2+}. In these ions the actual electronic structure depends upon the magnitude of the bonding interaction. If the bonding interaction is great, as it is in a cyanide complex, the electrons pair up in the t_{2g} level and the complex is a low-spin complex. The greater the bonding interaction the lower the energy of the $\pi(t_{2g})$ function and the greater the energy of the antibonding $\sigma^*(e_g)$ function. The electrons will be unpaired if the complex is high spin and the bonding interaction is weak. The spin state in these complexes depends upon the strength of the ligand field.

Molecular orbitals may be used to understand the relative stability

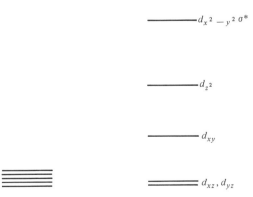

Figure 15-24 *Molecular orbital diagram for a square planar complex (σ bonding only).*

395

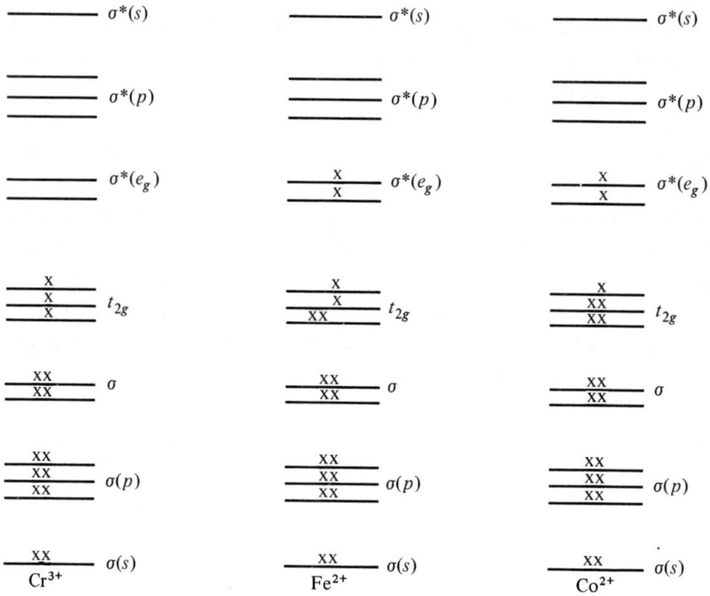

Ch. 15 The Ligand Field

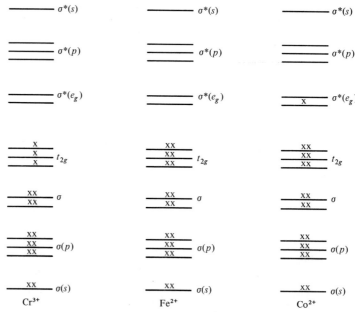

Figure 15-25 Molecular orbital diagrams for hydrates of Cr^{3+}, Fe^{2+}, and Co^{2+}.

Figure 15-26 Molecular orbital diagram for cyanides of Cr^{3+}, Fe^{2+}, and Co^{2+}.

of complexes such as the carbonyls. The enthalpy data for the metal carbonyls were given in Table 14-3.

As always the compound with a maximum number of bonding electrons and the fewest antibonding electrons is the most stable. The most favorable enthalpy of formation in Table 14-3 occurs at $Cr(CO)_6$. In this compound, all of the bonding levels are filled. As a result hexacarbonyls on the right and left of chromium are all less stable than $Cr(CO)_6$. In the case of iron where the two additional electrons would occupy the antibonding orbitals (if the molecule were octahedral), the stoichiometry and structure are changed from 1:6 and octahedral, to 1:5 and trigonal bipyramid. The molecular orbital diagram appropriate for a trigonal bipyramid is given in Fig. 15-27. The different structure and stoichiometry allow the use of a nonbonding level for the electrons rather than an antibonding level.

Although none of the diagrams given here are strictly appropriate to sandwich compounds, the relative stability of these compounds may be similarly understood. In the sandwich compounds, a change in structure is not possible. With more than six electrons on the metal, the stability of the complex drops rapidly. Sandwich compounds have not been prepared for the very early transition elements (Sc) or for the very late transition elements (Cu). Where the number of d electrons is greatly different from six, compounds of this type are not stable.

Compounds with benzene and cyclopentadienyl ligands (see pp. 354 and 362) are known as sandwich compounds. The metal ion lies between two planar ligands as in a sandwich.

For ions with few d electrons the very large majority of complexes are octahedral [that is, Mn(II), Fe(II and III), Co(III)]. For simple complexes containing eight d electrons, competition between the octahedral and square planar structures becomes really important because here again changes in stoichiometry and structure act to cut down the number of antibonding electrons. This tendency is opposed by electron repulsion and the spin correlation stabilization energy of the high-spin

 $d_{z^2}\sigma^*$

Figure 15-27 *Molecular orbital diagram for the d orbitals in a trigonal bipyramid complex.*

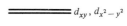 $d_{xy}, d_{x^2-y^2}$

 d_{xz}, d_{yz}

state. Only ligands that create a strong field are able to form square planar complexes with nickel(II).

Most copper(II) complexes have a structure that is midway between octahedral and square planar. Antibonding electrons must be used in either case. This, of course, weakens the bonding which is the main reason why copper compounds are so easily reduced to metallic copper.

II. The Strength of the Ligand Field

In Chapter 13 ligands were arranged in an order of increasing ease of polarization. Experimental measurement of the ligand field strength gives a series

$$CO > CN^- > NO_2^- > o\text{-phenanthroline} > \text{ethylenediamine} > NH_3 > H_2O > OH^- > F^- > NO_3^- > Cl^- > S^{2-} > Br^- > I^-$$

which is similar to but somewhat different from the sequence used in Chapter 13. The order given here is the result of a number of factors, including ease of polarization and π bonding capabilities. The ease of polarization deals strictly with the formation of σ bonds. A series of σ bonding ligands would follow ease of polarization. The tendency of a substance to form π bonds is responsible for the inversions that have occurred in the two series. The ability to form π bonds by accepting electrons from the metal would make a weak σ donor a much better ligand with metals containing a large number of d electrons. This is the case for CO which forms all low-spin complexes. At the same time the bonding of CO would be poor when combined with metals that contain few or no d electrons, such as Sc(III). The π bonding that stabilizes carbonyls was shown in Fig. 14-2. The same figure is applicable to CN^- and NO_2^-; except for CN^- and NO_2^- the σ donor properties are also good (CN^- is easily polarized). On the other hand ligands such as Cl^-, H_2O, or OH^- are good σ donors, but they are also π donors.

The difference in the molecular orbital diagrams for ligands of the CO type and for ligands of the Cl^-, H_2O type was given in Figs. 15-21 and 15-22. In complexes involving ligands which are good π acceptors the t_{2g} metal electrons occupy orbitals which are essentially bonding. On the other hand, in complexes containing good π donors the metal electrons occupy π antibonding orbitals. In the first case π bonding stabilizes the molecule when the t_{2g} level is occupied. With the π donor ligands the d electrons occupy π antibonding orbitals and the result is a destabilization of the molecule.

Sec. II The Strength of the Ligand Field

Table 15-4 *Classification of Ligands in the Spectrochemical Series*

Very strong σ donors and π acceptors	σ donors only	σ and π donors
CO, CN$^-$, NO$_2^-$, o-phenanthroline	Ethylenediamine, NH$_3$	H$_2$O, OH$^-$, F$^-$ NO$_3^-$, Cl$^-$, Br$^-$, I$^-$

The relative stabilities of the complex ammines and the hydrates of the first transition series are understandable on this basis. The σ bond strength in both the hydrates and the ammines increases across the series. However, because of π bonding capabilities the early elements form relatively more stable hydrates (because there are empty d orbitals on the metal to accept the p electrons of the oxygen). The hydrates of the late transition metal ions are destabilized because of the π donor tendency of water. In this case (Fig. 15-22) antibonding t_{2g} levels are occupied. Ammonia contains only σ donor electrons; therefore, an increase in stability follows an increased polarizing ability of the metal.

In terms of the above discussion, the series given on page 398 may be broken into the three groups shown in Table 15-4. This series is called the spectrochemical series. The name is derived from the fact that spacing of the d orbitals or ligand field strength influences the spectrum (or color) of a complex. The absorption spectrum of complexes is one method of determining the magnitude of the spacing or the ligand field strength.

QUESTIONS

1. What will be the number of σ bonding LCAOs possible in SO$_3$, SCl$_2$, SO$_2$, PCl$_5$, HCl?
2. How many independent ligand combinations appropriate for σ bonding may be written for SO$_3$, SO$_2$, PCl$_5$? Write as many of them as possible.
3. Set up the ligand skeleton for a number of different molecules: H$_2$S, PH$_3$, NO$_3^-$, and SO$_4^{2-}$.
4. Write the σ bonding and antibonding functions for the molecules in Question 3.
5. What is a nonbonding orbital?
6. Sketch the molecular orbitals for the molecules and ions in Question 3.
7. Construct the energy level diagrams for the molecules in Question 3.
8. Explain the difference between a high-spin complex and a low-spin complex.
9. What factors determine whether a complex will be high or low spin?
10. How does the presence of an electron in an antibonding orbital affect the stability of a complex?

11. Explain why only a few of the elements of the first transition series form sandwich complexes.
12. Explain the ammine complexing tendencies of the ions of the first transition series.
13. Which ion, NO_2^- or F^-, has the larger ligand field? Explain.
14. Suggest the stable geometry of ligands surrounding each of the following ions. Justify your choice.

   Ni^{2+}, Cr^{3+}, Co^{2+}, Zn^{2+}

15. Why is CuO easier to reduce than either NiO or ZnO?

PROBLEMS I
1. Draw a molecular orbital diagram for NH_3 labelling all orbitals according to their bonding, nonbonding, or antibonding character. Show which orbitals are occupied.
2. Omitting closed shell orbitals, show the molecular orbitals of FeF_6^{3-}, labelling the orbitals in as much detail as you can. Describe the extent to which the various $3d$ orbitals of iron are involved in the bonding.
3. Construct a molecular orbital diagram for square planar $Ni(CN)_4^{2-}$. Why is this ion low spin?

PROBLEMS II
1. Draw the octet rule structure for NO_3^-.
2. What is the shape of the NO_3^- ion?
3. What nitrogen orbitals form the three σ bonds in NO_3^-?
4. What orbital on nitrogen is involved in π bonding?
5. Sketch the π bonding molecular orbital.
6. Draw the molecular orbital diagram for NO_3^-.
7. How many valence atomic orbitals are used in constructing these molecular orbitals?
8. Fill the diagram in Problem 6 with the correct number of electrons.
9. Draw the molecular orbital diagram for NO_3.
10. On what atoms in NO_3 is the unpaired electron most likely to be found?
11. Is the combination $\psi_{p_z O_1} + \psi_{p_z O_2} + \psi_{p_z O_3}$ involved in the bonding? How?
12. What orbital can the combination $\psi_{p_z O_1} - \psi_{p_z O_2}$ overlap with?
13. What is a nonbonding orbital?
14. Why are nonbonding orbitals lower in energy than antibonding orbitals?

SUGGESTED READINGS
GRAY, H. B., *Electrons and Chemical Bonding*. New York: W. A. Benjamin Inc., 1965.

ORCHIN, M. and JAFFE, H. H., *The Importance of Antibonding Orbitals*. Boston: Houghton Mifflin Co., 1967.

Whereas U. V. absorption was highly constitutive and characterized systems rather than individual groups, the atomic and molecular spectra of infra-red absorption were found to be far less so. With certain qualifications that have gradually become understood and codified, I. R. spectra show directly what individual groups are present in the molecule. I cannot refrain from confirming the truth of a rumour which has returned to base. On a laboratory round I asked Mr. X: "Well, is that substance you showed me an acid?" "Sorry, I can't say, Professor. I haven't got the I. R. result back yet!"*

SIR ROBERT ROBINSON

Chapter 16 Physical Methods in Chemistry

The *d* orbital splitting discussed in the last chapter and the absorption spectra that result from the splitting provide a method for determining the structure of some of the complexes of transition metals. This, along with a discussion of the application of some other physical methods of determining structures, will be the subject of this chapter.

The application of spectra to the determination of the structure of molecules is unique to those cases where a metal ion contains an unfilled *d* shell. Compounds involving representative elements as well as those of group IB (in a $+1$ oxidation state) and group IIB require alternative methods.

I. The Structure and Spectra of Transition Metal Complexes

The detailed examination of the absorption spectrum of a material requires measurements with light over a range of wavelengths. This is most conveniently accomplished with a recording spectrophotometer. The fundamental parts of this instrument are indicated in Fig. 16-1. It consists of an intense light source and a prism or grating which

*R. A. Y. Jones, et. al., *The Techniques of NMR and ESR*, (London: United Trade Press, (1965).

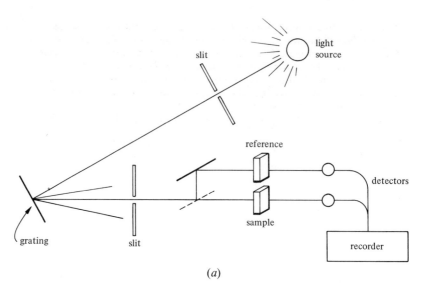

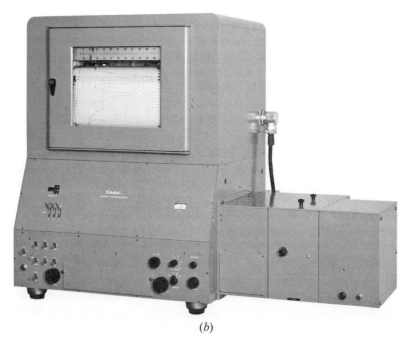

Figure 16-1 *Spectrophotometer. (a) Schematic diagram. (b) Recording spectrophotometer. (Courtesy of Cary Instruments, Monrovia, Calif.)*

Sec. I The Structure and Spectra of Transition Metal Complexes

separates the white light into different wavelengths. More complex instrumentation is used to split the light beam in two. One of the beams passes through the sample cell and the other, which is used as a standard reference, is generally passed through a similar cell containing pure solvent. The two beams are then compared electronically and either the intensity ratio or the absorbance is automatically plotted. Commercial instruments can be used to determine the spectrum of a compound from the infrared region of the spectrum into the ultraviolet region.

Absorbance is defined as log $(I_{\text{ref}}/I_{\text{sample}})$, the logarithm of the intensity ratio.

The visible absorption spectra of complexes of the transition metals result from an electronic transition between the split d levels. The splitting is the result of the interaction of the metal d orbitals with the ligand orbitals. As a result the splitting must depend upon the number of ligands involved as well as the structure of the compound. Consequently the splitting and the observed absorption spectrum must be different for compounds that involve a different number of ligands or different structures.

The characteristic spectral features of all of the ions of the first transition series have been studied in detail. The most interesting ion from this point of view is nickel(II) because of the variety of structures that are formed and the variable coordination number (four or six) in water solution. It is also of interest because many nickel(II) complexes were believed to have a tetrahedral structure for many years. Only after the idea of a ligand field was applied were many of these complexes shown to have an octahedral structure.

Many of the peaks in the absorption spectrum of a transition metal complex involve transitions of the d electrons. The transitions arise because the bonding in complexes splits the d orbitals into sets. In most octahedral complexes one of the transitions involves simply raising a d electron from t_{2g} to an e_g orbital. The energy difference between these orbitals is essentially responsible for the visible spectrum. Changes in color may be observed when ligands that are higher or lower in the spectrochemical series are substituted into the complex.

A plot of the energy levels of a particular ion versus the splitting of the t_{2g} and e_g orbitals (represented by the symbol Δ) is known as an Orgel diagram. The Orgel diagram for Ni(II) octahedral complexes is shown in Fig. 16-2. If we approximate some of the curves in this diagram by straight lines, the *position of the absorption bands* in an octahedral Ni(II) complex will be given by

$$\nu_1 = \Delta$$
$$\nu_2 = 36$$
$$\nu_3 = 21 + \Delta$$
$$\nu_4 = 61$$
$$\nu_5 = 36 + \tfrac{3}{2}\Delta$$

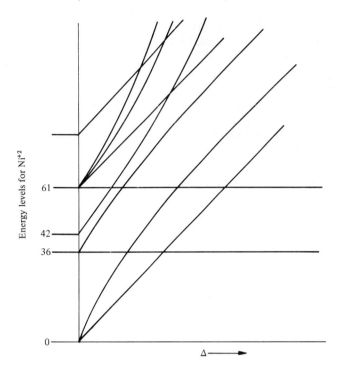

Figure 16-2 *Energy levels for octahedral complexes of Ni^{2+} as a function of Δ.*

Figure 16-3 shows the spectra of $Ni(H_2O)_6^{2+}$ and $Ni(NH_3)_6^{2+}$. By choosing an appropriate value for Δ the energies of the various observed bands can be calculated. In Table 16-1 the band positions obtained using the above relations are compared with the observed spectrum for these and other octahedral nickel complexes.

The observed spectra of a number of nickel complexes that are not octahedral are given in Table 16-2. The visible spectra of a tetrahedral and a square planar complex are given in Fig. 16-4. The characteristic features of the three different structures (octahedral, square planar, and tetrahedral) are summarized in Table 16-3.

Octahedral nickel(II) complexes are characterized by a large number of absorption bands. The four or five bands usually observed are weak (have a low intensity). The ratio of the energies of the first and third bands E_3/E_1 is between 1.6 and 1.8.

Only two or three of the bands influence the color of an octahedral complex. The first two long wavelength bands are usually in the near infrared. The color of the compound corresponds to the wavelength of light which is not absorbed. The color spectrum on Fig. 16-3 shows that green light is not absorbed in $Ni(H_2O)_6^{2+}$. The blue color of $Ni(NH_3)_6^{2+}$ results from the absorption of all wavelengths except those in the blue region.

Square planar nickel(II) complexes are characterized by as few as one band in the 550–330 mμ region of the spectrum. The bands in square planar complexes are fairly intense. Absorption in the blue region results in a red color for many square planar nickel(II) complexes.

The first band in a tetrahedral complex is usually found in the near infrared. Another important feature of tetrahedral spectra is the fine structure (irregularities) observed on the visible bands. (See Fig. 16-4.)

As an example of the application of spectra to structure determination, we shall consider the complexes of nickel(II) with ethylenimine (Az)

$$\underset{CH_2\!-\!CH_2}{\overset{\ddot{N}H}{\diagup\!\diagdown}}$$

The preparation of a number of complexes containing this ligand was reported in 1961. The complete spectrum of Ni(Az)$_6$(NO$_3$)$_2$ reported

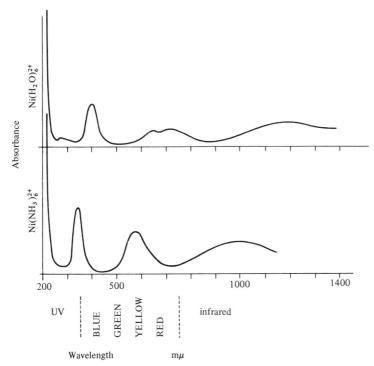

Figure 16-3 *The spectra of two octahedral nickel complexes.*

Table 16-1 *The Spectrum of Some Octahedral Nickel Complexes*

Compound	Absorption bands		Wavelength, mμ	Intensity,
	Calc., kcal mole^{-1}	Exptl., kcal mole^{-1}		
pyHNiBr$_3^*$	10	—		†
	36	32	900	
	31	30	960	
	61	57	505	
	51	50	575	
pyHNiCl$_3$	12	—		†
	36	36	800	
	33	33	870	
	61	62	465	
	54	58	495	
KNiF$_3$	22	—		†
	36	36	800	
	43	44	655	
	61	60	480	
	69	68	420	
Ni(H$_2$O)$_6^{2+}$	24	24	1175	2.0
	36	39	740	1.8
	45	44	650	1.5
	61	63	455	shoulder
	72	73	396	5.2
Ni(edta)$^{2-}$ ‡	28	32	900	3.1
	36	37	788	5.4
	49	49	588	8.4
	61	—		
	78	75	382	12.6
Ni(edta)(NH$_3$)$_2^{2-}$	28	29	980	16.6
	36	36	788	3.4
	49	49	582	10.9
	61	—		
	78	77	372	20.2
Ni(edta)$_n^{2-}$ ‡	28	29	995	25.0
	36	36	800	4.7
	49	50	579	8.1
	61	—		
	78	78	369	10.8
Ni(NH$_3$)$_6^{2+}$	30	31	930	
	36	38	760	
	51	50	572	
	61	—		
	81	78	355	

Sec. I The Structure and Spectra of Transition Metal Complexes

Table 16-1 (*Continued*)

Compound	Absorption bands		Wavelength, mμ	Intensity, ϵ
	Calc., kcal mole^{-1}	Exptl., kcal mole^{-1}		
Ni(en)$_3^{2+}$§	32	32	894	7.3
	36	36	807	5
	53	53	545	6.7
	61	—		
	84	83	345	8.6

*The abbreviation pyH stands for the pyridinium cation, $C_5NH_6^+$.
†Intensities of absorption bands in solids are difficult to measure and are not usually reported.
‡The abbreviation edta has been used for the anion of ethylenediamine tetraacetic acid which has the structure

$$\begin{array}{c} \text{structure shown} \end{array}$$

This ion can coordinate in four, five, or six positions.
§Similarly en stands for ethylenediamine which has the structure H_2N—CH_2—CH_2—NH_2. It coordinates as a bidentate ligand.

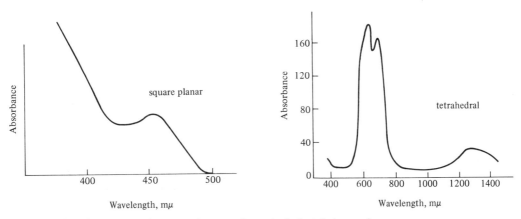

Figure 16-4 *The spectra of square planar and tetrahedral nickel complexes. (After F. A. Cotton and G. Wilkinson, Advanced Inorganic Chemistry, New York: Interscience, 1962.)*

Table 16-2 *The Spectra of Other Nickel Complexes*

Compound	Wavelength, mµ	Energy, kcal mole^{-1}	Intensity
Square planar	(Low spin)		
Ni(DMG)$_2^*$	475	60	120
	410	70	(520)†
Ni(DPG)$_2^*$	450	64	(350)†
Ni(OC$_6$H$_4$CH=N—CH$_3$)$_2$	450	53	130
Ni(CN)$_4^{2-}$	330	87	(250)†
	440	65	2
Tetrahedral	(High spin)		
Ni[(C$_6$H$_5$)$_3$AsO]$_2$Br$_2$	630	45	185
	1300	22	35
NiCl$_4^{2-}$	660	43	180
	860	33	
	1350	21	
NiBr$_4^{2-}$	700	41	
	930	30	
	1430	20	
Ni(II) in ZnO	570	50	
	740	39	

*DMG and DPG stand for the ligands dimethyl glyoxime and diphenyl glyoxime respectively.

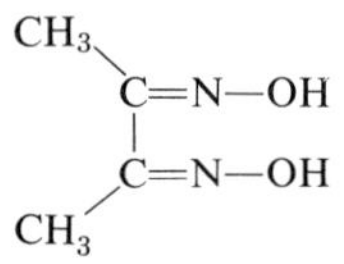

dimethyl glyoxime

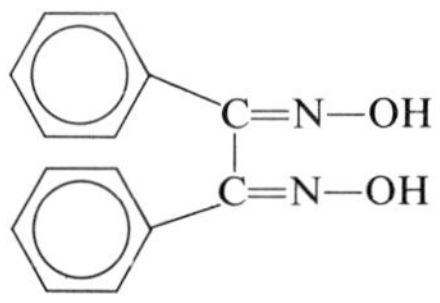

diphenyl glyoxime

†Shoulder on a large ultraviolet peak. The ultraviolet peak contributes to the observed intensity giving unusually large values.

in 1962 is given in Table 16-4. The spectrum has all of the characteristics of an octahedral complex. An octahedral structure may be assigned to some other ethylenimine complexes (see Table 16-5) on the basis of the two main peaks in the visible and near ultraviolet part of the spectrum. It is also apparent that the sulfate ion must be a bidentate ligand.

Although spectra alone can be used to determine structures, the method is even more useful when combined with magnetic data.

Table 16-3 *The Characteristics of Octahedral, Square Planar, and Tetrahedral Nickel(II) Complexes*

High-spin octahedral Ni(II) complexes
  (1) As many as four or five absorption bands
  (2) Low intensity $\epsilon = 1$–50
  (3) $E_3/E_1 = 1.6$–1.8
  (4) Absorption maxima fall in the ranges 1200–950, 825–750, 650–530, 400–350 mμ

Spin-paired square planar Ni(II) complexes
  (1) As many as three but sometimes only two or one band.
  (2) Medium intensity $\epsilon = 100$
  (3) Absorption maxima fall in the range 550–330 mμ

High-spin tetrahedral Ni(II) complexes
  (1) First band in the range 2000–1200 mμ
  (2) Visible band often in the red region resulting in an intense blue color
  (3) High intensity visible band $\epsilon = 200$
  (4) Visible band often split

Table 16-4 *The Spectrum of $Ni(Az)_6(NO_3)_2$*

Wavelength, mμ	Energy, kcal	Intensity, ϵ
953	30	6.7
585	49	7.2
364	79	11.5
299	96	—

Table 16-5 *Absorption Peaks for Some Nickel Complexes with Ethylenimine*

$Ni(Az)_4SO_4$		$Ni(Az)_4Cl_2$		$Ni(Az)_6Br_2$	
mμ	kcal	mμ	kcal	mμ	kcal
603	47	616	46	621	46
376	76	380	75	376	76

II. The Application of Magnetic Susceptibility to Structure

Often magnetic properties can aid in establishing the structure of a complex. Substances that contain unpaired electrons are paramagnetic. This property can be measured using a balance and a magnet. If a paramagnetic substance is suspended from the arm of a balance between the poles of a magnet, the observed weight will depend on the field strength. A paramagnetic substance will be pulled into the field. On the other hand the weight of a diamagnetic substance (with no unpaired electrons) will be nearly independent of the field strength. (A diamagnetic substance will be pushed out of the field, but only very slightly.)

Ch. 16 Physical Methods in Chemistry

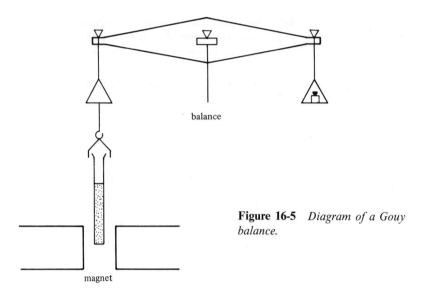

Figure 16-5 *Diagram of a Gouy balance.*

The apparatus that may be used to measure the magnetic susceptibility of a substance is illustrated in Fig. 16-5. The apparatus consists of a magnet and a balance. The sample is suspended from the arm of the balance. The measurement consists of weighing the sample twice, once in the field and once out of the field. The difference in weight is related to the number of unpaired electrons in a molecule of the substance.

The measurable force (weight difference) arises because an unpaired electron acts as a magnet and in a magnetic field the two possible orientations (spin up and spin down) will be different in energy. If this energy difference is larger than kT the electrons tend to line up, making the paramagnetic material into a magnet itself.

In some cases the magnetic susceptibility can be used to determine structure. For example, a spin-paired nickel complex (no unpaired electrons) must have a square planar structure except under unusual circumstances. Tetrahedral cobalt(II) must necessarily be high spin.

III. The Application of Infrared Spectroscopy to Structure

Although the visible region provides evidence for the structure of a coordination compound, the infrared spectrum is a valuable aid in structure studies of many more molecules. The transitions that occur in this region of the spectrum are not electronic but vibrational.

The atoms in a molecule undergo continuous vibrational motion.

These vibrations result in the absorption of characteristic frequencies in the infrared region of the spectrum. It is sufficient here to say that a particular grouping such as C≡O or C=O absorbs at a characteristic frequency and that this frequency varies only slightly from one molecular environment to another. As a result, the C≡O groups in the molecules listed in Table 16-6 absorb in the immediate vicinity of 2000 cm^{-1}. Of course, this is quite a simplification. No segment of a molecule is independent of any other segment; therefore, some care must be exercised in using infrared data according to this simple rule, but the method does allow a general (and fairly reliable) cataloging of characteristic groups. The characteristic frequency found for some additional groups is given in Table 16-7.

The applications of the method that may be found in the literature are too numerous to count. Only a few examples will be given here. In Fig. 16-6 the infrared spectrum of a free sulfate ion, a monodentate coordinated sulfate ion, and bidentate sulfate ion are shown. There are obvious differences that could be used to provide evidence that Ni(II) in Ni(Az)$_4$SO$_4$ is six coordinate and that the SO$_4^{2-}$ is bidentate.

Table 16-6 *Carbonyl Stretching Frequencies Observed in Infrared Spectra*

Complex	ν, cm^{-1}
BH$_3$CO	2165
Ni(CO)$_4$	2057
Fe(CO)$_5$	2028, 1994
Mn(CO)$_5^-$	1898, 1863
Mo(CO)$_6$	2000

Table 16-7 *The Normal Position of Certain Bands in the Infrared Spectrum*

Bond	Motion	Position, cm^{-1}
C—H	stretch	2850–2970
C—H*	stretch	3010–3040
O—H	stretch	3590–3650
O—H†	stretch	3450–3570
N—H	stretch	3030–3500
C=C	stretch	1620–1680
C=O	stretch	1550–1750
C—H	bending	1300–1470
C—O	stretch	1210–1410
C—N	stretch	1020–1360
C—F	stretch	1000–1400
C—Cl	stretch	600–800
C—Br	stretch	500–600

*If the carbon of a C-H bond is double bonded to another carbon atom the absorption normally falls in the higher energy region shown here.

†When the hydrogen of an O-H bond is involved in hydrogen bonding the absorption normally falls in the lower energy region shown here.

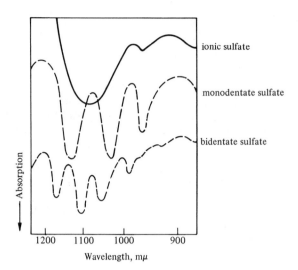

Figure 16-6 *Infrared spectrum of three types of sulfate ions.* (*After K. Nakamoto, Infrared Spectra of Inorganic and Coordination Compounds, New York: John Wiley, 1963.*)

The infrared spectra of *cis*- and *trans*-Co(en)$_2$Cl$_2^+$ are shown in Fig. 16-7. The less symmetrical cis isomer contains more vibrational absorption bands than the more symmetrical trans isomer. The spectra of the isomers of Co(NH$_3$)$_4$(NO$_2$)$_2^+$ are shown in Fig. 16-8. Again the cis isomer shows more bands than the trans isomer.

Another important application of infrared studies deals with the determination of the relative metal ligand bond strength. A variety of ammine and substituted ammine complexes shows an absorption

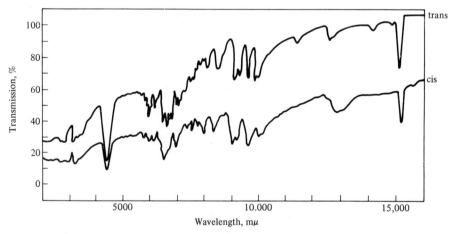

Figure 16-7 *Infrared spectra of the cis and trans isomers of* Co(en)$_2$Cl$_2$Cl. (*After F. A. Cotton and G. Wilkinson, Advanced Inorganic Chemistry, New York: Interscience, 1962.*)

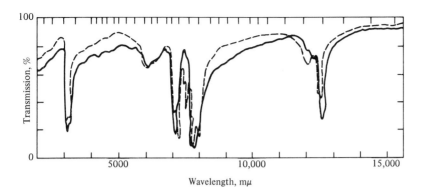

Figure 16-8 *Infrared spectra of cis (dotted curve) and trans (solid curve)* $Co(NH_3)_4(NO_2)_2^+$. *(After J. P. Faust and J. V. Quagliano, J. Am. Chem. Soc., 76:5346, 1954.)*

band due to the stretching vibration of the M-N bonds. The frequency observed for these bands decreases in the order Pt(IV), Pt(II), Co(III), Pd(II), Cr(III), Cu(II), Ni(II), Co(II). This order is in agreement with that expected from the discussion on bonding in Chapters 13, 14, and 15.

An obvious disadvantage to the identification of cis-trans isomers by qualitative examination of the infrared spectra is that both isomers are needed. Difficulties in isolating one of the isomers would then prevent the characterization of the other isomer by infrared methods. Other methods of establishing the structure of geometrical isomers do not always require the preparation of both isomers. One example is the dipole moment.

IV. Dipole Moments and Structure

The dipole moment of a complicated molecule containing three or more atoms is the vector sum of the bond moments. If the atoms are not all of one kind, there will be a net dipole moment unless the molecule is sufficiently symmetrical for the separate bond moments to cancel completely. The trans isomer of an octahedral complex (MA_4B_2) is highly symmetrical and cannot have a dipole moment. On the other hand, the cis isomer has the same symmetry as a water molecule and must have a dipole moment whatever the nature of the bonds.

Dipole moments have been used when other methods (short of X-rays) were not applicable. One example involves a five-coordinate nickel complex. Nickel forms the complex $Ni[P(C_2H_5)_3]_2Br_3$. The compound could conceivably be a nickel(III) complex salt containing the cation $Ni[P(C_2H_5)_3]_2Br_2^+$. A salt would very likely be soluble in

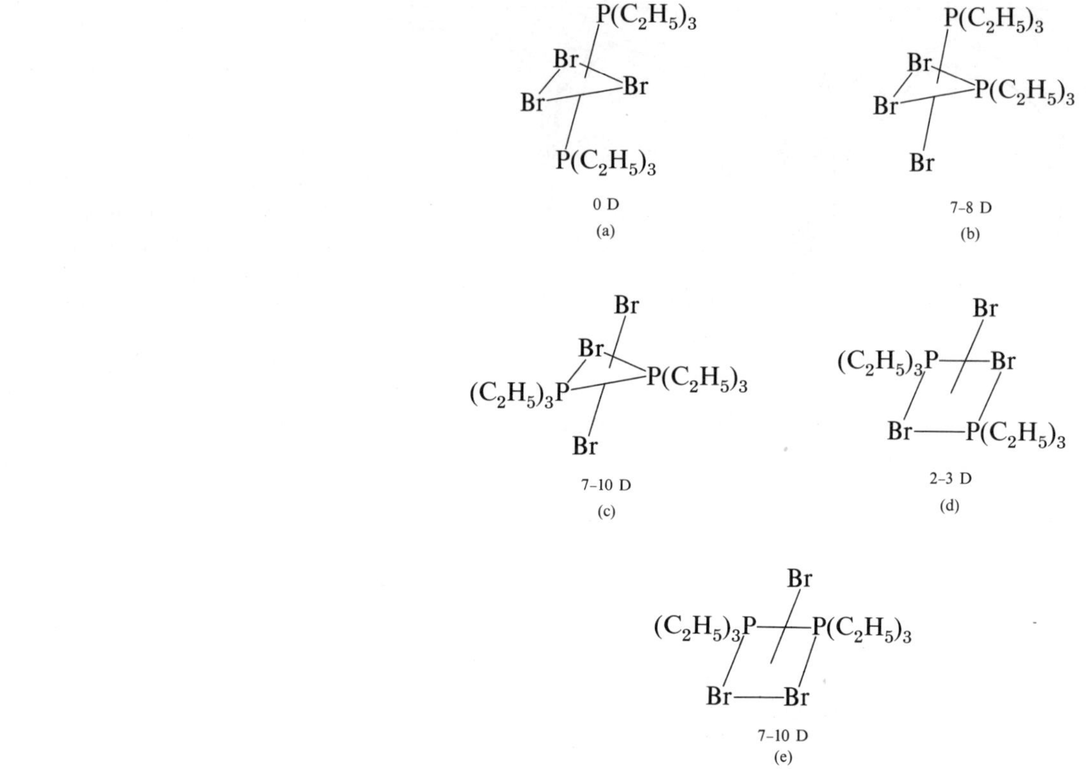

Figure 16-9 Possible structures for Ni[P(C$_2$H$_5$)$_3$]$_2$Br$_3$ with dipole moments (in debye, D) calculated for each structure.

water but not necessarily. However, the substance is soluble in benzene which indicates that the compound is not a salt. Salts are not usually soluble in organic solvents. Magnetic susceptibility measurements can be used to establish the oxidation state. Nickel(III) would show either one or three unpaired electrons. Nickel(II) must show either two or zero unpaired electrons corresponding to a tetrahedral or square planar structure respectively.

The measured magnetic susceptibility corresponds to one unpaired electron indicating a five-coordinate nickel-(III) complex. Infrared spectroscopy is not applicable to the determination of the structure of this molecule by simple means, and very few nickel(III) complexes are known so that the visible spectrum will not be easy to interpret. The possible structures of this complex are given in Fig. 16-9. The observed dipole moment of zero indicated that the structure is that shown in Fig. 16-9a, and the less symmetrical structures, Fig. 16-9b through 9e are excluded.

V. Mass Spectrometry

Any molecule can be ionized either by absorption of far ultraviolet light or by electron impact. If the ionization is conducted in the gas phase, the resulting positive ions can be accelerated in an electric field and deflected by a magnetic field. The pressure inside the instrument is kept low to avoid collision between the ions. Ions of different mass are deflected by different amounts and by either moving a detector or varying the electric or magnetic fields, the intensity of the beam of ions produced from a particular sample can be observed as a function of the mass. The instrument used for this purpose is known as a mass spectrometer. A schematic diagram of a mass spectrometer is shown in Fig. 16-10.

The first mass spectrometer was used to demonstrate the existence of isotopes. The instrument can be used to measure isotopic composition very accurately as well as study the processes by which ions are formed and the reactions they undergo. However, the principal use of a mass spectrometer is in the analysis of complicated mixtures. Each chemical compound present will show a characteristic pattern of fragment ions. Several typical patterns are shown in Fig. 16-11.

In favorable cases the heavier ions observed may provide a chemical analysis of a new compound. The mass spectrum of the first "compound of xenon and fluorine" prepared (Fig. 16-12) showed ions of mass 205 and 208. Since these correspond to XeF_4^+ the compound had to contain at least four fluorine atoms and was most probably XeF_4. This was, of course, confirmed by vapor density, chemical analysis, and infrared spectroscopy.

An example of a case where mass spectrometry can be used to obtain

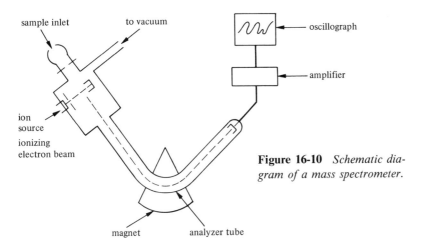

Figure 16-10 *Schematic diagram of a mass spectrometer.*

Ch. 16 Physical Methods in Chemistry

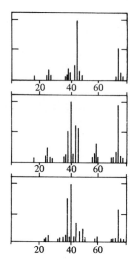

thiocyclobutane
CH₂—CH₂—CH₂—S

propylene sulfide
CH₃—CH—CH₂—S

allyl mercaptan
CH₂=CH—CH₂SH

Figure 16-11 *Some typical mass spectral patterns.*

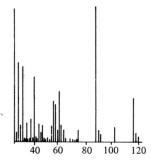

S—butyl ethyl sulfide
CH₃—CH₂—CH—S—C₂H₅
          |
          CH₃

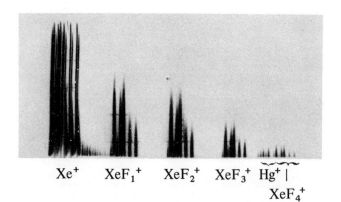

Figure 16-12 *Mass spectrum of a sample of* XeF₄. *(Courtesy of Argonne National Laboratory.)*

data on the structure of a complex is provided by the spectrum of $H_4N_2Fe_2(CO)_6$.

The compound $H_4N_2Fe_2(CO)_6$ was first reported in 1960. The analytical data and absorption spectra available at that time were insufficient to determine the correct number of hydrogen atoms present in the molecule. This feature of the formula was first demonstrated (1968) from mass spectral data. A series of ions that correspond to $H_4N_2Fe_2(CO)_n^+$ where the value of n ranges from 0 to 6 is apparent in the spectrum. In this case the fact that all the hydrogens are bonded to nitrogen and not carbon or oxygen was already clear from the infrared spectrum, but it is confirmed by the presence of $H_4N_2Fe_2^+$ ions in the mass spectrum. This particular ion is the most abundant ion produced.

VI. Nuclear Magnetic Resonance

Another modern instrumental technique for the examination of the structure of a new material is nuclear magnetic resonance. Since neutrons and protons have spins, all nuclei except those in which all the neutrons and protons are paired up will show magnetic properties, and the energies will be different for different orientations in the magnetic field. The splitting can be observed by matching the energy of a radio frequency field to the energy of the splitting. An instrument for doing this is known as an NMR spectrometer (Fig. 16-13). The precise matching (resonance) depends on the value of the magnetic field at the nucleus, and this is determined by the electronic structure of the molecule as well as the value of the external magnetic field. Nuclei of the same element in different chemical environments will show resonance at different values of the magnetic field. A low resolution NMR spectrum of ethyl alcohol is shown in Fig. 16-14. The peaks correspond to the resonance of hydrogen in three different chemical environments. This is in agreement with the structural formula of ethyl alcohol.

$$CH_3-CH_2-OH$$

Methyl groups in other organic molecules will show peaks near the position of the CH_3 peak of ethyl alcohol. Under high resolution a further splitting of the peaks (Fig. 16-15) provides even more data on the structure, but the interpretation requires quantum mechanical methods.

With a suitable instrument it is possible to observe nuclear magnetic resonance for other nuclei with odd numbers of protons or neutrons

The splitting in energy between the different possible nuclear orientations in a magnetic field is quite small compared to that for electrons, and the nuclei cannot be aligned at temperatures above about 2 K.

Ch. 16 Physical Methods in Chemistry

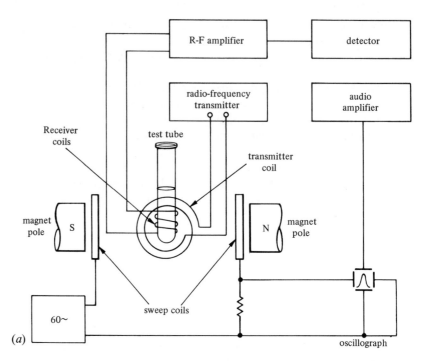

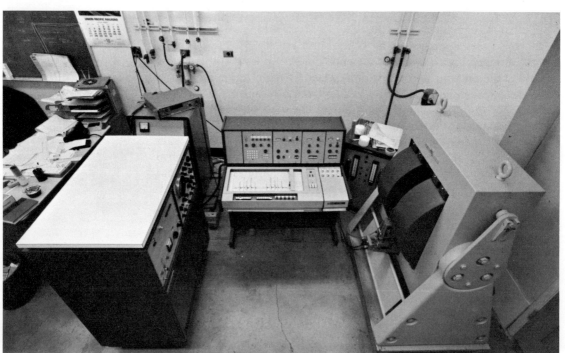

Figure 16-13 (a) Schematic diagram of an NMR spectrometer, (b) NMR spectrophotometer in use. (*Courtesy Varian Associates.*)

Sec. VI Nuclear Magnetic Resonance

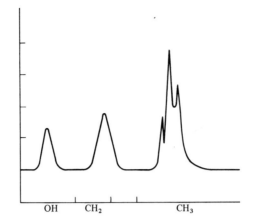

Figure 16-14 *NMR spectrum of ethanol.*

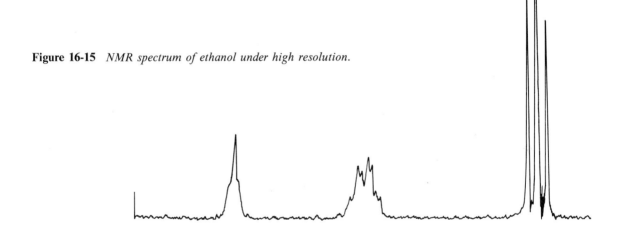

Figure 16-15 *NMR spectrum of ethanol under high resolution.*

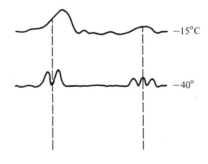

Figure 16-16 *Fluorine NMR peaks of* ClF_3 *at* -15 *and* $-40°C.$ *(After J. A. S. Smith, Lab. Pr. (London), 13(10):951, 1946.)*

419

such as ^{19}F and ^{17}O. With fluorine magnetic resonance it is possible to distinguish between the isomers of $Cr(NH_3)_3F_3$. The NMR spectrum of ClF_3 (Fig. 16-16) has two fluorine peaks in agreement with the unsymmetrical structure predicted by the methods of Chapter 8 (see page 204).

PROBLEMS I

1. The mass spectrum of $HCoC_4O_4$ shows the presence of peaks at masses of 172, 144, 143, 116, 115, 88, 87, 71, 60, 59, and 28. What are the formulas of the 115 and 116 peaks? What atom is the hydrogen bonded to? Explain the absence of a CoO^+ peak, $m = 75$.
2. The ^{31}P NMR spectrum of P_4S_3 shows two peaks with an intensity ratio of 3:1. Of the various octet rule structures possible for P_4S_3 find one which agrees with the NMR spectrum. (There are actually three with the required symmetry.)
3. There are two isomers of $PtCl_2(NH_3)_2$. Draw their structures. Which one will have a dipole moment?
4. In solid $CsCuCl_3$ each Cu atom is octahedrally surrounded by six chlorides. It shows d-d absorption peaks at 11,800 and 11,000 cm^{-1}. Two forms of $CuCl_4^{2-}$ are known with spectral bands given below. Which of these has a structure closest to tetrahedral? (Hint: Δ values are smaller for tetrahedral than for octahedral complexes.) The other $CuCl_4^{2-}$ species is known to be square planar.
    form A of $CuCl_4^{2-}$: 14,300, 13,100, and 10,900 cm^{-1}
    form B of $CuCl_4^{2-}$: 9,050, 7,900, 5,550, and 4,800 cm^{-1}
5. Br$^-$ is lower in the spectrochemical series than Cl$^-$. Only one form of $CuBr_4^{2-}$ is known, with d-d absorption peaks at 8,500, 7,570, 5,405, and 4,545 cm^{-1}. Which of the two $CuCl_4^{2-}$ species in Problem 3 does $CuBr_4^{2-}$ resemble?
6. A new compound has been prepared with the formula $C_2PSiOF_6H_3$. It shows only one infrared peak between 1,500 and 4,000 cm^{-1}, at 2,214 cm^{-1}. The normal stretching frequencies of H-M bands are H-C, $\sim$2,900; H-P, $\sim$2,400; H-Si, $\sim$2,250; and H-O, $\sim$3,500 cm^{-1}. How are the hydrogen atoms bonded?
7. The compound in Problem 6 also shows C-F, P-O, and Si-O stretching frequencies in its infrared spectrum. Draw a probable structure.
8. (a) Fe^{2+} reacts with MnO_4^- in aqueous acid solutions to give Fe^{3+} and Mn^{2+}. Write a balanced equation for this reaction.
    (b) What volume of 0.010 M $KMnO_4$ solution is required to just react with 15 ml of 0.050 M $FeSO_4$?
    (c) What indicator is used for this reaction?

PROBLEMS II

1. The lowest energy d-d peak for $Ni(H_2O)_6^{2+}$ is at 24 kcal mole^{-1}. What is the value of Δ in this ion?
2. Estimate Δ for a Ni(II) ion octahedrally surrounded by six Cl$^-$.

3. The value of Δ for Ni(II) in $C_5NH_6NiCl_3$ is 12 kcal mole^{-1}. Where will the lowest energy *d-d* peak in the spectrum of this salt be?
4. The salt in Problem 3 also has absorption peaks at 36 and 54 kcal mole^{-1}. Locate these transitions in Fig. 16-2.
5. To what colors of light do the peaks given in Problem 4 correspond?
6. What color of light is least absorbed by the salt $C_5NH_6NiCl_3$?
7. What color is $C_5NH_6NiCl_3$?
8. $Ni(H_2O)_6^{2+}$ is green. Predict the color of $Ni(NH_3)_6^{2+}$.
9. From Table 16-7 locate the position of the O-H stretch absorption band in liquid water.
10. To what energy does the band referred to in Problem 9 correspond?
11. Why does H_2O have a dipole moment?
12. What charge distribution in the H_2O molecule (bond angle 105°) corresponds to the observed dipole moment of 1.85 debye?
13. What mass spectral peaks will H_2O have?
14. How many NMR peaks should be observed for $H_2O(l)$?

> For they had observed from former battles that Gauls in general are most formidable and spirited in their first onslaught, while still fresh, and that, from the way their swords are made, as has been already explained, only the first cut takes effect; after this they at once assume the shape of a strigil, being so much bent both length-wise and side-wise that unless the men are given leisure to rest them on the ground and set them straight with the foot, the second blow is quite ineffectual.
>
> POLYBIUS
> (translation W. R. Paton)

Chapter 17 The Common Metals

For want of anything better, stone age man learned how to use rocks for tools and weapons. The difficulties involved in getting the desired shapes and the fragility of a sharp edge are obvious.

In contrast to silicate rocks, metals are generally malleable. The bonding in a metal crystal is not weakened seriously by a slight displacement, and when a mass of metal is struck with a hammer it does not ordinarily shatter. If the blow is strong enough, the shape of the metal is changed. Most metals can be pounded to a sharp edge, a process much more convenient than the chipping required for sharpening rocks. Thus the available metals have been of considerable importance to man from prehistoric times.

However, the common metals are found as oxides, and a chemical reaction (reduction) is required to obtain them in the metallic state. The metals which could be found in the free state, principally gold, platinum, silver, and copper were always highly valued. They are still the principal metals used in coins and in jewelry. All four of these metals are very malleable, so much so that an object formed from them is easily bent. This fact has seriously limited their usefulness; they are practically worthless for such practical things as fishhooks or the blades of knives.

Gold, platinum, and silver are so rare they will always be relatively expensive. Copper on the other hand is reasonably abundant, but it is usually found as copper(II). The reduction of copper oxide (CuO) to copper was the first major step in metallurgy, the art or science of obtaining metals in useful form.

I. The Metallurgy of Copper

Let us take a look at the thermodynamics of the reaction

$$CuO(s) + C(s) \longrightarrow CO(g) + Cu(s)$$

ΔH^0 for this reaction is $+11.2$ kcal mole^{-1}, but one of the products is a gas; therefore the entropy change is favorable, $\Delta S^0 = +42.7$ eu. The equilibrium constant at 25°C is 14. Thus the reaction is favorable (the equilibrium constant is greater than one at room temperature). The reaction proceeds at a reasonable rate at temperatures easily obtained in a wood fire, and the process was probably discovered by accident when a wood fire happened to be built on a particularly rich copper ore. We can tell from the analysis of copper objects of the ancient Egyptians that the regular preparation of copper from its ores dates from about 3500 B.C. Some samples of very impure copper were obtained by accident at times, and some of these were much harder than ordinary copper. Within 500 years bronze (the alloy of tin and copper) was being purposefully made. Bronze has a very desirable combination of properties in that it is very malleable at elevated temperatures but hard enough to hold a cutting edge at 25°C. To understand this we will have to examine the solubility of tin in solid copper as expressed by a phase diagram.

A. PHASE DIAGRAMS

The solubility of tin in solid copper is strictly limited. Copper crystallizes in the face centered cubic structure, and some copper atoms can be replaced by tin atoms up to the limit of concentration shown by the edge of the shaded region in Fig. 17-1. Note from this figure that the solubility of tin in copper increases as the temperature is increased. After reaching a maximum at about 17% at 550° the solubility decreases again.

If more tin is added to the copper, a different solid solution with a different crystal structure is obtained. Between room temperature and 350°C a hexagonal close packed structure containing about 38% tin can be obtained. A simple calculation

38 g Sn with 62 g Cu

0.320 mole Sn with 0.977 mole Cu

1.00 mole Sn with 3.05 moles Cu

indicates that this phase can be given the formula Cu$_3$Sn; however, as shown in Fig. 17-2 the composition is more variable than that found in a definite compound.

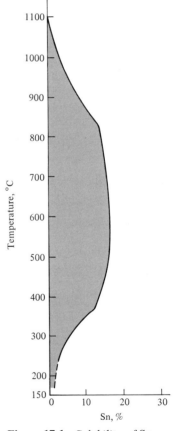

Figure 17-1 *Solubility of* Sn *in* Cu.

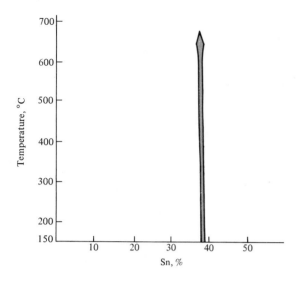

Figure 17-2 *Ranges of composition observed for* Cu_3Sn.

At 200°C and 1 atm pressure it is possible to have a solid composed of cubic closest packed crystals with a composition between 0 and 2% tin and a solid composed of hexagonal close packed crystals which ranges between 37.5 and 38.5% tin. A single crystalline form that contains between 2 and 37.5% tin cannot be obtained at equilibrium under these conditions. A solid containing 8% tin (at 200°C) consists of a mixture of two types of crystals. Some, the cubic close packed crystals, will be low in tin, while others, the hexagonal close packed crystals, will contain larger amounts (38%) of tin.

A good bronze will contain between 3 and 12% tin in the copper. This much tin will dissolve in the copper at high temperatures (400–800°C) resulting in a malleable single-phase system that is easily hammered into the desired shape. When the object is cooled small crystals of Cu_3Sn will form or start to form. The second crystals act to impart hardness to the finished object.

Figure 17-3 shows the full phase diagram for the Cu-Sn system. All the shaded regions correspond to single phases of more or less variable composition. In this diagram they are all solid phases except for the liquid phase at the top and to the right side of the diagram. The other characteristic feature of two component phase diagrams is the presence of horizontal lines connecting three different phases. The presence of a particular set of three phases together in equilibrium is possible only at the particular temperature shown. This is a consequence of Gibb's phase rule.

B. THE PHASE RULE

We know that a single phase of a pure material (liquid water or solid copper) can exist over a range of temperatures and pressures.

Sec. I The Metallurgy of Copper

This fact is expressed by the statement that a one-component system with one phase present has two degrees of freedom (the two independent variables, temperature and pressure). If the letters F and P are used for the number of degrees of freedom and the number of phases respectively, for any pure material

$$F = 2 + 1 - P$$

This expression states that there is a very simple relationship between the number of independent variables and the phases present. From this expression we see that three phases of a pure material can exist in equilibrium only if there are no degrees of freedom. There is only one pressure and one temperature at which the three phases of water or any substance can exist together at equilibrium. This corresponds to a single point on a pressure versus temperature diagram. Both the temperature and pressure are fixed by the requirement of equilibrium between the phases.

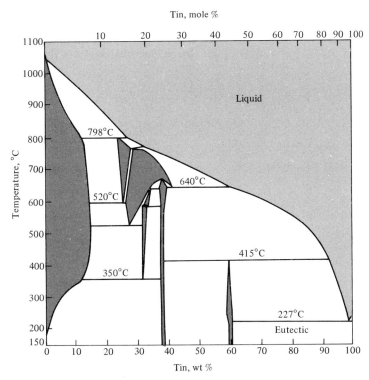

Figure 17-3 *Equilibrium diagram for the Cu-Sn system. (After W. Hume-Rothery, Electrons, Atoms, Metals, and Alloys, New York: Philosophical Library, 1955.)*

425

The condition for equilibrium between *two* phases of a pure material is represented by a line on a T versus P phase diagram. Pure water can boil at a number of different pressures, but at each pressure there is a definite equilibrium temperature. Similarly at a specified pressure, a pure material has a definite melting point.

The melting point of pure copper is given in Fig. 17-3 as 1083°C. The addition of tin lowers the melting point. The lowest temperature (the eutectic temperature) at which liquid can exist in equilibrium with mixtures of solid tin and copper alloys occurs at 227°C on the tin-rich side. The definite temperature observed for a eutectic (and other three-phase equilibria) in a two-component system is obviously analogous to the freezing point of a pure material. Hence there is one degree of freedom at a eutectic point. This is expressed in the phase rule as

$$F = 2 + 2 - P$$

The phase rule is a simple, unifying principle that holds for all systems in equilibrium. It was discovered by J. Willard Gibbs. To use the phase rule in general for systems of any number of components it can be written as

$$F = 2 + C - P$$

where C is the number of component materials.

C. THE PREPARATION OF PURE COPPER

The major current use of copper is in electrical wire. For this purpose very pure copper is desirable. Unfortunately the best copper ores (up to 20% CuO) have been exhausted. Currently most copper production is from copper sulfide (CuS) ores. In fact ores which assay less than 40 lb of copper sulfide per ton of ore are currently being used, and a number of extra steps are required in their treatment. The ore is first crushed very fine and then concentrated by flotation. In this process the powdered ore is stirred with air, water, and oil. Since the fragments of copper sulfide are concentrated in the oil, they are found in the oily froth that floats to the surface. The oil is then distilled off, leaving a concentrated ore for further processing.

Copper sulfide can be converted to copper oxide by heating in air.

$$2CuS + 3O_2 \longrightarrow 2CuO + 2SO_2(g)$$

Free copper is produced when the copper oxide formed reacts with copper sulfide. (Sulfide ion acts as a reducing agent.)

$$CuS + 2CuO \longrightarrow 3Cu + SO_2(g)$$

Copper produced in this way is called blister copper because of blisters caused by the release of dissolved sulfur dioxide (SO_2) during the crystallization. Blister copper can be made 99% pure, but for electrical uses it is further refined by electrolysis. Blister copper is used as an anode

$$Cu \longrightarrow Cu^{2+} + 2e^-$$

and copper of over 99.95% purity is deposited on the cathode

$$Cu^{2+} + 2e^- \longrightarrow Cu$$

The potential of the cell is adjusted so that impurities such as silver are not oxidized (they collect as a black sludge left when the anode dissolves), and also so that impurities such as iron are not reduced at the cathode but are left as positive ions in the electrolyte bath.

D. COPPER ALLOYS

We have seen how the addition of tin to copper gives a vast improvement in its physical properties for many uses. Actually any of a number of metals can be alloyed with copper for this purpose, and in modern practice the alloys of copper and zinc (brass) are most widely used. Zinc is much cheaper than tin, primarily because it is much more abundant. The ancient peoples, however, were limited by the fact that their metallurgical techniques were not satisfactory for the reduction of any metals from ores with electrode potentials more negative than about −0.15 volt. As Table 17-1 shows, tin, lead (Pb), and antimony (Sb) are the principal metals whose oxide ores could be reduced under conditions only slightly more extreme than those required for copper oxide. Unfortunately, these three elements are all quite rare. From the beginning of the bronze age the ores needed to make alloys of copper were in high demand and quite expensive. The importance of SnO_2, the tin ore, is emphasized by the fact that it is the only commodity for which the merchants of the time were consistently willing to risk voyages beyond the shelter of the Mediterranean Sea. The Phoenicians made regular voyages to Wales for the rich deposits of SnO_2 to be found there. Of course this made bronze weapons quite expensive, and this in turn favored a feudal organization of society. Only the leaders of the Greek army described in the Iliad were fully armed, and the course of battle was largely decided by conflicts between individual heroes. This situation had a strong romantic appeal for the Greeks of Homer's time, but the growth of Greek democracy was ultimately dependent upon the cheaper steel weapons introduced about 1400 B.C. which inaugurated the iron age.

Ch. 17 The Common Metals

Table 17-1 *Electrode Potentials and Abundances for Some Metals*

	E^{0}* volts	Cosmic abundance atoms per 10^6 Si atoms†
$Au^{3+} + 3e^- = Au$	1.50	0.18
$Ag^+ + e^- = Ag$	0.800	0.21
$Cu^{2+} + 2e^- = Cu$	0.340	1,000
$Sb_2O_3 + 6H_3O^+ + 6e^- = 2Sb + 9H_2O$	0.15	0.28
$Pb^{2+} + 2e^- = Pb$	−0.126	1.0
$Sn^{2+} + 2e^- = Sn$	−0.141	1.2
$Ni^{2+} + 2e^- = Ni$	−0.236	28,000
$Fe^{2+} + 2e^- = Fe$	−0.409	230,000
$Zn^{2+} + 2e^- = Zn$	−0.762	600
$SiO_2 + 4H_3O^+ + 4e^- = Si + 6H_2O$	−0.991	1,000,000
$V^{2+} + 2e^- = V$	−1.18	210
$Ti^{2+} + 2e^- = Ti$	−1.63	2,500
$Al^{3+} + 3e^- = Al$	−1.68	53,000
$Sc^{3+} + 3e^- = Sc$	−2.08	22
$Mg^{2+} + 2e^- = Mg$	−2.36	800,000

*Electrode potentials are calculated from Appendix IV or taken from *Handbook of Chemistry and Physics*, (Cleveland, Ohio: Chemical Rubber Publishing Co., 1964), p. D75.

†Abundances are from L. H. Aller, *The Abundance of the Elements*, (New York: Interscience, 1961).

II. The Metallurgy of Iron

If we examine the thermodynamics of the reaction

$$Fe_3O_4 + 4C \longrightarrow 3Fe + 4CO$$

	H^0, kcal mole^{-1}	S^0, eu
Fe_3O_4	−267.3	35.0
4C	4(0)	4(1.37)
3Fe	3(0)	3(6.52)
4CO	4(−26.42)	4(47.30)

for which $\Delta H^0 = +162$ kcal mole^{-1} and $\Delta S^0 = +168$ eu, we can see that a temperature of almost 1000 K ($\Delta H^0 - T\Delta S^0 = 0$ at $T = 960$ K or 690°C) is required before the favorable entropy can balance the endothermicity of the reaction. As implied above such a temperature is not reached in an ordinary wood fire; however, it is obtainable with a forced draft. It appears that the bellows was developed about 1800 B.C. and within 200 years it was being used regularly for the production of wrought iron.

A. WROUGHT IRON

Under the conditions described above iron is formed as a spongy mass of crystals enclosing all kinds of impurities, but on repeated hammering, even cold, it can be made into a coherent mass as most of the impurities are pressed out. The term wrought iron is derived from this hammering process. Wrought iron is the purest of the various kinds of iron readily formed from the ore. It has the same disadvantage as pure copper. In fact, wrought iron is softer than bronze. As a result iron did not have much social impact until the means were developed to add sufficient carbon to form steel. Some peoples could apparently make steel and temper it fairly reliably as early as 1200 B.C. Nevertheless the requirements for a good steel were not understood for thousands of years, during which time the properties of an exceptional sword could be as logically ascribed to magic as to anything else. Some understanding of the difficulties can be obtained from an examination of the iron-rich side of the Fe-C phase diagram (Fig. 17-4).

B. STEEL

At room temperature pure iron forms body centered cubic crystals. There is also a cubic closest packed form that is stable between 900 and 1400°C. Carbon dissolves in iron, not by replacing iron atoms but by filling the holes between them (interstices). The octahedral holes

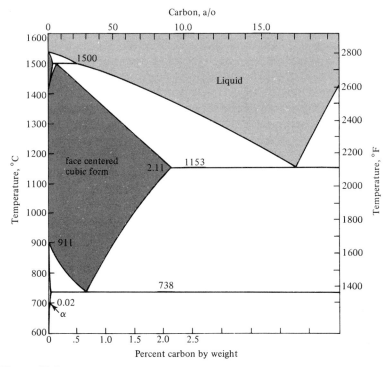

Figure 17-4 *Part of the* Fe–C *phase diagram. (After M. G. Benz and J. F. Elliott, Trans. AIME, 221:323, 1961.)*

in the cubic closest packed form are actually larger than any of the interstitial spaces in the body centered cubic crystals, and appreciable amounts of carbon dissolve only in the cubic close packed form. A steel with about 1% carbon can be easily forged at temperatures between 800 and 1300°C where a single solid phase is stable. On cooling, the single phase separates into crystals of nearly pure iron and the compound Fe_3C or, if it is cooled rapidly, a supersaturated solution of carbon in body centered cubic iron (called martensite) is obtained.

Martensite is harder than the equilibrium mixture of iron and Fe_3C. With careful control of the quenching process, the hardness of a steel object can be accurately controlled.

The word steel is used for any alloy of iron with significantly improved properties over those of pure iron. A good steel, containing only iron and carbon, will have 0.05–2.0% carbon. Material containing more carbon than 2% will start to melt without ever reaching a single solid phase region in which it can be worked. The alloys of over 2% carbon are known as cast iron, since the only way the alloys can be easily shaped is by casting from the melt.

C. CAST IRON AND THE BLAST FURNACE

The temperature required to produce a melt of iron that will dissolve 3–4% carbon and flow from the bottom of a furnace is most easily obtained in a large furnace with a really strong forced draft of air. Such a furnace for preparing cast iron is called a blast furnace. They were first developed in China where cast iron was quite widely used before A.D. 1. It is clear from excavations in Europe that cast iron was prepared there as well, but only by accident, and the material so produced was generally discarded immediately. It was not until 1440 that furnaces especially for the production of cast iron were built in Europe. This date is an important one for the industrial revolution, not as a technological advance but as an indication of when the sailing ship was used enough in trade that a relatively cheap and heavy material could be produced in quantities too large for local markets.

D. THE PRODUCTION OF STEEL

The original procedure for making steel from wrought iron, which spread from tribe to tribe at the beginning of the iron age, involved simply heating the solid metal in contact with charcoal in order to dissolve the desired amount of carbon. This process is known as cementation, and as late as 1700 essentially all the steel produced in Europe was made in this way. A much more uniform steel could be made by melting the wrought iron with the right amount of carbon in a crucible. This process was developed around 500 B.C. in India using crucibles capable of holding 2–5 lbs of steel. By A.D. 600 steel was made in China by the addition of wrought iron to molten cast iron, but again production was held to a relatively small scale.

After the year 1700 metallurgical developments occurred rapidly in

Europe and America. Many men were willing to try new methods, and all the Chinese and Indian advances had been rediscovered by 1800. The timing of the discoveries was relatively immaterial—they were successful when, but not before, the marketing conditions were ready. Good cast iron for boilers was already available when Watt developed the steam engine (1765), and cheap steel rails from the Bessemer process (1856) followed within 20 years of the success of railroads. Other steelmakers in the attempt to duplicate Bessemer's success and get around his patents developed the open hearth process within 10 years after this.

Since the most economical large scale reduction of iron ore yields cast iron, the problem in the production of cheap steel is that of removal of carbon instead of adding carbon. This is most easily accomplished by the use of an oxidizing agent. The Bessemer process uses air blown through the melt; the open hearth process uses iron ore; and the most recent developments involve the use of oxygen blown directly into the melt. In all three processes the basic technical problems were in obtaining furnace linings, and so forth, which would resist the very high temperatures developed.

E. ALLOY STEELS

The systematic study of iron alloys containing other metals dates from around 1800. Small amounts of metals such as chromium and nickel are present in or are deliberately added to most modern steel. Substantial amounts can impart special characteristics to the steel. High manganese alloys are nonmagnetic, chromium alloys (stainless) are noted for their corrosion resistance, and nickel steels generally have improved high-temperature properties. One of the most important specialty steels is that used in making tools for the machining of steel. Tool steel must be very hard, even when hot. It has been made in proportions very close to 77% iron, 18% tungsten, and 4% chromium ever since 1890.

III. More Active Metals

Zinc oxide can be reduced using carbon at 1000°C. At this temperature zinc is a gas that distills from the furnace. When zinc sulfide ore is used the sequence of reactions is

$$ZnS + \tfrac{3}{2}O_2 \longrightarrow ZnO + SO_2$$
$$ZnO + C \longrightarrow CO(g) + Zn(g)$$
$$Zn(g) \longrightarrow Zn(l) \longrightarrow Zn(s)$$

The process was developed before A.D. 1.

Although zinc is a useful metal as an alloying element (as in brass) or as a protective coating (galvanized iron) it is not special enough to have had any major historical importance.

The metals which are still more active than zinc could not be prepared until the electrolytic process of reduction was developed. A whole series of new elements (Mg, Ca, Sr, Ba, Na, K) were first prepared by Sir Humphrey Davy in the years 1807–1808.

One active metal can be used to displace another

$$2Mg + ZrCl_4 \longrightarrow 2MgCl_2 + Zr$$

Aluminum could be prepared in this manner for about $4 per lb from the reaction

$$AlCl_3 + 3Na \longrightarrow 3NaCl + Al$$

The development of a direct electrolytic process for the preparation of aluminum by Hall and Heroult independently in the 1800s was a major advance which brought the price of aluminum down to about $0.30 per lb and allowed the development of innumerable uses of the metal. Molten chlorides are used in most electrolytic processes, but aluminum chloride ($AlCl_3$) is much too volatile. C. M. Hall and P. L. T. Heroult found that fluoride was a satisfactory anion. The electrolytic bath of aluminum oxide (Al_2O_3) dissolved in molten cryolite (Na_3AlF_6), that they developed, is still used. Carbon anodes are used, and the electrode reactions are

$$Al^{3+} + 3e^- \longrightarrow Al$$
$$2O^{2-} + C \longrightarrow CO_2 + 4e^-$$

The production of magnesium is more standard; the simple electrolysis of molten $MgCl_2$ is

$$Mg^{2+} + 2e^- \longrightarrow Mg$$
$$2Cl^- \longrightarrow Cl_2 + 2e^-$$

An interesting sequence of reactions is used to obtain the $MgCl_2$ from sea water without consuming any chemicals more expensive than lime (CaO) or natural gas (CH_4).

$$Mg^{2+} + CaO + H_2O \longrightarrow Mg(OH)_2 + Ca^{2+}$$
$$Mg(OH)_2 + 2HCl \longrightarrow MgCl_2 + 2H_2O$$
$$4Cl_2 + 2CH_4 + O_2 \longrightarrow 8HCl + 2CO$$

Both aluminum and magnesium are fairly soft when pure. Pure aluminum can be rolled into very thin sheets which have largely displaced other foils such as tin foil from the market. Improved properties can be obtained by alloying aluminum and magnesium with other metals such as Mn, Zn, Cu, and Si. There are also a whole series of useful alloys containing just magnesium and aluminum. Magnalium (90% Al, 10% Mg) and Dowmetal (88% Mg, 12% Al) are two examples. Some of the properties of these alloys are represented in the Mg-Al phase diagram shown in Fig. 17-5.

There are a few applications where none of the steels or any Al-Mg alloys have been satisfactory. There is a great demand for a lightweight metal, useable at high temperatures. In looking for high-melting light elements, silicon must be considered first. However, silicon is too hard and brittle for most applications. The next reasonable possibilities are in the series Sc-Ti-V. Considering the abundance of these elements, it is not surprising that billions of dollars have been spent on the development of titanium metal. Titanium dioxide (TiO_2) is abundant and cheap and the development of cheaper ways to reduce it to the metal would open up tremendous markets. The metal is now made commercially by the reaction of titanium tetrachloride ($TiCl_4$) with magnesium or by the thermal decomposition of titanium tetraiodide (TiI_4). The properties of titanium are such that they complicate the preparation of titanium objects. Titanium is neither malleable nor liquid at even moderately high temperatures. The best current method of fashioning an object from titanium involves starting with powdered titanium pressed into a shape that is close to the one desired. Here again there is room for substantial improvement of the current practice.

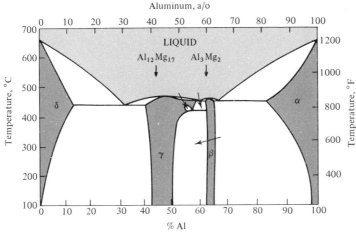

Figure 17-5 Al–Mg *phase diagram.* (*After J. B. Clark and F. N. Rhines, J. Metals, 9(Sec. 2) Trans. AIME, 209:6, Jan. 1957.*)

QUESTIONS

1. What range of composition is found for NaCl in water? for NaCl in ice?
2. What oxides in Table 4-1 should be as easily reduced as CuO?
3. What is meant by the term triple point for a pure material?
4. How does the vapor pressure of water vary with temperature?
5. Define eutectic.
6. What chemical steps would be required to get lead from PbS ores?
7. The term pearlite is used for a mixture of iron and Fe_3C with an overall composition of 0.95% carbon. What is the significance of this composition (see Fig. 17-4)?
8. Why does cast iron have a high percentage of carbon?
9. What are the two principal uses of tungsten?
10. Why is $AlCl_3$ more volatile than AlF_3? (Hint: see Chapter 6.)
11. How is lime (CaO) made? Why is it cheaper than NaOH?
12. Why are carbon and silicon less malleable than typical metals?
13. List the metals with melting points above 1600°C (Table 1-2).
14. Outline the two regions of the periodic table of elements often found naturally in the elemental form.

PROBLEMS I

1. What is the formula of the intermetallic compound which is 21% titanium, 79% copper?
2. Calculate ΔH^0 for the reaction $PbO + C \longrightarrow CO + Pb$ using data from Table 4-1.
3. Estimate the conditions required for the reaction $ZnS + 2ZnO \longrightarrow 3Zn(g) + SO_2(g)$ if $\Delta S^0 = 140.0$ eu.
4. What three phases in Fig. 17-5 are in equilibrium at 425°C?
5. What is the percentage of aluminum in Mg_2Al?
6. What is the composition of a solid which is 32% tin and 68% copper?
7. What three phases in Fig. 17-3 are in equilibrium at 350°C?
8. The addition of 8% tin to copper makes a marked improvement in the properties. Would a bronze of 16% tin be even better? Why?
9. Aluminum (mp 660°C) and zinc (mp 420°C) form one intermetallic compound, Al_2Zn, which melts at 580°C. The two metals do not dissolve appreciably in each other in the solid state. There are two eutectic temperatures in the Al-Zn system, one at 480°C and one at 370°C. Sketch the Al-Zn phase diagram.
10. For the process $Al(s) \longrightarrow Al(l)$, $\Delta H^0_{753 \text{ K}} = 2.6$ kcal mole^{-1} and $\Delta S_{753 \text{ K}} = 2.8$ eu. Calculate the equilibrium constant for this reaction at 480°C (753 K). Estimate the composition of the eutectic liquid at 480°C.

PROBLEMS II

1. Calculate the formula of the intermetallic compound which is 59–60% tin and 40–41% copper.
2. The melting point of tin is 232°C. From Fig. 17-3, how much copper should be dissolved in liquid tin to lower the freezing point 5°C?
3. What three phases in the Cu-Sn system are in equilibrium at 227°C?

4. Solid Cu_5Sn_4 decomposes at 415°C into Cu_3Sn and a liquid phase. What is the composition of the liquid?
5. A liquid phase of 7% copper in tin starts to crystallize at 400°C. What is the composition of the crystals formed?
6. What happens to the composition of the liquid described in Problem 5 as Cu_5Sn_4 crystals are formed?
7. A liquid of 7 g copper and 93 g tin is cooled to 228°C. At that point 13.0 g Cu_5Sn_4 have formed. What is the composition of the liquid that is left?
8. What phases are present in the system in Problem 7 after it is cooled to 226°C?
9. What is the eutectic temperature in the Fe-C system (Fig. 17-4)?
10. What is the composition of the last liquid to solidify in the Fe-C system?
11. A cast iron contains 4% carbon in iron. Describe two processes by which this material can be converted to steel.

SUGGESTED READINGS

Fisher, D. A., *The Epic of Steel*. New York: Harper and Row, 1963.

Hume-Rothery, W., *Electrons, Atoms, Metals and Alloys*. New York: Philosophical Library, 1955.

Wertime, T. A., *The Coming of the Age of Steel*. Chicago: University of Chicago Press, 1962.

There are several widespread, common misconceptions concerning the rare earths which we hope this symposium will tend to dispel.

First, rare earths are rare. It comes as a surprise to many people when they are told that cerium is reportedly more abundant in the earth's crust than tin, and that yttrium is more plentiful than lead; even the least abundant naturally occurring rare earths are far more plentiful than the platinum-group metals.*

F. H. SPEDDING

Chapter 18 Aluminum, Nickel, and Related Metals

The solubility relationships discussed in Chapter 13 provide a way of dividing the cations in the periodic table into groups. Chloride and sulfide may be used to remove the soft acid cations which include all the metals toward the bottom and right of the periodic chart and all of the group IB cations. The cations in the darkest section of Fig. 18-1 form insoluble chlorides and sulfides. If the nonmetals at the upper right of the periodic table are not included, the ions remaining after the soft cations are precipitated include only those that form substantially ionic salts. The cations with the lowest polarizing ability have hydroxides that are soluble at pH 9. These include all the alkali metals and all but the smallest (Be^{2+}) of the alkaline earth metal ions (11 metals in all). The next group of 47 elements includes most of the first transition series, some of the second and third transition series, as well as all of the lanthanum and actinium series of elements. Figure 18-1 shows the location of all the elements included in this group which we shall call the aluminum-nickel group. There are some striking similarities in the chemistry of the cations of these elements so it is quite appropriate to group them together in one chapter.

*Used by permission from F. H. Spedding and A. H. Daane, *The Rare Earths* (New York: John Wiley, 1961).

Sec. I Chemical Properties of the Entire Group

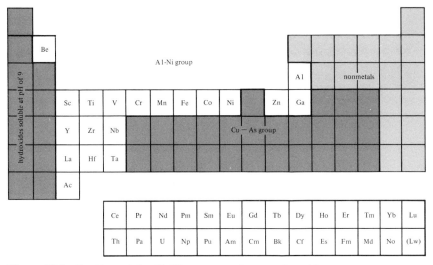

Figure 18-1 *Position of the Al–Ni group in the periodic table.*

I. Chemical Properties of the Entire Group

All the elements of the Al–Ni group have hydroxides that are insoluble in water at pH 7. The cations are all acids and their enthalpies of hydration are substantial. The electric field around the ions is so great that the water molecules in the first hydration sphere tend to lose a proton, resulting in ionization equilibria such as

$$Fe(H_2O)_6^{3+} \rightleftharpoons Fe(H_2O)_5OH^{2+} + H^+$$

$Fe(H_2O)_6^{3+}$ is actually a stronger acid than acetic acid with $K_a = 6.8 \times 10^{-3}$.

There is a good correlation between the acidity and polarizing ability of cations. This was discussed in Chapters 11, 13, and 14, see especially Table 14-8 which shows also the pH range over which the hydroxides are insoluble. As shown in Table 14-8 some of the hydroxides are soluble in excess hydroxide ion. The cations of the Al–Ni group may be divided on the basis of their reaction with excess hydroxide ions.

A. THE ALUMINUM SUBGROUP

The amphoteric cations of the Al–Ni group constitute the aluminum subgroup. Amphoterism is characteristic of those cations with polarizing ability (PA) between *3.0 and 4.4*. Of the Al–Ni group elements,

437

only Al^{3+}, Be^{2+}, and Ga^{3+} have polarizing abilities in this range, but Zn^{2+} and Cr^{3+} are included in the aluminum subgroup as well.

1. Aluminum

As shown in the E_h-pH diagram in Fig. 18-2, aluminum is found almost exclusively in the $+3$ oxidation state. The amphoteric character of aluminum is shown on the diagram by the regions I and II where the predominate species are the positive hydrated Al^{3+} ion and the negatively charged $Al(OH)_4^-$ respectively.

The oxygen/aluminum radius ratio of 2.80 is close to the transition point between octahedral and tetrahedral coordination. Aluminum(III) is found coordinated in both. Crystalline Al_2O_3 is basically a cubic closest packing of oxide ions with two thirds of the octahedral holes occupied by aluminum ions. It is a hard material that is valuable as an abrasive. The refractive index of Al_2O_3 is high, and when it is colored by appropriate impurities it is valued as a gem (sapphire, ruby, and so forth).

See Chapter 14 for coordinating tendencies of many of the ions in this chapter.

The only common ligand which competes effectively with hydroxide ion for coordination with aluminum is fluoride. The whole series of complex ions from AlF^{2+} to AlF_6^{3-} can be prepared. Other salts of aluminum(III) such as $Al_2(SO_4)_3$ and $AlCl_3$ can be prepared, but they react violently with water to give hydrates or $Al(OH)_3$. These reactions as well as

$$AlCl_3 + CH_3-O-CH_3 \longrightarrow AlCl_3 \cdot CH_3OCH_3$$

and

$$AlCl_3 + Cl^- \longrightarrow AlCl_4^-$$

are examples of the acidic character of aluminum salts. When no other bases are available a salt such as $AlCl_3$ will react with itself to give the dimeric molecules, Al_2Cl_6

$$\begin{array}{ccccc} Cl & & Cl & & Cl \\ & \diagdown & & \diagup & \\ & Al & & Al & \\ & \diagup & & \diagdown & \\ Cl & & Cl & & Cl \end{array}$$

Because Al_2Cl_6 is not very polar, it dissolves in most organic solvents and serves as a very effective acid catalyst for a great variety of reactions in organic chemistry.

2. Beryllium

As a result of the similar polarizing abilities there is a very marked diagonal resemblance between the elements of the second and third

Sec. I Chemical Properties of the Entire Group

rows. Like aluminum, beryllium forms stable complexes primarily with hydroxide and fluoride ions. There are very few uses for beryllium, partly because it is quite rare and partly because it is extremely poisonous. However, because of the high enthalpy of formation of many of the compounds of beryllium, it is of interest as a potential rocket propellant. The E_h-pH diagram for beryllium is shown in Fig. 18-3.

3. Gallium

The chemistry of gallium is so similar to that of aluminum that it hardly merits any separate comment. The only substantial difference is that it is somewhat easier to reduce to the metal. ($E^0_{Ga} = -0.55$ compared to $E^0_{Al} = -1.68$ volts, see Fig. 18-4).

4. Chromium

Freshly precipitated $Cr(OH)_3$ does dissolve appreciably in NaOH solution; therefore, chromium is ordinarily included in the aluminum subgroup. However, the thermodynamic data show that the equilibrium solubility of Cr(III) is negligible even at a pH of 15. $Cr(OH)_3$ and Cr_2O_3 are easily dissolved in base if an oxidizing agent is present to convert Cr(III) to Cr(VI). The chemistry of chromium will be considered in more detail in Chapter 21.

5. Zinc

As is shown by the E_h-pH diagram in Fig. 18-5, Zn(II) is amphoteric. The bonding in zinc salts is considerably more covalent than that of Al(III), or in other words, Zn(II) is a much softer acid. Thus ZnS can be precipitated from aqueous solutions and complexes of Zn^{2+} with Cl^-, Br^-, and I^- are easily prepared.

$$Zn^{2+} + 4Cl^- = ZnCl_4^{2-}$$

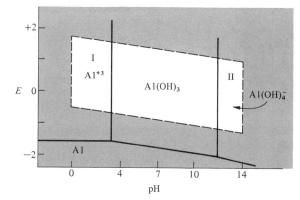

Figure 18-2 E_h-pH *diagram for aluminum.*

Ch. 18 Aluminum, Nickel, and Related Metals

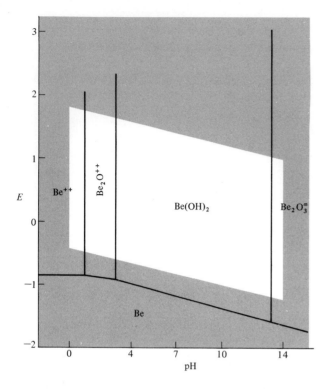

Figure 18-3 E_h-pH *diagram for beryllium.*

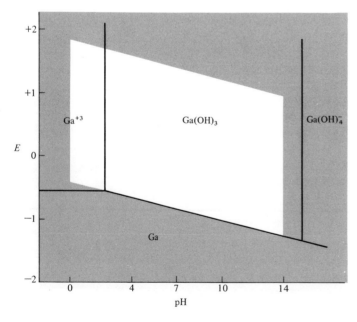

Figure 18-4 E_h-pH *diagram for gallium.*

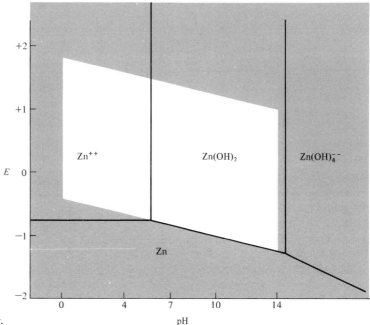

Figure 18-5 E_h-pH *diagram for zinc.*

The induced polarizing ability of Zn^{2+} is evidently significant even with OH^- as a ligand since the polarizing ability, $(Z + S)/(r + 0.50)^2$, of Zn^{2+} is not quite large enough (2.92) for a clear prediction of amphoteric behavior.

B. THE NICKEL SUBGROUP

The remaining elements of the Al-Ni group are not amphoteric in their normal oxidation states. They are classified as the nickel subgroup. It is convenient to consider the nickel subgroup in two sections: the IIIB family and the rest of the subgroup. All the IIIB elements are very similar chemically. The family includes the 15 lanthanide elements in which the $4f$ level is being filled.

1. THE IIIB FAMILY OF ELEMENTS

a. *Occurrence and Discovery.* The story of these elements begins in 1788 when a soldier collected some samples of a black heavy mineral near the small Swedish town of Ytterby. This mineral, ytterbite or gadolinite, with a formula of the type $Be_2FeY_2Si_2O_{10}$, is similar to coal in appearance but it is much more dense. By 1794, J. Gadolin had isolated the oxide of yttrium (Y_2O_3) from this mineral and started the processes that led to the eventual identification of four separate elements (yttrium, terbium, erbium, and ytterbium), all named after this one small place. Yttrium is the predominate +3 cation in gadolinite, but all the naturally occurring rare earths are present to some extent. All the heavier rare earths (atomic numbers 65 through 71, with ionic radii 0.93 through 1.00 Å) were originally isolated from this source.

An even rarer complex silicate, cerite, $[(Ca,Fe)Ce_3(OH)_3Si_3O_{10}]$ yielded first cerium (Ce) and then all the larger rare earths (atomic numbers 57–64, with ionic radii 1.02–1.15 Å) except promethium (Pm) (atomic number 61) which does not occur naturally.

The rare earth elements can easily substitute for each other because they all exhibit a +3 oxidation state and a similar size. They range in size from 0.93 Å (Lu^{3+}) to 1.15 Å (La^{3+}). The rare earth ions are too large for the octahedral holes of a close packed array of oxide ions. The radius ratios of the largest and smallest cations ($1.40/1.15 = 1.22$ and $1.40/.93 = 1.51$) lie within or close to the range (1.00–1.37) in which eight-fold coordination is expected. Since Ca^{2+} is eight-fold coordinate in silicates, it is not surprising that there are traces of the rare earths in all silicate rock that contains calcium. Compensation for the extra positive charge is made by additional substitutions such as O^{2-} for F^- or Al^{3+} for Si^{4+} or by incomplete occupancy of all the cation sites. In a similar way CaF_2 can dissolve substantial amounts (over 50% in some cases) of rare earth fluorides.

The third anion (in addition to fluoride and silicate) with which the rare earths are associated in commercially significant quantities is phosphate. Phosphate is not a common ion geologically but apparently, because of the match in charge, several minerals are known that are predominately rare earth phosphates. Monazite is a solid solution of principally $ThSiO_4$, $CePO_4$, and $LaPO_4$.

b. Chemistry of the +3 State. All the rare earth elements are included in group IIIB of the periodic table because they exist principally in the +3 oxidation state. The +3 ions are good polarizing cations because of the high charge. They are found coordinated to either oxygen or fluorine. Both hydroxides and fluorides (formulas approximately $MF_3 \cdot \frac{1}{2}H_2O$) are easily precipitated from solution. None of the +3 ions have electronic structures that would lead to induced polarizing ability; consequently complexes with easily polarized anions are not formed. The +3 cations are, like Al^{3+}, too acidic to precipitate as the sulfide. The sulfides react with water to give the hydroxide

$$La_2S_3 + 6H_2O \longrightarrow 2La(OH)_3 + 3H_2S$$

Although there are only a few reliable thermodynamic values for any of the rare earth compounds, they vary in a regular manner with ionic radius so that in most cases good values can be obtained by interpolation or extrapolation. This is illustrated in Table 18-1. Electrode potentials and solubility products calculated from these values show similar regular trends but, since the activity of a +3 ion is not closely related to its concentration, the values commonly given have only qualitative significance.

c. Other Oxidation States. Scandium, yttrium, and lanthanum are not stable in any oxidation states except 0 and +3. However there are a fair number of compounds of the rare earths containing an incomplete $4f$ subshell that can exist in an oxidation state of +2 or +4. Only the largest of the +3 ions show any tendency to lose an f electron. Ce^{4+} is reasonably stable in water solutions, and the ion is a commonly used oxidizing agent. The standard electrode potential for the Ce^{3+}-Ce^{4+} couple (around +1.61 volts) is close to the limit of kinetic stability with respect to oxidation of water. This makes Ce^{4+} a stronger oxidizing agent than Cl_2 ($E^0 = 1.36$ volts). The next smaller element, praseodymium, does form solid PrO_2, but the Pr^{3+}-Pr^{4+} potential is about +2.9 volts so that Pr^{4+} in solution would be an even

Table 18-1 *Thermodynamic Properties of Group IIIB (H^0, kcal mole^{-1}; S^0, eu)**

Metal	Ionic radius	M^{3+}		M_2O_3		MF_3		MCl_3		$M(OH)_3$		M
		H^0	S^0	H^0	S^0	H^0	S^0	H^0	S^0	H^0	S^0	S^0
$_{21}$Sc	0.81	−149		−456.2	18.4	−367		−221				(8.2)
$_{39}$Y	0.93	−168	(−55)	−455	23.7	−410.7	(26.2)	−232.7	(41.5)	−338		10.6
$_{71}$Lu	0.93	−160		−449	26.0	(−392)		−228				12.2
$_{70}$Yb	0.94	−161	(−53)	−436	31.8	−376		−224				(15.0)
$_{69}$Tm	0.95	−161		−451	(36.5)			(−230)				17.4
$_{68}$Er	0.96	−162	(−50)	−454	36.6			−229.1	(36)	−340		17.5
$_{67}$Ho	0.97	−164		−450	37.8			(−233)				18.0
$_{66}$Dy	0.99	−166		−446	35.8			(−238)				17.9
$_{65}$Tb	1.00	−168		−437	(37.5)			(−242)				17.5
$_{64}$Gd	1.02	−169	(−50)	−434	36.0	−404		−240.1	(45)			15.8
$_{63}$Eu	1.03	−169	(−50)		(35)			(−247)				(17.0)
$_{62}$Sm	1.04	−170		−434	36.1	−405		−245	(36)			16.6
$_{61}$Pm	1.06	−170		(−433)								(17.2)
$_{60}$Nd	1.08	−171	(−49)	−432	36.9			−249.0	(34)	(−332)	(32)	17.5
$_{59}$Pr	1.09	−173	(−48)	−437	(37.9)	−401		−252	(34)			17.5
$_{58}$Ce	1.11	−174	(−47)	−434.9	36.0	−416	27.5	−253	(35)			16.7
$_{57}$La	1.15	−176	(−46)	−428	30.6	−405		−255.9	(44.5)			13.6

*The values given in parentheses are estimates. In addition to the references in Appendix IV, values given in this table are taken from *Advances in Chemistry Series*, Vol. 71, *Lanthanide/Actinide Chemistry*, (Washington, D. C.: American Chemical Society, 1967), E. M. Larsen, *Transitional Elements* (New York: W. A. Benjamin, 1965), and R. L. Montgomery, *U. S. Bur. Mines, Rept. Invest.*, **5468** (1959).

stronger oxidizing agent than F_2. The +4 oxidation state of terbium (Tb) is of borderline stability. In this case the exceptionally high spin correlation stabilization energy of seven unpaired $4f$ electrons contributes to the stability of the +4 oxidation state. Although TbO_2 can be prepared, Tb^{4+} is too strong an oxidizing agent to exist in solution.

A similar situation is encountered for the +2 oxidation state. The most stable of the M^{2+} ions is that of europium (Eu^{2+}). This ion has a $4f^7$ electron configuration. The electrode potential of this ion

$$Eu^{3+} + e^- = Eu^{2+} \qquad E^0 = -0.43 \text{ volt}$$

Lutetium, (Lu) is smaller than ytterbium and thulium, but Lu^{2+} is not stable. There would be a 5d electron present in Lu^{2+}.

is near the borderline for kinetic stability in water. There is apparently enough extra spin correlation stabilization energy for six $4f$ electrons over five so that Sm^{2+} can be prepared, but it reacts rapidly with water to liberate hydrogen. The other two rare earth elements for which +2 compounds can be prepared in the absence of water are thulium (Tm) and ytterbium (Yb). These ions are the smallest of all the lanthanide ions. In summary, cerium and europium are the only rare earths with an oxidation state other than +3 which can be made in the presence of water.

2. Uranium and Thorium

The elements with inner $5f$ electrons might be expected to behave in a manner similar to the rare earths, but oxidation states higher than +3 are quite accessible in the actinide series. Thorium (Th) is found almost exclusively in the +4 oxidation state, whereas uranium (U) can occur in compounds with oxidation states all the way from +3 to +6. The +6 state, represented by the uranyl ion (UO_2^{2+}) is by far the most stable.

The most significant thorium ore, monazite, may contain as much as 9% thorium. The best uranium ore, pitchblende, is predominately an oxide of uranium(IV) and uranium(VI). The demand for uranium for atomic power is great and many other uranium bearing minerals can be profitably mined. Thorium is used in gasoline lantern mantles. Before its radiological hazard was recognized uranium compounds were used in yellow and orange glazes for pottery. At the present time the principal use of both of these elements is as nuclear fuel (see Chapter 19).

3. Families IVB and VB

The chemistry of titanium, zirconium, and hafnium is almost exclusively the formation of an insoluble +4 oxide and hydroxide. The fluorides are also stable, and hydrofluoric acid is one of the best reagents for dissolving the dioxides. The chlorides and sulfides (MCl_4

and MS_2) can be prepared, but they react with water to form insoluble oxides or hydroxides. This is, of course, typical behavior of cations that are hard acids.

Except for the difference in stoichiometry niobium (Nb) and tantalum (Ta) (group VB) are similar chemically to the IVB elements. Tantalum chloride ($TaCl_5$) is readily hydrolyzed and hydrofluoric acid is almost the only reagent which will dissolve niobium pentoxide (Nb_2O_5) and tantalum pentoxide (Ta_2O_5).

$$Nb_2O_5 + 12HF \longrightarrow 2NbF_6^- + 2H^+ + 5H_2O$$

Tantalum metal is a strong reducing agent

$$Ta_2O_5 + 10H^+ + 10e^- = 2Ta + 5H_2O \qquad E^0 = -0.81 \text{ volt}$$

but it is so well protected by an oxide coating that it vies with platinum as an inert metal in applications requiring extreme corrosion resistance. Tantalum is even less reactive than platinum toward some reagents such as aqua regia (HNO_3 plus HCl).

4. THE 3d TRANSITION SERIES

Much of the chemistry of the 3d transition series has been discussed in previous chapters. Therefore we need only discuss a few additional features here.

Reasonably reliable electrode potentials for the $M^{3+} + e^- = M^{2+}$ half-reactions are known or can be estimated for the full series of elements from titanium through nickel. They are shown plotted against the number of d electrons present in the $+2$ ion in Fig. 18-6.

The data in this figure can be interpreted if we recall that spin correlation and ligand field effects combine to make the three-d-electron configuration, illustrated in Fig. 18-7, particularly stable. A fourth electron will be easily removed from Cr^{2+} regardless of whether it is in the antibonding e_g level or in the t_{2g} level. For this reason the electrode potentials do not increase from V to Cr. There is a similar drop between Mn and Fe that is related to the high spin correlation stabilization energy for the half-filled d subshell. The sixth d electron present in Fe^{2+} is readily removed. Thus Mn^{2+} and Fe^{2+} are almost equally good reducing agents but are considerably weaker than V^{2+} and Cr^{2+}. They are, however, much stronger reducing agents than the $+2$ ions of the later elements in the series. In fact, Ni^{2+} is very difficult to oxidize and Cu^{3+} is very rare.

The involvement of d orbitals in chemical bonding similarly complicates other trends across the 3d transition metals. The general trend is toward decreasing size as the atomic number increases, but the

Ch. 18 Aluminum, Nickel, and Related Metals

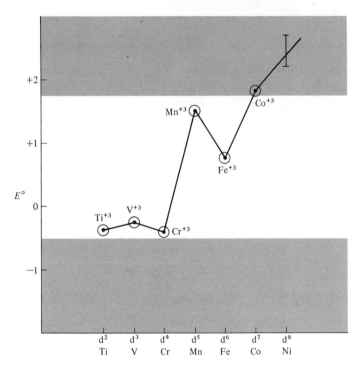

Figure 18-6 *Electrode potentials for the oxidation of M^{2+}. The half-reaction is generally $M^{3+} + e^- = M^{2+}$.*

Figure 18-7 *The stable configuration of three d electrons in octahedral complexes.*

There are significant quantities of vanadium in some iron, gold, and uranium ores.

addition of antibonding electrons (into the e_g^* orbitals) causes an increase in bond lengths. The effect of antibonding electrons on size predominates in both the $Ti^{2+} - Cr^{2+} - Mn^{2+}$ and the $Ni^{2+} - Cu^{2+} - Zn^{2+}$ sequences with the result that Mn^{2+} is larger than Ti^{2+}, and Zn^{2+} is larger than Ni^{2+}.

The position of cupric ion (Cu^{2+}) in the sequence of sizes is particularly interesting. The ninth d electron occupies the d_{z^2} orbital and it is especially antibonding for the two ligands toward which this orbital is oriented. Four of the ligands in Cu^{2+} complexes are generally closer than those in the corresponding Ni^{2+} complexes and two are at longer distances than those in Zn^{2+}.

a. Vanadium. Pure vanadium is the hardest of all metals. Like chromium it will take a very high polish. Since chromium is much cheaper and more abundant, it is usually used instead of vanadium. The principal use of vanadium is in iron alloys, which are prepared by reducing a mixture of vanadium oxide and iron oxide with carbon. The vanadium oxide used is usually a by-product of another mining venture, although substantial amounts are also recovered from the ash of oil-burning furnaces. The analysis of the vanadium content of oils is a useful way to distinguish between samples from different oil fields. The behavior of the various oxidation states of vanadium is summarized in Fig. 18-8.

Vanadium is quite different chemically from the other group VB elements because of its smaller size. Both the acidic behavior of V(V) and the stability of lower oxidation states are consequences of the small size. Both of these points are clearly evident in the E_h-pH diagram for vanadium shown in Fig. 18-8. Compounds can be prepared with vanadium in all the oxidation states from 0 to +5 except for +1. Classification of vanadium in the nickel subgroup or the aluminum subgroup depends on the oxidation state involved since V(II) and V(III) are basic, V(IV) is amphoteric, and V(V) is on the borderline between amphoteric and acidic behavior.

b. *Manganese.* Manganese is the first of the 3d transition elements in which the +2 oxidation state is truly stable in water. (The elements to the left of Mn exist in higher oxidation states.) A strong oxidizing agent will convert Mn(II) to the bright purple permanganate ion, MnO_4^-.

$$MnO_4^- + 8H^+ + 5e^- = Mn^{2+} + 4H_2O \qquad E^0 = 1.51 \text{ volts}$$

Potassium permanganate is a very commonly used oxidizing agent. In acid solution it is reduced to Mn^{2+} but in basic or neutral solution manganese dioxide (MnO_2) is produced.

$$MnO_4^- + 2H_2O + 3e^- = MnO_2(s) + 4OH^- \qquad E^0 = 0.597 \text{ volts}$$

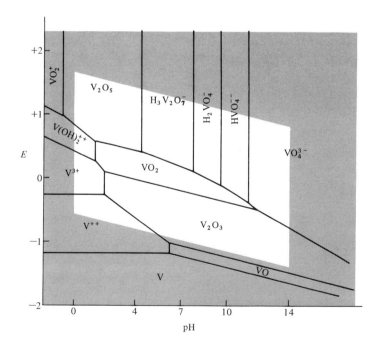

Figure 18-8 E_h-pH *diagram for vanadium.*

Permanganate ion is useful as an indicator as well as a reactant in titrations because the purple color of the ion is so intense that a single drop of 0.02 M potassium permanganate is easily seen in 100 ml of solution. Manganese(VII) exists as the anion MnO_4^- over the entire pH range. Manganese(VII) is accordingly highly acidic. Manganese(II) is basic, but manganese dioxide is so insoluble in both acid and base that the acid-base properties of the substance cannot be observed. Manganese dioxide is a common manganese ore and a useful oxidizing agent that can be used for the preparation of chlorine from hydrochloric acid solutions.

$$MnO_2 + 4H^+ + 2Cl^- \longrightarrow Cl_2 + Mn^{2+} + 2H_2O$$

The intermediate oxidation states in which manganese appears (+3, +5, and +6) are all unstable with respect to disproportionation so they do not appear on the E_h-pH diagram for manganese shown in Fig. 18-9. They can all be prepared under suitable conditions in compounds such as MnO_4^{2-}, MnO_4^{3-} and $Mn(CN)_6^{3-}$.

c. Iron. In discussing the chemistry of iron in aqueous solution only two oxidation states, +2 and +3, need to be considered. However, the chemistry of iron is complicated by the formation of a large variety of complex ions. It is not difficult to write formulas for the complex ions formed, since iron shows a coordination number of six almost exclusively.

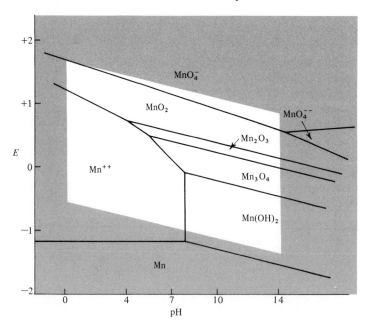

Figure 18-9 E_h-pH *diagram for manganese.*

Fe(H$_2$O)$_6^{3+}$ (a high spin d^5 ion) is almost colorless. Most solutions containing Fe(III) are yellow due to the presence of complex ions such as Fe(OH)(H$_2$O)$_5^{2+}$ or FeCl(H$_2$O)$_5^{2+}$. The reddish brown color of Fe(OH)$_3$ and the color of rust arise from similar spectral transitions. The really intense colors shown by some iron compounds are due either to electron transfer from an easily reduced ligand such as Br$^-$ or SCN$^-$ to Fe(III) or to electron transfer between two iron atoms in different oxidation states. FeO$_{1.05}$ and magnetite (Fe$_3$O$_4$) are both almost black because of the presence of both Fe(II) and Fe(III). The deep blue colors of both Prussian blue and Turnbull's blue are due to an anionic lattice of overall composition, Fe$_2$(CN)$_6^-$, in which half of the iron atoms are in each oxidation state. For applications in which the color of Fe(III) compounds is a problem, it is sometimes possible to add sodium fluoride to the solution to form colorless FeF$_6^{3-}$ ions.

All the complex ions in the series from Fe(SCN)(H$_2$O)$_5^{2+}$ *to* Fe(SCN)$_6^{3-}$ *are bright red in color.* FeBr$_3$ *has a very dark red color.*

d. *Nickel.* Nickel oxide (like FeO) usually contains enough Ni(III) to give it an intense black color, and Ni(OH)$_2$ can be oxidized in basic solution to a composition approaching NiO(OH). This is the basis for the anode reaction in the Edison storage battery.

anode   Ni(OH)$_2$ + OH$^-$ = NiO(OH) + H$_2$O + e^-

cathode   Cd(OH)$_2$ + 2e^- = Cd + 2OH$^-$

There are however, only a few additional compounds of Ni(III).

Nickel is found in solution almost exclusively in the +2 oxidation state. Octahedral Ni(II) complexes have been discussed in detail in previous chapters. An octahedral structure is not the only stable structure for Ni(II) complexes. Square planar complexes are also observed, and they exist because the polarizing ability of Ni(II) depends upon the spin state of the complex. There are two spin states common for Ni(II) complexes, one the high-spin state (that is, a maximum number of unpaired spins)

Ni(II) $\underset{3d}{\text{⸬ ⸬ ⸬ · ·}}$, $\underset{4s'}{\text{—}}$ $\underset{4p}{\text{— — —}}$

contains two unpaired electrons; the other, the low-spin state

Ni(II) $\underset{3d}{\text{⸬ ⸬ ⸬ ⸬ —}}$, $\underset{4s'}{\text{—}}$ $\underset{4p}{\text{— — —}}$

has all the electrons paired. The molecular structure that results if the nickel is low spin is not octahedral but square planar. This arrangement gives the minimum interaction between nickel d electrons and ligand electron pairs. Low-spin Ni(II) is more highly polarizing than high-spin

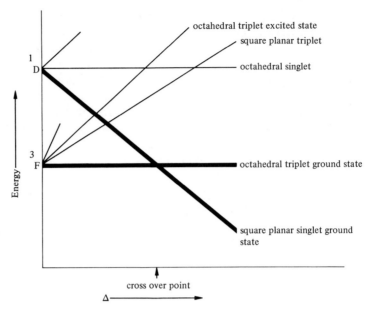

Figure 18-10 *A few of the octahedral and square planar states of d^8 complexes.*

Ni(II) because the d electrons provide even less shielding in the direction of the ligands since the $d_{x^2-y^2}$ orbital is completely empty.

The excess or induced polarizing ability available through such a distortion of structure depends upon the ligand. As the ligand group becomes easier to polarize, the metal responds by becoming more highly polarizing. At some point in a ligand sequence such as F^-, Cl^-, O^{2-}, S^{2-}, CN^-, there will be a change in spin state of Ni(II) and the low-spin square planar or distorted octahedral configuration becomes the most stable.

This situation can be represented by an energy level diagram such as that shown in Fig. 18-10. In this diagram the state with the lowest energy is the ground state. The section on the left side of the cross over point (indicated by the arrow) represents low ease of polarization of the ligand and the section at the right high ease of polarization.

QUESTIONS

1. Which of the rare earth +3 ions is the most basic? Why?
2. Table 3-7 gives the same ionic radius for Sc^{3+} and In^{3+}, yet In_2S_3 is too insoluble for indium to be considered in the Al-Ni group. Account for this difference.
3. In general there is an increase in size in going down a family, but the

pairs Zr^{4+}-Hf^{4+} and Nb^{5+}-Ta^{5+} show radii very close to the same. Account for this behavior.
4. List the number of d electrons present in each of the following:
   (a) Ni(II), Cr(IV), Mn(III), Co(III), V(II)
   (b) Fe^{3+}, Cr^{3+}, Cr^{2+}, Al^{3+}, Zn^{2+}
   (c) CrO_4^{2-}, MnO_4^{2-}, MnO_2, $TiCl_3$, $ScCl_3$
   (d) $Fe(CN)_6^{4-}$, $NiCl_4^{2-}$, $Cr(NH_3)_6^{3+}$, $Co(NH_3)_5Cl^{2+}$, Fe_3O_4
5. Why don't cobalt and nickel have any stable oxidation states higher than $+3$?
6. What chemical reactions occur in the rusting of iron?
7. Write chemical equations for:
   (a) dissolving $Sc(OH)_3$ in acid
   (b) dissolving $Ni(OH)_2$ in NH_3
   (c) dissolving $Al(OH)_3$ in NH_3
   (d) dissolving HfO_2 in HF solution
   (e) the reaction of $Ni(OH)_2$ with S^{2-}
   (f) the reaction of Fe_3O_4 with H_2
   (g) the hydrolysis of La^{3+}
8. In each of the lists of four ions or molecules below, pick the one specified at the left.
   (a) the strongest oxidizing agent: La^{3+}, Ce^{3+}, Pr^{3+}, Nd^{3+}
   (b) the strongest acid: La^{3+}, Sm^{3+}, Ho^{3+}, Lu^{3+}
   (c) the strongest acid: $La(OH)_3$, $Al(OH)_3$, $Eu(OH)_2$, $In(OH)_3$
   (d) the strongest oxidizing agent: Cr^{3+}, Cr^{2+}, Mn^{3+}, Ni^{3+}
   (e) the strongest reducing agent: Ti^{2+}, Cr, Fe^{2+}, Ni
   (f) the most stable complex: $Mn(NH_3)_6^{2+}$, $Fe(NH_3)_6^{3+}$, $Fe(NH_3)_6^{2+}$, $Cr(NH_3)_6^{2+}$
   (g) the most soluble in NaOH: FeO, NiO, ZnO, Fe_2O_3
   (h) the least soluble: MnS, Al_2S_3, CoS, CaS
9. Why do Mn^{2+} and Fe^{3+} have less color than most transition metal ions?

PROBLEMS I

1. Calculate the electrode potentials for the following half-reactions (Table 18-1):
   (a) $La^{3+} + 3e^- = La$
   (b) $Sc(OH)_3 + 3e^- = Sc + 3OH^-$
   (c) $Lu^{3+} + 3e^- = Lu$
2. Calculate the solubility products of $La(OH)_3$ and $Eu(OH)_3$.
3. The solubility product of $Al(OH)_3$ is 5×10^{-33}. Calculate the concentration of Al^{3+} in a saturated neutral solution (pH = 7.0).
4. The equilibrium constant for the reaction $Fe^{3+} + SCN^- = FeSCN^{2+}$ is 1×10^{-3}. Calculate the concentration of $FeSCN^{2+}$ when 10^{-2} mole of NaSCN are added to 1 liter of a 0.10 M solution of $Fe(NO_3)_3$.
5. The pK_a for Fe^{3+} is 2.6. Calculate the pH of a 0.10 M solution of $Fe(NO_3)_3$. Is the neglect of the $Fe(OH)^{2+}$ ion in Problem 4 justified?
6. Write a balanced equation for the reaction of Fe^{2+} with MnO_4^-. What volume of 0.010 M $KMnO_4$ solution is needed for the titration of 15 ml of 0.050 M $FeSO_4$?

7. For the reaction $Al(OH)_3 + OH^- \longrightarrow Al(OH)_4^-$, $K = 40$. What is the minimum concentration of NaOH solution which can be used to dissolve 0.10 mole $Al(OH)_3$ in 1 liter?
8. Complete and balance the three following half-reactions:
   (a) $\_I_2 \longrightarrow \_IO_3^-$
   (b) $\_I^- + \_Cu^{2+} \longrightarrow \_CuI$
   (c) $\_MnO_4^- \longrightarrow \_Mn^{2+}$
9. Combine two of the half-reactions in Problem 8 to give the balanced chemical equation for the reaction of MnO_4^- with I_2.
10. The ionization constant of Fe^{3+} as an acid is 3×10^{-3}. Calculate the pH of a solution prepared by dissolving 0.05 mole of NaOH in 1 liter of 0.25 M $Fe(NO_3)_3$ solution.
11. If 0.40 mole of NaOH are added to 1.00 liter of 0.25 M $Fe(NO_3)_3$, some of the iron precipitates out as $Fe(OH)_3$, $K = 10^{-36}$. Calculate the pH of the remaining solution.

PROBLEMS II

1. The ionic radius of La^{3+} is 1.22 Å. Calculate the polarizing ability of La^{3+}.
2. Will La_2O_3 dissolve in 1 M NaOH?
3. Balance the half-reaction:

$$\_La_2O_3 + \_H_2O + \_e^- \longrightarrow \_La + \_OH^-$$

4. Calculate ΔG^0 for the half-reaction in Problem 3, given the values:

	G^0
H_2O	-56.69 kcal mole^{-1}
OH^-	-37.59 kcal mole^{-1}
La_2O_3	-407 kcal mole^{-1}

5. Calculate E^0 for the half-reaction in Problem 3.
6. Write the Nernst equation for this half-reaction.
7. How does the electrode potential for the half-reaction in Problem 3 vary with pH?
8. At what pH will this half-reaction have a potential of -2.29 volts (assuming La_2O_3 and lanthanum are present as pure solids)?
9. H^0 for La^{3+} is -176 kcal mole^{-1}. Calculate ΔH^0 for the half-reaction $La^{3+} + 3e^- \longrightarrow La$.
10. Using the entropy for lanthanum metal from Table 18-1 and the estimate $S^0(La^{3+}) = -46$ eu, calculate ΔS^0 for the half-reaction in Problem 9.
11. Calculate ΔG^0 for the half-reaction in Problem 9.
12. Calculate E^0 for the half-reaction $La^{3+} + 3e^- = La$.
13. Draw the E_h-pH diagram for lanthanum.
14. List the potential and pH for the point of intersection of the three lines in the E_h-pH diagram for lanthanum.
15. Estimate the equilibrium constant for the reaction $\frac{1}{2}La_2O_3 + 3H^+ \longrightarrow La^{3+} + \frac{3}{2}H_2O$.

SUGGESTED READINGS

LARSEN, E. M., *Transitional Elements*. New York: W. A. Benjamin, Inc., 1965.

MACKAY, K. M., and R. A. MACKAY, *Modern Inorganic Chemistry*. London: Intertext, 1968.

MOELLER, T., *The Chemistry of the Lanthanides*. New York: Reinhold Pub. Corp., 1963.

> Nature is neutral. Man has wrested from nature the power to make the world a desert or to make the deserts bloom. There is no evil in the atom; only in men's souls.
>
> ADLAI STEVENSON
> Speech, Hartford, Conn.,
> Sept. 18, 1952

Chapter 19 The Structure of Nuclei

I. The Formation of Atoms

From measurements and estimates of the composition of the earth, meteorites, other planets, and stars, it is possible to calculate the overall elemental composition of at least a major part of our galaxy. The relative abundance of each element in this overall composition is called the cosmic abundance of the element. The cosmic abundance of an element is customarily expressed as the number of atoms of that element relative to 10^6 atoms of silicon. Figure 19-1 and Table 19-1 show the values for a large number of elements on a logarithmic scale. These data do not show any relationship to the electronic structure of the elements. There is, however, a fairly general alternation between elements with even and odd atomic number. The elements with even atomic number are more abundant. The relative abundance of the elements is primarily due to the structure of the nuclei.

As indicated in Fig. 19-1, hydrogen is by far the most abundant element. From the cosmic abundance we can calculate that 93% of all atoms in this part of the universe are hydrogen. Hydrogen is the principal element in all the main sequence stars (stars on the same luminosity temperature curve as the sun). The outer layers of the sun are 59% hydrogen by weight, 86% hydrogen by atoms.

Sec. 1 The Formation of Atoms

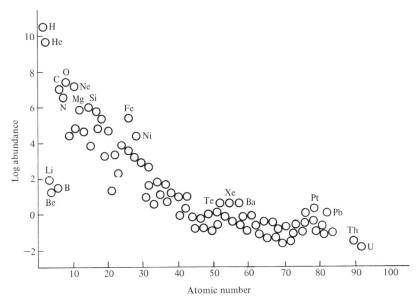

Figure 19-1 *Cosmic abundances of the elements.*

Helium is the second most abundant element in the sun. In fact, the source of the sun's energy is the nuclear reaction in which helium is formed from hydrogen. The reaction $4\,^1\text{H} \longrightarrow\,^4\text{He}$ is extremely exothermic ($\Delta H = -6.14 \times 10^8$ kcal mole^{-1}).

There are several pathways by which hydrogen can be converted to helium; however, the two processes outlined in Table 19-2 can account for the energy production in the main sequence stars. In both processes most of the energy appears as gamma rays (γ) and kinetic energy of the nuclei, but the neutrinos (ν) produced carry off about 2% of the energy. Both the carbon cycle and the proton-proton chain process require the extremely high temperature of the main sequence stars (about 10^7 K).

Although ^{4}He is the most stable of the very light elements, continued reaction does take place with the production of heavier elements. One mechanism by which ^{12}C may be produced involves the simultaneous collision of three helium nuclei.

$$3\,^4\text{He} \longrightarrow\,^{12}\text{C} \qquad \Delta H = -1.67 \times 10^8 \text{ kcal mole}^{-1}$$

The reaction cannot occur in a stepwise fashion through the intermediate ^{8}Be because this isotope of beryllium is unstable. The production of ^{12}C from ^{4}He apparently occurs at a significant rate in the interior of red giant stars (densities about 10^5 g cm^{-3}, temperatures

In designating individual isotopes it is customary to use the mass number (the integer closest to the atomic weight of the isotope) as a superscript to the left.

Energy changes as large as this are quite easily determined by measurements of mass. A mole of ^{4}He weighs 4.00260 g, whereas four moles of H (4×1.007825) weighs 4.031300 g; therefore the energy change may be obtained from $E = mc^2$, using the value of 2.14×10^{10} kcal g^{-1} for c^2.

455

Ch. 19 The Structure of Nuclei

Table 19-1 *Logarithms of the Abundances of Elements**

IA	IIA	IIIB	IVB	VB	VIB	VIIB	VIII			IB	IIB	IIIA	IVA	VA	VIA	VIIA	Inert gases
10.5																	9.7
2.0	1.3											1.4	7.1	6.6	7.4	4.5	7.2
4.8	5.9											4.7	6.0	3.9	5.8	4.8	5.4
3.3	4.7	1.4	3.4	2.3	3.9	3.6	5.4	3.3	4.4	3.0	2.8	1.0	1.7	0.6	1.8	1.1	1.7
0.8	1.2	1.0	1.0	0.0	0.4		−0.1	−0.7	−0.2	−0.7	0.0	−0.8	0.1	−0.6	0.6	−0.1	0.6
−0.3	+0.6	−0.4	−1.1	−0.8	−0.9	−0.6	−0.1	−0.3	0.2	−0.8	−0.7	−1.0	0.0	−1.0			
				−0.2	−0.8	−0.1		−0.6	−1.0	−0.4	−1.2	−0.4	−1.1	−0.7	−1.4	−0.7	−1.4
				−1.5		−1.8											

*Data are from L. H. Aller, *The Abundance of the Elements* (New York: Interscience, 1961) and H. Suess, and H. C. Urey, *Revs. Mod. Phys.*, **28**:53 (1956).

Table 19-2 *Pathways for the Nuclear Reaction $4\,^1H \longrightarrow {}^4He$*

Carbon cycle	Proton-proton chain
$^{12}C + {}^1H \longrightarrow {}^{13}N + \gamma$	$^1H + {}^1H \longrightarrow {}^2H + e^+ + \nu$
$^{13}N \longrightarrow {}^{13}C + e^+ + \nu$	$^2H + {}^1H \longrightarrow {}^3He + \gamma$
$^{13}C + {}^1H \longrightarrow {}^{14}N + \gamma$	$^3He + {}^3He \longrightarrow {}^4He + 2\,{}^1H$
$^{14}N + {}^1H \longrightarrow {}^{15}O + \gamma$	
$^{15}O \longrightarrow {}^{15}N + e^+ + \nu$	
$^{15}N + {}^1H \longrightarrow {}^{12}C + {}^4He$	
$2e^+ + 2e^- \longrightarrow 4\gamma$	

about 10^8 K). It is believed that further reactions of helium and carbon under similar conditions account for the formation of the elements from oxygen through titanium (atomic number 22). Some typical nuclear reactions for the production of these elements are shown in Table 19-3.

The conversion of lighter elements to ^{56}Fe is also exothermic. These reactions require temperatures over 10^9 K where entropies are significant, even compared to nuclear energies, and as a result an equilibrium is possible. The abundance of the elements vanadium through nickel can be interpreted in terms of an equilibrium established at about 4×10^9 K.

For example, the isotopic abundance of the isotopes of iron is shown in Table 19-4. If we consider the nuclear reaction

$$2\,{}^{56}Fe = {}^{54}Fe + {}^{58}Fe$$

the equilibrium constant expression is

$$K = \frac{[{}^{54}Fe][{}^{58}Fe]}{[{}^{56}Fe]^2} = \frac{(0.0033)(0.0582)}{(.9166)^2} = 2.3 \times 10^{-4}$$

From the atomic weights (mass change 0.0031 g mole^{-1}) we can

Table 19-3 *Some Typical Nuclear Reactions for the Synthesis of the Elements of Atomic Numbers 8 Through 22*

$^{12}C + {}^4He$	$\longrightarrow {}^{16}O + \gamma$
$^{12}C + {}^{12}C$	$\longrightarrow {}^{23}Na + {}^1H$
$^{12}C + {}^{12}C$	$\longrightarrow {}^{20}Ne + {}^4He$
$^{20}Ne + \gamma$	$\longrightarrow {}^{16}O + {}^4He$
$^{20}Ne + {}^4He$	$\longrightarrow {}^{24}Mg + \gamma$
$^{24}Mg + {}^4He$	$\longrightarrow {}^{28}Si + \gamma$
$^{44}Ca + {}^4He$	$\longrightarrow {}^{48}Ti + \gamma$
$^{21}Ne + {}^4He$	$\longrightarrow {}^{24}Mg + n$

Table 19-4 *Abundances and Atomic Weights of the Iron Isotopes*

Isotope	Mass number	Atomic weight	Abundance, %
^{54}Fe	54	53.9396	5.82
^{56}Fe	56	55.9349	91.66
^{57}Fe	57	56.9354	2.19
^{58}Fe	58	57.9333	0.33

calculate that $\Delta H^0 = 6.6 \times 10^7$ kcal mole^{-1} for the reaction, and from the nuclear spins we can estimate $\Delta S^0 = 0$ eu. Then from the equation

$$K = 10^{-\Delta H^0/4.5757T} 10^{\Delta S^0/4.5757}$$

we can calculate an equilibrium temperature of 4.2×10^9 K.

The standard enthalpy change of the preceding reaction was calculated from atomic weights. Another example will be used to show the details of this calculation. For the reaction

$$3\ ^{56}\text{Fe} = 2\ ^{59}\text{Co} + ^{50}\text{Cr}$$

we have an initial mass of $3 \times 55.9349 = 167.8047$-g mole^{-1}. The weight of the products is $2 \times 58.9332 + 49.9461 = 167.8125$ g mole^{-1}, an increase of 0.0078 g mole^{-1}. The conversion factor between grams and kilocalories is 2.14×10^{10} kcal g^{-1} so this reaction is endothermic by 1.67×10^8 kcal mole^{-1} or 1.67×10^{11} cal mole^{-1}.

The entropy change for this reaction is +8.3 eu, and the equilibrium constant at 4.2×10^9 K may be calculated

$$K = 10^{-\Delta H^0/4.5757T} 10^{\Delta S^0/4.5757} = 10^{-8.70} 10^{1.82}$$

$$K = 10^{-6.88} = 1.3 \times 10^{-7}$$

The ground state of ^{59}Co is eightfold degenerate, whereas the ground states of ^{56}Fe and ^{50}Co are nondegenerate. The reaction results in a larger number of possible orientations of nuclear spin. Therefore, the entropy increases. $R \ln 8 = 4.13$ eu, so the entropy of a ^{59}Co nucleus is 4.13 eu.

Because the reaction is endothermic the equilibrium constant is small and we expect the abundance of ^{50}Cr and ^{59}Co to be considerably less than that of ^{56}Fe. The equilibrium constant calculated from the measured cosmic abundances

$$\frac{[^{50}\text{Cr}][^{59}\text{Co}]^2}{[^{56}\text{Fe}]^3} = 1.1 \times 10^{-7}$$

agrees with the calculated value to within 20%.

The equilibrium temperature may be obtained from the equilibrium constant calculated using the cosmic abundance of the isotopes and the equation $\Delta G = -RT \ln K$.

When comparing the stability of a number of different isotopes, it is convenient to use packing fractions in place of the atomic weight of isotopes. The packing fraction for an isotope is obtained by dividing the weight of the isotope by the mass number. For ^{56}Fe the packing fraction is

$$\frac{55.9349}{56} = 0.99884$$

Any isotope that is less stable than ^{56}Fe will have a larger packing fraction (0.99887 for ^{59}Co). Isotopes that are less stable than ^{12}C will have a packing fraction greater than 1.00000. For ^{4}He the packing fraction is

$$\frac{4.00260}{4} = 1.00065$$

The packing fraction curve (Fig. 19-2) is a convenient summary of the energy (or enthalpy) of the various nuclei. Any balanced nuclear reaction will be exothermic in the direction toward the nuclei with the lower packing fraction. Figure 19-3 is an enlargement of the region near the minimum of the packing fraction curve. The packing fraction of ^{56}Fe is smaller than the packing fraction of any other isotope, and ^{56}Fe is the most stable single nucleus. In the equilibrium (at around 4×10^9 K) that is apparently established temporarily in the interior

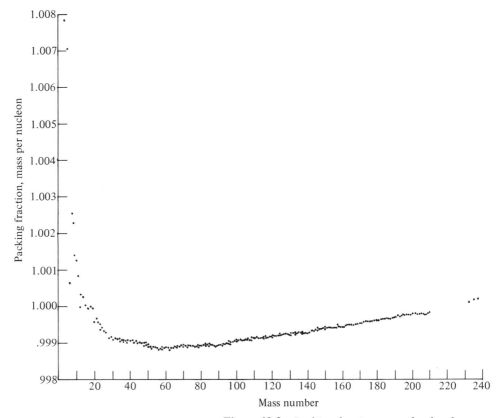

Figure 19-2 *Packing fraction curve for the elements.*

Ch. 19 The Structure of Nuclei

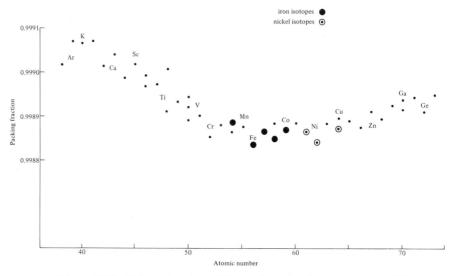

Figure 19-3 *Enlarged region of the packing-fraction curve around* ^{56}Fe.

of certain stars, ^{56}Fe will be the predominant nucleus. There will also be substantial amounts of other nuclei of the elements from chromium through zinc with packing fractions less than 0.99890.

The course of the further evolution of stars is less certain, but in some, if not all, cases there are catastrophic explosions. In such novae and supernovae explosions a substantial amount of the previously synthesized material is returned to interstellar space to be used in the formation of new stars and planetary systems.

There is excellent evidence that the elements beyond zinc are produced in low abundance by the successive capture of neutrons followed by appropriate radioactive decay reactions.

II. The Stability of Nuclei

A. ENERGY LEVELS IN NUCLEI

There are two levels of protons in ^{7}Li (requiring 2.3×10^8 kcal mole^{-1} and 3.7×10^8 kcal mole^{-1} for removal) just as there are two levels of electrons. This information may be obtained by bombarding ^{7}Li with protons of sufficient energy to remove a proton, just as electrons can be used to ionize atoms. In contrast to the electron subshells, the magnetic properties of the ^{7}Li nucleus indicate that the higher energy proton level is a *p* level rather than an *s* level. From this we see that the ordering of energy levels in a nucleus is not the same as the ordering of electron levels.

It is now known that the correct ordering of subshells for neutrons or protons is 1s, 2p, 3d, 2s, 4f, 3p, 5g, and so forth. This order, which is just the reverse of the ordering of electronic energy levels in that $2p < 2s$, $3d < 3p < 3s$, and so forth, was proposed on theoretical grounds in the 1930s. It can be rationalized as follows: The attractive forces between nuclear particles are extremely short ranged and there is no particular attraction to the center of the nucleus; therefore, levels with the most penetration have a particularly high, rather than particularly low, energy.

Figure 19-4 is an enlargement of the packing fraction curve for the

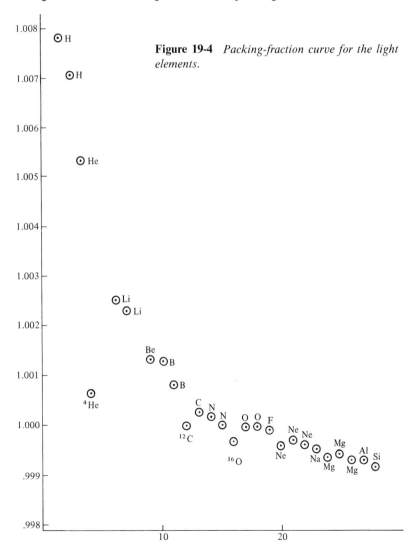

Figure 19-4 *Packing-fraction curve for the light elements.*

Ch. 19 The Structure of Nuclei

Table 19-5 *The Order of Filling of Neutron and Proton Subshells*

$1s$	$2p_{3/2}$	$2p_{1/2}$	$3d_{5/2}$	$2s$	$3d_{3/2}$	$4f_{7/2}$	$3p_{3/2}$	$4f_{5/2}$	$3p_{1/2}$	$5g_{9/2}$
2	4	2	6	2	4	8	4	6	2	10
2		6	8		20	28				50

$5g_{7/2}$	$4d_{5/2}$	$6h_{11/2}$	$4d_{3/2}$	$3s_{1/2}$	$6h_{9/2}$	$5f_{7/2}$	$5f_{5/2}$	$7i_{13/2}$	$4p_{3/2}$	$4p_{1/2}$	$6g_{9/2}$	$5d_{5/2}$	$7i_{11/2}$
8	6	12	4	2	10	8	6	14	4	2	10	6	12
				82						126			

light elements, and the stability of the nuclei, ^{4}He and ^{16}O is evident. (The packing fraction of these isotopes is below the more or less smooth curve of values for other elements.) In ^{4}He there is a filled $1s^2$ subshell of both protons and neutrons, and in ^{16}O there is a closed subshell of protons with a $1s^2\ 2p^6$ configuration.

At this point it is no longer desirable to continue the analogy between the stability of nuclei and the stable electron configurations. The most stable nuclei are those with 2, 6, 8, 20, 28, 50, 82, and 126 neutrons or protons. The proposed ordering of the subshells is shown in Table 19-5. In filling these subshells, neutrons and protons are considered separately. If there is a large energy difference between the lowest unoccupied proton level and the lowest unoccupied neutron level, a proton can change into a neutron or vice versa. The process by which this occurs is known as beta (β) decay.

B. BETA DECAY

Free neutrons are unstable with respect to a proton and an electron and the reaction $^1n \longrightarrow {}^1H + e^- + \nu$ *is exothermic and occurs spontaneously.*

The process of β decay involves the conversion of a neutron into a proton and an electron (e^-) or the conversion of a proton into a neutron and a positron (e^+). The process occurs when a proton is in a level higher in energy than an unoccupied neutron level. In this case the transformation of a proton into a neutron is exothermic. Thus the atomic weight of ^{12}N is 12.01871, and it spontaneously changes into ^{12}C according to the equation

$$^{12}\text{N} \longrightarrow {}^{12}\text{C} + e^+ + \nu$$

Some additional examples are shown in Table 19-2. Even if there is insufficient energy to produce a positron, it is possible for a proton to change into a neutron by the capture of an orbital electron.

$$^7\text{Be} + e^- \longrightarrow {}^7\text{Li} + \nu$$

The reverse reaction, in which a neutron is changed into a proton, commonly follows the capture of a neutron by a nucleus. This is represented by the sequence

$$^{62}\text{Ni} + {}^1n \longrightarrow {}^{63}\text{Ni} + \gamma$$

$$^{63}\text{Ni} \longrightarrow {}^{63}\text{Cu} + e^- + \nu$$

Some additional examples of the synthesis of heavier elements by neutron capture followed by β decay are shown in Table 19-6. In the examples given in Tables 19-2 and 19-6, an electron or positron must be included to balance the electrical charges, and it is necessary to include a neutrino to allow for the conservation of energy, momentum, and spin. The three processes found (emission of e^- and ν, emission of e^+ and ν, and electron capture with neutrino emission) are together classified as β-decay processes

Since the interconversion of neutrons and protons occurs so readily, it is useful to consider them as two states of a single particle, *the nucleon*. It is convenient to refer to the mass number of a nucleus as the number of nucleons present. The distinction between protons and neutrons can be made by assigning them values of $\frac{1}{2}$ and $-\frac{1}{2}$ for a quantum number very much like spin. This quantum number is called the isobaric spin. In this terminology the ^{4}He nucleus is considered to be made up as a $1s$ orbital filled with four nucleons. A single orbital will hold four nucleons since there are four combinations of isobaric spin and ordinary spin, $(\frac{1}{2}, \frac{1}{2})$, $(\frac{1}{2}, -\frac{1}{2})$, $(-\frac{1}{2}, \frac{1}{2})$, and $(-\frac{1}{2}, -\frac{1}{2})$.

For an odd number of nucleons there will be one division into neutrons and protons which is most stable. This is illustrated in Fig. 19-5. All nuclei of this mass number with more neutrons than the most stable number will be radioactive and emit negative β particles, e^-. (^{125}Sn and ^{125}Sb are the examples shown in Fig. 19-5). Nuclei with too few neutrons will decay by positron emission or by electron capture. (^{125}I, ^{125}Xe, ^{125}Cs and ^{125}Ba are shown in Fig. 19-5.)

Table 19-6 *Typical Processes for the Production of Heavier Elements by Neutron Capture*

A	$^{59}\text{Co} + {}^1n \longrightarrow {}^{60}\text{Co} + \gamma$
	$^{60}\text{Co} \longrightarrow {}^{60}\text{Ni} + e^- + \nu + 2\gamma$
B	$^{238}\text{U} + {}^1n \longrightarrow {}^{239}\text{U} + \gamma\text{'s}$
	$^{239}\text{U} \longrightarrow {}^{239}\text{Np} + e^- + \nu + \gamma\text{'s}$
	$^{239}\text{Np} \longrightarrow {}^{239}\text{Pu} + e^- + \nu + \gamma\text{'s}$

Ch. 19 The Structure of Nuclei

In general, the single stable nucleus will contain more neutrons than protons because the electrostatic repulsions between protons tend to raise the energy of all proton levels. The only exceptions to this are ^{1}H and ^{3}He. The repulsion between protons is also responsible for the lower stability of the elements with very high atomic numbers. Similarly proton-proton repulsion also accounts for the energy released in alpha (α) decay and in nuclear fission.

In the case of elements with an even number of nucleons (Fig. 19-6) the situation is more complicated, and there are often two or even three stable nuclei with the same mass number. As shown in the figure, decay occurs in this case to ^{128}Xe and ^{128}Te.

A factor that is important in explaining the stability of nuclei is the tendency for protons and neutrons within a subshell to have opposed spins. This is the exact opposite of what is found for electrons. The only significant force between electrons is electrostatic repulsion, and on the average when they have parallel spins they stay further apart. Neutrons and protons must also satisfy the Pauli exclusion principle, but in nuclei the state in which spins are opposed (they tend to come together) is favored energetically. Well over half of all known stable nuclei have an even number of both protons and neutrons, and all of the nucleons are paired up to give a total spin for the nucleus of zero. As shown in Fig. 19-6 for ^{128}I, an odd-odd nucleus may decay to either of two stable nuclei by both electron emission and electron capture.

Nuclei with odd numbers of both protons and neutrons are exceptionally unstable. Only four odd-odd nuclei are known which are not radioactive: ^{2}H, ^{6}Li, ^{10}B, and ^{14}N.

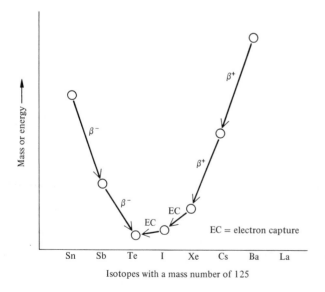

Figure 19-5 *Typical β decay possibilities for an odd mass number.*

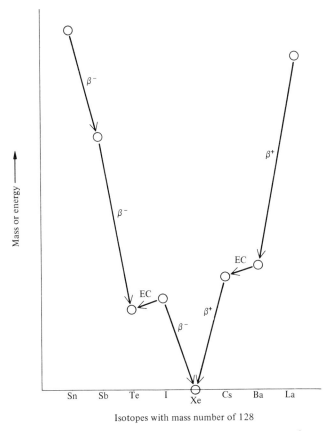

Figure 19-6 *Typical β decay possibilities for an even mass number.*

III. Rates of Radioactive Decay

Electron capture is the only radioactive decay process where laboratory conditions can be used to change measurably the rate of reaction. Ordinarily the decay of any particular radioactive atom is completely independent of the number of other atoms around it. If each ^{12}N atom in a sample of 10^{18} atoms has a 50% chance of decaying within 0.011 sec, the last ^{12}N atom left in the sample will also have a 50% chance of decaying in the next 0.011 sec. *As long as the sample is fairly large, very close to half of the sample will decay in 0.011 sec and only about one quarter of the sample will be left after 0.022 sec.* In this way the time required for half of a sample to decay, the half-life, is a very good measure of the rate at which any radioactive decay process occurs.

Figure 19-7 shows a graph of the decay as a function of time. The curve is a typical exponential function

$$N = N_0 e^{-\lambda t}$$

where N is the number of atoms remaining and N_0 is the initial number of atoms.

It is usually much easier to measure the disintegration rate than the number of radioactive atoms remaining. However, since the disintegration rate is proportional to the number of atoms remaining, it also decreases by a factor of two during each half-life. This is illustrated by the experimental data on the counting rate shown in Fig. 19-8 for a sample of ^{128}I. The proportionality constant that relates the disintegration rate to the number of atoms present is called the rate constant. For radioactive decay it is customary to use the symbol λ for the rate constant. Thus

$$\text{Rate} = \lambda N = \lambda N_0 e^{-\lambda t} = (\text{initial rate}) e^{-\lambda t}$$

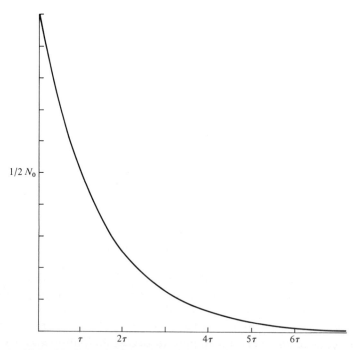

Figure 19-7 *Plot of the exponential decay of a typical radioactive isotope.*

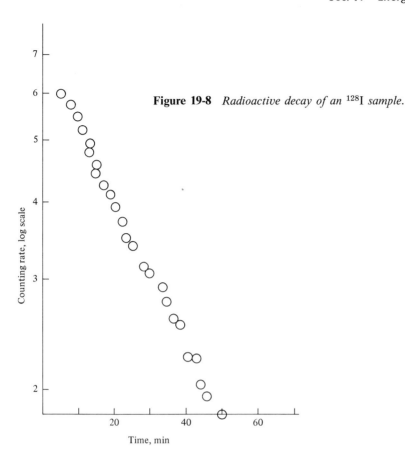

Figure 19-8 *Radioactive decay of an ^{128}I sample.*

The relationship between the rate constant and the half-life, τ, is simply

$$\lambda = \frac{0.693}{\tau}$$

or

$$\tau = \frac{0.693}{\lambda}$$

IV. Energy Requirements for Human Life

It is clear from the ΔH values for nuclear reactions, (measured in millions of electron volts (Mev) or containing exponential terms of 10^7 or 10^8 in kcal mole^{-1}) that enormous quantities of energy are involved. The fusion reactions in the sun maintain the surface of the sun at a high temperature so that tremendous quantities of energy are radiated. The earth and its atmosphere intercept only a small fraction of this radiation and only about two thirds of this energy reaches the earth's surface. Nevertheless, sunlight supplies about 8×10^{20} kcal year^{-1} to the earth's surface.

Almost a third of this energy goes into the evaporation of water. About one part in 10^4 of this energy is recovered as potential energy of water above sea level in rivers and lakes, and only 1% of this is actually converted into electrical power. In spite of these large losses, the total hydroelectric power production of 3×10^{15} kcal year^{-1} is a very important part of the economic picture in many nations.

The conversion of solar energy to other forms via natural processes, such as photosynthesis, is considerably more efficient than the conversion to hydroelectric power. Under optimum conditions an energy yield from dried algae corresponding to 1.2×10^{11} kcal year^{-1} km^2 can be obtained. Over the entire earth's surface this would amount to 6×10^{19} kcal year^{-1}. Over much of the earth's surface plant growth is limited by the shortage of water, and there is a critical shortage of the nutrients nitrogen and phosphorus in most of the oceans, but the total energy yield from photosynthesis (4×10^{16} kcal year^{-1} on land and 3×10^{17} kcal year^{-1} in the oceans) is substantial.

About one quarter of the plant material produced by photosynthesis on land, an energy equivalent of 1×10^{16} kcal year^{-1}, is currently being used for food and fuel. There has been and still is somewhere between 1×10^{19} and 5×10^{19} kcal available in easily recoverable fossil fuels—coal and oil.

The use of coal as a fuel in western civilization dates from about 1600. The year 1600 was a particularly important one for the industrial revolution and the subsequent history of mankind, as it marks the time when coal was first used on a substantial scale in the British Isles. The use of coal demanded, and at the same time made possible, a mechanized system of transportation. The first railroads dating from 1812 were primarily for the transportation of coal from the mines to ports from which it could be economically transported to the large coastal cities.

The importance of the change in method of transportation can be made clear by a consideration of the transport of food. Two oxen could move 1 ton of grain, corresponding to 2.5×10^6 kcal about 25 km day^{-1}, while consuming about 2.5×10^4 kcal in feed. If they had to go over 2000 km, there would be essentially nothing left to deliver. Thus, before 1700 each area was dependent on its own food supply, and with such unpredictable regional weather there were often disastrous famines in areas where the population levels were only about one third of the present population.

The process of industrialization has brought whole countries to a standard of living incomprehensible to even the nobility of 1600, but this has been at the expense of a phenomenal per capita energy consumption. For the population of the United States this amounts to roughly 10^8 kcal year^{-1} per person. This is over 100 times the per capita food requirement. Even if the world population were stabilized at 4 billion people, this standard of consumption would require 4×10^{17} kcal year^{-1} which is enough to exhaust the economical world reserves of fossil fuels in 20–100 years.

Fortunately nuclear energy is also available on earth. At the present time the only fully developed method for using this energy involves nuclear fission with thermal neutrons, a process which is applicable only to a few very heavy nuclei with odd mass numbers. These include ^{235}U, ^{233}U, and ^{239}Pu. An energy of about 1.2×10^8 kcal mole^{-1} is required to break the nuclei into fragments. The addition of a neutron to ^{238}U

$$^{238}\text{U} + {}^1n \longrightarrow {}^{239}\text{U} \qquad \Delta H = -1.11 \times 10^8 \text{ kcal mole}^{-1}$$

does not provide enough energy for fission unless the neutron has a considerable kinetic energy. However, if there is already an unpaired neutron present in the nucleus, the energy of pairing the two neutrons is available and fission occurs.

$$^{235}\text{U} + {}^1n \longrightarrow [{}^{236}\text{U} + 1.47 \times 10^8 \text{ kcal mole}^{-1}] \begin{array}{c} \nearrow^{15\%} {}^{236}\text{U} + \gamma \\ \searrow_{85\%} \text{fission fragments} \end{array}$$

There are an enormous number of ways that a heavy nucleus can break up into two fragments with about half the initial mass. No one reaction accounts for over 6% of the fission events, and appreciable amounts of the elements from Zn ($Z = 30$) to Tb ($Z = 65$) are found among the products. The yields of different products from the slow neutron-induced fission of ^{235}U are shown in Fig. 19-9. A typical example of a balanced fission process

$$^{235}_{92}\text{U} + {}^1_0n \longrightarrow {}^{99}_{40}\text{Zr} + {}^{134}_{52}\text{Te} + 3\,{}^1_0n + \text{energy}$$

must be regarded as only a sample of over 100 similar reactions, all of which occur with various probabilities, but it does illustrate several typical points. Two or three free neutrons are emitted which are capable of starting more fission events in a chain process that can be

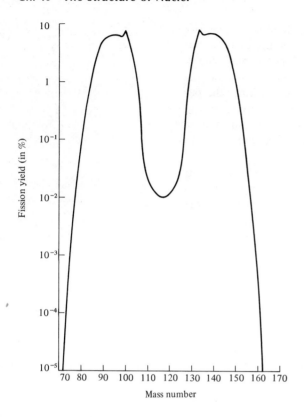

Figure 19-9 Yields of products from slow neutron-induced fission of ^{235}U. (After G. Friedlander, J. W. Kennedy, and J. M. Miller Nuclear and Radiochemistry, New York: John Wiley, 1966.)

controlled or allowed to go faster and faster. The isotopes produced have very high neutron-proton ratios and they are generally highly radioactive (β^- emission). In all cases the energy release averages a substantial 4.6×10^9 kcal mole^{-1} (200 Mev per fission event).

Although ^{235}U is a relatively rare isotope of uranium, it can be concentrated for use in nuclear reactors. Fortunately, from the standpoint of energy reserves, some of the extra neutrons can be used to convert the more common heavy isotopes such as ^{238}U into fissionable material. Thus ^{238}U is converted into ^{239}Pu (see Table 19-6) and ^{232}Th into ^{233}U

$$^{232}\text{Th} + {}^1n \longrightarrow {}^{233}\text{Th} \longrightarrow {}^{233}\text{U} + 2e^- + 2\nu$$

With breeder reactors designed to make use of these processes, it is possible to make use of most of the energy available in accessible deposits of uranium and thorium ores. On the basis of estimates carefully made by the U. S. Atomic Energy Commission, these deposits will provide an energy reserve of 4×10^{21} kcal. This should be sufficient even for a population of 10–20 billion people for several hundred years. There is also a high probability that before our reserves of

At the current United States per capita consumption, this energy supply would last a stabilized population of 4 billion for 2,800 years.

fissionable material are exhausted mankind will know how to use the energy from the fusion of deuterium in nondestructive ways. The following sequence of reactions for this purpose has already shown considerable promise in the laboratory:

$$^3H + {}^2H \longrightarrow {}^4He + {}^1n \qquad \Delta H = -40.6 \times 10^8 \text{ kcal mole}^{-1}$$

$$^1n + {}^6Li \longrightarrow {}^3H + {}^4He \qquad \Delta H = -10.9 \times 10^8 \text{ kcal mole}^{-1}$$

The energy equivalent of the deuterium present in the oceans amounts to an energy equivalent of over 3.5×10^{26} kcal.

It is hopeful that the development of the peaceful use of the fission process along with the development of the fusion process will end the division of the world into the "have" and "have not" nations.

QUESTIONS
1. What feature of the nuclear structure of tin accounts for the fact that it is more abundant than its neighbors?
2. Which elements are rarer than gold? (See Table 19-1)
3. What is a nucleon?
4. What is a neutrino?
5. What changes in the electron configuration of an atom are expected after the capture of a K electron by the nucleus?
6. How is ^{233}U made?
7. Why can ^{233}U be used in a nuclear reactor whereas ^{232}Th cannot?
8. Why are most fission products β^- emitters?
9. Why is iron an abundant element?
10. The heaviest element with a cosmic abundance greater than 0.8 is zirconium. Explain this in terms of the closed neutron shell at 50 neutrons.
11. Locate the stable isotopes with 82 neutrons in a table of isotopes (*Handbook of Chemistry and Physics,* (Cleveland, Ohio; Chemical Rubber Co., 1964) p. B6). How does this compare to cosmic abundances?

PROBLEMS I
1. Write balanced nuclear equations for the following decay processes:
    (a) α particle emission by ^{234}U
    (b) the formation of ^{234}Th from ^{238}U
    (c) the decay of ^{16}F
    (d) the decay of ^{16}N
    (e) the electron capture decay of ^{59}Ni
2. Write balanced nuclear reactions for
    (a) the production of ^{12}C by α particle bombardment of ^9Be
    (b) neutron capture by ^{127}I
    (c) the formation of plutonium
    (d) several steps in the carbon cycle in the sun

3. Calculate the energy released in the β decay of ^{16}F atomic weight 16.01171. The atomic weight of ^{16}O is 15.99491.
4. Calculate the energy released in the process $3\ ^4\text{He} \longrightarrow\ ^{12}\text{C}$.
5. Nuclear energies are often expressed in million electron volts (Mev). If 1 kcal mole^{-1} = 23.05 ev, calculate the energy in Problem 3 in Mev.
6. The calorie is defined as 4.184 joules. Express the energy in Problem 3 in joules mole^{-1}.
7. Given the data below for chromium isotopes, calculate ΔH^0 and ΔS^0 for the nuclear reaction $3\ ^{52}\text{Cr} = ^{50}\text{Cr} + 2\ ^{53}\text{Cr}$.

Isotope	Abundance, %	Atomic weight	Nuclear spin	Nuclear entropy, eu
^{50}Cr	4.31	49.9461	0	0
^{52}Cr	83.76	51.9405	0	0
^{53}Cr	9.55	52.9407	$\frac{3}{2}$	2.7
^{54}Cr	2.38	53.9389	0	0

8. At what temperature will the equilibrium constant for the reaction in Problem 7 match the observed isotopic abundances?
9. ^{53}Mn decays by electron capture with a half-life of 10^6 years. How will this decay have affected the ratio of chromium isotopes since the time of equilibrium nucleosynthesis of these isotopes?
10. The atomic weight of ^{60}Ni (26.23% of all Ni atoms) is 59.9332. Calculate the value of the equilibrium constant for the reaction $2\ ^{56}\text{Fe} \longrightarrow\ ^{60}\text{Ni} + ^{52}\text{Cr}$ at 4.2×10^9 K. $\Delta S^0 = 0.0$ eu. How does this compare to the observed cosmic abundances of these isotopes?

PROBLEMS II

1. Write a balanced nuclear equation for the α decay of ^{221}Fr.
2. The only stable nucleus of mass number of 26 is ^{26}Mg. Why is ^{26}Na radioactive?
3. Write a balanced nuclear equation for the decay of ^{26}Na.
4. What kind of radioactivity is shown by ^{26}Al?
5. Show how ^{26}Al can be prepared by proton capture by ^{25}Mg.
6. The atomic weight of ^{26}Al is 25.9869. Calculate the packing fraction.
7. The atomic weight of ^{26}Mg is 25.9826. Calculate the energy released in the decay of ^{26}Al.
8. The half-life of ^{26}Al is 7.4×10^6 years. Calculate the decay constant λ.
9. In a sample of 6×10^{23} atoms of ^{26}Al, how many will decay in 1 sec?
10. How much heat is produced by the decay of this ^{26}Al sample every second?
11. How much of the ^{26}Al sample in Problem 9 will be left after 1.48×10^7 years?
12. What will be the decay rate of the ^{26}Al from Problem 9 in disintegrations per second in the sample left after 1.48×10^7 years?
13. How much of the ^{26}Al from Problem 9 will be left after 1×10^9 years?

There are approximately $\pi \times 10^7$ seconds in 1 year.

SUGGESTED READINGS

COTTRELL, W. F., *Energy and Society.* New York: McGraw Hill, 1955.
HARVEY, B. G., *Nuclear Chemistry.* Englewood Cliffs, N.J.: Prentice Hall, 1965.
PAULING, L., *College Chemistry.* San Francisco: Freeman, 1964, Chapter 29.
THIRRING, H., *Energy for Man.* Bloomington, Ind.: Indiana University Press, 1958.

Into the same rivers we step and we do not step.
Everything flows and nothing abides; everything gives way and nothing stays fixed.
You cannot step twice into the same river, for other waters are continually flowing on.
Cool things become warm, the warm grows cool; the moist dries, the parched becomes moist.
It is in changing that things find repose.
People do not understand how that which is at variance with itself agrees with itself. There is a harmony in the bending back, as in the case of the bow and the lyre.*

HERACLITUS

Chapter 20 Chemical Reactions

I. Rates of Reaction and the Rate Law

Whenever the rate of a chemical reaction is studied it is found to be dependent on such things as the nature of the reactants, the concentration of the reactants, the temperature, and so forth. The influence of the reaction conditions on the rate, as determined by experiment, provides the primary experimental evidence upon which a study of the reaction mechanism is based. The reaction mechanism gives the details of a reaction in terms of a reaction path.

To establish the complete dependence of the rate upon concentration the effect of varying the concentration of each reacting species must be determined. When this has been done a rate law may be written. A rate law expresses the dependence of the rate on the concentration of each substance involved. For the reaction

$$a\text{A} + b\text{B} \longrightarrow c\text{C} + d\text{D}$$

the rate law might have the form

$$\text{Rate} = k[\text{A}]^n[\text{B}]^m \qquad (20.1)$$

---

*Used by permission from the translation of P. Wheelwright, *Heraclitus*, (Princeton, N.J.: Princeton University Press, 1959).

In this expression the concentration of the reacting species are raised to a power. The power is called the order of the reaction. For a reaction that is first-order with respect to substance A, $n = 1$. A second-order reaction with respect to A is one for which $n = 2$. The same possibilities also exist for B. The overall order of a reaction is given by the sum of the orders with respect to the individual reactants, that is, $n + m$.

Mathematical manipulations with first-order rate laws are particularly simple, and we have already considered them in connection with radioactive decay. When a rate law has the form

$$\text{Rate} = k[A] \tag{20.2}$$

a plot of log [A] versus t will be a straight line as shown in Fig. 20-1. Slightly more complex mathematical procedures are available for other orders of reaction.

The constant k in the rate law is known as the rate constant. The value of the rate constant for a reaction can be obtained from a measurement of the rate under a particular set of conditions once the form of the rate law is known, or from an appropriate plot of some function of the concentration versus time. For a first-order reaction k is the slope of a $\ln_e$ [A] versus time plot, or it may be obtained from the half-life

$$k = 0.693/\tau_{1/2} \tag{20.3}$$

The rate laws observed for a number of reactions are given in Table 20-1.

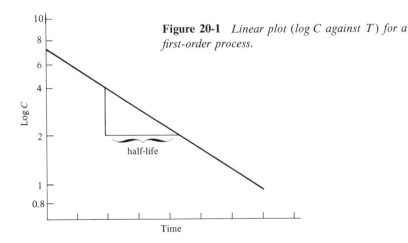

Figure 20-1 *Linear plot (log C against T) for a first-order process.*

Table 20-1 *Rate Expressions for Some Typical Reactions*

Reaction	Rate law
$2BH_3CO \longrightarrow B_2H_6 + 2CO$	Rate = $k[BH_3CO]^2[CO]^{-1}$
$2N_2O_5 \longrightarrow 4NO_2 + O_2$	Rate = $k[N_2O_5]$
$2NO + O_2 \longrightarrow 2NO_2$	Rate = $k[NO]^2[O_2]$
$H + Br_2 \longrightarrow HBr + Br$	Rate = $k[H][Br_2]$
$NH_4^+ + CNO^- \longrightarrow (NH_2)_2CO$	Rate = $k[NH_4^+][CNO^-]$
$C_2H_6 \longrightarrow C_2H_4 + H_2$	Rate = $k[C_2H_6]$
$2H_3O^+ + 3I^- + H_2O_2 \longrightarrow 4H_2O + I_3^-$	Rate = $k[H_3O^+][I^-][H_2O_2]$
$CHCl_3 + Cl_2 \longrightarrow CCl_4 + HCl$	Rate = $k[CHCl_3][Cl_2]^{1/2}$

II. Reaction Mechanisms and the Rate Law

A reaction mechanism is a detailed description of the path of the reaction. Some simple reactions take place in a single reaction but most involve a sequence of reactions. Even the simple reaction of hydrogen and iodine to form hydrogen iodide does not take place in a single step, and neither does the combination of hydrogen and bromine.

The reaction

$$2BH_3CO \longrightarrow B_2H_6 + 2CO$$

probably involves at least two steps. The mechanism of this reaction most probably involves the production of the intermediate species BH_3

$$BH_3CO \rightleftharpoons BH_3 + CO$$

followed by

$$BH_3CO + BH_3 \longrightarrow B_2H_6 + CO$$

Oxidation-reduction reactions such as

$$Tl(I) + 2Co(III) \longrightarrow Tl(III) + 2Co(II)$$

must also occur in a sequence such as

$$Tl(I) + Co(III) \rightleftharpoons Tl(II) + Co(II)$$
$$Tl(II) + Co(III) \longrightarrow Tl(III) + Co(II)$$

The rate law is fundamental to the understanding of mechanisms such as those illustrated because at least a portion of the mechanism is represented by the rate law. In order for a reaction to occur, collision between the reactants is necessary. The assembly of the reactants in the rate determining step with the required energy forms the activated complex of the reaction. The activated complex then goes on to dissociate into the products of the reaction.

The energy required for the formation of the activated complex may come from the kinetic energy of the collision.

The species that appear in the rate law are those that are involved in the rate determining step and in the formula of the activated complex. This principle will be illustrated with a few specific cases.

The rate law for the decomposition of nitrogen pentoxide

$$2N_2O_5 \longrightarrow 4NO_2 + O_2$$

is

$$\text{Rate} = k[N_2O_5]$$

The formula of the activated complex is

$$(N_2O_5)^\ddagger$$

and the rate determining step of the decomposition passes through this activated complex.

The rate law for the reaction

$$2BH_3CO \longrightarrow B_2H_6 + 2CO$$

is

$$\text{Rate} = k[BH_3CO]^2[CO]^{-1}$$

The appearance of the negative power in the rate law implies that a fragment of the molecule is split out before the rate determining step. The formula for the activated complex is $(B_2H_6CO)^\ddagger$.

The mechanism proposed for this reaction is accordingly

$$BH_3CO \rightleftharpoons BH_3 + CO$$

$$BH_3CO + BH_3 \longrightarrow (B_2H_6CO)^\ddagger \longrightarrow B_2H_6 + CO$$
$$\text{rate determining step}$$

The step labelled as rate determining proceeds through the activated complex. When the rate of a reaction is measured at different tempera-

477

The significance of the Arrhenius equation is described in the next section.

tures, it is found that the temperature dependence of the rate constant is given by the Arrhenius equation

$$k = Ae^{-E_a/RT} \qquad (20.4)$$

where A is a constant and the quantity E_a is called the activation energy. As the temperature rises, the rate of reaction will increase. Values of the constant A and the activation energy E_a can be assigned for each reaction. A plot of the logarithm of the rate versus $1/T$ will be a straight line. The slope of this line will be $E_a/2.303R$; and in this way an overall activation energy can be determined for any reaction or set of reactions.

The significance of the Arrhenius expression can be understood from a detailed consideration of the various equilibria between products, reactants, and activated complexes.

III. The Theory of Activated Complexes

A. THE ACTIVATED COMPLEX

An activated complex is a molecule in the process of reacting. The complex often contains the bonds that are being formed and the bonds that will be broken; however, the total bonding in the complex is less than that in either the products or the reactants. In the plot of the energy of a reaction as a function of the progress of the reaction, as shown in Fig. 20-2, the activated complex is at the energy maximum.

In the reaction of a hydrogen atom with a hydrogen molecule

$$H + H_2 \longrightarrow H_2 + H$$

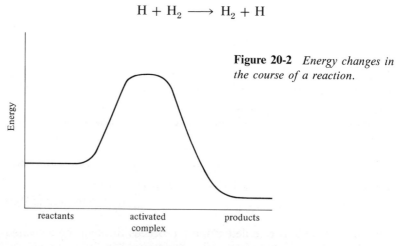

Figure 20-2 *Energy changes in the course of a reaction.*

Sec. III The Theory of Activated Complexes

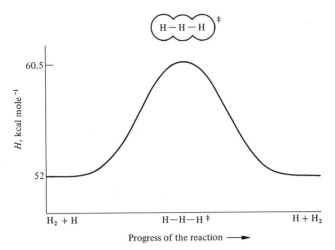

Figure 20-3 *Energy changes in the reaction of H_2 with H atoms.*

the activated complex (Fig. 20-3) is nearly linear. In the complex both H-H distances are about 15% longer than the distance in the H_2 molecule (0.85 Å compared to 0.74 Å).

Another example of an activated complex is shown in Fig. 20-4. In this reaction

$$I^- + C_4H_9Cl \longrightarrow C_4H_9I + Cl^-$$

the iodide ion displaces the chloride ion through an activated complex in which the carbon participates in five bonds.

The unusual bonding present in activated complexes is illustrated by both of these examples. Hydrogen does not form two covalent bonds, and carbon does not form five covalent bonds. The activated complexes in Figs. 20-3 and 20-4 exist only momentarily in the process of the reaction. In both cases the total bonding energy is less than that in either the products or the reactants.

The fact that the activated complex of any reaction has the maximum energy of any of the species formed during the reaction has some important consequences. An activated complex can move along the

Figure 20-4 *An activated complex with five-coordinate carbon.*

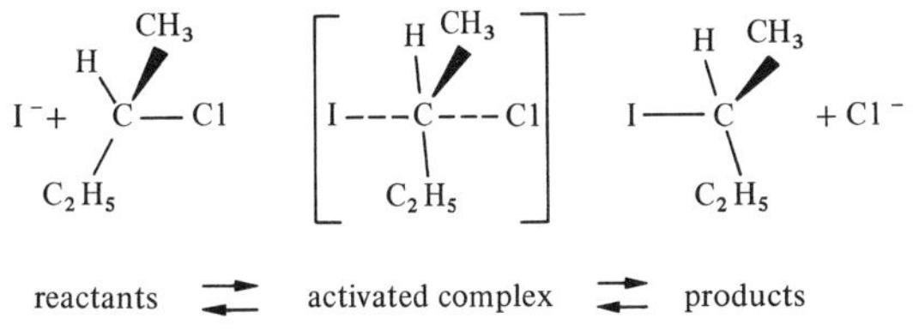

progress of the reaction coordinate in one direction or the other. Any activated complex that is formed will be moving either in the direction of products or back toward the reactants. Every activated complex contributes to the rate of the reaction in one direction or the other.

The energy that the activated complex has over and above the energy of the reactants is the activation energy (E_a). The activation energy is a quantity of extreme significance to the interpretation of rates. No reaction can take place unless this amount of energy is supplied to the reactants. If the activation energy of a reaction is high, there will be very little of the activated complex present at low temperatures, and the reaction will be very slow regardless of the enthalpy term for the complete reaction. An extreme example of this behavior is provided by nuclear fusion reactions such as

$$^3He + {}^3He \longrightarrow {}^4He + 2\,{}^1H$$

for which $\Delta H^0 = -2.96 \times 10^8$ kcal mole^{-1} and yet the reaction is negligibly slow at low temperatures. In this case the activated complex is two very close ^{3}He nuclei. For the two ^{3}He nuclei to react they must approach within about 10^{-12} cm. Because of the electrostatic repulsion of the two nuclei, this process requires an activation energy of about 1.3×10^7 kcal mole^{-1}. Although this is small compared to the exothermicity of the nuclear reaction, the nuclei almost never get close enough to react at normal temperatures. Temperatures in excess of 10^5 K are required.

B. ENTROPY AND ENTHALPY OF ACTIVATION

It is intuitively obvious that the rate of a reaction will be proportional to the number of molecules that are in the process of reacting at any particular time. The rate can be expressed in terms of a constant (k^*) and a concentration ($C^\ddagger$) of activated complex

$$\text{Rate} = k^* C^\ddagger \tag{20.5}$$

if the concentration of activated complex is known.

There is one set of conditions for which the calculation of the concentration of activated complexes present ($C^\ddagger$) is always simple. When equilibrium has been established between the reactants and products, there is no net reaction. Molecules of the activated complex are being formed from the reactants and products and they are in equilibrium with both reactants and products. Equilibrium considerations can be applied, and the amount of activated complex present can be calculated.

At the beginning of a reaction when the concentration of products is small, the activated complex cannot be in equilibrium with the products. In general, that portion of the activated complex that forms from products and decomposes to reactants is missing in the early stages of the reaction. Except in unusual cases this has no influence on the amount of the activated complex that is formed from reactants and decomposes to products. The amount of activated complex present that decomposes to form products is usually the amount in equilibrium with the reactants (that is, formed from the reactants), and the following derivation (of equations 20.6 through 20.12) is based upon this assumption.

When the activated complex is in equilibrium with the reactants, the equilibrium can be represented by the chemical equation

$$a\text{A} + b\text{B} + c\text{C} = \text{C}^\ddagger \tag{20.6}$$

The equilibrium constant for this reaction is

$$K^\ddagger = \frac{\text{C}^\ddagger}{[\text{A}]^a[\text{B}]^b[\text{C}]^c} \tag{20.7}$$

Rearranging we obtain

$$\text{C}^\ddagger = K^\ddagger [\text{A}]^a[\text{B}]^b[\text{C}]^c \tag{20.8}$$

This expression may be substituted for $\text{C}^\ddagger$ in equation 20.5

$$\text{Rate} = k^* \text{C}^\ddagger$$

to give

$$\text{Rate} = k^* K^\ddagger [\text{A}]^a[\text{B}]^b[\text{C}]^c \tag{20.9}$$

In this rate law the product $k^* K^\ddagger$ is the rate constant. The equilibrium constant $K^\ddagger$ can be rewritten in terms of the free energy or the enthalpy and entropy.

$$\text{Rate} = k^* 10^{-\Delta G^\ddagger/2.303RT} [\text{A}]^a[\text{B}]^b[\text{C}]^c \tag{20.10}$$

$$\text{Rate} = k^* 10^{-\Delta H^\ddagger/2.303RT} 10^{\Delta S^\ddagger/2.303R} [\text{A}]^a[\text{B}]^b[\text{C}]^c \tag{20.11}$$

Thus we see that the rate constant k in

$$\text{Rate} = k[\text{A}]^a[\text{B}]^b[\text{C}]^c$$

may be written as

$$k = k^* 10^{-\Delta H^{\ddagger}/2.303RT} 10^{\Delta S^{\ddagger}/2.303R} \qquad (20.12)$$

To show the relationship between equation 20.12 and the Arrhenius equation (20.4), equation 20.12 may be rewritten as

$$k = k^* 10^{\Delta S^{\ddagger}/2.303R} 10^{-\Delta H^{\ddagger}/2.303RT} \qquad (20.13)$$

This is equivalent to the Arrhenius expression if

$$A = k^* 10^{\Delta S^{\ddagger}/2.303R} \qquad (20.14)$$

and

$$\Delta H^{\ddagger} \cong E_a$$

The rate constant may thus be calculated from a knowledge of A and E_a or k^*, $\Delta S^{\ddagger}$, and $\Delta H^{\ddagger}$. The value of k^* has not been specified above, but the quantum mechanical treatment of the motion of decomposition of the activated complex gives a simple equation for k^*. It depends on the gas constant R, Avogadro's number N_0, Planck's constant h, and the absolute temperature T

$$k^* = \left(\frac{RT}{N_0 h}\right) = (2.08 \times 10^{10})T \qquad (20.15)$$

Using this value for k^* equations 20.13 and 20.14 become

$$k = (2.08 \times 10^{10} T) 10^{\Delta S^{\ddagger}/2.303R} 10^{-\Delta H^{\ddagger}/2.303RT} \qquad (20.16)$$

and

$$A = (2.08 \times 10^{10} T) 10^{\Delta S^{\ddagger}/2.303R} \qquad (20.17)$$

The entropy of activation of a reaction can be evaluated from the constant A, or it can be obtained directly from the value of the rate constant and the activation energy. The second method may be illustrated by rearranging equation 20.16 to give

$$\frac{k}{2.08 \times 10^{10} T} = e^{\Delta S^{\ddagger}/R} e^{-\Delta H^{\ddagger}/RT} \qquad (20.18)$$

$$4.5757 \log \frac{k}{2.08 \times 10^{10} T} = \Delta S^{\ddagger} - \frac{\Delta H^{\ddagger}}{T} \qquad (20.19)$$

$$\Delta S^{\ddagger} = \frac{\Delta H^{\ddagger}}{T} + 4.5757 \log \frac{k}{2.08 \times 10^{10} T} \qquad (20.20)$$

Sec. III The Theory of Activated Complexes

For example, the reaction $H_2 + I_2 \longrightarrow 2HI$ follows the rate law

$$\text{Rate} = k[H_2][I_2]$$

with $k = 3.53 \times 10^{-4}$ at 575 K and an activation energy (which is approximately $\Delta H^\ddagger$) of 39,900 cal mole^{-1}. The value of $k/(2.08 \times 10^{10} T)$ is 2.95×10^{-17}, and we have from equation 20.20

$$\Delta S^\ddagger = \Delta H^\ddagger/T + 4.5757 \log(2.95 \times 10^{-17})$$

$$\Delta S^\ddagger = 69.3 \text{ eu} + 4.5757(-16.53)$$

$$\Delta S^\ddagger = 69.3 - 75.7 = -6.4 \text{ eu}$$

The values of the activation energy, and values of A and $\Delta S^\ddagger$ for some reactions are given in Table 20-2.

As an example of the utility of the preceding discussion we shall consider the structure of the activated complex formed between H_2 and I_2. There are several possible structures for this activated complex. They are shown in Fig. 20-5.

Structure (c) can be ruled out first because the enthalpy of formation of a mole of hydrogen atoms (52 kcal mole^{-1})

$$\tfrac{1}{2}H_2 \longrightarrow H \qquad \Delta H = 52 \text{ kcal mole}^{-1}$$

Table 20-2 *Activation Energies and Related Quantities for Typical Reactions*

Reaction	A	$\Delta H^\ddagger$, kcal mole^{-1}	$\Delta S^\ddagger$, eu
$H_2 + I_2 \longrightarrow 2HI$	5×10^{11} M^{-1} sec^{-1}	39.9	-6.4
$2C_2H_4 \longrightarrow C_4H_8$	7×10^7 M^{-1} sec^{-1}	37.7	-23
$NH_4^+ + CNO^- \longrightarrow (NH_2)_2CO$	4×10^{12} M^{-1} sec^{-1}	23.2	$+1$
$N_2O_5 \longrightarrow 2NO_2 + \tfrac{1}{2}O_2$	5×10^{13} sec^{-1}	24.7	$+4$
$2NO_2 \longrightarrow N_2O_4$	5×10^{11} M^{-1} sec^{-1}	0	-5
$H_2P_2O_7^{2-} + H_2O \longrightarrow 2H_2PO_4^-$	3×10^{13} sec^{-1}	29	$+3$
$5Br^- + BrO_3^- + 6H^+ \longrightarrow 3Br_2 + 3H_2O$	3.5×10^{12} M^{-3} sec^{-1}*	15.8	-1
$CH_3COCH_3 + Br_2 \xrightarrow{H^+} CH_3COCH_2Br + HBr$	2×10^{13} sec^{-1}†	19.9	$+2$

*Rate $= k[Br^-][BrO_3^-][H^+]^2$
†Rate $= k[H^+]$, measured in 91% acetone, 9% water

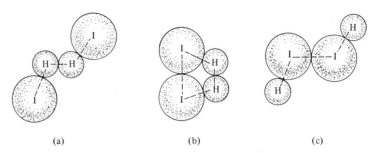

Figure 20-5 *Three conceivable activated complexes for the $H_2 + I_2$ reaction.*

is greater than the activation energy (40 kcal mole^{-1}, see Table 20-2).

The entropy change in a reaction in which two gaseous molecules combine should be substantial and it should be negative. For the cyclic activated complex (Fig. 20-5b) an even more negative value ($\Delta S^{\ddagger} \approx -15$ eu) would be expected. To give agreement with the measured value, $\Delta S^{\ddagger} = -6.4$ eu, the activated complex should have the structure that shows the least amount of order (Fig. 20-5a). That the cyclic structure in Fig. 20-5b is definitely incorrect, was proven by a series of experiments performed in 1967.

To form the activated complex shown in Fig. 20-5a it is necessary to break the I_2 bond completely before the configuration of the activated complex is reached. The mechanism should be written as

$$I_2 = 2I$$

$$H_2 + I + I \longrightarrow (H_2I_2)^{\ddagger} \longrightarrow 2HI \quad \text{rate determining step}$$

In this particular example the stoichiometry of formation of the activated complex

$$H_2 + I_2 \rightleftharpoons H_2I_2^{\ddagger}$$

is identical to that of the overall reaction

$$H_2 + I_2 = 2HI$$

so that the powers in the rate law

$$\text{Rate} = k[I_2][H_2]$$

happen to be the same as the coefficients in the balanced chemical equation. This is just a coincidence. In any reaction the powers in the rate law are determined from the equilibrium expression for the activated complex (equation 20.7)

$$K^{\ddagger} = \frac{C^{\ddagger}}{[A]^a[B]^b[C]^c}$$

IV. Complex Reaction Mechanisms

A. CATALYSIS

According to the rate expression

$$\text{Rate} = k[CH_3CO_2C_2H_5][H_2O]$$

the hydrolysis of ethyl acetate

$$CH_3CO_2C_2H_5 + H_2O \rightleftharpoons CH_3CO_2H + CH_3CH_2-OH$$

proceeds through an uncharged activated complex

$$(CH_3CO_2C_2H_5H_2O)^{\ddagger}$$

If the reaction is carried out in basic solution, the rate expression is found to be more complex. In this case the rate law is

$$\text{Rate} = k[CH_3CO_2C_2H_5][H_2O] + k'[CH_3CO_2C_2H_5][H_2O][H^+]^{-1}$$

The second term is most simply explained in terms of a second activated complex, pictured in Fig. 20-6, in which the ethyl acetate molecule is attacked by a hydroxide ion. In this case the hydroxide ion acts as a catalyst by introducing a new path for the reaction.

The oxidation of carbon monoxide

$$2CO + O_2 \longrightarrow 2CO_2$$

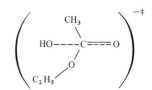

Figure 20-6 *Activated complex for attack of* OH$^-$ *on ethyl acetate.*

is catalyzed by nitrogen dioxide (NO$_2$). The addition of nitrogen dioxide provides a new and faster path for the reaction. The rate of the catalyzed reaction is given by the expression

$$\text{Rate} = k[CO][NO_2]$$

The activated complex is

$$(CNO_3)^{\ddagger}$$

and the mechanism is simply

$$CO + NO_2 \longrightarrow (CNO_3)^{\ddagger} \longrightarrow CO_2 + NO$$

$$NO + \tfrac{1}{2}O_2 \rightleftharpoons NO_2$$

The separate steps of the faster reactions following the rate determining step are not shown in detail. The rate law only provides information on the rate determining step. Additional experiments are needed to determine the mechanism for the oxidation of nitric oxide (NO). The net reaction

$$CO + NO_2 + \tfrac{1}{2}O_2 \longrightarrow CO_2 + NO_2$$

shows that there is no net change in the amount of nitrogen dioxide present. A substance such as nitrogen dioxide, which increases the rate of a reaction without being consumed or altered in the process, is called a catalyst. The function of a catalyst is to introduce a new pathway for the reaction, and in this sense it is correct to consider hydroxide ion a catalyst for the hydrolysis of ethyl acetate even though some of it may react with the acetic acid produced in the reaction.

Thermodynamic arguments were used many years ago to demonstrate that a catalyst cannot change the conditions for equilibrium. A catalyst merely changes the rate of a reaction. If a catalyst speeds up the forward reaction by providing an alternate pathway, it must provide the same pathway for the reverse reaction, and the rate of the reverse reaction must increase proportionately. Therefore, nitrogen dioxide must increase the rate of dissociation of carbon dioxide even though the equilibrium favors the production of carbon dioxide so strongly that it is very difficult to observe any dissociation of carbon dioxide. Consequently the reverse reaction cannot be observed experimentally. With carbon dioxide as the reactant, the mechanism must be reversed,

$$K = \frac{[CO_2]^2}{[CO]^2[O_2]} = 1.5 \times 10^{91}$$

$$moles^{-1}\ liter.$$

$$NO_2 \rightleftharpoons NO + \tfrac{1}{2}O_2$$

$$NO + CO_2 \longrightarrow CO + NO_2 \quad \text{rate determining step}$$

and the rate expression must have the form

$$\text{Rate} = k'[CO_2][NO_2][O_2]^{-1/2}$$

The balanced equation representing the formation of the activated complex from carbon dioxide is

$$CO_2 + NO \longrightarrow (CNO_3)^{\ddagger} \longrightarrow CO + NO_2$$

The rate expressions for the forward and reverse reactions arise from considering $(CNO_3)^\ddagger$ in equilibrium with the two sets of reactants. The equilibrium expression for the reaction can be obtained by equating the two rate expressions

$$k[CO][NO_2] = k'[CO_2][NO_2][O_2]^{-1/2}$$

$$k^2[CO]^2 = k'^2[CO_2]^2[O_2]^{-1}$$

$$\frac{k^2}{k'^2} = K = \frac{[CO_2]^2}{[O_2][CO]^2}$$

B. SUCCESSIVE ACTIVATED COMPLEXES

In some reactions the form of the rate law leads to the possibility that either of two different activated complexes may be involved in the rate determining step. For example, the experimentally determined rate expression for the reaction

$$H_2 + Br_2 = 2HBr$$

has the form

$$\text{Rate} = \frac{k[H_2][Br_2]^{3/2}}{[Br_2] + k'[HBr]}$$

If the concentration of HBr is negligible the rate expression reduces to

$$\text{Rate} = k[H_2][Br_2]^{1/2}$$

In this case the activated complex is $(H_2Br)^\ddagger$ and the rate determining step is

$$Br + H_2 \longrightarrow HBr + H$$

Conversely when the concentration of HBr is large and the concentration of Br_2 is small the rate law reduces to

$$\text{Rate} = (k/k')[H_2][Br_2]^{3/2}[HBr]^{-1}$$

Under these conditions the activated complex will have the stoichiometry $(HBr_2)^\ddagger$. The full mechanism for the reaction is

$$Br_2 \rightleftharpoons 2Br$$

$Br + H_2 \longrightarrow HBr + H$ sometimes rate determining

$H + Br_2 \longrightarrow HBr + Br$ sometimes rate determining

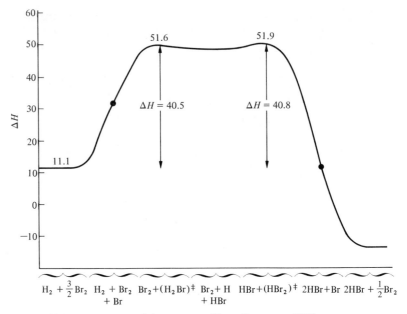

Figure 20-7 *Energetics of the reaction* $H_2 + Br_2 \longrightarrow 2HBr$.

The reaction of hydrogen with bromine is called a chain reaction because once a bromine atom is formed, the two steps

$$Br + H_2 \longrightarrow HBr + H$$
$$H + Br_2 \longrightarrow HBr + Br$$

can occur alternately hundreds of times.

The energetics of this typical chain reaction are summarized in Fig. 20-7. The maxima shown in this figure represent the activated complexes $(H_2Br)^{\ddagger}$ and $(HBr_2)^{\ddagger}$. The corresponding activation energies are 40.5 and 40.8 kcal mole^{-1}. The species present as the reaction progresses are labelled along the bottom of the figure.

The first step in the reaction involves the decomposition of Br_2. The alternative first step of the reaction ($H_2 \longrightarrow 2H$) is not energetically favored because the enthalpy of formation of hydrogen atoms (52.1 kcal mole^{-1}) is larger than the measured activation energy (40.6 kcal mole^{-1}). Similarly in the reverse reaction (between HBr and Br_2) the decomposition of HBr is not the first step because the bond energy of HBr is 87 kcal mole^{-1}.

The energetics of the $H_2 + I_2$ reaction are shown in Fig. 20-8. In this case there is only one activated complex and, as a result, the reaction is not a chain reaction. The maximum corresponds to an activation energy of 40 kcal mole^{-1}.

In this system the reaction that would lead to a chain reaction

$$I + H_2 \longrightarrow HI + H$$

is a possible reaction. However, a chain reaction does not occur at an appreciable rate because the formation of HI does not liberate a sufficient amount of energy (HI bond energy is 71 kcal mole^{-1}) to break the H-H bond (bond energy 104 kcal mole^{-1}).

Most high-temperature gas-phase reactions are chain reactions. The formation of CH_3Cl from CH_4 proceeds by way of a chain reaction. It involves the methyl radical (CH_3). The mechanism of the reaction is

$$Cl_2 \longrightarrow 2Cl$$

$Cl + CH_4 \longrightarrow CH_3 + HCl$ sometimes rate determining

$CH_3 + Cl_2 \longrightarrow CH_3Cl + Cl$ sometimes rate determining

The reaction of H_2 with Cl_2 also proceeds through a similar mechanism. Although the reaction is negligibly slow at 300 K, it is sufficiently exothermic ($\Delta H = -44$ kcal mole^{-1}) to raise the temperature of the reaction vessel substantially. Therefore, if the reaction is started by dissociating Cl_2 molecules with intense ultraviolet light or by heating part of the reaction vessel to around 500 K, the heat produced will cause the reaction to proceed faster and faster, and an explosion results. In the explosion of a mixture of H_2 and Cl_2 the temperature rises to about 2000 K and the reaction will proceed essentially to completion in less than 10^{-2} seconds. The explosion of a mixture of H_2 and O_2

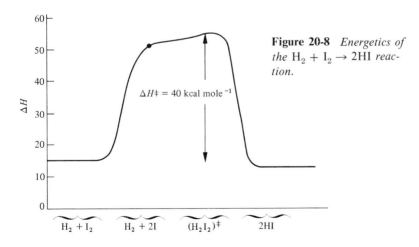

Figure 20-8 *Energetics of the $H_2 + I_2 \rightarrow 2HI$ reaction.*

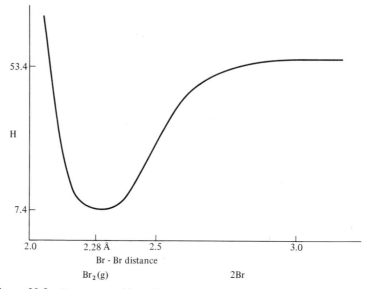

Figure 20-9 *Energetics of Br$_2$ dissociation.*

is even more violent (2600 K, 10^{-3} seconds) partly because this reaction is even more exothermic and partly because the reaction

$$H + O_2 \longrightarrow OH + O$$

produces two reactive chain carriers (an oxygen atom and the hydroxyl free radical in this case) for each chain carrier that reacts.

C. DISSOCIATION OF DIATOMIC MOLECULES AND RECOMBINATION OF ATOMS

In spite of the fact that the dissociation of Br$_2$ and the recombination of Br atoms

$$Br_2 \rightleftharpoons 2Br$$

is very fast, both reactions have been studied in detail. The activation energy for the dissociation of Br$_2$ is the same, within experimental error, as the enthalpy of the reaction. The energetics pictured in Fig. 20-9 show that there is no energy maximum between reactants and products; Br$_2$ molecules dissociate immediately once the molecule has sufficient energy. Once a Br$_2$ molecule has sufficient energy to dissociate there is nothing to keep the two atoms together. Conversely when two Br atoms collide, they cannot remain bonded very long unless some other atom or molecule is nearby to accept some of the energy. In the dissociation of Br$_2$ and the recombination of Br atoms, the energy transfer is the rate determining step. Reactions such as this are the only commonly encountered exception to equation 20.11.

In order for Br$_2$ to dissociate, energy must be supplied through the collision of Br$_2$ molecules with an atom or another molecule. There are no chemical requirements on the atom or the molecule. The rate of reaction is found to include a term for every molecule present during the reaction. For example, in the presence of Ar and N$_2$, the rate is

$$\text{Rate}_f = k_1[\text{Br}_2][\text{Br}_2] + k_2[\text{Br}_2][\text{Br}] + k_3[\text{Br}_2][\text{Ar}] + k_4[\text{Br}_2][\text{N}_2]$$

Under these conditions the rate of recombination of bromine atoms is given by

$$\text{Rate}_r = k_{-1}[\text{Br}]^2[\text{Br}_2] + k_{-2}[\text{Br}]^2[\text{Br}] + k_{-3}[\text{Br}]^2[\text{Ar}] + k_{-4}[\text{Br}]^2[\text{N}_2]$$

It is customary to abbreviate rate laws of this form where any arbitrary molecule may take part. The symbol (M) is used for the effective total concentration of atoms and molecules. In this particular case the equations may be written as

$$\text{Rate}_f = k_5[\text{Br}_2][\text{M}]$$

$$\text{and Rate}_r = k_{-5}[\text{Br}]^2[\text{M}]$$

V. S$_N$2 Substitution

Some activation energy data for the substitution of one ligand by another on a tetrahedral carbon atom are given in Table 20-3. Reactions of this type, called substitution reactions, are distinguished by a capital letter S. The N and the 2 in S$_N$2 refer to substitution by reaction with a base (N = nucleophile, another word for a Lewis base or electron pair donor) in a second order reaction (2). The incoming

Table 20-3 *Reactions of LiCl with RBr in Acetone*

R equals	Methyl (—CH$_3$)	Ethyl (—CH$_2$CH$_3$)	n-Propyl (—CH$_2$(C$_2$H$_5$))	i-Propyl (—CH(CH$_3$)$_2$)
$\Delta H^{\ddagger}$, kcal mole^{-1}	15.7	17.6	17.5	18.8
$\Delta S^{\ddagger}$, eu	-16	-18	-19	-22
k (at 25°C), liter moles^{-1} sec^{-1}	$10^{-2.22}$	$10^{-4.01}$	$10^{-4.20}$	$10^{-5.89}$

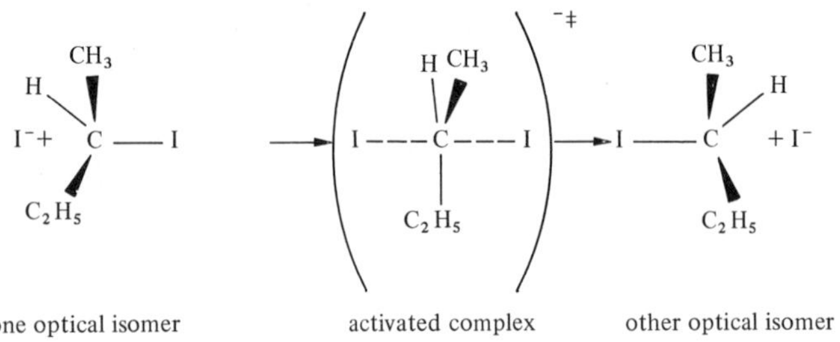

Figure 20-10 *Inversion of configuration of C_4H_9I.*

one optical isomer → activated complex → other optical isomer

group occupies the fifth coordination position of the carbon in the activated complex shown in Fig. 20-10.

The shape of the activated complex shown in Fig. 20-10 is based upon the fact that an inversion in the configuration about the carbon takes place during the reaction. This inversion is a very important characteristic of an S_N2 reaction. In this mechanism the three groups that are retained pass through a planar configuration. When the three groups are all different, as they are in C_4H_9I (Fig. 20-10), the substitution of I^- for I^- does not give the same molecule back. The molecule that results is the mirror image of the original. A molecule which is not identical to its mirror image is optically active. The configuration of the carbon atom where reaction occurs is inverted in an S_N2 reaction.

Activated complexes in which the carbon is five-coordinate will be more crowded than the tetrahedral structure of the reactant. The crowding will increase the energy of the activated complex, and thus affect the rate of reaction. This is known as steric hindrance. The larger the groups present, the greater the steric hindrance and the slower the reaction. This is demonstrated by the data given in Table 20-3 for the reactions of different organic bromides with chloride ion. As the hydrogen atoms in methyl bromide (CH_3Br) are progressively replaced by larger methyl (CH_3) groups, the activation energy increases and the reaction becomes progressively slower.

The variation in the reaction rate with the leaving group is even more interesting. Methyl iodide reacts faster with lithium chloride (LiCl) in acetone than does methyl bromide, and methyl chloride is the slowest of the three. There is evidently a preference for the larger more easily polarized halide ions in the activated complex. The relative rates of S_N2 reactions with other leaving groups are given in Table 20-4. These data indicate that in the activated complex carbon is a soft acid. It can show extra polarizing ability if it is induced by an easily polarized ligand. It is, of course, not surprising that the electron distribution of the activated complex can be distorted because the bonding is much more diffuse. Since the activated complex is symmetrical with respect to the entering and leaving groups, it is not surprising that the same softness of the activated complex shows up when the nucleophile is varied.

Table 20-4 *Relative Rates of S_N2 Substitution Reactions for Various Leaving Groups*

Leaving group	Log(relative rate)
$-OSO_2C_6H_5$	1.0
$-I$	0.2
$-Br$	0.0
$-Cl$	-1.6
$-ONO_2$	-2
$-F$	-4
$-OCH_3$	-6.5

The observed rate of S_N2 displacement reactions provides a method of determining the relative softness of the entering ligand. The standard order of nucleophiles, given in Table 20-5, is arranged with the softest nucleophiles at the top. There are some slight discrepancies between this order and that determined with soft inorganic acids, some of which are related to the differences in solvents used.

The softness of carbon in the activated complex is in sharp contrast to the relative thermodynamic stability of the carbon compounds. Owing to its small size, carbon should be a hard acid. In line with this expectation the equilibrium for a reaction such as

$$CH_3I + Cl^- = CH_3Cl + I^-$$

lies well over to the right.

Table 20-5 *An Average Order of Nucleophiles in S_N2 Reactions at Carbon*

Nucleophile (entering ligand)	Average relative S_N2 displacement rates
$C_4H_9S^-$	2×10^{-7}
$S_2O_3^{2-}$	1×10^6
CN^-	1×10^5
I^-	1×10^5
SCN^-	6×10^4
OH^-	2×10^4
Br^-	1×10^4
Cl^-	1×10^3
C_5H_5N	1×10^3
CH_3COO^-	5×10^2
NO_3^-	30
H_2O	1

Table 20-6 *Halide Exchange for Methyl Halides in Acetone Solution*

Reaction	ΔH, kcal mole^{-1}	ΔS, eu	$\Delta H^{\ddagger}$, kcal mole^{-1}	$\Delta S^{\ddagger}$, eu
$CH_3I + LiBr$	0	+1	15.7	−8.5
$CH_3I + LiCl$	∼−3	∼−4	15.4	−17.5
$CH_3Br + LiBr$	0	0	15.2	−12
$CH_3Br + LiCl$	∼−3	∼−4	15.1	−18.5
$CH_3Cl + LiCl$	0	0	19.5	−13

The substitution of Cl$^-$ and I$^-$ for Br$^-$ in CH_3Br provides another example. From a thermodynamic standpoint, CH_3Cl is the favored product, but because of the softness of the activated complex, CH_3Br reacts faster with I$^-$ than with Cl$^-$. The thermodynamic data available for some of these reactions are summarized in Table 20-6.

VI. Heterogeneous Reactions

The term heterogeneous reaction is applied to reactions that occur at the boundary between two phases. There are a great many important reactions of this type including all phase transitions and most of the reactions of solids. The rates of heterogeneous reactions are always dependent on the surface area available for reaction. Zinc will react faster with acids, and sugar will dissolve faster in water, if the particles are very finely divided.

In some cases, the rate of a heterogeneous reaction is limited by the rate at which the reactants arrive at the surface. The rate of condensation of a gas on a cold surface is often almost equal to the rate at which the condensable molecules strike the surface. Electrode reactions at small electrodes often reach a limiting rate equal to the rate of arrival of the reactant by diffusion to the electrode. This is the basis of the method of analysis of solutions known as polarography.

Some half-reactions are intrinsically slow because they involve a high-energy activated complex. In Chapter 5 we mentioned that the production of H_2 from water solutions

$$2H_3O^+(aq) + 2e^- \longrightarrow H_2(g) + 2H_2O$$

is normally so slow that many reducing agents strong enough to liberate hydrogen do so only at a negligible rate. This phenomenon, known as overvoltage, depends greatly on the chemical nature of the surface on which the bubbles of hydrogen are formed. For example, zinc dissolves readily in acids because the potential of the half-reaction

$$Zn = Zn^{2+} + 2e^- \qquad E^0 = -0.763 \text{ volt}$$

is so negative that even an unfavorable activated complex can be formed. Nevertheless the reaction goes much faster if the zinc is in contact with copper or platinum. As illustrated in Fig. 20-11, most of the hydrogen is then produced on the copper or platinum surface. The same catalytic effect can be produced by tiny crystals of copper formed on the zinc surface by the reaction

$$Zn + Cu^{2+} \longrightarrow Cu + Zn^{2+}$$

In all these cases the reaction proceeds through an activated complex where hydrogen atoms are bonded to the metal surface. Copper and

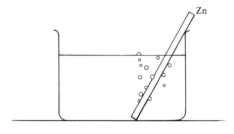
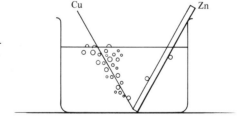
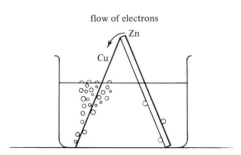

Figure 20-11 *Hydrogen evolution from zinc in sulfuric acid:* $Zn + 2H^+ \to H_2 + Zn^{2+}$.

Table 20-7 *Melting Points of Some Typical Molecules with C-C Double Bonds and the Corresponding Hydrogenated Compounds*

Unsaturated compound*	Melting point, °C	Saturated compound†	Melting point, °C
4-Octene, C_8H_{16}	−105	Octane, C_8H_{18}	−56
9-Octadecene, $C_{18}H_{36}$	−30	Octadecane, $C_{18}H_{38}$	28
Maleic acid, $C_4H_4O_4$	131	Succinic acid, $C_4H_6O_4$	182
7-Hexadecenoic acid, $C_{15}H_{29}COOH$	33	Hexadecanoic acid, $C_{15}H_{31}COOH$	63
2-Methylpenten-3-one, $C_6H_{10}O$	70	2-Methylpentane-3-one, $C_6H_{12}O$	114
Glyceryl trioleate, $C_3H_5(O_2CC_{17}H_{33})_3$	−5	Glyceryl tristearate, $C_3H_5(O_2CC_{17}H_{35})_3$	73

*Unsaturated compounds contain C-C double bonds.
†Saturated compounds are fully hydrogenated.

platinum are catalysts because the Cu-H and Pt-H bonds are stronger than the Zn-H bond at the surface of the solid metals.

The $2H_3O^+(aq) + 2e^- \longrightarrow H_2(g) + 2H_2O$ half-reaction is used as a standard for electrode measurements. This is possible only because a platinum electrode of large surface area allows the half-reaction to proceed at an appreciable rate at potentials negligibly different from 0.000 volt. In a similar way platinum and palladium metals will catalyze many other reactions of hydrogen. They are very widely used to catalyze the reaction of hydrogen with compounds containing C-C double bonds.

$$C_3H_6 + H_2 \overset{Pt}{\rightleftharpoons} C_3H_8$$

In a molecule such as C_3H_6 the π-π overlap in the C-C double bond prevents rotation about the bond axis. As a result, a double bond present in the middle of a long chain of carbon atoms considerably reduces the flexibility of the molecule. This in turn may interfere with the formation of the best crystal structure, and a few double bonds in a complex organic molecule often lower the melting point substantially (Table 20-7). The hydrogenation of a vegetable oil (over a platinum or palladium catalyst) converts it into a saturated fat with a melting point above room temperature.

QUESTIONS
1. What is a zero-order reaction? A first-order reaction?
2. What is an activated complex? Give an example.
3. What is the significance of the number 2.303?
4. The reaction of Fe^{2+} with H_2O_2 has the rate law given below. What is the activated complex?

$$\text{Rate} = k[Fe^{2+}][H_2O_2]$$

5. Can a catalyst be found to increase the amount of ammonia present in the equilibrium $3H_2 + N_2 \rightleftharpoons 2NH_3$?
6. Define $\Delta H^{\ddagger}$.
7. Finely divided lead sparks when exposed to air. Explain.

PROBLEMS I
1. Growing wood contains enough ^{14}C to give a disintegration rate of 15.3 disintegrations per second per gram (dis sec^{-1} g^{-1}) of carbon. What is the age of a beam from a pyramid which shows 6.0 dis sec^{-1} g^{-1}? The half-life of ^{14}C is 5770 years.
2. The decomposition of N_2O_5 has a half-life of 207 min at 45°C. Calculate the rate constant at this temperature and give its units.
3. Evaluate R/N_0h.

4. In acid solutions the reaction of I^- with H_2O_2 goes through the activated complex $(H_3IO_2)^{\ddagger}$. Write the rate law.
5. The rate law for the reaction of Fe^{3+} with iodide is given below. Write the formulas, including the correct charge, for the two activated complexes indicated by the rate law.

$$\text{Rate} = \frac{k[I^-]^2[Fe^{3+}]}{1 + k'[Fe^{2+}]/[Fe^{3+}]}$$

6. The reaction $(CH_3)_3CCl + H_2O \longrightarrow (CH_3)_3COH + HCl$ proceeds by the mechanism given below. Does the rate depend on the concentration of H_2O? Write the rate expression. This is called an S_N1 reaction. Explain the term S_N1.

$$(CH_3)_3CCl \longrightarrow (CH_3)_3C^+ + Cl^- \quad \text{rate determining step}$$
$$(CH_3)_3C^+ + H_2O \longrightarrow (CH_3)_3COH + H^+$$

7. The rate constant for the reaction $2N_2O \longrightarrow 2N_2 + O_2$ is 2.6×10^{-11} at $300°C$ and 2.1×10^{-10} at $330°C$. Calculate the activation energy.
8. For the reaction $2NOCl \longrightarrow 2NO + Cl_2$, $\Delta H^{\ddagger}$ is 24 kcal mole^{-1} and $\Delta S^{\ddagger}$ is -5 eu. Calculate the rate constant at $100°C$.
9. An enzyme in the human body may increase the rate of a chemical reaction a million-fold. How much must it lower $\Delta H^{\ddagger}$ to account for this effect?

PROBLEMS II

1. The reaction $2NO + O_2 \longrightarrow 2NO_2$ has the rate law given below. What is the order of the reaction with respect to NO?

$$\text{Rate} = k[NO]^2[O_2]$$

2. What is the overall order of the reaction in Problem 1?
3. The decomposition of gaseous C_2H_5Cl is first order. Write the rate law for the decomposition of C_2H_5Cl.
4. After 40 min at 800 K only 25% of a sample of C_2H_5Cl is left. What is the half-life for the decomposition?
5. What fraction of the C_2H_5Cl will be left after 100 min at 800 K?
6. Calculate the rate constant for the decomposition of C_2H_5Cl.
7. Write the activated complex for the decomposition of C_2H_5Cl.
8. Write the activated complex for the reaction $2NO + O_2 \longrightarrow 2NO_2$.
9. The activated complex for the reaction $2I^- + H_2O_2 + 2H^+ \longrightarrow 2H_2O + I_2$ is $(H_2O_2I^-)^{\ddagger}$. Write the rate law for this reaction.
10. The rate law for the reaction $2Fe^{2+} + Tl^{3+} \longrightarrow 2Fe^{3+} + Tl^+$ is found to have the form given below. What is the formula of the activated complex for the step which is rate determining when $[Fe^{3+}]$ is negligibly small?

$$\text{Rate} = \frac{k[\text{Fe}^{2+}]^2[\text{Tl}^{3+}]}{[\text{Fe}^{3+}] + k'[\text{Fe}^{2+}]}$$

11. The mechanism given below has been proposed for the reaction in Problem 10. Which step is rate determining when (Fe^{3+}) is negligibly small?

$$\text{Fe}^{2+} + \text{Tl}^{3+} = \text{Fe}^{3+} + \text{Tl}^{2+}$$

$$\text{Tl}^{2+} + \text{Fe}^{2+} = \text{Fe}^{3+} + \text{Tl}^+$$

SUGGESTED READINGS

CAMPBELL, J. A., *Why Do Chemical Reactions Occur?* Englewood Cliffs, N. J.: Prentice Hall, 1965.

KING, E. L., *How Chemical Reactions Occur.* New York: W. A. Benjamin, 1963.

SYKES, A. G., *Kinetics of Inorganic Reactions.* London: Pergamon Press, 1966.

A factor of great importance in determining the behavior of complex ions is the rate at which substitution reactions for them take place. It is, in fact, of greater importance for many of the observations which are made than is the factor of stability. A useful classification of complex ions can be based on the speed at which they adjust to equilibrium with respect to substitution reactions.*

H. TAUBE

Chapter 21 The Robust Complexes

I. The Rate of Ligand Exchange Reactions

A. ROBUST AND LABILE COMPLEXES

The equilibrium reactions of complex ions in solution are often very involved because of the large number of different complexes that can be formed. For example, in the reaction of $Fe(H_2O)_6^{3+}$ with thiocyanate ion (SCN^-), complexes containing up to six coordinated thiocyanate ligands can be formed.

$$Fe(H_2O)_6^{3+} + SCN^- \rightleftharpoons Fe(SCN)(H_2O)_5^{2+} + H_2O$$
$$Fe(SCN)(H_2O)_5^{2+} + SCN^- \rightleftharpoons Fe(SCN)_2(H_2O)_4^+ + H_2O$$
$$Fe(SCN)_2(H_2O)_4^+ + SCN^- \rightleftharpoons Fe(SCN)_3(H_2O)_3 + H_2O$$
$$Fe(SCN)_3(H_2O)_3 + SCN^- \rightleftharpoons Fe(SCN)_4(H_2O)_2^- + H_2O$$
$$Fe(SCN)_4(H_2O)_2^- + SCN^- \rightleftharpoons Fe(SCN)_5(H_2O)^{2-} + H_2O$$
$$Fe(SCN)_5(H_2O)^{2-} + SCN^- \rightleftharpoons Fe(SCN)_6^{3-} + H_2O$$

In addition there may be several distinct complexes of the same

*Used by permission from H. Taube, *Chem. Revs.*, **50**, 68 (1952); © 1952 by Williams and Wilkins Co., Baltimore, Md.

stoichiometry such as the cis and trans forms of $Fe(SCN)_2(H_2O)_4^+$ (Fig. 21-1). As a result there will be a large number of different complex ions present when equilibrium is established.

Including the geometrical isomers, there is a total of ten octahedral thiocyanate complexes of Fe(III).

Two of the ten thiocyanate complexes of Fe(III) are fairly easy to study because they can be prepared relatively free of the others. In solutions containing a low concentration of thiocyanate the ion $Fe(SCN)(H_2O)_5^{2+}$ predominates, whereas the ion $Fe(SCN)_6^{3-}$ is the predominate species when the thiocyanate concentration is high. Although the complex of intermediate stoichiometry $[Fe(SCN)_3(H_2O)_3]$ can not be isolated in solution, solid $Fe(SCN)_3$ can be prepared pure. When $Fe(SCN)_3$ is dissolved in water, all of the possible iron thiocyanate complexes as well as hydrated iron are formed within a second.

The rate of formation of the first complex $[Fe(SCN)(H_2O)_5]^{2+}$ can be measured by modern techniques. The rate law for the reaction

$$Fe(H_2O)_6^{3+} + SCN^- \rightleftharpoons Fe(SCN)(H_2O)_5^{2+} + H_2O$$

is

$$\text{Rate} = k_1[Fe^{3+}][SCN^-] + k_2[Fe^{3+}][SCN^-][H^+]^{-1}$$

with $k_1 = 127$ liter moles^{-1} sec^{-1} at 25°C. The enthalpy of activation for the reaction is 13 kcal mole^{-1}. The rate constant is so large (or the activation energy is so low) that equilibrium is established in less than 1 second even in dilute solutions where the reaction is expected to be the slowest. The form of the rate law shows that thiocyanate is present in the activated complex. This indicates that the anion is bonded to the iron in the activated complex, so the activated complex involves seven-coordinate Fe(III).

In contrast to the reaction of Fe(III), the reaction of $Cr(H_2O)_6^{3+}$ with thiocyanate

$$Cr(H_2O)_6^{3+} + SCN^- \rightleftharpoons Cr(SCN)(H_2O)_5^{2+} + H_2O$$

involves an activated complex that is primarily five-coordinate rather than seven-coordinate. The principal terms in the rate law are

$$\text{Rate} = k_1[Cr(H_2O)_6^{3+}] + k_2[Cr(H_2O)_6^{3+}][H^+]^{-1}$$

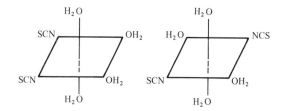

Figure 21.1 *Cis and trans isomers of $Fe(SCN)_2(H_2O)_4^+$.*

There is also a significant difference in the value of the rate constants and activation energies for the two reactions. For the Cr(III) reaction

$$k_1 = 1.8 \times 10^{-6} \text{ sec}^{-1} \quad \text{and} \quad \Delta H^\ddagger = 25.7 \text{ kcal mole}^{-1}$$

The activation energy for this reaction is so large (compare with $\Delta H^\ddagger = 13$ kcal mole^{-1} for Fe(III)) that the formation of Cr(SCN)(H$_2$O)$_5^{2+}$ is very slow at room temperature. Instead of a fraction of a second, almost 7 days are required before the reaction begins to approach equilibrium. As is the case for Fe(III) the equilibrium mixture is very complicated, but because of the slow rate in this case it is possible to isolate each individual complex ion by fractional precipitation or ion exchange techniques. The term "robust" is applied to complexes in which each complex can be separated and studied. However, it must be emphasized that the term robust carries no implication of thermodynamic stability. A very stable complex will ordinarily dissociate slowly, but a complex is not classified as robust unless the rate of formation is also slow.

The term "labile" is used for complexes that undergo rapid ligand substitution, such as the thiocyanate complexes of Fe(III).

B. THE EFFECTS OF SIZE, CHARGE, AND ELECTRONIC STRUCTURE ON THE RATE OF LIGAND EXCHANGE

The complexes of Co(III) and Cr(III) are the most widely studied robust complexes. There are, however, a number of interesting robust complexes of other elements. In attempting to predict or interpret data in order to determine which complexes are robust, it is desirable to examine as many of the elements in the periodic table as possible. The techniques used to study the kinetics of fast reactions have been developed to the point where rate data for the substitution of O^{2-}, OH^-, or H_2O by water are widely available for a large number of elements. Data are now available for even the extremely fast ($k \approx 10^9$ sec^{-1}) exchange reaction of water molecules coordinated to an ion such as Ca^{2+}.

The rate of slow exchange reactions can be determined using ^{18}O as a tracer.

$$Ca(H_2O)_n^{2+} + H_2O \rightleftharpoons Ca(H_2O)_n^{2+} + H_2O$$

Like other chemical properties we can expect the rate of ligand exchange to correlate with the size, charge, and electronic structure of the cation involved. From the data in Table 21-1 we see a drastic decrease in the rate of exchange across a row, that is, from Na(I) to Cl(VII), and a substantial increase in rate down each group. Thus both decreased charge and increased size favor faster rates of ligand exchange. As a result, there are a number of robust oxo anions (ClO$_4^-$,

Table 21-1 *Rates of Exchange of O^{2-}, OH^-, or H_2O Ligands by Water for d^0 and d^{10} Ions*

(1) very fast exchange: $k > 10^{+8}$ sec^{-1}
(2) fast exchange: $10^3 < k < 10^8$
(3) moderately slow exchange: $10^{-3} < k < 10^3$
(4) robust complexes: $k < 10^{-3}$ sec^{-1}

IA	IIA	IIIB	IVB	VB	VIB	VIIB	VIII	VIII	VIII	IB	IIB	IIIA	IVA	VA	VIA	VIIA	Inert gases
H^+ 1																	
Li^+ 1	Be^{2+} 3												CO_3^{2-} 3	NO_3^- 4			
Na^+ 1	Mg^{2+} 2											Al^{3+} 3	SiO_4^{4-} 3	PO_4^{3-} 4	SO_4^{2-} 4	ClO_4^- 4	
K^+ 1	Ca^{2+} 1				CrO_4^{2-} 3						Zn^{2+} 2	Ga^{3+} 3					
Rb^+ 1	Sr^{2+} 1										Cd^{2+} 1	In^{3+} 2					
Cs^+ 1	Ba^{2+} 1										Hg^{2+} 1	Tl^{3+} 1					

Table 21-2 *Exchange of Bound Water Molecules with Free Water in the 3d Transition Metals*

	+2 Ions		+3 Ions	
	Ion	$k\ \text{sec}^{-1}$	Ion	$k\ \text{sec}^{-1}$
d^0	Ca^{2+}	10^9		
d^1				
d^2			V^{3+}	1×10^3
d^3	V^{2+}	200	Cr^{3+}	3.3×10^{-6}
d^4	Cr^{2+}	7×10^9		
d^5	Mn^{2+}	3×10^7	Fe^{3+}	3×10^3
d^6	Fe^{2+}	3×10^6	Co^{3+}	$<1 \times 10^3$
d^7	Co^{2+}	1×10^6		
d^8	Ni^{2+}	2.7×10^4		
d^9	Cu^{2+}	8×10^9		
d^{10}	Zn^{2+}	5×10^7	Ga^{3+}	8×10^7

PO_4^{3-}, and so forth) in the upper right hand corner of the periodic table.

The rate of the ligand exchange reaction for the transition metal ions depends upon both the charge and electronic configuration of the ion. For all of the +2 ions in the series from Ca^{2+} through Zn^{2+}, the rate of water exchange is fairly rapid (V^{2+} with a d^3 configuration is the slowest). All of these +2 ions are labile.

It is apparent from Table 21-2 that the rate of exhange is slowest for the +3 cation with a $3d^3$ electron configuration. The reason for this exceptional behavior of octahedral d^3 complexes is evident from a comparison of the molecular orbital diagrams in Fig. 21-2. Three d electrons is a particularly favorable arrangement in an octahedral complex. For ions with this electron configuration there is less stabilization in an activated complex with a coordination number of either five or seven (see Fig. 21-2) than there is in an octahedral arrangement.

Low-spin d^6 complexes are expected to be robust for the same reason. Most of the complexes of Fe^{2+} (a d^6 ion) are high spin and quite labile; however, a few such as $Fe(CN)_6^{4-}$ are low spin and they undergo ligand exchange quite slowly. The +3 ions in the first transition series with d^3 and d^6 configurations are Cr(III) and Co(III).

Another example of an electron configuration that favors the formation of robust complexes is the d^8 configuration in square planar complexes. Low-spin, square-planar d^8 systems are also stabilized by ligand field splitting. There are a few square planar complexes of Ni(II) that are robust; however, the principal examples of robust square planar complexes are formed by the d^8 ions of the $4d$ and $5d$ transition series. These ions, Pd(II), Pt(II), and Au(III) will be discussed in detail in Chapter 24. Here we shall examine a special consequence of slow ligand exchange reactions by considering one aspect of the chemistry of chromium and then we shall consider the mechanism of redox reactions.

C. THE AMPHOTERISM OF CHROMIUM(III)

One of the distinctive features of the chemistry of Cr(III) is the amphoteric character of $Cr(OH)_3$. To interpret this behavior we must first look at the crystal structure of a solid hydroxide such as $Al(OH)_3$. The basic unit of the structure, shown in Fig. 21-3, consists of two close packed layers of hydroxide ions with two thirds of the octahedral holes occupied by Al^{3+} ions. In a structure such as this each OH^- is shared by two Al atoms, and the sandwich type units are held together by hydrogen bonds.

Since the individual sandwich type units can hydrogen-bond water molecules, $Al(OH)_3$ normally precipitates with variable amounts of water between the layers so that the precipitate does not have the long range order characteristic of true crystals. In fact, crystals of $Al(OH)_3$ suitable for X-ray analysis are only formed over a long period of geologic time. As prepared in the laboratory, $Al(OH)_3$ is an amorphous gelatinous solid. Ferric hydroxide is a similar gelatinous precipitate that presumably has the same type of structure. It is reasonable to expect that freshly precipitated $Cr(OH)_3$ will have this structure. However, the formation of even a single sandwich layer of the hydroxide of Al(III), Fe(III), or Cr(III) requires ligand exchange. To form the layer hydroxide structure, metal-oxygen bonds must be broken. Water must be released from the primary coordination sphere and

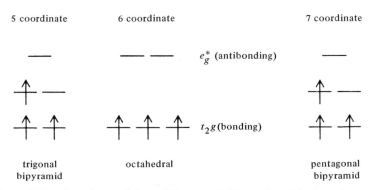

Figure 21-2 *Complexes of three different stoichiometries with three d electrons.*

Ch. 21 The Robust Complexes

M-OH-M linkages must be formed in its place. For Fe(III) and Al(III) the ligand exchange is rapid, but for Cr(III) the exchange is very slow. On the other hand the precipitate formed by the process

$$Cr(H_2O)_6^{3+} + 3OH^- \longrightarrow Cr(OH)_3(H_2O)_3 + 3H_2O$$

only requires the transfer of a proton which is practically instantaneous.

The polarizing ability of Cr(III) is 2.45

Chromium(III) is not small enough to be truly amphoteric, and a well-aged $Cr(OH)_3$ precipitate is insoluble in concentrated sodium hydroxide. However, the formation of the Cr-OH-Cr linkage is very

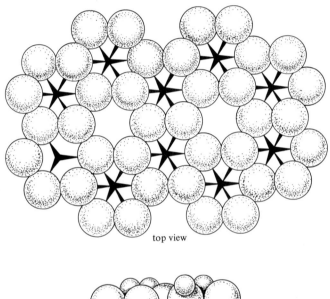

top view

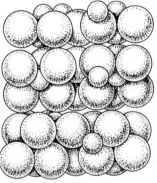

side view

Figure 21-3 *A layer of the* $Al(OH)_3$ *crystal structure. (After R. W. G. Wyckoff, Crystal Structures, New York: John Wiley, 1964.)*

Sec. II Mechanisms of Redox Reactions

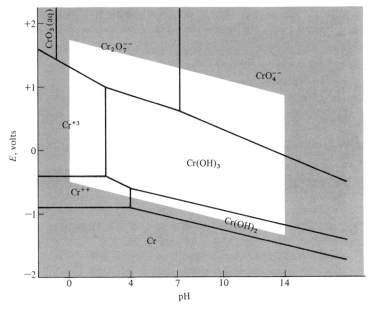

Figure 21-4 E_h-pH *diagram for chromium.*

slow, and freshly precipitated $Cr(OH)_3$ will dissolve in NaOH hours or even days after it is precipitated. The actual behavior of freshly precipitated $Cr(OH)_3$ is quite different from that indicated on the E_h-pH diagram shown in Fig. 21-4. This is a case where the solubility of a substance depends upon kinetics rather than true equilibrium.

II. Mechanisms of Redox Reactions

The reactions that occur between positively charged ions are always catalyzed by anions. The anions may or may not be directly coordinated to the metal atoms and if the cation-anion complex is labile, it may be impossible to decide if a particular ion is present as a ligand in the activated complex.

In cases where at least one of the reactants or products is robust, three separate possibilities can be distinguished. Either two, one, or zero ligands may be shared in the activated complex. In cases where ligands are shared the activated complex is singly or doubly "bridged." When no ligands are shared the reaction proceeds through an outer sphere activated complex. Examples of reactions that involve each type of activated complex are known.

A double bridge is probably involved in the air oxidation of Cr(II). The mechanism of the reaction has been formulated as

$$2Cr(H_2O)_6^{2+} + O_2 \longrightarrow 2Cr(H_2O)_4(OH)_2^{2+} + 2H_2O \quad \text{rate determining step}$$

$$Cr(H_2O)_4(OH)_2^{2+} + Cr(H_2O)_6^{2+} \longrightarrow 2H_2O + \left((H_2O)_4Cr\underset{\underset{H}{O}}{\overset{\overset{H}{O}}{\diamond}}Cr(H_2O)_4\right)^{4+}$$

Double bridging is involved in the reaction between Cr(II) and Cr(IV). This reaction is unusual in that both cations in the product are robust. As a result the configuration of the activated complex for the second step is fully maintained in the product.

$$Cr(H_2O)_6^{2+} + Cr(H_2O)_4(OH)_2^{2+} \longrightarrow$$

$$2H_2O + \left((H_2O)_4Cr\underset{\underset{H}{O}}{\overset{\overset{H}{O}}{\diamond}}Cr(H_2O)_4\right)^{4+\ddagger} \longrightarrow \left((H_2O)_4Cr\underset{\underset{H}{O}}{\overset{\overset{H}{O}}{\diamond}}Cr(H_2O)_4\right)^{4+} + 2H_2O$$

Singly bridged activated complexes are more common for metals in low oxidation states. These reactions provide a way of preparing monohalo complexes of Cr(III). The reaction sequence may be represented by the equations

$$Cr(H_2O)_6^{2+} + Co(NH_3)_5(Cl)^{2+} \longrightarrow ((NH_3)_5Co-Cl-Cr(H_2O)_5)^{4+\ddagger} + H_2O$$
$$\longrightarrow Cr(H_2O)_5Cl^{2+} + 5NH_3 + Co^{2+} + H_2O$$

In this case the Co(II) produced is quickly hydrated to $Co(H_2O)_6^{2+}$ (predominately) in dilute solutions. Another reaction that involves a singly bridged activated complex is the conversion of polychloro Cr(III) complex ions to the monochloro complex $Cr(H_2O)_5Cl^{2+}$.

$$Cr(H_2O)_3Cl_3 + Cr(H_2O)_6^{2+} \longrightarrow ((H_2O)_3Cl_2Cr-Cl-Cr(H_2O)_5)^{2+\ddagger} + H_2O$$
$$\longrightarrow Cr^{2+} + 3H_2O + 2Cl^- + Cr(H_2O)_5Cl^{2+} + H_2O$$

Finally a great many of the reactions of anions, for example, the reaction of the two anions ferrocyanide and permanganate

$$Fe(CN)_6^{4-} + MnO_4^- \longrightarrow Fe(CN)_6^{3-} + MnO_4^{2-}$$

proceeds through the outer sphere mechanism (no ligands shared in the activated complex).

The outer sphere activated complex is favored in complexes that contain complex organic molecules such as phenanthroline (abbreviated phen). In complexes of this type the π structure of the rings

provides a pathway for electrons that allows them to move away from or toward the atom that is being oxidized or reduced. This type of mechanism has been proposed for the reaction

$$\text{Fe(phen)}_3^{2+} + \text{Ce(H}_2\text{O)}_n(\text{OH})_2^{2+} \longrightarrow$$
$$\text{Fe(phen)}_3^{3+} + \text{OH}^- + \text{Ce(H}_2\text{O)}_n(\text{OH})^{2+}$$

A. HIGHER OXIDATION STATES OF CHROMIUM

Chromium(III) is easily oxidized to CrO_4^{2-} in either basic or acidic solution if a sufficiently strong oxidizing agent such as Ce(IV) is used. The rate expression for the oxidation of Cr(III) with Ce(IV) in sulfuric acid solutions is

$$\text{Rate} = k[\text{CeSO}_4^{2+}]^2[\text{CrSO}_4^+][\text{CeSO}_4^+]^{-1}$$

This reaction apparently takes place through a stepwise oxidation that can be represented as

Cr(III) + Ce(IV) $\rightleftharpoons$ Cr(IV) + Ce(III)

Cr(IV) + Ce(IV) $\rightleftharpoons$ Cr(V) + Ce(III)   rate determining step

Cr(V) + Ce(IV) $\rightleftharpoons$ Cr(VI) + Ce(III)

The water of hydration has been omitted in this mechanism.

The activated complex formed in the rate determining step may be considered to be composed of Cr(IV) and Ce(IV). The step between Cr(IV) and Cr(V) is similarly rate determining in the reaction between $HCrO_4^-$ and a one-electron reducing agent such as Fe^{2+}. The rate of this reaction is given by

$$\text{Rate} = k[\text{H}^+]^3[\text{Fe}^{2+}]^2[\text{Fe}^{3+}]^{-1}([\text{HCrO}_4^-] + k_2[\text{Cr}_2\text{O}_7^{2-}])$$

Similarly for the reaction between Cr(VI) and V(IV) the rate is

$$\text{Rate} = k[\text{V(OH)}_2^{2+}]^2[\text{V(OH)}_4^+]^{-1}[\text{HCrO}_4^-]$$

Table 21-3 *Properties of the Various Oxidation States of Chromium*

	II	III	IV	V	VI
Oxidation number					
Coordination number	6	6	6	4	4
Stability	borderline	stable	disproportionates	disproportionates	stable
Color	blue	green, violet		green	yellow, orange
Acidity	basic	basic (kinetically amphoteric)	strongly hydrolyzed amphoteric	acidic	acidic
Rate of ligand exchange	very fast	robust	moderately fast	moderately slow	moderately slow
Strength as reducing agent	very strong	good	strong	strong	none
E^0	-0.41	$+1.33$	(~ 1.2)*	(~ 1.3)*	
Strength as oxidizing agent	weak	weak	strong	strong	strong
E^0	-0.90	-0.41	(~ 1.6)*	(~ 1.2)*	$+1.33$

*Estimates for one-electron changes in acidic solutions.

The Cr(IV) $\rightleftharpoons$ Cr(V) reaction step is rate determining in all of these reactions because the coordination number of chromium drops from six to four at this point. The species present in solution are probably $Cr(H_2O)_4(OH)_2^{2+}$ and CrO_4^{3-}.

Chromium(IV) and (V) are too unstable to be prepared in quantity, and reaction rates provide the only reliable information on their properties. Both Cr(IV) and Cr(V) are stronger oxidizing agents than Cr(VI) and at the same time stronger reducing agents than Cr(III).

The oxidizing power of Cr(IV) may be illustrated by a phenomenon that is called an induced reaction. $HCrO_4^-$ is not a strong enough oxidizing agent to oxidize Mn^{2+} to Mn(IV)(in MnO_2), but it is strong enough to oxidize AsO_3^{3-} to AsO_4^{3-}

$$7H_2O + 5H^+ + 3H_3AsO_3 + 2HCrO_4^- \longrightarrow 3H_2AsO_4^- + 2Cr(H_2O)_6^{3+}$$

However with both arsenate(III) and excess Mn(II) present, the reaction does proceed according to the equation.

$$4H_2O + 2H^+ + HCrO_4^- + H_3AsO_3 + \tfrac{1}{2}Mn^{2+} \longrightarrow$$
$$Cr(H_2O)_6^{3+} + H_2AsO_4^- + \tfrac{1}{2}MnO_2$$

The first step in this reaction is presumably

$$Cr(VI) + As(III) \longrightarrow Cr(IV) + As(V)$$

This is followed by the rapid reactions

$$Cr(IV) + Mn(II) \longrightarrow Mn(III) + Cr(III)$$

$$2Mn(III) \longrightarrow Mn(II) + Mn(IV)$$

From data of this type an estimate of the oxidizing ability of all the oxidation states of chromium can be made. These are given in Table 21-3 along with some other properties of intermediate chromium species. Sufficient CrO_4^{3-} can be obtained in strongly basic solutions for the observation of its green color. Even for Cr(IV) which is never present in more than trace amounts during the course of a reaction, a substantial amount of information is available.

In a water solution containing Cr(VI) three species will be in equilibrium. They are the yellow chromate ion (CrO_4^{2-}), the orange acid chromate ion ($HCrO_4^-$), and the orange dichromate ion ($Cr_2O_7^{2-}$). The equilibrium reaction

$$2HCrO_4^- \rightleftharpoons Cr_2O_7^{2-} + H_2O$$

is concentration-dependent, as are all dimerization reactions. In dilute solution $HCrO_4^-$ predominates whereas $Cr_2O_7^{2-}$ is the principal species in concentrated solution. These two ions react at different rates with hydroxide. $HCrO_4^-$ reacts instantaneously, but the reaction

$$Cr_2O_7^{2-} + 2OH^- \longrightarrow 2CrO_4^{2-} + H_2O$$

involves the breakage and formation of a Cr-O bond and several minutes are required before equilibrium is reached. The rate law and rate constants are given in Table 21-4.

Most chromates are soluble in water; the common exceptions to this rule being $BaCrO_4$, $SrCrO_4$, $PbCrO_4$, $CaCrO_4$, and Ag_2CrO_4. All of

Table 21-4 *Rate of the Reaction of $Cr_2O_7^{2-}$ with Base*

$$Cr_2O_7^{2-} + 2OH^- \longrightarrow 2CrO_4^{2-} + H_2O$$

Rate (at 13°C) = $k_1[Cr_2O_7^{2-}][OH^-] + k_2[Cr_2O_7^{2-}] + k_3[Cr_2O_7^{2-}][H^+]$

$k_1 = 1 \times 10^2$ liter mole^{-1} sec^{-1}

$k_2 = 0.012$ sec^{-1}

$k_3 = 8 \times 10^3$ liter mole^{-1} sec^{-1}

these are yellow except Ag_2CrO_4 which is red. The formation of red Ag_2CrO_4 from a yellow solution is commonly used to indicate the end point in titration of Cl^- with $AgNO_3$. $PbCrO_4$ and $BaCrO_4$ are used as yellow pigments. Ammonium dichromate $((NH_4)_2Cr_2O_7)$ is an interesting compound in that the cation is a reducing agent and the anion is an oxidizing agent. When a small pile of $(NH_4)_2Cr_2O_7$ is ignited (as with a magnesium ribbon), the reaction

$$(NH_4)_2Cr_2O_7 \longrightarrow N_2 + Cr_2O_3 + 4H_2O$$

occurs. This reaction is exothermic; the material glows red and the gases produced blow sparks and green clouds of Cr_2O_3 upward giving the appearance of a miniature volcano.

B. COBALT

Cobalt(II) is a d^7 ion. Since a tetrahedral structure is not particularly unfavorable for a high-spin d^7 electron configuration (the molecular orbital diagram for a tetrahedral complex is given in Fig. 21-5), high-spin Co(II) is found in both octahedral and tetrahedral complexes. The aqueous solution chemistry of Co(II) is particularly colorful because the equilibrium between pink $Co(H_2O)_6^{2+}$ and a blue tetrahedral complex such as $CoCl_4^{2-}$ is easily shifted. A change in temperature or concentration is sufficient.

While tetrahedral Co(II) complexes can be made, ligands that form strong σ bonds will favor the octahedral geometry and, if Δ is large enough, low-spin complexes are obtained. A low-spin Co(II) complex is easily oxidized. The e_g electron shown in Fig. 21-6b can be removed by a variety of oxidizing agents. High- and low-spin Co(III) complexes are known.

The high spin state complexes are formed with ions such as O^{2-} and F^-. High-spin CoF_6^{3-} can be prepared and a precipitate of $Co(OH)_2$ (blue) is readily oxidized to $CoO(OH)$

$$OH^- + Co(OH)_2 \longrightarrow CoO(OH) + H_2O + e^- \qquad E^0 = 0.17 \text{ volt}$$

These high-spin Co(III) compounds undergo ligand exchange reactions readily.

Low-spin Co(III) complexes are obtained with ligands that are high

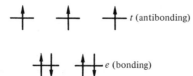

Figure 21-5 *Electronic configuration of a d^7 tetrahedral complex.*

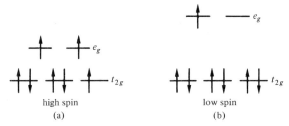

Figure 21-6 *Electronic configuration of a d^7 octahedral complex.*

in the spectrochemical series (large Δ). With NH_3 the half-reaction is

$$(Co(NH_3)_6)^{3+} + e^- \rightleftharpoons Co(NH_3)_6^{2+} \qquad E^0 = 1.17 \text{ volts}$$

The standard potential for this reaction is well within the water range. $Co(NH_3)_6^{3+}$ has a low spin d^6 electron configuration. It is a robust complex.

When hydrogen peroxide is used to oxidize $Co(NH_3)_6^{2+}$, it is possible to obtain a high yield of $Co(NH_3)_5(H_2O)^{3+}$. The mechanism for the reaction probably involves the formation of a single bridge formed by hydrogen peroxide

$$2Co(NH_3)_6^{2+} + H_2O_2 \rightleftharpoons (NH_3)_5Co\!-\!\overset{H}{\underset{|}{O}}\!-\!\overset{H}{\underset{|}{O}}\!-\!Co(NH_3)_5^{4+} + 2NH_3$$

$$\downarrow$$

$$((NH_3)_5Co\!-\!\overset{H}{\underset{|}{O}}\!-\!\overset{H}{\underset{|}{O}}\!-\!Co(NH_3)_5^{4+})^\ddagger \longrightarrow 2(NH_3)_5CoOH^{2+}$$

The water molecule in $Co(NH_3)_5(H_2O)^{3+}$ can be replaced by a variety of ligands. Complexes such as $Co(NH_3)_5Cl^{2+}$ may be readily prepared. By appropriate variation in the concentration of ligand and oxidizing agent, it is possible to prepare the *cis-* and *trans-*dichlorotetramine complexes and many other complex ions of Co(III).

In the absence of a small highly charged ligand or ligands high in the spectrochemical series, the oxidation of Co(II) is almost impossible. $Co(H_2O)_6^{3+}$ is such a strong oxidizing agent ($E^0 = +1.84$ volts) that it rapidly oxidizes water producing O_2

$$4Co(H_2O)_6^{3+} + 2H_2O \longrightarrow 4H^+ + 4Co(H_2O)_6^{2+} + O_2$$

The most widely studied inorganic redox reactions are the reactions between the complex ions of Co(III) and the Cr(II) ion. The various ligand transfer mechanisms are particularly easy to distinguish in these reactions since both the oxidizing agent and one of the products are robust. Cis-Co(NH$_3$)$_4$Cl$_2^+$ reacts with Cr(II) according to two mechanisms

$$Co(NH_3)_4Cl_2^+ + Cr(H_2O)_6^{2+} \longrightarrow \begin{pmatrix} & Cl & \\ (NH_3)_4Co & & Cr(H_2O)_4^{3+} \\ & Cl & \end{pmatrix}^{\ddagger} + 2H_2O \longrightarrow$$

$$4NH_3 + Co^{2+} + Cr(H_2O)_4Cl_2^+ + 2H_2O$$

$$\longrightarrow \left((NH_3)_4ClCo-Cl-Cr(H_2O)_5^{3+}\right)^{\ddagger} + H_2O \longrightarrow$$

$$4NH_3 + Cl^- + Co^{2+} + Cr(H_2O)_5Cl^{2+} + H_2O$$

In this case both the singly and doubly bridged activated complexes are important and substantial amounts of both products are formed.

QUESTIONS

1. Define "ligand exchange reaction."
2. Which of the following complex ions are robust: CrO_4^{2-}, SO_4^{2-}, CoF_6^{3-}, $Co(CN)_6^{3-}$, Cr^{2+}, Cr^{3+}, Fe^{3+}?
3. Write the rate expression for the reaction of $Cr(H_2O)_6^{2+}$ with $Co(NH_3)_4Cl_2^+$ (see above).
4. Suggest a mechanism for the disproportionation of Cr(IV).
5. List four soluble chromates and four insoluble chromates.
6. Why isn't $Cr(H_2O)_2(OH)_4^-$ shown on the chromium E_h-pH diagram when it is easily prepared in 1 M NaOH solution?
7. List the factors which determine the relative stability of Co(II) and Co(III).
8. Suggest some robust negatively charged complex ions which could be considered for use in tracking the movement of underground water. (Cations are strongly adsorbed on soil and are not satisfactory.)

PROBLEMS I

1. From the data in Table 21-3 estimate G^0 for Cr(IV) and Cr(V).
2. Calculate equilibrium constants for the reactions below.

$$4Co(H_2O)_6^{3+} + 2H_2O = 4H^+ + 4Co(H_2O)_6^{2+} + O_2$$

$$4Co(NH_3)_6^{3+} + 2H_2O = 4H^+ + 4Co(NH_3)_6^{2+} + O_2$$

3. The pK_a of Cr^{3+} is 3.8. Calculate the pH of a 0.050 M solution of Cr(NO$_3$)$_3$.

4. From the data on p. 502 calculate the temperature at which k_1 for H_2O exchange on Cr(III) is 1.0×10^{-4}.
5. What is the half-life of $Cr(H_2O)_6^{3+}$ toward ligand exchange at 25°C in an acid solution?
6. Calculate $\Delta S^{\ddagger}$ for the reaction $Fe(H_2O)_6^{3+} + SCN^- = FeSCN(H_2O)_5^{2+} + H_2O$ using the data on p. 501.

PROBLEMS II
1. The reaction $Co(NH_3)_5Br^{2+} + H_2O \longrightarrow Co(NH_3)_5H_2O^{3+} + Br^-$ has the rate law given below. What is the order of the reaction?

$$\text{Rate} = k[Co(NH_3)_5Br^{2+}][OH^-]$$

2. At 25°C the rate constant for the reaction in Problem 1 is $k = 1.0\ M^{-1}\ \text{sec}^{-1}$. What is the half-life of $Co(NH_3)_5Br^{2+}$ in a buffer of pH = 7?
3. If the activation energy for the reaction in Problem 1 is 24 kcal mole^{-1}, what will the rate constant be at 60°C?
4. Which of the following activated complexes are consistent with the rate law in Problem 1?

$(Co(NH_3)_5^{3+})^{\ddagger}$ $(Co(NH_3)_4NH_2Br^+)^{\ddagger}$

$(Co(NH_3)_5H_2OBr^{2+})^{\ddagger}$ $(Co(NH_3)_5OHBr^+)^{\ddagger}$

5. How many d electrons are present in Cr(IV)?
6. Calculate $Z + S$ for Cr(IV).
7. From the data in Table 3-7, estimate the ionic radius of Cr(IV).
8. Estimate the polarizing ability of Cr(IV).
9. Predict the acid base behavior of Cr(IV) from its polarizing ability.
10. Estimate the radius ratio $r_{O^{2-}}/r_{Cr^{4+}}$.
11. Predict the coordination number for Cr(IV).
12. Repeat Problems 5–11 with Cr(V) in place of Cr(IV).
13. The rate law for the reaction $3VO^{2+} + HCrO_4^- + H^+ \longrightarrow 3VO_2^+ + Cr^{3+} + H_2O$ is given below. What is the formula of the activated complex?

$$\text{Rate} = k[VO^{2+}]^2[HCrO_4^-][VO_2^+]^{-1}$$

14. What chromium species reacts with VO^{2+} in the rate determining step?

SUGGESTED READINGS
HARRIS, G. M., *Chemical Kinetics.* Boston: D. C. Heath, 1966.
LANGFORD, C. H., and T. R. STENGLE, *Ann. Rev. Phys. Chem.*, **19**: 193 (1968).
SYKES, A. G., *Adv. Inorg. Chem. Radiochem.*, **10**: 153 (1967).

> Nitrogen in some available form we have long known as absolutely essential to plant growth but the problem has been to "fix" air nitrogen, to convert it into soluble compounds acceptable to growing plants. There is no lack of "free" nitrogen, free as air, for 20,000,000 tons of this gas rest on every square mile of earth.*
>
> HARRY N. HOLMES

Chapter 22 The Chemistry of Nitrogen

I. Elemental Nitrogen

Although nitrogen had been prepared earlier the discovery is usually credited to Daniel Rutherford in 1772 because he was the first to distinguish nitrogen from carbon dioxide. The name nitrogen comes originally from the presence of nitrogen in sodium nitrate ($NaNO_3$), a compound that had been used for thousands of years under names derived from the Greek, nitron, and Egyptian, ntry.

Lavoisier proposed the name azote (a-, without, plus zoe, Greek for life) for nitrogen because it does not support respiration.

Nitrogen can be prepared from air. Nearly any metal is satisfactory as a reducing agent to remove the oxygen, although some require heating. The commercial preparation of nitrogen uses the difference in boiling point of oxygen (bp, $-218.4°C$) and nitrogen (bp, $-209.86°C$) for the separation.

Nitrogen is most important as an atmospheric diluent. The dilution of oxygen reduces the rate of reactions involving oxygen.

Elemental nitrogen is characterized by its chemical inertness. The reactivity of N_2 is, of course, related to the strength of the bonds in the molecule. The total bond energy of the N_2 molecule may be obtained from the enthalpy of formation of N^* atoms.

$$1/2 N_2 \xrightarrow{\Delta H = 113} N \xrightarrow{SCSE = 48} N^*$$

$$N_2 \longrightarrow 2N^* \quad \Delta H = 322 \text{ kcal mole}^{-1}$$

*Out of the Test Tube (New York: Emerson Books, Inc., 1943)

Table 22-1 *Calculation of Bond Energies*

Compound	H^0, kcal mole^{-1}	Bond	Bond energy, kcal mole^{-1}
$NH_3(g)$	−11.0	N-H	109
$N_2H_4(g)$	22.8	N-N	71
$NF_3(g)$	−29.8	N-F	83
$N_2F_4(g)$	−1.7	N-N	68

This (322 kcal mole^{-1}) is an exceptionally large value even for three bonds.   :N:::N:

An *estimate* of the N-N single bond energy in the N_2 molecule may be obtained from the data for NH_3, N_2H_4, NF_3, and N_2F_4 given in Table 22-1. The data for NH_3 and NF_3 are used to obtain the N-H and N-F bond energies. These values are then used to estimate the N-N bond energy in N_2H_4 and N_2F_4. From this estimate less than one third of the total bond energy in nitrogen is due to σ bonding.

The energy of an N-N π bond must be about 126 kcal mole^{-1} to account for the bonding in N_2. The fact that this is an exceptionally strong bond accounts for the stability and inertness of N_2.

II. Compounds of Nitrogen

A. IONIC COMPOUNDS OF NITROGEN

Nitrogen forms simple compounds (nitrides, imides, and amides) with a number of metallic elements. The electronic structure of the nitride, imide, and amide ions are shown in Fig. 22-1. The metal-nitrogen bonds in these compounds are generally quite polar because of the high electronegativity of nitrogen. These compounds have high melting and boiling points (AlN sublimes at 2000°C), or they decompose before the melting point is reached (Na_3N decomposes at 300°C).

The formulas and the enthalpy of formation of a number of nitrides are given in Table 22-2. Definite trends are evident from these data. In general the stability of the nitrides decreases down a group. As a result Na_3N is unstable when heated and Cs_3N is difficult to prepare.

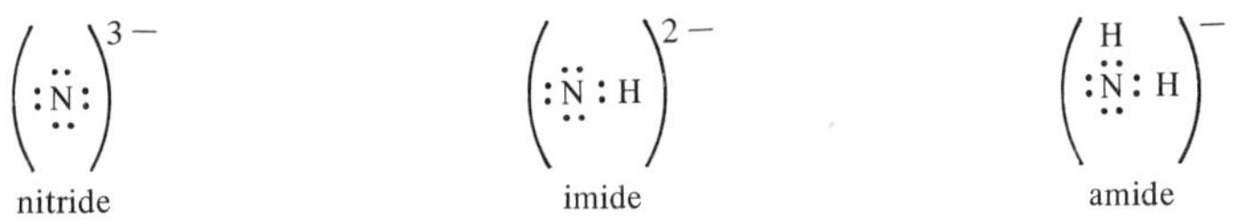

$$\left(:\ddot{\underset{..}{N}}:\right)^{3-} \qquad \left(:\ddot{\underset{..}{N}}:H\right)^{2-} \qquad \left(:\overset{H}{\underset{..}{\ddot{N}}}:H\right)^{-}$$

    nitride   imide   amide

Figure 22-1 *Electronic structures of some nitrogen anions.*

Ch. 22 The Chemistry of Nitrogen

Table 22-2 *Enthalpy of Formation of Nitrides (kcal mole^{-1} of nitrogen atoms)*

IA	IIA	IIIB	IVB	VB	VIB	VIIB	VIII			IB	IIB	IIIA	IVA	VA	VIA	VIIA	Inert gases
												BN −60	C_2N_2 +37	N_2 0.0	NO −22	NF_3 −28	
NH_3 −11																	
Li_3N −47	Be_3N_2 −68											AlN −76	Si_3N_4 −44	P_3N_5 −71.4	N_4S_4 +32	NCl_3 +55	
	Mg_3N_2 −55																
	Ca_3N_2 −52	ScN −75	TiN −80	VN −52	CrN −29	Mn_4N −30	Fe_4N −3	Co_3N −2	Ni_3N 0	Cu_3N +18	Zn_3N_2 −3	GaN −25	Ge_3N_4 −4				
	Sr_3N_2 −47	YN −72	ZrN −87	Nb_2N −61						Ag_3N +61	Cd_3N_2 −19	InN −5			NSe +42		
	Ba_3N_2 (−43)	LaN −71	HfN −88	Ta_2N −65													

518

On going from group IA to group IIA there is an increase in stability. In rows 4, 5, and 6 the group IV nitrides show exceptional stability.

From the relative acidity of water and ammonia we can deduce that the amide ion (NH_2^-) is more basic than the hydroxide ion. Because of the higher charge on the imide (NH^{2-}) and nitride (N^{3-}) ions, they are even stronger bases than amide ions. As a result all amides, imides, and nitrides react with water to form ammonia.

$$Mg_3N_2 + 6H_2O \longrightarrow 3Mg(OH)_2 + 2NH_3 \quad \Delta H = -220 \text{ kcal mole}^{-1}$$

$$AlN + 3H_2O \longrightarrow Al(OH)_3 + NH_3 \quad \Delta H = -34 \text{ kcal mole}^{-1}$$

$$LiNH_2 + H_2O \longrightarrow Li^+ + OH^- + NH_3 \quad \Delta H = -40 \text{ kcal mole}^{-1}$$

In some cases, although the reaction is highly exothermic

$$2TiN + 3H_2O \longrightarrow Ti_2O_3 + 2NH_3 \quad \Delta H = -48 \text{ kcal mole}^{-1}$$

the insoluble oxide forms a protective coating on the solid nitride. In general the reaction is extremely slow when a highly insoluble oxide is formed.

The nitrides of many of the transition metals listed in Table 22-2 do not represent the expected stoichiometry. Compounds such as Fe_4N are more metallic than saltlike in their properties. They are known as interstitial compounds because they consist of a metallic lattice in which nitrogen atoms occupy the spaces between the metal atoms.

From the trends shown in Table 22-2 it is possible to estimate the enthalpy of formation of nitrides and understand the fact that some nitrides such as gold nitride (Au_3N) are highly explosive, and others such as thallium nitride (TlN) probably cannot be prepared. The values given for yttrium nitride (YN) and hafnium nitride (HfN) in this table are estimates based on the observed trends. If the trend in group IIIA is continued, it seems quite unlikely that TlN exists. The most stable nitrides fall along a line from beryllium to zirconium.

Compounds such as BN, S_4N_4, and P_3N_3 do not, of course, contain N^{3-} ions. These compounds are covalent nitrides.

B. COVALENT NITROGEN COMPOUNDS

There are many covalent nitrogen compounds in addition to the covalent nitrides. These include ammonia, the amino acids, and the oxides of nitrogen.

1. Ammonia

Ammonia is probably the most important industrial nitrogen compound. It was originally prepared by the decomposition of proteins or urea

$$H_2O + NH_2CONH_2 \xrightarrow{\Delta} CO_2 + 2NH_3$$
urea

However, the demand for fixed nitrogen for fertilizer, dyes, and explosives soon exhausted this supply. The extensive deposits of sodium nitrate in the desert regions of Chile were the principal source of fixed nitrogen in the last part of the nineteenth century. A number of industrial processes for fixing nitrogen were developed in the early years of this century, but by far the most successful was Haber's synthesis of ammonia in 1913. The equilibrium reaction for the preparation of ammonia was discussed in Chapter 9. The Haber process accounted for the production of about 10^7 tons of ammonia in the United States in 1966.

The reaction in the Haber process is $N_2 + 3H_2 = 2NH_3$.

2. Amino Acids and Proteins

Plants and animals require fixed nitrogen for the formation of amino acids and nucleic acids.

Amino acids contain an acid carboxyl group ($-\overset{\overset{O}{\|}}{C}-OH$) and a basic amino group ($-\overset{H}{\underset{|}{N}}-H$). The 20 amino acids that occur with the most frequency are listed in Table 22-3.

In a solution of an amino acid, the acid grouping (carboxyl group) is ionized and the amino group is protonated.

$$H-\overset{H}{\underset{H^+}{\underset{|}{N}}}-CH_2-\overset{\overset{O}{\|}}{C}-O^-$$
glycine

This form of an amino acid is known as a zwitterion.

The alpha amino acids are important because of the variety of side chains that they may have and because they may be linked together to form a carbon-nitrogen bond which is called a peptide bond. The products are referred to as dipeptides, tripeptides, and so forth. The dipeptide glycylglycine may be formed in the reaction

$$NH_2-CH_2-\overset{\overset{O}{\|}}{C}-OH + NH_2-CH_2-\overset{\overset{O}{\|}}{C}-OH \longrightarrow$$
glycine   glycine

$$NH_2-CH_2-\overset{\overset{O}{\|}}{C}-\overset{H}{\underset{|}{N}}-CH_2-\overset{\overset{O}{\|}}{C}-OH + H_2O$$
glycylglycine

Table 22-3 *The 20 Amino Acids Found in Proteins*

Name	Formula	pK_a	pK_b
Glycine	$CH_2(NH_2)COOH$	9.78	11.65
Alanine	$H_3CCH(NH_2)COOH$	9.87	11.65
Serine	$HOCH_2CH(NH_2)COOH$	9.15	11.79
Threonine	$H_3CHCOHCH(NH_2)COOH$		
Cysteine	$HSCH_2CH(NH_2)COOH$		
Cystine	$HOOC(NH_2)CHCH_2S—SCH_2CH(NH_2)COOH$	8.00	11.95
Methionine	$H_3CSCH_2CH_2CH(NH_2)COOH$	9.21	11.72
Valine	$(H_3C)_2CHCH(NH_2)COOH$	9.72	11.71
Leucine	$(H_3C)_2CHCH_2CH(NH_2)COOH$	9.74	11.67
Isoleucine	$H_3CCH_2CH(CH_3)CH(NH_2)COOH$	9.76	11.68
Proline	(pyrrolidine-2-carboxylic acid ring structure)	10.60	12.0
Hydroxyproline	(4-hydroxypyrrolidine-2-carboxylic acid ring structure)	9.73	12.08
Aspartic acid	$HOOCCH_2CH(NH_2)COOH$	3.86	11.93
Glutamic acid	$HOOCCH_2CH_2CH(NH_2)COOH$	4.07	11.90
Arginine	$H_2N(HN=)CNHCH_2CH_2CH_2CH(NH_2)COOH$	12.48	4.96
Lysine	$H_2NCH_2CH_2CH_2CH_2CH(NH_2)COOH$	10.53	5.05
Histidine	(imidazole)—$CH_2CH(NH_2)COOH$	9.18	7.90
Phenylalanine	(phenyl)—$CH_2CH(NH_2)COOH$	9.24	11.42
Tyrosine	HO—(phenyl)—$CH_2CH(NH_2)COOH$	9.11	11.80
Tryptophan	(indole)—$CH_2CH(NH_2)COOH$	9.39	11.62

The tripeptide glycylglycylglycine may be formed by the reaction of a third molecule of glycine with glycylglycine

$$NH_2-CH_2-\underset{\underset{O}{\|}}{C}-\underset{\underset{H}{|}}{N}-CH_2-\underset{\underset{O}{\|}}{C}-OH + NH_2-CH_2-\underset{\underset{O}{\|}}{C}-OH \longrightarrow$$

$$NH_2-CH_2-\underset{\underset{O}{\|}}{C}-\underset{\underset{H}{|}}{N}-CH_2-\underset{\underset{O}{\|}}{C}-\underset{\underset{H}{|}}{N}-CH_2-\underset{\underset{O}{\|}}{C}-OH + H_2O$$

Proteins are high molecular weight polypeptides, each with a definite sequence of amino acids. In the biosynthesis of a protein the individual amino acids are activated and ordered by first reacting with complex molecules (one for each amino acid) known as transfer RNA (ribonucleic acid).

Because there are 20 different amino acids, the number of possible combinations of amino acids in proteins is very great; this fact is responsible for the diversity of structure and function shown by natural proteins.

3. Oxides of Nitrogen

The most important group of simple nitrogen compounds are the

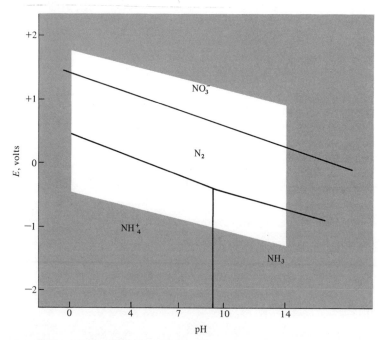

Figure 22-2 E_h-pH *diagram for nitrogen.*

oxides and the acids derived from them. The nitrate ion is the only oxygen compound of nitrogen to appear on the E_h-pH diagram given in Fig. 22-2. All other oxides of nitrogen are thermodynamically unstable.

The chemical reactions of the nitrogen oxides will be examined in the next section. Here we will be concerned with introducing a method for calculating the energy of the bond in the nitrogen oxide.

When only σ bonding is present the calculation of bond energies is straightforward. However the π bonding in nitrogen oxides and the oxyacids of nitrogen introduces some additional complications.

From the molecular orbital diagram in Fig. 22-3 or the octet rule structure, we see that there is a total of $2\frac{1}{2}$ bonds in NO. The five shared electrons correspond to one σ bond and $1\frac{1}{2}$ π bonds. If we use the symbols $\alpha_{\text{N-O}}$ and $\beta_{\text{N-O}}$ to represent the energy of the σ and π bonds respectively, the total bonding in NO can be represented by

$$\text{Total bond energy} = \alpha_{\text{N-O}} + 1\tfrac{1}{2}\beta_{\text{N-O}}$$

Given the enthalpy of formation of NO is 22 kcal mole^{-1} and the enthalpy of formation of the hypothetical atoms N* and O* the total bond energy can be calculated using the cycle shown in Fig. 22-4.

The total bond energy in any nitrogen-oxygen compound can be calculated in a similar manner. For example, there are three σ bonds and one π bond in the nitrate ion,

Octet rule structure for NO

3 N-5-O 3

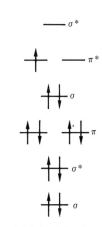

Figure 22-3 *Molecular orbital diagram for NO.*

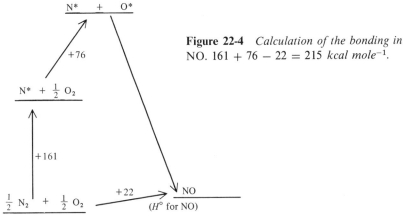

Figure 22-4 *Calculation of the bonding in NO. 161 + 76 − 22 = 215 kcal mole^{-1}.*

Ch. 22 The Chemistry of Nitrogen

The bonding in the nitrate ion can be represented by

$$\text{Total bond energy} = 3\alpha_{\text{N-O}} + C\beta_{\text{N-O}}$$

where the constant C is included to allow for the fact that the π bonding p orbital of the nitrogen overlaps with p orbitals on all three oxygen atoms simultaneously or, alternatively, that resonance is involved. The net contribution of the π bond in nitrate to the total bond energy should be greater than it is in NO. In this case the electrons are more delocalized than they are in NO. To a reasonable degree of approximation the stabilization in cases such as this can be included by multiplying β by the square root of the number of equivalent ligands that are involved in the bonding. In this case $C = \sqrt{3}$.

A large number of nitrogen-oxygen compounds that involve π bonding are known and the *N-O bond energy is not expected to be exactly the same in two different compounds* but an average value can be estimated. One way of estimating the average value of $\alpha_{\text{N-O}}$ and $\beta_{\text{N-O}}$ is to consider a large number of compounds. The total bond energy of each of the compounds in Table 22-4 is written in terms of $\alpha_{\text{N-O}}$ and $C\beta_{\text{N-O}}$. A value for the total bond energy in any compound can then be obtained if a single value of $\alpha_{\text{N-O}}$ and $\beta_{\text{N-O}}$ (along with $\alpha_{\text{N-N}}$, $\alpha_{\text{N-H}}$ and $\alpha_{\text{O-H}}$ where needed) can be assigned. The calculated values shown in the table were obtained using $\alpha_{\text{N-O}} = 65$ kcal mole^{-1}, $\beta_{\text{N-O}} = 101$ kcal mole^{-1}, $\alpha_{\text{N-N}} = 70$ kcal mole^{-1}, $\alpha_{\text{N-H}} = 109$ kcal mole^{-1}, and $\alpha_{\text{O-H}} = 119$ kcal mole^{-1}.

The experimental values listed in the table are obtained from H^0 and H^* values using a Born Haber cycle calculation. The agreement between the experimental and calculated energies indicates that the

The π bonding in gaseous HNO$_3$ has been estimated as half way between a bond distributed over two positions and one distributed equally over three positions.

Table 22-4 *A Comparison of Calculated and Experimental Nitrogen-Oxygen Bond Energies (binding present, kcal mole^{-1})*

Compound	H^0	Exptl.	Calc*	
NH$_2$OH	−8	401	402 = $2\alpha_{\text{N-H}} + \alpha_{\text{O-H}} + \alpha_{\text{N-O}}$	
NO	22	215	217 =	$\alpha_{\text{N-O}} + 1.5\beta_{\text{N-O}}$
NO$_3$	17	372	370 =	$3\alpha_{\text{N-O}} + \sqrt{3}\beta_{\text{N-O}}$
N$_2$O$_3$	20	530	509 = $\alpha_{\text{N-N}}$	$+3\alpha_{\text{N-O}} + (1 + \sqrt{2})\beta_{\text{N-O}}$
N$_2$O$_4$	2	624	616 = $\alpha_{\text{N-N}}$	$+4\alpha_{\text{N-O}} + 2\sqrt{2}\beta_{\text{N-O}}$
HNO	24	265	275 = $\alpha_{\text{N-H}}$	$+\alpha_{\text{N-O}} + 1.0\beta_{\text{N-O}}$
HNO$_2$	−18	383	373 = $\alpha_{\text{O-H}}$	$+2\alpha_{\text{N-O}} + \sqrt{1.5}\beta_{\text{N-O}}$
HNO$_3$	−32	473	474 = $\alpha_{\text{O-H}}$	$+3\alpha_{\text{N-O}} + \sqrt{2.5}\beta_{\text{N-O}}$
NO$_2^-$		278	272 =	$2\alpha_{\text{N-O}} + \sqrt{2}\beta_{\text{N-O}}$
NO$_3^-$		383	370 =	$3\alpha_{\text{N-O}} + \sqrt{3}\beta_{\text{N-O}}$

*Calculations made for $\alpha_{\text{N-O}} = 65$ kcal mole^{-1} and $\beta_{\text{N-O}} = 101$ kcal mole^{-1}.

Table 22-5 *Bond Energy Parameters $\alpha_{N\text{-}X}$ and $\beta_{N\text{-}X}$ for Bonds to Nitrogen*

	C	N	O	F
α	85	70	65	83
β	90	126	101	
	Si	P		Cl
α	98	92		64

energy of the bonds in these molecules may be represented by the α and β parameters. *This method of estimating bond energies provides a direct method for obtaining the total bond energy of a compound or a method of estimating H^0.*

An important feature of the data given in Table 22-4 is the fact that exceptionally strong π bonding between nitrogen and oxygen must be postulated to account for it. This is a feature that accounts for a number of aspects of nitrogen chemistry. Table 22-5 gives the bond energy parameters for nitrogen bonded to most of the other nonmetals.

III. Chemical Reactions

From its position in the periodic table nitrogen is expected to form compounds in which the oxidation state ranges from -3 to $+5$. Only the oxidation states -3, 0, and $+5$ (see Fig. 22-2) are thermodynamically stable; however, the decomposition of compounds involving other oxidation states is often quite slow and, as a result, there are a large number of nitrogen oxides.

A. REACTIONS OF NITROGEN OXIDES

The relative stability of the nitrogen oxides can be used to illustrate the importance of nitrogen-oxygen π bonding. Nitrogen pentoxide (N_2O_5), in which nitrogen has an oxidation number of $+5$, can be prepared, but it decomposes rapidly to nitrogen dioxide and oxygen according to the equation

$$2N_2O_5(g) = 4NO_2(g) + O_2(g)$$

The entropy change that accompanies this reaction is easily understood from the change in the number of moles. The enthalpy change can be understood if we examine the electronic structures of N_2O_5, NO_2 and O_2.

The electronic structure of N_2O_5 is given in Fig. 22-5. In this mole-

$2N_2O_5(g) = 4NO_2(g) + O_2(g)$
$\Delta H = +26.3 \text{ kcal mole}^{-1}$
$\Delta S = 108.4 \text{ eu}$
$K_{25°C} = 2.5 \times 10^4$

Ch. 22 The Chemistry of Nitrogen

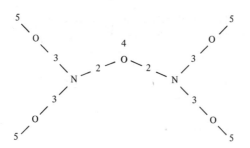

Figure 22-5 *An octet rule structure for* N_2O_5.

Electronic structure of NO_2

One resonance form of NO_3 *is*

:Ö:
Ö::N:Ö:

There is another form of NO_3 *that can be formed from NO and* O_2. *This molecule presumably has the structure*

2 3
4 O-4-N-2-O-3-O 5

In this structure there are a total of three σ bonds and $1\frac{1}{2}$ *π bonds.*

:N=N=Ö:

cule each nitrogen atom is involved in three σ bonds and one π bond.

In one of the products of the decomposition (NO_2) the seven bonding electrons form two N-O σ bonds and $1\frac{1}{2}$ π bonds. In oxygen there is one σ bond and 2 halves of π bonds. The π bonding in NO_2 and O_2 stabilizes the products, and the decomposition of N_2O_5 is not highly endothermic.

The simplest mechanism for the decomposition of N_2O_5 would involve the N-O bonds in the N-O-N link. If one of these bonds breaks the products would be NO_3 and NO_2. In NO_3 nine electrons should be shared to satisfy the octet rule. However, a central nitrogen atom cannot share more than eight electrons, so one p nonbonding orbital on the three oxygens is left unfilled. The unpaired electron on an oxygen can form an additional bond without reducing the number of N-O bonding electrons. As a result NO_3 is a very reactive molecule. In the next step of the decomposition a collision between NO_3 and NO_2 results in the formation of an O_2NO—ONO bond.

$$NO_2 + NO_3 \longrightarrow \left(\begin{array}{c} O \\ N-O\cdots O-N \\ O \end{array} \right)^{\ddagger} \longrightarrow NO + O_2 + NO_2$$

This is the rate determining step in the decomposition of N_2O_5.

The subsequent oxidation of NO to NO_2, which is much faster, presumably proceeds through the reaction of NO with another NO_3 radical.

$$NO + NO_3 \longrightarrow 2NO_2$$

The mechanism for the decomposition of N_2O_5 is

$$N_2O_5 \longrightarrow NO_2 + NO_3$$
$$NO_2 + NO_3 \longrightarrow (N_2O_5)^{\ddagger} \longrightarrow NO + O_2 + NO_2$$
rate determining step
$$NO + NO_3 \longrightarrow 2NO_2$$

The activated complex involved in the reaction has the stoichiometry $(N_2O_5)^{\ddagger}$. Even though the structure of the activated complex (Fig. 22-6) is quite different from the structure of N_2O_5, the reaction follows a simple first order rate law

$$\text{Rate} = k[N_2O_5]$$

The activation energy is 24.6 kcal mole^{-1}.

The unpaired electron in NO_3 is responsible for the extreme reactivity of the molecule. This is usually the case; however, there are examples of molecules containing unpaired electrons that are not highly reactive. NO and NO_2 are not highly reactive. If these molecules were to form another bond, the N-O π bonding would be substantially weakened and there would be little net energy gain. Consequently the dimerization of NO_2 is only slightly exothermic because the additional N-N σ bond is formed at the expense of a considerable amount of N-O π bonding.

At room temperature and 1 atm pressure NO_2 and N_2O_4 are present in equilibrium in a sample of NO_2. The equilibrium is a particularly easy one to observe since NO_2 is a reddish brown gas and the dimer is practically colorless. A sample of the gas becomes noticeably paler when cooled to 0°C.

NO dimerizes even less readily than NO_2. Again π bonding is destroyed upon dimer formation.

The slight tendency for NO and NO_2 to combine results in a blue compound that has the structure

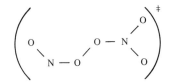

Figure 22-6 *The structure of the activated complex for the decomposition of* N_2O_5.

$NO_2 + NO_2 \rightleftharpoons N_2O_4$
$\Delta H = -13.64 \text{ kcal mole}^{-1}$
$\Delta S = -42.00 \text{ eu}$
$K_{25°C} = 0.146$

$2NO \rightleftharpoons N_2O_2$
$\Delta H \approx -1 \text{ kcal mole}^{-1}$
$\Delta S \approx -30 \text{ eu}$
$K_{25°C} \approx 10^{-6}$

An alternative structure

$$O=N-O-N=O$$

is less stable because α_{N-O} is smaller than α_{N-N} and because there are fewer positions over which the π bonding electrons may be distributed (there is less resonance energy).

The difference between the nitrogen-nitrogen and the nitrogen-oxygen σ and π bond energy is responsible for many of the commonly observed reactions in nitrogen chemistry. Ammonium nitrate decomposes to nitrous oxide (N_2O) and water when heated. During the reaction one of the oxygens of the nitrate ion is replaced by an amide group. This type of reaction results in the formation of

Ch. 22 The Chemistry of Nitrogen

an amide of nitric acid (NH$_2$NO$_2$). The amide of a number of acids can be made; however, this compound can not be isolated. Once the nitrogen-nitrogen single bond is formed, the loss of a second water molecule allows the formation of a second nitrogen-nitrogen bond.

$$NH_4^+ + NO_3^- \longrightarrow (NH_3 + HNO_3) \longrightarrow \left(H_2N-N\underset{O}{\overset{O}{\diagup\!\!\!\!\diagdown}} \right) + H_2O$$

$$\longrightarrow N=N=O + H_2O$$

There are many similar reactions in which a nitrogen-nitrogen bond is formed. Whenever this results in a grouping such as

$$\underset{N-N}{H\diagdown\diagup OH} \quad \text{or} \quad \underset{N-N}{H\diagdown\diagup O}$$

the reaction goes further to give a double or triple bond.

This sequence of reactions occurs in the preparation of nitrogen from ammonium nitrite.

$$NH_4NO_2 \xrightarrow{\Delta} (H_2NNO + H_2O) \longrightarrow N_2 + 2H_2O$$

and in the formation of azide ion (N$_3^-$) from nitrous oxide and the amide ion

$$NH_2^- + N_2O \longrightarrow \left(\underset{N_-}{H_2N-N=O} \right) \longrightarrow N=N=N^- + H_2O$$

A similar mechanism is involved in the formation of hyponitrous acid (N$_2$(OH)$_2$)

$$NH_2OH + HNO_2 \longrightarrow \left(HO-NH-N\underset{O}{\overset{O}{\diagup\!\!\!\!\diagdown}} \right) + H_2O \longrightarrow$$

$$HO-N=N-OH + H_2O$$

Hydroxylamine (NH$_2$OH) contains nitrogen in the -1 oxidation state. The corresponding compound in which a hydroxide group is replaced by amide is known as hydrazine (N$_2$H$_4$). Both hydroxylamine and hydrazine can be prepared from ammonia in a three-step sequence.

$$NH_3 + Cl \longrightarrow NH_2 + HCl$$

$$NH_2 + Cl_2 \longrightarrow NH_2Cl + Cl$$

$$NH_2Cl + NH_3 \longrightarrow N_2H_4 + HCl$$

$$NH_3 + Cl \longrightarrow NH_2 + HCl$$

$$NH_2 + Cl_2 \longrightarrow NH_2Cl + Cl$$

$$NH_2Cl + H_2O \longrightarrow NH_2OH + HCl$$

B. NITRATION AND RELATED REACTIONS

Nitrogen in the +5 (V) oxidation state (HNO_3) is a strong oxidizing agent as well as a strong acid. The preparation of nitric acid was discussed in Chapter 10. The intermediates involved in the preparation of nitric acid (NO and NO_2) are formed when the acid acts as an oxidizing agent.

The nitrate ion must be classified as a robust complex. Most of the ligand exchange reactions of nitrogen(V) that have been studied go through a two-coordinate intermediate instead of a four-coordinate complex. The intermediate ion NO_2^+ is formed under very acid conditions

$$HNO_3 + H^+ \rightleftharpoons NO_2^+ + H_2O$$

The ion is a very reactive Lewis acid that can be used to form nitrates from alcohols and nitro compounds of benzene and similar organic molecules

$$R-OH + NO_2^+ \longrightarrow R-ONO_2 + H^+$$

$$C_6H_6 + NO_2^+ \longrightarrow C_6H_5NO_2 + H^+$$

The structural formulas of several organic nitrates are given in Fig. 22-7. Glyceryl trinitrate (misnamed trinitroglycerine) is the principal component of dynamite.

Nitrous acid, a weak acid with $K_a = 4.5 \times 10^{-4}$, has the formula HNO_2. It decomposes in acid solutions in a matter of minutes to NO and NO_2. The nitrite ion (NO_2^-) is relatively stable towards disproportionation. The alkali metal nitrites are easily prepared by the thermal decomposition of the nitrates.

$$2NaNO_3 \longrightarrow O_2 + 2NaNO_2$$

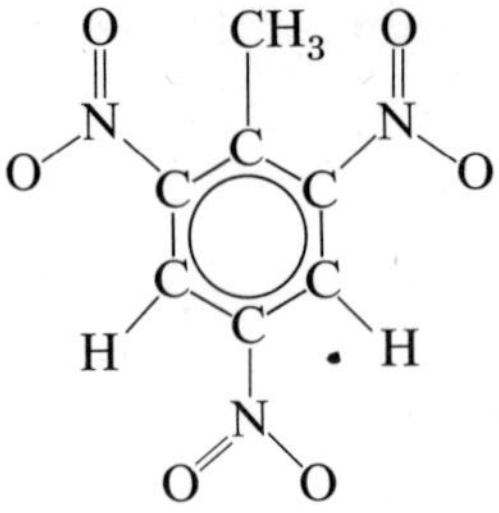

Ch. 22 The Chemistry of Nitrogen

trinitrotoluene

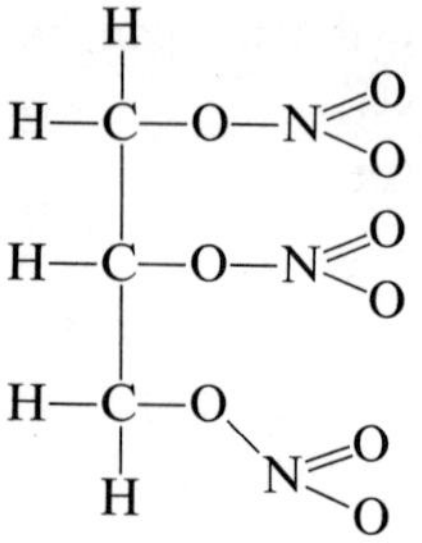

glyceryl trinitrate

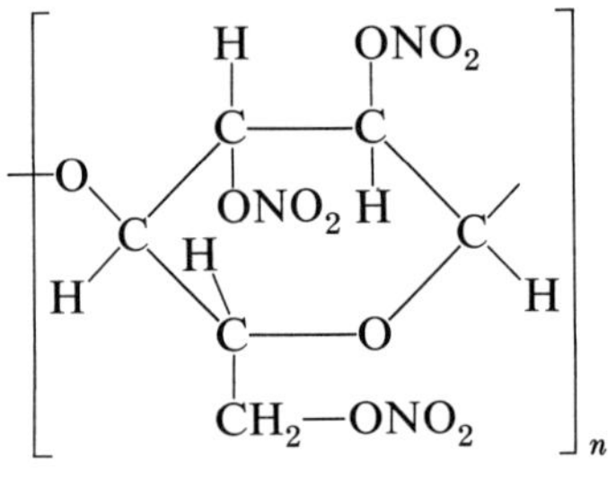

nitrocellulose

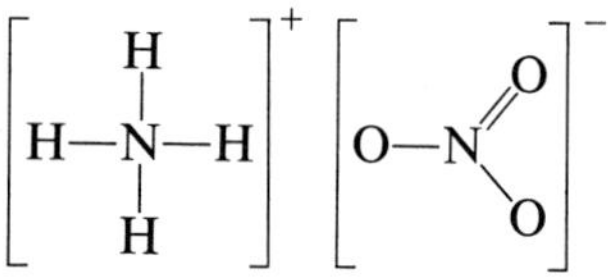

ammonium nitrate

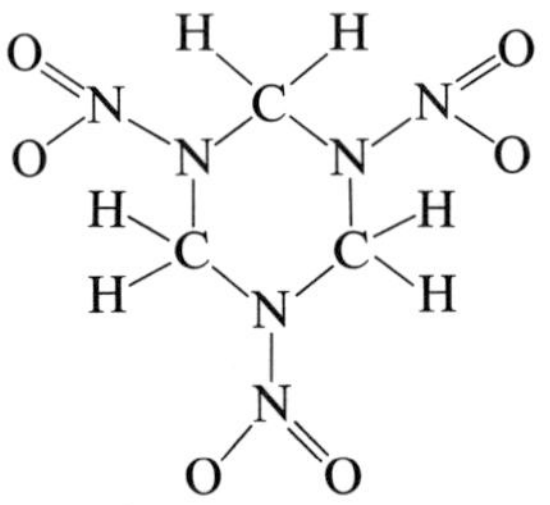

cyclotrimethylenetrinitramine

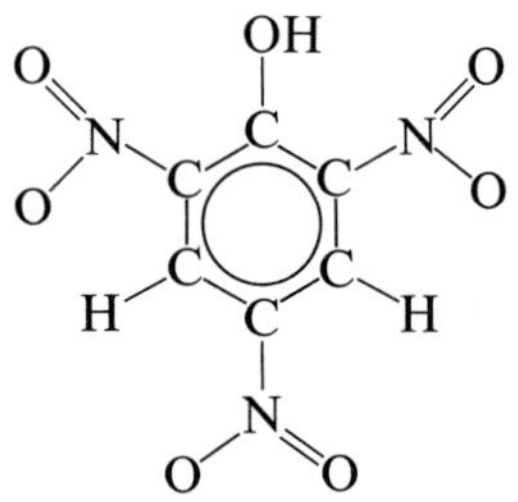

picric acid

Figure 22-7 *Several common explosives.*

QUESTIONS
1. Which of the following elements will react with nitrogen at high temperature: Na, Mg, Al, Si, P?
2. Under what conditions can Na_3N be formed?
3. Write an equation to represent the reaction of an amide and an imide with water.

Sec. III Chemical Reactions

4. How much π bonding is present in NO_3^-?
5. Estimate the bond angle in NO_2 and N_2O.
6. Write a set of reactions for forming the azide ion, N_3^-.
7. Why is lead azide an explosive?
8. Write a set of equations for the formation of NH_4NO_3 from N_2.
9. What is an amino acid?
10. Write the equation for amide formation between acetic acid and glycine.
11. What products would you expect from the reaction of nitrous acid with glycine?
12. Why is the formula of nitric acid HNO_3 instead of H_3NO_4?
13. Write a mechanism for the formation of nitroglycerine. Suggest reasons why sulfuric acid is used as a catalyst.
14. Draw the molecular orbital diagram for NO_3^- showing the correct number of valence electrons.
15. Complete and balance the following chemical equations:

$$\_Fe + \_NO_3^- + \_H^+ \longrightarrow$$

$$\_NH_2OH + \_HNO_2 \longrightarrow$$

$$\_Cu^{2+} + \_NH_3 \longrightarrow$$

$$\_NO_2 + \_H_2O \longrightarrow$$

$$\_Ag(NH_3)_2^+ + \_H^+ \longrightarrow$$

16. Do you think H_2N_2 would be a stable molecule? Explain. Under what conditions would it be most stable?
17. When hydrazine is oxidized by one-electron oxidizing agents, N_2H_3 radicals are formed which should be able to dimerize to N_4H_6. Suggest a mechanism for the rapid decomposition of N_4H_6 to N_2 and NH_3.
18. Calculate H^0 for NI_3 using Pauling's expression, $\alpha_{A-B} = (\alpha_{A-A}\alpha_{B-B})^{1/2} + 30(\Delta\chi)^2$ to estimate the N-I bond energy.

PROBLEMS I

1. The enthalpy of formation of AlN is -76 kcal mole^{-1}. What is the Al-N bond energy?
2. If the Na-N bond energy is about 84 kcal mole^{-1}, estimate H^0 for Na_3N.
3. Using the result from Problem 2, calculate ΔH for the decomposition of Na_3N. If $\Delta S = 45$ eu for this reaction, at what temperature will Na_3N decompose?
4. From the data in Table 22-5 calculate H^0 for NF_3.
5. H^0 for N_2 is 0.00. Calculate β_{N-N} assuming $\alpha_{N-N} = 70$ kcal mole^{-1}.
6. Calculate H^0 for NO_2 using $\alpha_{N-O} = 65, \beta_{N-O} = 101$ kcal mole^{-1}.
7. Calculate the pH of a 0.10 M solution of glycine.
8. Given the electrode potentials below, calculate the equilibrium constant for the disproportionation of HNO_2 into NO_3^- and NO.

$$NO + 2H_2O = NO_3^- + 4H^+ + 3e^- \quad E^0 = +0.96 \text{ volt}$$

$$NO + H_2O = HNO_2 + H^+ + e^- \quad E^0 = +1.00 \text{ volt}$$

531

9. What is the empirical formula of a compound which is 60.5% O, 26.4% N, and 13.1% Li?
10. Calculate the enthalpy (H^0) of HNO_3(g) given $\alpha_{N-O} = 65$, $\beta_{N-O} = 101$, and $\alpha_{O-H} = 119$ kcal mole^{-1}.

PROBLEMS II

1. The following rate constants were observed for the gas phase decomposition of N_2O_5.* Calculate the activation energy for this reaction.

$$k = 7.87 \times 10^{-7} \text{ sec}^{-1} \text{ at } 273 \text{ K}$$
$$k = 1.35 \times 10^{-4} \text{ sec}^{-1} \text{ at } 308 \text{ K}$$
$$k = 4.87 \times 10^{-3} \text{ sec}^{-1} \text{ at } 338 \text{ K}$$

2. Calculate $\Delta S^{\ddagger}$ for the reaction $2N_2O_5(g) \longrightarrow 4NO_2(g) + O_2$.
3. H^0 for N_2O_5(g) is 2.7 kcal mole^{-1}, and for NO_2(g) $H^0 = 7.9$ kcal mole^{-1}. Calculate ΔH^0 for the decomposition of N_2O_5.
4. H^0 for NO_3 is $+17$ kcal mole^{-1}. Calculate ΔH^0 for the reaction $N_2O_5 \longrightarrow NO_2 + NO_3$. How does this compare to $\Delta H^{\ddagger}$ for the decomposition of N_2O_5?
5. H^0 for NH_3 is -11.02 kcal mole^{-1}. Calculate α_{N-H}.
6. If the N-N single bond energy, α_{N-N} is 70 kcal mole^{-1}, calculate H^0 for N_2H_4.
7. Calculate H^0 for N_2H_2 assuming $\beta_{N-N} = 126$ kcal mole^{-1}.
8. Using the value from Problem 7, estimate ΔH^0 for the reaction $N_2H_2 \longrightarrow N_2 + H_2$.
9. Draw the octet rule structure for NO.
10. Calculate H^0 for NO assuming $\alpha_{N-O} = 65$ and $\beta_{N-O} = 101$ kcal mole^{-1}.
11. What is the strength of the π bond in NO_3?
12. How does the result for Problem 11 compare to the value $\sqrt{3}\, \beta_{N-O}$?

*F. Daniels and E. H. Johnston, *J. Am. Chem. Soc.*, **43**: 53(1921).

Her eyes in torture fix'd, and anguish drear,
Hot, glaz'd, and wide, with lid-lashes all sear,
Flash'd phosphor and sharp sparks, without one cooling tear.

JOHN KEATS
"Lamia"

Chapter 23 The Chemistry of Phosphorus

I. Elemental Phosphorus

Elemental phosphorus was discovered in 1669. Although earlier alchemists (Paracelsus, 1540) reported effects that may have been due to elemental phosphorus, Hennig Brandt was the first person to prepare (in 1669) the element in quantity. At the time of Brandt's discovery white phosphorus was unique in that it glowed on exposure to air regardless of any prior exposure to light.

In the original method of preparation, a phosphorus(V) compound (impure sodium metaphosphate) was reduced with carbon from charred organic materials at a temperature of about 1000°C.

Because of the curious nature of phosphorus, Brandt was able to sell his process for around 200 Thalers ($200). Others who bought his process or developed their own made even more by revealing the secrets of the preparation to the nobility.

$$NaNH_4HPO_4 \cdot H_2O \xrightarrow{\Delta} NaPO_3 + NH_3 + 2H_2O$$

$$8NaPO_3 + 10C \xrightarrow{\Delta} 2Na_4P_2O_7 + P_4 + 10CO$$

A substance that glows in the dark after exposure to light is phosphorescent. Phosphorus is chemiluminescent.

Larger yields are achieved when silicon dioxide (SiO_2) is added to combine with the metals present (that is, Na or Ca).

$$4NaPO_3 + 2SiO_2 + 10C \longrightarrow P_4 + 10CO + 2Na_2SiO_3$$

$$4Ca_5F(PO_4)_3 + 30C + 18SiO_2 \longrightarrow 3P_4(g) + 30CO(g) + 2CaF_2 + 18CaSiO_3$$

The routine manufacture of white phosphorus for about $5.00 per g dates from about 1680. The substance remained an expensive curiosity until better sources were discovered about 100 years later.

By 1743 phosphorus had been isolated from a variety of plant and animal materials. It was first obtained from a mineral source [from $Pb_5Cl(PO_4)_3$] in 1780. The general use of phosphorus in matches dates from about this time. The mineral apatite, $Ca_5F(PO_4)_3$, is the predominate source of phosphorus today.

At 1000°C phosphorus consists of an equilibrium mixture of P_4 and P_2 molecules. As the gas is cooled the equilibrium is shifted and the substance exists almost exclusively as P_4 molecules which persist in the liquid and in solid white phosphorus (mp, 44°C; bp, 280°C).

The reactivity and physical properties of white phosphorus (low melting and boiling points and solubility in organic solvents) are the consequence of the molecular structure (P_4) of the solid. White phosphorus bursts into flame upon prolonged exposure to air. The substance is highly poisonous. It is readily absorbed through the skin and serious burns result.

When exposed to light or heated under pressure in the absence of air, white phosphorus changes to a red or violet form. Red phosphorus contains long chains of covalently bonded phosphorus atoms. It is insoluble in organic solvents and the melting point is quite high (590°C).

When either form of phosphorus is subjected to a pressure of around 35,000 atm or is heated at somewhat lower pressures, a third crystalline form may be produced. The crystal structure of black phosphorus consists of layers of covalently bonded phosphorus atoms, each with three nearest neighbors.

Neither red phosphorus nor black phosphorus reacts with air at an appreciable rate at room temperature. However, both forms will ignite at temperatures that can be produced by friction. Red phosphorus and P_4S_3 have replaced white phosphorus in the manufacture of matches.

The phosphorus-phosphorus bonds in white phosphorus are hydrolyzed in hot basic solution according to the equation

$$3H_2O + P_4 + 3OH^- \longrightarrow PH_3 + 3H_2PO_2^-$$

This reaction is very slow at a pH of 7, and white phosphorus can be conveniently stored under water. Thermodynamically the disproportionation of phosphorus is favored at all pHs and elemental phosphorus does not appear in the E_h-pH diagram (Fig. 23-1).

The chemical behavior of phosphorus is similar to that of nitrogen in many ways except that the phosphorus-phosphorus π bonding in phosphorus compounds is much weaker. This difference accounts for

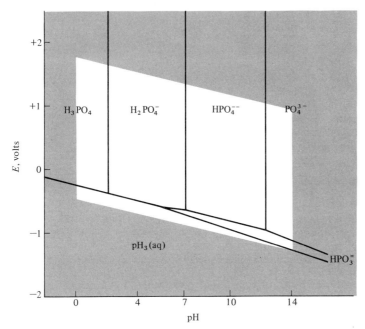

Figure 23-1 E_h-pH *diagram for phosphorus.*

the occurrence of P_4 molecules in contrast to the N_2 molecules of nitrogen. In addition phosphorus does not form oxides that are comparable to the oxides of nitrogen (NO, NO_2, and N_2O). As a result phosphorus is not found in a large variety of oxidation states. The only common oxidation states of phosphorus in compounds are -3, $+3$, and $+5$.

II. Compounds of Phosphorus

A. PHOSPHIDES

Phosphorus will react with some metals to form phosphides in which the phosphorus is in the -3 oxidation state (Na_3P, Ca_3P_2). These compounds react with water to form phosphine (PH_3).

The transition metals form some phosphides (FeP, Ni_2P, Mo_2P_2) in which the oxidation numbers are not clear. The overall properties of the compounds formed with the transition metals are more like those of a metal or semiconductor than a salt. The compounds are typically insoluble in water and dilute acids. They dissolve in an oxidizing acid, such as nitric acid, to produce H_3PO_4.

$$Ni_2P + 13H^+ + 9NO_3^- \longrightarrow 2Ni^{2+} + H_3PO_4 + 9NO_2 + 5H_2O$$

To balance an equation such as this, first write a balanced half-reaction such as

$$Ni_2P + 4H_2O \longrightarrow 2Ni^{2+} + H_3PO_4 + 5H^+ + 9e^-$$

Ch. 23 The Chemistry of Phosphorus

Table 23-1 *The Enthalpy of Formation of Some Phosphides (kcal mole^{-1})*

Compound	H^0	Compound	H^0
MnP	−23	Cu_3P	−54
Fe_3P	−39	CuP_2	−24
Fe_2P	−38	AgP_2	−11
FeP	−29	AgP_3	−16
FeP_2	−43	Zn_3P_2	−113
Co_2P	−47	Cd_3P_2	−27
CoP	−69		
Ni_3P	−54	GaP	−21
Ni_2P	−44	InP	−21
NiP_2	−40		
Ni_5P_2	−104		

Many of the transition metal phosphides have negative enthalpies of formation. The data in Table 23-1 indicate a maximum bond energy in the neighborhood of cobalt and nickel. Obviously something more than electronegativity difference is involved in the bonding in these compounds since the bond energies are comparable to that in Ca_3P_2 (α_{Ca-P} = 76 kcal mole^{-1}).

The difference in stability between the nitrides and phosphides of cobalt and nickel suggests that the empty $3d$ orbitals on phosphorus may be involved in the bonding. The involvement of d orbitals in the bonding between a metal atom and phosphorus is one aspect of the ease of polarization of phosphorus. As a result phosphorus compounds are particularly good ligands for soft cations. Many phosphides can be precipitated from aqueous solution by the addition of phosphine.

For the reaction
$3Cd^{2+} + 2PH_3 = Cd_3P_2 + 6H^+$
$\Delta H^0 = 24.4$ *kcal mole*$^{-1}$
$\Delta S \approx +5$ eu
$K \approx 10^{-17}$

$$3Cd^{2+} + 2PH_3(g) \longrightarrow Cd_3P_2 + 6H^+$$

Ammonia, of course, complexes with many metals, but it never loses H$^+$ to form nitrides.

$$3Cd^{2+} + 2NH_3(g) = Cd_3N_2 + 6H^+ \qquad \Delta H^0 = +115 \text{ kcal mole}^{-1}$$

$$Cd^{2+} + 4NH_3(g) = Cd(NH_3)_4^{2+} \qquad \Delta H^0 = -45 \text{ kcal mole}^{-1}$$

B. PHOSPHINE

Because phosphorus in the −3 oxidation state is easily polarized, it forms stable complexes with soft acids (cations with induced polarizing ability). With the softer cations, phosphine (PH$_3$) is a stronger base than ammonia.

The cations that do complex with phosphine generally react further to form phosphides. Therefore most of the studies of complexes containing phosphorus ligands have been conducted with substituted phosphines such as triethyl phosphine [$P(C_2H_5)_3$] or triphenyl phosphine [$P(C_6H_5)_3$]. The spectra and spin states of the complexes of these ligands indicate that phosphorus compounds create a high ligand field. In this respect as well as in the stabilization of low oxidation states of metals, phosphorus ligands are very similar to carbon monoxide.

In Table 10-4 the P-H bond energy in PH_3 was given as 85 kcal mole^{-1}. This value is very nearly the average of the P-P (67 kcal mole^{-1}) and H-H (104 kcal mole^{-1}) bond energies. Accordingly the enthalpy of formation of PH_3 is close to 0, (actually $+1.3$ kcal mole^{-1}). Since the entropy effect also favors the elements,

$$\tfrac{3}{2}H_2 + P(s) \longrightarrow PH_3(g) \qquad \Delta S = -22.0 \text{ eu}$$

the equilibrium constant for the formation of phosphine is small ($K = 10^{-6}$ at 25°C). At 1 atm pressure PH_3 is unstable with respect to decomposition; however, the reaction is very slow at ordinary temperatures. Because of the slow decomposition rate (the overvoltage effect), PH_3 is stable in water solution. (The E_h-pH diagram is given in Fig. 23-1.)

Phosphine reacts vigorously with air to give phosphorus(V) oxide (P_4O_{10}).

$$4PH_3 + 8O_2 \longrightarrow P_4O_{10} + 6H_2O$$

C. BOND ENERGIES IN PHOSPHORUS COMPOUNDS

One value of the P-Cl bond energy may be used to represent the total bond energy of the two stable chlorides of phosphorus, PCl_3 and PCl_5. The applicability of one value of the bond energy is illustrated in Table 23-2 where the values of 84 and 87 are the bond energies

Table 23-2 *Bond Energy Calculations for Phosphorus Chlorides (kcal mole^{-1})*

	H^0 for valence state atoms	Bonding*	H^0 Calc*	H^0 Exptl.
$PCl_3(g)$	186	252	−66	−68.6
$PCl_3(l)$	186	261	−75	−76.4
$PCl_5(g)$	244	336	−92	−89.6
$PCl_5(s)$	244	348	−104	−106.0

*The values $\alpha_{P-Cl} = 84$ and 87 are used for the gaseous and condensed phases, respectively.

Ch. 23 The Chemistry of Phosphorus

in the gaseous and condensed phases respectively. A value of $\alpha_{P-Br} = 75$ is similarly satisfactory for both PBr_3 and PBr_5.

The phosphorus-fluorine bond energy in phosphorus trifluoride (PF_3)($\alpha_{P-F} = 125$) is not at all satisfactory for phosphorus pentafluoride (PF_5). The value of H^0 for phosphorus pentafluoride is -381 kcal mole^{-1} and the total bond energy in this molecule is 575 kcal mole^{-1}. This amounts to 75 kcal mole^{-1} more than $4\alpha_{P-F}$. The octet rule cannot be used to describe the bonding in phosphorus pentafluoride satisfactorily.

Use H^0 and H^* for atoms and a Born Haber cycle for this calculation.

Similarly a single parameter cannot be used to represent the bonding in phosphorus-nitrogen and phosphorus-oxygen compounds. In both cases the total bond energy is larger than that given by the single parameter, α.

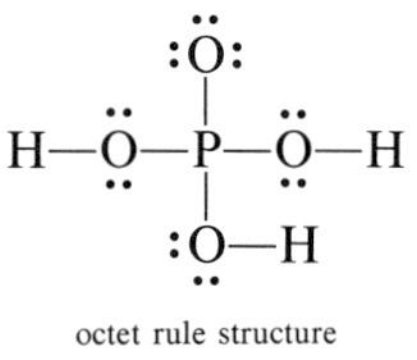

In order to obtain an expression that can be used to represent the total bonding in phosphorus-oxygen compounds, the bonding must be examined in more detail. The octet rule structure shown in Fig. 23-2, in which 14 valence electrons are shared, does not require that the phosphorus atom form π bonds. However the short phosphorus-oxygen bond lengths shown in Fig. 23-3 suggest that there is some double bond character in the phosphorus-oxygen bonds in some molecules. Since the s and p orbitals of phosphorus are used for σ bonding, π bonding in phosphorus-oxygen compounds (as well as phosphorus-nitrogen and phosphorus-fluorine) must involve the $3d$ orbitals of phosphorus.

octet rule structure

Figure 23-2 *The σ bonding in H_3PO_4.*

hydrogen phosphite ion

phosphate ion

pyrophosphate ion

phosphorous acid

phosphoric acid

triphosphate ion

phosphorus III

phosphorus V

polyphosphates

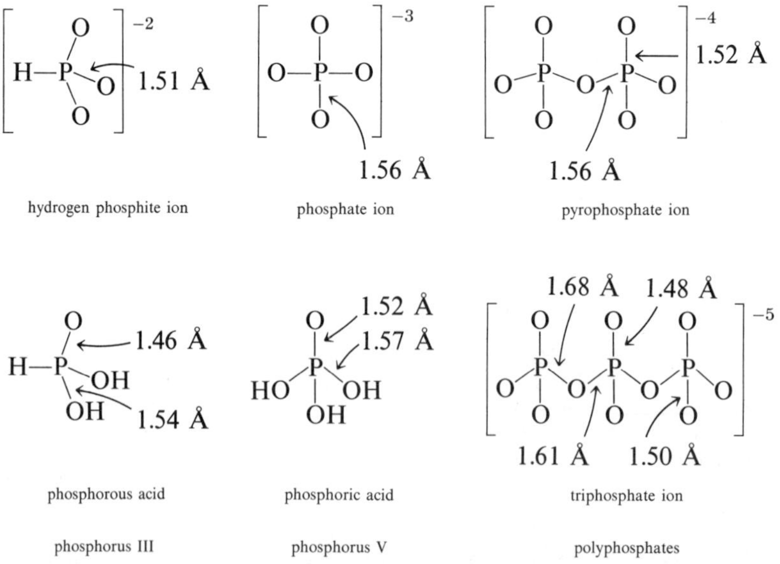

Figure 23-3 Bond lengths of P-O bonds.

When the d-p π overlap is significant, the total bond energy can be expressed in the form

$$\text{Total bond energy} = 4\alpha_{\text{P-O}} + \gamma_{\text{P-O}}$$

The parameter γ takes d-p π bonding into account. The value of γ will be large for compounds where the π bonding is particularly important and small or zero when the π bonding is negligible. In phosphorus-oxygen compounds the value of $\gamma_{\text{P-O}}$ for a particular compound will depend to a large extent upon the number of bonds formed by the oxygen atom. The value of γ and hence the amount of d-p π bonding is largest where oxygen forms bonds to one phosphorus atom only (oxo ligands). It is smaller for hydroxo ligands, and it is smallest when the oxygen atom is in a linkage such as P-O-P.

The value of $\alpha_{\text{P-O}}$ and $\gamma_{\text{P-O}}$ can be estimated from the data available in much the same way that α and β were estimated for nitrogen. Some of the data used to obtain these estimates for ionic phosphates and some other phosphorus compounds is presented in Table 23-3.

The bond energy data, shown in Table 23-4, for phosphorus and the elements of groups VA, VIA, and VIIA show the expected trends. Note that d-p π bonding is not important ($\gamma = 0$) in compounds containing phosphorus and the heavier elements. Most of the chemistry of phosphorus in the positive oxidation states is summarized by the phosphorus-nonmetal bond energy data in Table 23-4. The compounds formed between phosphorus and the heavier nonmetals react vigorously with water or alcohol because of the exceptional stability of phosphorus-oxygen bonds. The reactions of phosphorus trichloride with water and alcohol may be represented by the equations

$$PCl_3 + 3H_2O \longrightarrow H_2HPO_3 + 3HCl$$

$$PCl_3 + 3CH_3OH \longrightarrow P(OCH_3)_3 + 3HCl$$

Phosphorus trichloride reacts in a slightly different manner with acids, such as acetic acid

$$PCl_3 + 3CH_3COOH \longrightarrow H_2HPO_3 + 3CH_3COCl$$

This reaction is used for the formation of CH_3COCl and the chlorides of similar acids.

The first stage of the hydrolysis of phosphorus pentachloride is highly exothermic. The enthalpy of the reaction can be estimated from the bond energies of the substances involved. The enthalpy of the reaction is approximately equal to the energy released in the formation

Ch. 23 The Chemistry of Phosphorus

Table 23-3 *Thermodynamic Data for Phosphorus-Oxygen d-p π Bonding*

Compound	H^0	Bond energies used in calculation, kcal mole^{-1}	Calculated
P_4O_6	−392		α_{P-O} 104
$POBr_3$	−110	$\alpha_{P-Br} = 75$	γ_{P-O} 37
$POCl_3$	−143	$\alpha_{P-Cl} = 87$	40
		average for single oxygen atom $\gamma_{P-O} =$	38
P_4O_{10}	−713		51
H_3PO_2	−144	$\alpha_{P-H} = 85, \alpha_{O-H} = 126*$	47
H_3PO_3	−230	$\alpha_{P-H}, \alpha_{O-H}$	64
HPO_3	−227	α_{O-H}	64
H_3PO_4	−306	α_{O-H}	71
$NaH_3P_2O_7$	−603	$\alpha_{Na-O} = 140, \alpha_{O-H}$	82
$Na_2H_2P_2O_5$	−506	$\alpha_{Na-O}, \alpha_{P-H}$	85
$Na_2H_2P_2O_7$	−663	$\alpha_{Na-O}, \alpha_{O-H}$	92
Na_2HPO_3	−338	$\alpha_{Na-O}, \alpha_{O-H}$	94
$Na_3HP_2O_7$	−711	$\alpha_{Na-O}, \alpha_{O-H}$	96
$Ca(H_2PO_4)_2$	−744	$\alpha_{Ca-O} = 144, \alpha_{O-H}$	90
$Ba(H_2PO_4)_2$	−750	$\alpha_{Ba-O} = 141, \alpha_{O-H}$	94
KH_2PO_4	−375	$\alpha_{K-O} = 139, \alpha_{O-H}$	96
		average for dihydrogenphosphate(V) $\gamma_{P-O} =$	93
Na_2HPO_4	−417	$\alpha_{Na-O}, \alpha_{O-H}$	102
$CaHPO_4$	−435	$\alpha_{Ca-O}, \alpha_{O-H}$	106
		average for monohydrogenphosphate(V) $\gamma_{P-O} =$	104
Na_3PO_4	−460	α_{Na-O}	105
$Ca_3(PO_4)_2$	−989	α_{Ca-O}	118
		average for orthophosphates $\gamma_{P-O} =$	112

*$\alpha_{O-H} = 119$ kcal mole^{-1} in the gas phase, but the larger value, 126 kcal mole^{-1}, is applicable to the solids under consideration in this table.

of hydrogen-chlorine and phosphorus-oxygen bonds minus the energy required to break two of the oxygen-hydrogen bonds in water and one phosphorus-chlorine bond in phosphorus pentachloride. The exothermicity of the reaction is due in part to the stability of the phosphorus-oxygen bond.

$$PCl_5 + H_2O \longrightarrow POCl_3 + 2HCl$$

$$\Delta H \cong -2\alpha_{H-Cl} - \alpha_{P-O} - \gamma_{P-O} + 2\alpha_{O-H} + \alpha_{P-Cl}$$
$$\Delta H \cong -206 - 104 - 72 + 238 + 87 = -57 \text{ kcal mole}^{-1}$$

The reaction is even more exothermic than this because of the reaction of hydrogen chloride with water.

$$H_2O(l) + HCl(g) \longrightarrow H_3O^+ + Cl^- \quad \Delta H^0 = -17.89 \text{ kcal mole}^{-1}$$

When the bond energies approximately cancel the reaction is shifted to the right because of the reaction of hydrogen chloride with water.

The strength of phosphorus-oxygen bonds leads to a rather unusual structure for phosphorous acid (H_3PO_3). The structure expected by analogy with nitrous acid (HNO_2) and arsenious acid (H_3AsO_3) should have a lone pair of electrons on the phosphorus. This structure is shown in Fig. 23-4a. In comparing the structure 23-4a to 23-4b we see that in structure b there is one less oxygen-hydrogen bond and an increase of one phosphorus-hydrogen bond. Since an oxygen-hydrogen bond is stronger than a phosphorus-hydrogen bond (119 compared to 85 kcal mole^{-1}) structure (a) would be 34 kcal mole^{-1} more stable than structure (b) on this basis alone. However, this energy difference is counterbalanced by an increased amount of d-p π bonding in structure (b). Structure (b) is stabilized by d-p π bonding by about 52 kcal mole^{-1}, and this is the observed structure of the acid. Only two of the three hydrogens present in phosphorous acid ionize in solution.

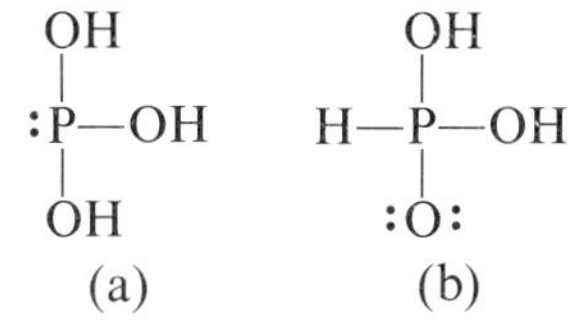

A phosphorus(I) acid, hypophosphorous acid (H_3PO_2), can be prepared, although it is quite unstable. With two phosphorus-hydrogen bonds this acid is actually about as strong as H_2HPO_3 (H_3PO_3) and H_3PO_4 ($K_a = 1.0 \times 10^{-2}$). The ionization constants of the oxoacids of phosphorus along with those of pyrophosphoric acid ($H_4P_2O_7$) are given in Table 23-5.

The exceptional strength of phosphorus-oxygen bonds is also responsible for the fact that phosphorus in the +5 (V) oxidation state

$$\text{P—Cl} + \underset{H}{\text{O—H}} \longrightarrow \text{P—O} + \text{H—Cl}$$
$$\underset{H}{}$$

$$\Delta H \approx -\alpha_{P-O} - \alpha_{H-Cl} + \alpha_{O-H} + \alpha_{P-Cl}$$
$$= -104 - 103 + 119 + 87 =$$
$$-1 \text{ kcal mole}^{-1}.$$

$$\begin{array}{cc}
\text{OH} & \text{OH} \\
| & | \\
:\text{P—OH} & \text{H—P—OH} \\
| & | \\
\text{OH} & :\text{O}: \\
(a) & (b)
\end{array}$$

Figure 23-4 *Possible structures for H_3PO_3.*

Table 23-4 *α and γ Parameters in Phosphorus Compounds (kcal mole^{-1})*

All of the data used to obtain the estimates of γ cannot be listed here.

α	$\alpha_{P-N} = 95$	$\alpha_{P-O} = 104$	$\alpha_{P-F} = 125$
γ	$\gamma_{P-N} = 50$	$\gamma_{P-O} = 72$*	$\gamma_{P-F} = 75$
α	$\alpha_{P-P} = 67$	$\alpha_{P-S} = 73$	$\alpha_{P-Cl} = 87$
γ	$\gamma_{P-P} = 0$	$\gamma_{P-S} = 0$	$\gamma_{P-Cl} = 0$
α			$\alpha_{P-Br} = 75$
γ			$\gamma_{P-Br} = 0$
α			$\alpha_{P-I} = 66$
γ			$\gamma_{P-I} = 0$

*P-O d-p π bonding varies considerably from one compound to another, see the text.

Ch. 23 The Chemistry of Phosphorus

Table 23-5 *Ionization Constants of Phosphorus Acids*

Acid	K_1	K_2	K_3	K_4
HH_2PO_2	1.0×10^{-2}			
H_2HPO_3	1.6×10^{-2}	7×10^{-7}		
H_3PO_4	7.1×10^{-3}	6.2×10^{-8}	4.5×10^{-13}	
$H_4P_2O_7$	1.4×10^{-1}	1.1×10^{-2}	2.9×10^{-7}	3.6×10^{-9}

is the stable form of phosphorus over most of the E_h-pH diagram.

As indicated above when phosphorus or phosphine is burned in air, a white smoke of phosphorus(V) oxide is formed. With a limited amount of air it is possible to stop the reaction after phosphorous(III) oxide (P_4O_6) has been produced. The structure of these oxides (shown in Fig. 23-5) is based on a tetrahedron of phosphorus atoms linked by P-O-P bonds. The bond lengths indicate that π bonding is involved in the short P-O bonds in P_4O_{10}. Both oxides react vigorously with water according to the equations

γ_{P-O} is larger for the P-O-H linkage than it is for the P-O-P linkage.

$$P_4O_6 + 6H_2O \longrightarrow 4H_2HPO_3 \qquad \Delta H^0 = -120 \text{ kcal mole}^{-1}$$

$$P_4O_{10} + 6H_2O \longrightarrow 4H_3PO_4 \qquad \Delta H^0 = -95 \text{ kcal mole}^{-1}$$

The equilibrium in reactions of this type lies to the right because of the increased possibility of d-p π bonding in the products.

Any compound containing phosphorus(V) in which an oxygen atom is partially π bonded to two atoms (as in P-O-P) is said to contain a high-energy phosphate bond. Simple examples of a high-energy phosphate bond are found in the pyrophosphate ion and phosphopyruvic acid.

$$\left(H-O-\underset{\underset{O}{|}}{\overset{\overset{O}{\|}}{P}}-O-\underset{\underset{O}{|}}{\overset{\overset{O}{\|}}{P}}-O-H \right)^{2-} \qquad \left(HO-\underset{\underset{O}{|}}{\overset{\overset{O}{\|}}{P}}-O-\overset{\overset{O}{\|}}{C}-\overset{\overset{O}{\|}}{C}-CH_3 \right)^{-}$$

pyrophosphate $\qquad\qquad$ phosphopyruvic acid

The principal energy transfer and storage agent in many biological systems, adenosine triphosphate (ATP), contains two high-energy P-O-P linkages (Fig. 23-6). The free energy change for the hydrolysis of ATP to adenosine diphosphate (ADP) at a pH of 7 is about -8 kcal mole^{-1}.

In ADP one $-\!\!\left(\!\!\begin{array}{c}O\\|\\P-O\\|\\O\end{array}\!\!\right)^{-}\!\!-$ group is hydrolyzed from ATP.

$$R-P_3O_{10}^{4-} + H_2O \rightleftharpoons RP_2O_7^{3-} + H_2PO_4^{-}$$

$$\Delta G^0 = -8 \text{ kcal mole}^{-1}$$

ATP is suitable for energy storage because phosphates undergo ligand exchange slowly (they are robust complexes). Thus, the hydrolysis reaction is very slow in spite of a favorable free energy change. As a result, ATP can be retained in a cell for a considerable period of time without an appreciable loss due to spontaneous hydrolysis. The free energy is available, however, for reactions within the cell. For example, the series of reactions in the contraction of a muscle can be represented schematically as

ATP^{4-} + relaxed muscle $\rightleftharpoons$ ADP^{3-} + $H_2PO_4^-$ + contracted muscle

The endothermic steps in a biosynthesis may be coupled to the hydrolysis of ATP. An example involves the synthesis of sucrose from glucose and fructose.

$$C_6H_{12}O_6 + C_6H_{12}O_6 \longrightarrow C_{12}H_{22}O_{11} + H_2O$$
$$\text{glucose} \qquad \text{fructose} \qquad \text{sucrose}$$

$$\Delta G° = +7 \text{ kcal mole}^{-1}$$

The free energy change for this reaction is unfavorable. Sucrose may be formed through the intermediate glucose-1-phosphate in a reaction sequence that involves ATP.

$$ATP + \text{glucose} \longrightarrow \text{glucose-1-phosphate} + ADP$$

$$\text{Glucose-1-phosphate} + \text{fructose} \longrightarrow \text{sucrose} + \text{phosphate}$$

The free energy change for the overall reaction is negative

$$C_6H_{12}O_6 + C_6H_{12}O_6 + C_{10}H_{12}O_{13}N_5P_3^{4-} \longrightarrow$$
$$\text{glucose} \qquad \text{fructose} \qquad \text{ATP}$$

$$C_{12}H_{22}O_{11} + C_{10}H_{12}O_{10}N_5P_2^{3-} + H_2PO_4^-$$
$$\text{sucrose} \qquad \text{ADP} \qquad \text{phosphate}$$

$\Delta G° = -1 \text{ kcal mole}^{-1}$

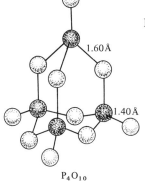
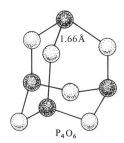

Figure 23-5 *The structure of phosphorus oxides.*

P_4O_{10} $\qquad$ P_4O_6

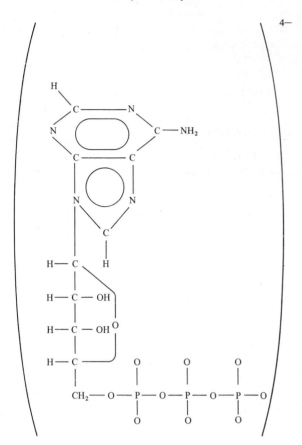

Figure 23-6 *The structure of adenosine triphosphate (ATP).*

In a more complicated synthesis such as the formation of a fatty acid, ATP may be required for each reaction in a whole series of steps. Thus about 50 molecules of ATP are hydrolyzed in the formation of one molecule of palmitic acid.

$$7H^+ + 8CH_3COO^- + 50ATP^{4-} + 50H_2O \longrightarrow$$
$$CH_3(CH_2)_{14}COO^- + 14O_2 + 50ADP^{3-} + 50H_2PO_4^-$$

D. USES OF PHOSPHORUS

Phosphorus is an essential nutrient for both plants and animals. It is one of three elements (K, N, and P) that is normally added to soil in the form of fertilizer. Some 3×10^7 tons of calcium phosphate (primarily as apatite) are mined each year for use in fertilizers. Calcium phosphate that is to be used in a fertilizer is usually treated with sulfuric acid

$$Ca_3(PO_4)_2 + 2H_2SO_4 \longrightarrow Ca(H_2PO_4)_2 + 2CaSO_4$$

to give "superphosphate," or with phosphoric acid to increase the solubility.

$$Ca_3(PO_4)_2 + 4H_3PO_4 = 3Ca(H_2PO_4)_2$$

The calcium dihydrogen phosphate produced with phosphoric acid is free from calcium sulfate and is sold as "triple superphosphate."

There are a great many other uses for phosphorus and phosphorus compounds that are important, but the quantities used in these applications do not approach the quantities used in the fertilizer and match industries. For example, P_4O_{10} is widely used as a drying agent, trisodium phosphate is a common cleaning agent, and phosphates are used in making buffer solutions.

E. PHOSPHATE BUFFER SOLUTIONS

Equilibrium calculations on phosphate systems are more complicated than any of those discussed in previous chapters because phosphoric acid is a triprotic acid. A solution of phosphoric acid at equilibrium will contain all the phosphate species (H_3PO_4, $H_2PO_4^-$, HPO_4^{2-}, and PO_4^{3-}). When the concentration of some of the phosphate species such as HPO_4^{2-} and PO_4^{3-} may be neglected, the calculation is identical to those for simple systems presented in Chapter 11. At a pH near 3 the concentration of both H_3PO_4 and $H_2PO_4^-$ is large, and the pH of the solution can be calculated from the equilibrium expression

$$K_1 = 7.1 \times 10^{-3} = \frac{[H^+][H_2PO_4^-]}{[H_3PO_4]}$$

In some cases, for example the calculation of the pH of a 0.05 M solution of NaH_2PO_4, it is necessary to consider two of the ionization equilibria. In this calculation there are six equations and six unknowns. There are three equilibrium expressions

$$K_1 = 7.1 \times 10^{-3} = \frac{[H^+][H_2PO_4^-]}{[H_3PO_4]}$$

$$K_2 = 6.2 \times 10^{-8} = \frac{[H^+][HPO_4^{2-}]}{[H_2PO_4^-]}$$

$$K_w = 10^{-14} = [H^+][OH^-]$$

and three stoichiometric equations

$$[Na^+] = 0.050$$
$$[H_3PO_4] + [H_2PO_4^-] + [HPO_4^{2-}] = 0.050$$
$$[Na^+] + [H^+] = [OH^-] + [H_2PO_4^-] + 2[HPO_4^{2-}]$$

The last of these equations is known as the charge balance equation. It is obtained by setting the total positive charge present (per liter of solution) equal to the negative charge carried by the anions present. This is almost always the easiest way to keep track of the stoichiometry of the hydronium and hydroxide ions. It is possible to simplify the equations by making appropriate substitutions and neglecting the species present in small concentrations

$$\text{Using } [HPO_4^{2-}] = x$$
$$[Na^+] \approx [H_2PO_4^-] + 2[HPO_4^{2-}]$$
$$0.050 = [H_2PO_4^-] + 2x$$
$$[H_2PO_4^-] = 0.05 - 2x$$
$$[H_3PO_4] = 0.05 - x - (0.05 - 2x) = x$$

and

$$K_1 K_2 = \frac{[H^+][H_2PO_4^-]}{[H_3PO_4]} \frac{[H^+][HPO_4^{2-}]}{[H_2PO_4^-]}$$

$$K_1 K_2 = \frac{[H^+](0.05 - 2x)}{x} \frac{[H^+]x}{(0.05 - 2x)} = [H^+]^2$$

$$[H^+]^2 = K_1 K_2$$

the hydronium ion concentration is

$$[H^+] = \sqrt{K_1 K_2} = \sqrt{(7.1 \times 10^{-3})(6.2 \times 10^{-8})}$$
$$= 2.10 \times 10^{-6} \, M$$

To avoid errors it is best to be certain that all six equations are written out in full before attempting to apply any simplifying assumptions. Similarly when it is not apparent which phosphate species might be neglected, the problem can be set up correctly by writing all three ionization equilibria

$$K_1 = 7.1 \times 10^{-3} = \frac{[H^+][H_2PO_4^-]}{[H_3PO_4]}$$

$$K_2 = 6.2 \times 10^{-8} = \frac{[H^+][HPO_4^{2-}]}{[H_2PO_4^-]}$$

$$K_3 = 4.5 \times 10^{-13} = \frac{[H^+][PO_4^{3-}]}{[HPO_4^{2-}]}$$

Sec. II Compounds of Phosphorus

and including all four phosphate species in the stoichiometric equation

Total phosphate concentration

$$= [H_3PO_4] + [H_2PO_4^-] + [HPO_4^{2-}] + [PO_4^{3-}]$$

For example, if 0.2 mole of Na_3PO_4 and 0.04 mole of P_2O_5 are dissolved in 2.0 liters of water, the full stoichiometric equations are:

$$[Na^+] = 0.2 \times \tfrac{3}{2} = 0.30 \ M$$

$$[H_3PO_4] + [H_2PO_4^-] + [HPO_4^{2-}] + [PO_4^{3-}] = \frac{(0.2 + .08)}{2} = 0.14 \ M$$

$$[Na^+] + [H^+] = [H_2PO_4^-] + 2[HPO_4^{2-}] + 3[PO_4^{3-}] + [OH^-]$$

Together with the equilibrium expressions

$$K_1 = 7.1 \times 10^{-3} = \frac{[H^+][H_2PO_4^-]}{[H_3PO_4]}$$

$$K_2 = 6.2 \times 10^{-8} = \frac{[H^+][HPO_4^{2-}]}{[H_2PO_4^-]}$$

$$K_3 = 4.5 \times 10^{-13} = \frac{[H^+][PO_4^{3-}]}{[HPO_4^{2-}]}$$

$$K_w = 10^{-14} = [H^+][OH^-]$$

they form a set of seven equations and seven unknowns. The concentration of each species is

$$[Na^+] = 0.300$$
$$[PO_4^{3-}] = 0.0169$$
$$[HPO_4^{2-}] = 0.123$$
$$[H_2PO_4^-] = 6.50 \times 10^{-6}$$
$$[H_3PO_4] = 3.0 \times 10^{-15}$$
$$[OH^-] = 0.00306$$
$$[H^+] = 3.27 \times 10^{-12}$$

The two compounds, Na_2HPO_4 and KH_2PO_4, are easily purified and they are ideal components for a standard buffer solution that may be used to calibrate pH meters. The pH of a solution containing 0.0249 M

H_2PO_4 and 0.0249 M Na_2HPO_4 is 6.685 at 25°C. The pH of this buffer calculated from $K_2 = 6.2 \times 10^{-8}$ is 7.21. The discrepancy between the calculated and observed value is the result of using concentration as an approximation for activity. The interaction of a triply charged ion such as PO_4^{3-} with the other species present in a solution is so large that any value given for the third ionization constant of an acid is at best only an order of magnitude estimate.

QUESTIONS

1. Which of the following compounds is the most stable: Na_3P, Mg_3P_2, Zn_3P_2, Co_2P, PCl_3?
2. Which of the compounds in Question 1 will dissolve in H_2O?
3. Which of the following species will disproportionate: P_4, HPO_3^{2-}, HPO_4^{2-}, H_3PO_4, $H_2PO_2^-$, $H_2PO_3^-$?
4. Which of the species in Question 3 is the strongest base?
5. Compare NH_3 and PH_3 in terms of basicity.
6. Which of the following is the strongest acid: H_3P, H_3PO_2, H_3PO_3, H_3PO_4? Why? Why is there so little difference between the acidity of H_3PO_3 and H_3PO_4?
7. Write the equation for the reaction of $Ca_3(PO_4)_2$ with H_3PO_4.
8. Write equations for the preparation of H_3PO_4 from P_4.
9. Which of the following phosphates are soluble in water: K_3PO_4, $Mg_3(PO_4)_2$, $AlPO_4$, $Co_3(PO_4)_2 \cdot 8H_2O$, $CePO_4$?
10. Which of the phosphates in Question 9 will dissolve in acid?
11. Why is a solution of trisodium phosphate basic?
12. Draw the structure of the complex ion formed from calcium ion and the pyrophosphate ion.
13. What factors affect the P-O bond energy?
14. Is P_2O_5 an acid? Explain.

PROBLEMS I

1. Calculate ΔH^0 for the reaction of $Zn(OH)_2$ ($H^0 = -147$) with PH_3 ($H^0 = +1$ kcal mole^{-1}). See Table 23-1.
2. Given the H^0 values: PF_3, -220 and PF_5, -381 kcal mole^{-1}, calculate $\alpha_{P\text{-}F}$ and $\gamma_{P\text{-}F}$.
3. Estimate H^0 for PF.
4. Show that the H^0 values: $PCl_3(g)$, -69 and $PCl_5(g)$, -90 can be approximated using $\alpha_{P\text{-}Cl}$ alone.
5. Calculate the equilibrium constant for the decomposition of PH_3 from the appropriate thermodynamic data.

	H^0, kcal mole^{-1}	S^0, eu
PH_3	1.3	50.2
H_2	0	31.2
P_4 (white)	0	39.28
P (red)	-4.2	5.45

6. Calculate the pH of a buffer solution prepared by dissolving 0.03 mole of KH_2PO_4 in 1.00 liter of 0.10 M H_3PO_4.
7. Calculate the pH of a 0.04 M solution of KH_2PO_3.
8. At what pH would you expect the first equivalence point in the titration of $H_4P_2O_7$ with NaOH?
9. What ratio of Na_2HPO_4 and KH_2PO_4 should be used to give a buffer solution with a pH of 7.0? (neglect activity coefficients)
10. Rework Problem 9 assuming the total salt concentration is such as to give $\gamma_{H^+} = \gamma_{H_2PO_4^-} = 0.84$ and $\gamma_{HPO_4^{2-}} = 0.41$.
11. What volume of 0.10 M NaOH should be added to 50 ml of 0.050 M Na_2HPO_4 to give a buffer solution with a pH of 11.50? The experimental value is 11.1 ml. Explain any discrepancy.

PROBLEMS II
1. H^0 for PBr_3 is -44 kcal mole^{-1}. Calculate α_{P-Br}.
2. Draw the octet rule structure for PBr_5.
3. How many P-Br bonds are present in PBr_5?
4. Estimate H^0 for PBr_5 using α_{P-Br} from Problem 1. The experimental value for solid PBr_5 is -64.5 kcal mole^{-1}.
5. Draw the octet rule structure for $POBr_3$.
6. If $\alpha_{P-O} = 104$ kcal mole^{-1}, calculate the energy of the four single bonds in $POBr_3$.
7. If $H^0 = -110$ for solid $POBr_3$, calculate γ_{P-O} in $POBr_3$.
8. Calculate the concentration of Na^+ in a 0.20 M solution of Na_2HPO_4.
9. What is the total phosphorus concentration in the solution in Problem 8?
10. Write the charge balance equation for a solution of Na_2HPO_4.
11. If the concentrations of $[H_3PO_4]$, $[H^+]$, and $[OH^-]$ are all very small, show that the equations from Problems 8, 9, and 10 require that $[H_2PO_4^-] = [PO_4^{3-}]$.
12. Write the equilibrium constant expression for the ionization of $H_2PO_4^-$.
13. Write the equilibrium constant expression for the ionization of HPO_4^{2-}.
14. Multiply the two equilibrium expressions in Problems 12 and 13 to get an equation of the form $K_2K_3 = \_\_\_\_$.
15. What is the pH of a 0.20 M solution of Na_2HPO_4?
16. What is the pH of a 0.20 M solution of KH_2PO_4?

The material referred to as antimony was actually Sb_2O_3. The other compounds prepared by this procedure would be $SbCl_3$ and $SbOCl$.

Pound together one pound of Antimony, half a pound of common salt, and five pounds of broken bricks; place in a retort and distill a yellow oil, when all the spirits will pass away. Pour into a fresh vessel, and remove its oiliness; there remains a powder, which, spread on a stone in a humid place, and you will have a humid balm which is of great efficacy against foul wounds.*

BASIL VALENTINE (JOHANN THÖLDE)
The Triumphal Chariot of Antimony (1604)

Chapter 24 Heavy Metals

Many of the compounds formed by the remaining elements in group VA are similar to those formed by nitrogen and phosphorus. The number of possible oxidation states is again limited for arsenic and antimony as it is in the case of phosphorus.

I. Arsenic

Elemental arsenic exists in two forms. The stable gray or metallic form has a crystal structure similar to black phosphorus; there is also an unstable yellow form consisting of As_4 molecules. The element burns in air to form the oxide. It reacts with the halogens and some other nonmetals.

In these respects the chemistry of arsenic resembles that of phosphorus very closely; however, there is a substantial difference in the stability of the compounds formed by the two elements because of the difference in the energy of the bonds. The arsenic-hydrogen bond energy (77 kcal mole^{-1}) is lower than the phosphorus-hydrogen bond energy (85 kcal mole^{-1}) and arsine (AsH_3) decomposes to the elements more readily than phosphine. The arsenic-oxygen bond energy is also lower than the phosphorus-oxygen bond energy and, as a result, elemental arsenic is not hydrolyzed (see Chapter 23). The E_h-pH

For the hydrolysis of the arsenic

$\Delta H = \alpha_{As-As} + \alpha_{O-H} - \alpha_{As-O} - \alpha_{As-H}$

$\Delta H = 60 + 126 - 94 - 77$

$= 15$ kcal mole^{-1}

*Used by permission from H. M. Leicester and H. S. Klickstein, *A Source Book in Chemistry 1400–1900* (Cambridge, Mass.: Harvard University Press, 1952).

diagram for arsenic (Fig. 24-1) is substantially different from that for phosphorus because of the difference in bond energies.

Arsenic and phosphorus form many compounds with the same coordination number and oxidation states. This aspect of group similarities leads to some important consequences in the case of biologically important phosphorus compounds. Arsenic can often substitute for phosphorus in biologically important compounds but, because of the chemical difference between the elements, the substitution of arsenic for phosphorus often leads to serious consequences. The toxicity of arsenic insecticides and herbicides is the basis of the principal use of the element.

Perhaps the main chemical interference of arsenic in biological systems involves the synthesis of nucleic acids. In line with this, one of the most serious effects of arsenic poisoning in humans is degeneration of the lining of the digestive tract. Cell division, for which the synthesis of nucleic acids is a necessary step, occurs more rapidly in the lining of the digestive tract than in any other tissue in an adult mammal.

Any arsenic compound as well as metallic arsenic can be handled safely, but great care should be taken with soluble arsenic compounds. Any compound of the element including the highly insoluble sulfides

Organic compounds of arsenic (such as salvarsan) were once used extensively in medicine. They have now been replaced by somewhat less hazardous antibiotics.

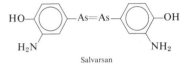
Salvarsan

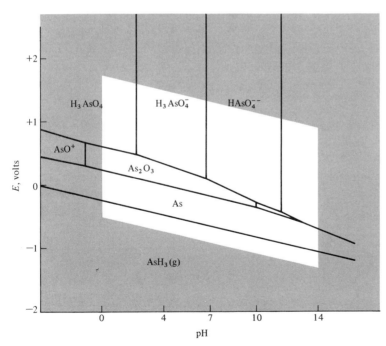

Figure 24-1 E_h-pH *diagram for arsenic.*

Table 24-1 *Ionization Constants of the Oxo Acids of Arsenic*

Acid	K_1	K_2	K_3
H_3AsO_4	2.5×10^{-4}	5.6×10^{-8}	3×10^{-13}
H_3AsO_3	6×10^{-10}	$\cong 10^{-14}$	

must be classed as highly toxic if ingested or inhaled, and volatile arsenic compounds such as AsH_3 (boiling point, $-62.5\,°C$) and $AsCl_3$ (boiling point, $130\,°C$) are extremely hazardous. Men handling As_2O_3 industrially wear special clothing that is changed frequently because repeated contact with As_2O_3 will cause dermatitis.

Arsenic is widely distributed in nature. The principal ores are FeAsS, $CuAsS_3$, AsS, and As_2S_3. The element is recovered as As_4O_6 vapor when the sulfide ore is roasted. The vapor condenses as a cubic molecular solid that consists of dimers of As_2O_3. This form is thermodynamically unstable. The thermodynamically stable form of arsenic(III) oxide is a monoclinic covalent solid (mp, $315\,°C$; bp, $457\,°C$).

Mining operations are not carried out specifically for arsenic because the quantity of arsenic(III) oxide recovered in refining operations of other metals is sufficient for all requirements.

The ionization constants of the oxo acids of arsenic are given in Table 24-1. As expected, H_3AsO_4 is slightly weaker than H_3PO_4. H_3AsO_3 is a very much weaker acid than H_3PO_3 because H_3AsO_3 does not contain a hydrogen-arsenic bond. Arsenic(III) oxide can be considered to be amphoteric because As_2O_3 dissolves in hydrochloric acid with the formation of complex ions such as $AsCl_2(H_2O)^+$ as well as $H_4AsO_3^+$.

II. Antimony

A method for preparing metallic antimony from the sulfide ore had been discovered by 3000 B.C. By that time some medical uses of antimony compounds had been developed.

It is not clear just when metallic antimony was recognized as a distinct metal because the ancients did not make a clear distinction between the various low-melting heavy metals (Sb, Bi, Sn, and Pb). Pliny (A.D. 90), in describing the purification of Sb_2S_3 for medical purposes, states that the heating must be carefully regulated so that the material does not turn into lead.

Some elemental antimony is found in nature. The element can be obtained when the sulfide ore (stibnite, Sb_2S_3) is roasted to the oxide

$$Sb_2S_3 + 4\tfrac{1}{2}O_2 \longrightarrow Sb_2O_3 + 3SO_2$$

and the oxide is reduced with carbon

$$Sb_2O_3 + 3C \longrightarrow 2Sb + 3CO$$

The sulfide can also be directly reduced with iron.

$$Sb_2S_3 + 3Fe \longrightarrow 2Sb + 3FeS$$

Antimony, like water, has the unusual property of having a lower density as a solid than as a liquid. The expansion that takes place on freezing antimony is important in casting because an exact reproduction of even small details of the mold is assured. As a result antimony is an important constituent (about 15%) of type metal.

The only stable oxidation states of antimony shown in the E_h-pH diagram in Fig. 24-2 are 0, +3, and +5. In stibine (SbH_3) the antimony may be considered to be in the -3 oxidation state. Stibine is formed when a solution of antimony(III) is reduced with zinc metal.

In contrast to the tetrahedral coordination found in H_3AsO_4, antimony(V) is almost always found with an octahedral coordination. Because of the larger coordination number and decreased polarizing ability the antimony(V) acid is much weaker than arsenic or phosphoric acid.

The hexacoordinate $Sb(OH)_6^-$ ion exists in strongly basic solution. Below a pH of 4 a white precipitate of hydrated antimony(V) oxide that is only slightly soluble in water is formed. In contrast to the highly exothermic reaction of P_4O_{10} with water the reaction of Sb_2O_5 with water is endothermic.

$P_4O_{10}(c) + 6H_2O \longrightarrow 4H_3PO_4(aq)$ $\Delta H = -107.6$ kcal mole^{-1}

$As_4O_{10}(c) + 6H_2O \longrightarrow 4H_3AsO_4(aq)$ $\Delta H = -12.8$ kcal mole^{-1}

$2Sb_2O_5(c) + 14H_2O \longrightarrow 4HSb(OH)_6(aq)$ $\Delta H = +7.1$ kcal mole^{-1}

Figure 24-2 E_h-pH *diagram for antimony.*

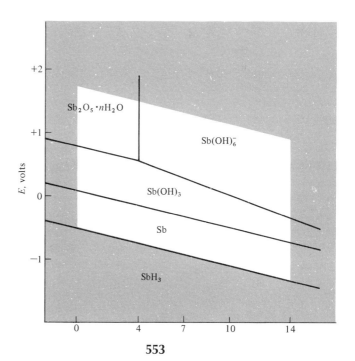

III. Tin

A coordination number greater than four is characteristic of all the elements from tin through iodine. Tin(IV) is six-coordinate in solid tin(IV) oxide and in ions such as $Sn(OH)_6^{2-}$ and $Sn(H_2O)(OH)_5^-$. This effect leads to some similarities in the chemical behavior of adjacent elements in the same row. Like Sb_2S_3, SnS will dissolve in 6 M HCl or in excess ammonium sulfide solution. The equilibrium constant for the reaction

$$Sn(OH)_2 + OH^- + H_2O = Sn(OH)_3(H_2O)^-$$

is 2.5×10^{-5}, so that $Sn(OH)_2$ is appreciably soluble in basic solution.

Tin and antimony behave similarly through the major separations of the qualitative analysis scheme.

Compounds of tin and antimony are generally distinguished by their redox behavior. Compounds containing antimony are expected to be better electron acceptors or stronger oxidizing agents than those containing tin because of the larger nuclear charge and smaller size of antimony. The electrode potential, given in Table 24-2, indicates that this is the case. Antimony compounds will oxidize tin metal, for example:

$$2Cl^- + 2SbCl_2^+ + 3Sn \longrightarrow 2Sb + 3SnCl_2$$

The two forms of $SnO_2 \cdot H_2O$ are prepared by the reactions

$$SnCl_4 + 3H_2O \longrightarrow$$
$$SnO_2 \cdot H_2O + 4HCl$$

and

$$Sn(OH)_6^{2-} + 2H^+ \longrightarrow$$
$$SnO_2 \cdot H_2O + 3H_2O$$

respectively.

There are two known forms of $SnO_2 \cdot H_2O$ but neither of them has been sufficiently well crystallized for characterization with X-rays. It is interesting that this case of isomerism, one of the first to be clearly established (Berzelius, 1811), is still not understood. The more stable (and less soluble) form of $SnO_2 \cdot H_2O$ is represented in the E_h-pH diagram shown in Fig. 24-3. This form is easily prepared by the action of strong oxidizing agents on metallic tin. It is presumably the form of the oxide that coats and protects the surface of metallic tin from corrosion.

Tin is one of the few metals that is not appreciably attacked by the

Table 24-2 *Standard Electrode Potentials*

	E^0, volts
Tin	
$Sn^{2+} + 2e^- \longrightarrow Sn$	-0.141
$SnO_2 + 4H^+ + 2e^- \longrightarrow Sn^{2+} + 2H_2O$	-0.093
Antimony	
$Sb(OH)_3 + 3H^+ + 3e^- \longrightarrow Sb + 3H_2O$	$+0.091$
$Sb_2O_5 \cdot nH_2O + 4H^+ + 4e^- \longrightarrow 2Sb(OH)_3 + (n-1)H_2O$	$+0.79$

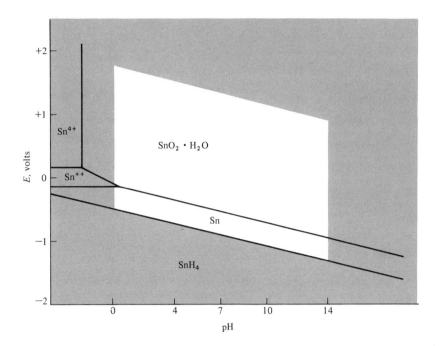

Figure 24-3 E_h-pH diagram for tin.

acids and complexing agents normally found in foods. The use of tin-coated steel for tin cans has long been the major use of the metal. Two other important uses of tin are in copper alloys (bronze) and in low-melting alloys such as type metal and solder. An alloy of 83% tin, 8.5% antimony, and 8.5% copper is used extensively in bearings.

IV. Lead and Bismuth

The metal-oxygen bond energies for the group IVA and VA elements of the third through the sixth rows are shown in Table 24-3. The decrease in the metal-oxygen bond energies between the third and fourth row elements is apparent from the table, and there is another

Table 24-3 *Metal-Oxygen Bond Energies for Some Group IVA and VA Elements*

$\alpha_{\text{Si-O}} = 119$	$\alpha_{\text{P-O}} = 104$
decrease between 3rd and 4th rows	
$\alpha_{\text{Ge-O}} = 94$	$\alpha_{\text{As-O}} = 94$
$\alpha_{\text{Sn-O}} = 92$	$\alpha_{\text{Sb-O}} = 91$
decrease between 5th and 6th rows	
$\alpha_{\text{Pb-O}} = 66$	$\alpha_{\text{Bi-O}} = 67$

decrease between the fifth and sixth rows. Two breaks in metal-oxygen bond energies divide these elements into three groups according to the softness of their cations. Silicon(IV) and phosphorus(V) should be classed as moderately hard cations that show a marked preference for oxide and fluoride ligands. Germanium, tin, lead, arsenic, antimony, and bismuth form typical soft cations.

The decrease in bond energy down each group can be explained on the basis of the polarizing ability of the cations. Polarizing ability decreases down a group so the magnitude of the interaction decreases down a group. There is also a large difference in size of the valence orbitals (between lead for example and oxygen) so the orbital overlap decreases down a group.

Oxygen is moderately effective at inducing the polarizing ability of the heavy metals in these two groups, and there are a number of similarities in the solubility of lead and bismuth salts.

Another consequence of the decrease in bond energies down the group is the ease with which the oxides and hydroxides of the metals are reduced. The electrode potentials, in volts, at pH 0, are

Compare these values with the lighter elements of the group.

$$Pb^{2+} + 2e^- \longrightarrow Pb \qquad E^0 = -0.126$$

$$PbO_2 + 2e^- + 4H^+ \longrightarrow Pb^{2+} + 2H_2O \qquad E^0 = 1.46$$

$$HBiO_3 + 2e^- + 3H^+ \longrightarrow BiO^+ + 2H_2O \qquad E^0 = 1.7$$

$$BiO^+ + 3e^- + 2H^+ \longrightarrow Bi + H_2O \qquad E^0 = 0.32$$

Alternatively lead(IV) and bismuth(V) are extremely strong oxidizing agents.

A. ACID-BASE BEHAVIOR

Another consequence of the low lead-oxygen and bismuth-oxygen bond energies and the lower polarizing ability of the ions is that lead and bismuth ions and the compounds formed from these ions should be weaker acids than the corresponding tin and antimony compounds. A pH greater than 14 is required for the dissolution of Bi_2O_5. (Sb_2O_5 dissolves as $Sb(OH)_6^-$ even in dilute acid.) $Pb(OH)_2$ is amphoteric with an equilibrium constant for the reaction

$$Pb(OH)_2 + H_2O + OH^- = Pb(OH)_3(H_2O)^-$$

equal to 500. (The corresponding equilibrium constant for $Sn(OH)_2$ is 40,000.) Similarly Bi(III) is less acidic than Sb(III) and Bi_2O_3 is a stronger base than Sb_2O_3. The equilibrium constant for the reaction

$$Bi_2O_3 + 2H^+ = 2BiO^+ + H_2O$$

is 5.5×10^6. The solubility of Bi_2O_3 is over 1 mole liter^{-1} at pH 3. In contrast to this, the equilibrium constant for the reaction

$$Sb_2O_3 + 2H^+ = 2SbO^+ + H_2O$$

is only 3.4×10^{-8}. As a result the concentration of SbO^+ in a saturated solution is small even at pH 0.

B. COLORS AND CHARGE TRANSFER SPECTRA

The molecular orbital diagram for lead tetrachloride ($PbCl_4$) is shown in Fig. 24-4. There are empty $5d$ orbitals on the lead and the antibonding molecular orbitals are also predominately on the lead atom. The absorption of a quantum of energy can result in the transfer of an electron from the nonbonding level on the chlorine atoms to a higher energy empty orbital on the lead. This type of electronic transition is known as a charge transfer transition because it involves the transfer of an electron from one part of the molecule to another (chlorine to lead in this case). The absorption bands that result from a charge transfer transition are usually very intense, and the position of the bands depends on the electron accepting properties of the central atom and the donor properties of the ligand.

If the lowest energy charge transfer peak in a lead compound is well into the ultraviolet region of the spectrum, as it is in lead chloride ($PbCl_2$), the compound will be colorless. As ligands that are better electron donors are substituted for chlorine, the peak moves farther into the visible region. The corresponding colors observed range from yellow and red to brown or black. This sequence is well illustrated by the data in Table 24-4.

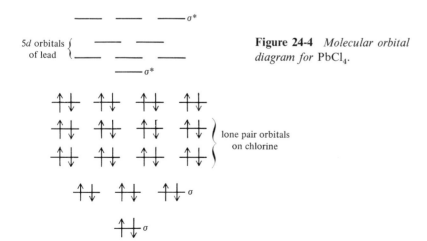

Figure 24-4 *Molecular orbital diagram for* $PbCl_4$.

Table 24-4 *The Color of Some Salts of Lead, Bismuth, Tin, and Antimony*

Salt	Color	Salt	Color
$SnCl_4$	white	$SbCl_5$	yellow
$SnCl_2$	white	$SbCl_3$	white
$SnBr_4$	white	$SbBr_3$	white
$SnBr_2$	pale yellow	SbI_3	yellow or red
SnI_4	red	Sb_2S_3	black (amorphous material is orange)
SnI_2	red		
SnS	yellow	Sb_2Se_3	gray or black
SnSe	gray or black	$BiCl_3$	white
$PbCl_4$	yellow	$BiBr_3$	yellow
$PbCl_2$	white	BiI_3	black
$PbBr_2$	white	Bi_2S_3	brown
PbI_2	yellow		
PbS	black		

The electron donor ability of a ligand depends on the electronegativity or the ease of polarization of the ligand. All of the fluorides of the elements in families IIIA through VIIA are white or colorless. The color deepens as chloride, bromide, and iodide ion are substituted in that order or as the oxidation state of the metal increases. Even oxides may be colored if the oxidation state is high enough or if oxide ion can induce the polarizing ability of the cation. The oxides of antimony(V) and bismuth(V) are yellow, and a whole series of lead oxides are colored.

Lead(II) oxide occurs in a number of different crystal forms that vary in color from yellow to red. Some black materials are also known that have a composition close to Pb_2O_3. Red lead (Pb_3O_4) is prepared by heating PbO in air to about 450°C. The strong oxidizing power of this substance (two thirds of the lead atoms are in the +4 oxidation state) leads to its very extensive use in paints for steel. The Pb_3O_4 pigment maintains a good oxide coating on the steel.

The use of Pb_3O_4 in paints (about 30,000 tons of lead a year) is a relatively small but important use. Other lead pigments such as $PbCrO_4$ (yellow) and $Pb(OH)_2 \cdot 2PbCO_3$ (white lead) have been used in paints. The use of lead compounds in paints has been sharply reduced by the use of less toxic materials such as TiO_2 (white), CdS (yellow), and $ZnCrO_4 \cdot 4Zn(OH)_2$ (yellow).

V. A Measure of the Induced Polarizing Ability of a Cation

The induced polarizing ability of a cation results from the deformation of the cation by an easily polarized ligand. The net result of this distortion is increased covalency in the bond and relatively stronger bonding to ligands with the largest ease of polarization. This effect should be reflected in the enthalpy data for compounds containing these cations, and in particular in the data for the displacement of one ligand by another.

The enthalpy of the reaction of a neutral chloride with water provides a convenient quantitative measure of the induced polarizing ability of each of the cations within a family. To compare the ions in different families, it is necessary to consider the differences in stoichiometry between families. The simplest way to obtain comparable numbers is to divide the enthalpy of the reaction with water by the number of chlorine atoms present in the formula. The available data are shown in Table 24-5.

Down a family of elements the induced polarizing ability of the cation increases and the displacement reaction of chloride by water or hydroxide ion becomes more endothermic. The region of the periodic table that contains the cations for which the reaction of the chloride with water is endothermic is outlined with a heavy rule in Table 24-5. This region includes the cations of the sixth row from osmium (Os) through bismuth (Bi) and a scattering of elements in the center of the periodic table (Rh, Pd, Ag, and Cu). These cations are easily distorted from a spherical shape. They all form stable anionic chloro complex ions.

A similar measure of the induced polarizing ability of a cation is provided by the enthalpy of the reaction of the sulfides with water. We noted in Chapter 18 that the sulfides of hard cations are readily hydrolyzed

All the chlorides of elements outside the outlined region in Table 24-5 are soluble in water. Within the region all the +1 chlorides, $CuCl$, $AgCl$, $AuCl$, $TlCl$, and Hg_2Cl_2 are insoluble, but the behavior of higher chlorides is more complex. $PbCl_2$ is insoluble in cold water, but quite soluble at $100°C$. $HgCl_2$ dissolves as $HgCl_2$ molecules. $OsCl_2$ and $PtCl_2$ are insoluble. These metal ions form robust complexes, and the solid chlorides do not form at an appreciable rate when chloride is added. As a result the only chlorides that precipitate from a solution of all of these metals on the addition of HCl are $AgCl$, Hg_2Cl_2, $TlCl$, and $PbCl_2$.

$$Al_2S_3 + 6H_2O \longrightarrow 2Al(OH)_3 + 3H_2S(aq) \qquad \Delta H^0 = -108 \text{ kcal mole}^{-1}$$

The reaction of lead sulfide is endothermic

$$PbS + 2H_2O \longrightarrow Pb(OH)_2 + H_2S$$

The enthalpy of the reaction of the sulfides of a number of cations with water is given in Table 24-6 (in terms of the enthalpy mole^{-1} of sulfur). These data represent the oxygen-sulfur competition of the cations in the periodic table.

The induced polarizing ability of the cations with a filled or nearly

Table 24-5 *Enthalpy of Reaction of Chlorides with Water (kcal mole^{-1} of chlorine atoms)*

IA	IIA	IIIB	IVB	VB	VIB	VIIB	VIII			IB	IIB	IIIA	IVA	VA	VIA	VIIA	Inert gases
																FCl −35	
												BCl$_3$ −25	CCl$_4$ −22	NCl$_3$ −22	Cl$_2$O −16	Cl$_2$ −0.3	
			TiCl$_4$ −16		CrCl$_2$ −9.3 CrCl$_3$ −16	MnCl$_2$ −8.5	FeCl$_2$ −10 FeCl$_3$ −12	CoCl$_2$ −9.2	NiCl$_2$ −10	CuCl +6.5 CuCl$_2$ −7.7	ZnCl$_2$ −8.8	AlCl$_3$ −17	SiCl$_4$ −19	PCl$_3$ −23 PCl$_5$ −26	SCl$_2$ +1.7	
			ZrCl$_4$ −13		MoCl$_4$ −19		RhCl$_2$ +1.3 RhCl$_3$ +1.3		PdCl$_2$ +5.0	AgCl +16	CdCl$_2$ −2.2	GaCl$_3$ −15		AsCl$_3$ −7.7	SeCl$_4$ −16	BrCl −2.2	
												InCl$_3$ −5.5	SnCl$_2$ −2.2 SnCl$_4$ −10	SbCl$_3$ −4.1 SbCl$_5$ −8.0			
					WCl$_4$ −22 WCl$_6$ −23				PtCl$_2$ +2.4	AuCl$_3$ +4.4	Hg$_2$Cl$_2$ +15 HgCl$_2$ +11	TlCl +7.5	PbCl$_2$ +2.8 PbCl$_4$ −2.8	BiCl$_3$ +1.7	TeCl$_4$ −5.6	ICl 3.7	

Table 24-6 *Enthalpy of Reaction of Sulfides with Water (kcal mole^{-1} of sulfur atoms)*

IA	IIA	IIIB	IVB	VB	VIB	VIIB	\multicolumn{3}{c}{VIII}	IB	IIB	IIIA	IVA	VA	VIA	VIIA	Inert gases
H_2S 0.0																	
	BeS −34											B_2S_3 −28	CS_2 −1.3				
Na_2S* −24	MgS* −26											Al_2S_3 −18	SiS_2 −25	P_2S_3 −21			
K_2S −18	CaS −13					MnS +10	FeS +14	CoS +17	NiS +17	Cu_2S +38 / CuS +32			GeS +14 / GeS_2 +16	As_2S_3 +20			
Rb_2S 33	SrS 20				WS_2 +14	MoS_2 +22 / MoS_3* +13	RuS_2 +57			Ag_2S +59	CdS +32	In_2S_3 +19	SnS +14	Sb_2S_3 +15			
Cs_2S 35	BaS 21	La_2S_3* 13						IrS_2 +54	PtS +61		HgS +51	Tl_2S +28	PbS +26	Bi_2S_3 +24			

*Calculations are made for the predominant species at a pH of 6. Where this involves a charged species, the value H^0 of $H^+ \equiv -5.7$ is used to make all the values roughly comparable.

full inner d subshell is evident in both Tables 24-5 and 24-6. The chloride data show the effect of oxidation number more clearly. The cations with low oxidation states (Cu^+, Ag^+, and so forth) are the most easily distorted cations.

This is not clear from the sulfide data because the oxidation state of a metal ion in a sulfide is not given reliably by the formula for the sulfide. Sulfur-sulfur bonding may occur. (Pyrite, FeS_2, contains Fe^{2+} and S_2^{2-} ions.)

Sulfide ion reduces many cations to lower oxidation states. A material of the stoichiometry, CuS, is precipitated when H_2S is added to a Cu(II) solution. However, examination of the crystals with X-rays shows that two thirds of the copper atoms are in the tetrahedral coordination sphere expected for Cu(I) instead of the octahedral coordination sphere expected for Cu(II). In addition the S_2^{2-} ion is present. The reduction of some of the Cu(II) to the more highly polarizing Cu(I) state is partly responsible for the fact that CuS is much less soluble than either NiS or ZnS.

VI. The Chemistry of the IB Cations

A. COPPER

Cu^+, Ag^+, and Au^+ form insoluble chlorides. They show either two-fold coordination in complex ions ($CuCl_2^-$, $Ag(NH_3)_2^+$, and so forth) or four-fold tetrahedral coordination.

Copper(I) is stabilized by easily polarized anions such as sulfide or iodide. The reaction between copper and elemental sulfur produces Cu_2S. CuI_2 cannot be prepared since the redox reaction

$$2Cu^{2+} + 4I^- \longrightarrow 2CuI + I_2$$

occurs when the ions are brought together. Some Cu(I) compounds such as CuI (brown) and Cu_2S (black) are colored due to charge transfer transitions. Cuprous compounds that contain weaker reducing agents than I^- (CuBr, CuCl, and CuCN) are colorless. The red color of Cu_2O is apparently due to the presence of some Cu(II) in the crystals.

The charge transfer peaks of $CuCl_2$ and $CuCl_4^{2-}$ that appear in the ultraviolet region of the spectrum overlap into the visible region and cause an intense yellow color. The d^9 electron configuration of Cu(II) should result in a d-d absorption spectrum. Absorption bands are also observed in the infrared region of the spectra of these compounds. The position of these peaks provides a way of distinguishing between complexes of different geometry (see Chapter 16).

The ion $CuCl_4^{2-}$ can exist in either a planar structure or a distorted tetrahedral structure. $CuCl_5^{3-}$ is also known. The structure of this ion is trigonal bipyramid. The most common structure for Cu(II) complexes is square planar with two additional ligands weakly held in the other two octahedral positions.

With ligands that produce a greater ligand field (those higher in the spectrochemical series) the *d-d* absorption bands of Cu(II) appear in the red end of the visible region of the spectrum. Accordingly $Cu(H_2O)_6^{2+}$ and $Cu(NH_3)_6^{2+}$ are blue.

B. SILVER

Metallic silver is a weaker reducing agent than copper. Silver occurs naturally as the element and in the compounds argentite (Ag_2S) and horn silver (AgCl).

The electrode potential is just slightly lower than that of palladium.

The standard method of analysis for chloride ions uses the reaction

$$Ag^+ + Cl^- \longrightarrow AgCl$$

The quantity of chloride present in a sample is determined either by measuring the volume of a silver nitrate solution required to react with the chloride or by weighing the solid silver chloride produced.

The chloride ions in silver chloride may be displaced by a number of ligands. For example, silver chloride dissolves in a water solution of ammonia. Ammonia can in turn be displaced by bromide ions and all of these ligands can be displaced by thiosulfate ($S_2O_3^{2-}$) or sulfide (S^{2-}).

The order in which the displacement reaction occurs can be determined by calculating the concentration of silver(I) in equilibrium with a solid (using the value of K_{sp}) or the concentration of silver(I) in equilibrium with a 0.10 M concentration of complex ion (from the dissociation constant of the complex). The value of the silver ion concentration in equilibrium with a number of solids and complexes is shown in Table 24-7. The values are arranged in the table according to a decreasing silver ion concentration. The ligands at the bottom of the table are stronger bases with silver(I) than those at the top. Once again we see that for an easily distorted cation such as silver(I) the strongest bases are those that are easily polarized.

Silver(I) is a stronger oxidizing agent than copper(I). The charge transfer absorption bands observed in the spectra of silver(I) compounds generally lie at a lower energy than those observed for copper(I) compounds. In contrast to cuprous bromide (CuBr), silver bromide has a light yellow color. Silver chloride looks white but absorbs light strongly in the near ultraviolet region. Both silver chloride and silver bromide are light sensitive substances that form free silver on exposure to strong sunlight.

$$2AgCl \xrightarrow{h\nu} 2Ag + Cl_2$$

Table 24-7 *Order of Ligands in Displacement Reactions of Silver Salts and Complexes*

Ligand		Complex or solid formed	K†	Concentration Ag^{+*}, M
SO_4^{2-}	$1M$	Ag_2SO_4	$K_{sp} = 1.6 \times 10^{-5}$	4×10^{-3}
Ac^-	$1M$	$AgAc(s)$	$K_{sp} = 4 \times 10^{-3}$	4×10^{-3}
CO_3^{2-}	$1M$	$Ag_2CO_3(s)$	$K_{sp} = 4 \times 10^{-12}$	2×10^{-6}
CrO_4^{2-}	$1M$	$Ag_2CrO_4(s)$	$K_{sp} = 1.4 \times 10^{-12}$	1.2×10^{-6}
O^{2-}	pH-14	$Ag_2O(s)$	$K_{sp} = 1.6 \times 10^{-8}$	1.6×10^{-8}
NH_3	$1M$	$Ag(NH_3)_2^+$	$K_d = 6 \times 10^{-8}$	6×10^{-9}
Cl^-	$1M$	$AgCl(s)$	$K_{sp} = 1.7 \times 10^{-10}$	1.7×10^{-10}
CNS^-	$1M$	$AgSCN(s)$	$K_{sp} = 1 \times 10^{-12}$	1×10^{-12}
Br^-	$1M$	$AgBr(s)$	$K_{sp} = 5 \times 10^{-13}$	5×10^{-13}
$S_2O_3^{2-}$	$1M$	$Ag(S_2O_3)_2^{3-}$	$K_d = 1 \times 10^{-13}$	1×10^{-14}
I^-	$1M$	$AgI(s)$	$K_{sp} = 8.5 \times 10^{-17}$	8.5×10^{-17}
CN^-	$0.1M$	$Ag(CN)_2^-$	$K_d = 1 \times 10^{-20}$	1×10^{-17}
S^{2-}	$0.1M$	$Ag_2S(s)$	$K_{sp} = 10^{-50}$	3×10^{-25}

*For complex ions, the concentration of Ag^+ is calculated as that in equilibrium with the complex ion at $0.10M$.

†K_{sp} is the equilibrium constant or the solubility product. K_d is the equilibrium constant for the complete dissociation of a complex ion.

It has been possible to make continual improvements in the sensitivity of silver bromide films to the point where the absorption of about ten photons can lead to the development of a silver bromide particle. In the photographic process a film is exposed briefly to light. The silver bromide grains that received enough energy to be sensitized will be largely reduced to metallic silver in a few minutes in a developing bath. After stopping this reaction, the unreacted grains of silver bromide are dissolved in a solution of sodium thiosulfate (hypo). In making a positive print from the negative another piece of film is exposed to light transmitted by the negative. On developing, this print will show dark regions where the original film was not illuminated. The light absorbed by a grain of silver bromide on the film, or the energy transferred from a dye molecule absorbed on the grain produces a latent image. The latent image consists of a tiny chemical change at some spot on the grain which makes it react rapidly with a reducing agent. Absolutely pure silver bromide does not form a latent image. Some sulfur compounds such as those present in natural gelatin must be present.

Under ordinary conditions silver does not exist in a $+2$ oxidation state. The compound AgO can be prepared, but X-ray analysis shows that it contains equal quantities of silver(I) and silver(III).

Silver(III) is a typical low spin d^8 ion that forms square planar complexes almost exclusively. Silver(III) is such a strong oxidizing agent that its compounds are quite unstable.

C. GOLD

Gold is normally found as the free metal. The only natural minerals of any importance in which gold is found combined are tellurides such as $AuTe_2$ and $AgAuTe_2$.

A simple mechanical separation with running water can be used to separate the larger particles of gold from lighter particles of sand and gravel. Somewhat smaller particles can be separated because they dissolve in liquid mercury.

The density of gold is 19.3 g cm^{-3}.

The development of the cyanide process in about 1890 contributed greatly to the world's supply of gold and silver. Using this process, gold can be profitably concentrated from poorer ores. Many of the tailings from earlier gold mines were reworked between 1890 and 1925 using the cyanide process.

The dissolution of gold requires a combination of an oxidizing agent and a complexing agent. Aqua regia ($H^+ + NO_3^- + Cl^-$) can be used, but nitric acid is much too expensive for treating tons of ore. The metals will dissolve in a solution of cyanide (a complexing agent) that is exposed to oxygen.

$$4Au + 8CN^- + O_2 + 2H_2O \longrightarrow 4Au(CN)_2^- + 4OH^-$$

$$4Ag + 8CN^- + O_2 + 2H_2O \longrightarrow 4Ag(CN)_2^- + 4OH^-$$

Large quantities of pulverized ore are leached with a solution of sodium cyanide saturated with air. Silver and gold are then recovered from the solution by electrolysis or displacement by a cheaper metal.

$$Zn + 2Au(CN)_2^- \longrightarrow Zn(CN)_4^{2-} + 2Au$$

When gold is dissolved in aqua regia a yellow planar complex ion containing Au(III) is obtained. The ion $AuCl_4^-$ undergoes ligand exchange reactions slowly (like other d^8 low-spin complexes). The complex ion reacts with ammonia to give explosive gold-nitrogen compounds such as $Au(NH)Cl$ or $Au(NH)NH_2$. The sulfide Au_2S_3 can be precipitated from cold solutions. In hot acidic solutions treatment with H_2S reduces $AuCl_4^-$ to metallic gold.

Both Au_2S_3 and metallic gold dissolve slowly in ammonium polysulfide and more rapidly in sodium polysulfide solutions.

$$Au_2S_3 + S^{2-} \longrightarrow 2AuS_2^-$$

$$2Au + 3S_2^{2-} \longrightarrow 2AuS_2^- + 2S^{2-}$$

VII. Zinc, Cadmium, Mercury, and Thallium

Going down the family of IIB elements there is a decrease in polarizing ability as the size of the cation increases. There is also a great increase in induced polarizing ability. As a result the solubility of the sulfides decreases down the group. All three elements form complex ions with the halides.

The dissociation constants of the halide complexes (Table 24-8) become progressively smaller as either the cation or ligand size is increased.

The chemistry of mercury and thallium exhibit some unique features. The ions Hg^{2+} and Tl^{3+} are isolectronic with Pb^{4+}. They are softer acids than Pb^{4+} because of the lower oxidation number. These ions are the softest ions known in their respective oxidation states.

Mercuric sulfide (HgS) is a very insoluble substance. The equilibrium constant for the reaction

$$HgS + 2H^+ \longrightarrow Hg^{2+} + H_2S(aq)$$

is 3×10^{-33}. A combination of chloride ion as the complexing agent and a strong oxidizing agent is required to dissolve mercuric sulfide.

$$HgS + 4Cl^- + 8H^+ + 8NO_3^- \longrightarrow$$
$$8NO_2 + HgCl_4^{2-} + SO_4^{2-} + 4H_2O$$

Thallium (I) is the stable form of thallium over most of the E_h-pH diagram (compare mercury and lead, Figs. 24-5 and 24-6).

The solubility product of thallium hydroxide (TlOH) is 8×10^{-2}, consequently the hydroxide ion concentration (3×10^{-1}) of a saturated solution of thallium hydroxide corresponds to a pH of 13.5. Thallium hydroxide is a strong base that is considerably more soluble in water than barium hydroxide. There are other soluble hydroxides at the right of the periodic table ($As(OH)_3$, and so forth), but the others all give acidic solutions.

Thallium(I), like lead(II), is a soft acid. Both TlCl and Tl_2S are

Table 24-8 *Dissociation Constants of Some Complex Ions of Group IIB Metals*
$$MX_4^{2-} \longrightarrow M^{2+} + 4X^-$$

$ZnCl_4^{2-}$	$K_d = 10^{+2.1}$	$ZnBr_4^{2-}$	$K_d = 10^{1.3}$	ZnI_4^{2-}	$K_d = 10^{-2.3}$
$CdCl_4^{2-}$	$K_d = 10^{-2.9}$	$CdBr_4^{2-}$	$K_d = 10^{-3.7}$	CdI_4^{2-}	$K_d = 10^{-5.6}$
$HgCl_4^{2-}$	$K_d = 10^{-16.2}$	$HgBr_4^{2-}$	$K_d = 10^{-21}$	HgI_4^{2-}	$K_d = 10^{-30.3}$

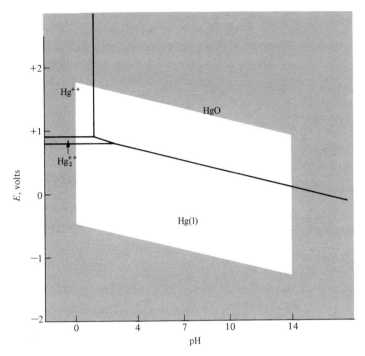

Figure 24-5 E_h-pH *diagram for mercury.*

insoluble. They are more soluble than might be expected because of a favorable entropy change. Although the enthalpy term for the reaction TlCl $\longrightarrow$ Tl$^+$ + Cl$^-$ is unfavorable ($\Delta H^0 = 10.1$ kcal mole^{-1}), the entropy change of 16.9 eu is favorable.

VIII. The Platinum Metals

The six elements ruthenium (Ru), rhodium (Rh), palladium (Pd), osmium (Os), iridium (Ir), and platinum (Pt) occur together naturally in alloys. Usually platinum is the principal element present. Occasionally natural samples that are over 50% osmium or iridium are found. All of the elements show considerable chemical similarity (they are all very weak reducing agents), and they are usually classified together as the platinum metals.

Platinum is the most important of these metals. It is widely used for laboratory ware (crucibles and electrodes) because it is chemically inert and can be used to catalyze the reactions of hydrogen. Palladium is very similar to platinum chemically and is also widely used as a catalyst. Hydrogen is exceptionally soluble in metallic palladium.

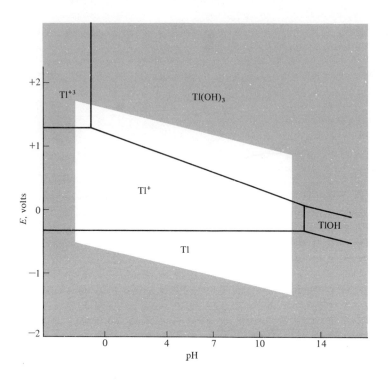

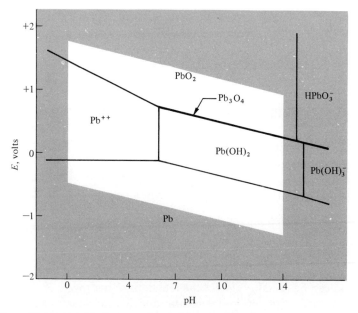

Figure 24-6 E_h-pH *diagrams for thallium and lead.*

Sec. VIII. The Platinum Metals

Table 24-9 *Oxidation States of the Platinum Metals*

	d^0	d^1	d^2	d^3	d^4	d^5	d^6	d^7	d^8
Ru	+8	+7	+6		+4	+3	+2		
Os	+8	(+7)	+6		+4	+3	(+2)		
Rh				(+6)		+4	+3		(+1)
Ir				+6		+4	+3		(+1)
Pd					(+6)		+4		+2
Pt					(+6)		+4		+2

Palladium and platinum are usually found in compounds in the +2 or +4 oxidation state. The complexes of both oxidation states are robust. The structure of the +2 compounds is square planar. The +4 compounds are octahedral. Because the ligand exchange reactions of both d^8 and low-spin d^6 compounds are slow, a great many complexes of palladium and platinum (such as $Pt(NH_3)_4^{2+}$, $PtCl_3(NH_3)^-$, $Pd(NH_3)_2Cl_2$, $Pt(CNS)_3CO^-$) can be isolated.

All the platinum metals precipitate as insoluble sulfides with hydrogen sulfide in acid solutions; however, some of the robust complexes, especially those of ruthenium(III) and iridium(III) react very slowly with hydrogen sulfide.

The density of osmium is greater than that of any other metal (22.5 g cm^{-3}) and it is the only metal that reacts readily with nitric acid to form a volatile oxide.

$$Os + 8H^+ + 8NO_3^- \longrightarrow 8NO_2 + OsO_4 + 4H_2O$$

Osmium tetroxide (OsO$_4$) vapor is extremely poisonous, but fortunately it has an extremely strong unpleasant odor that can serve as a warning. Osmium tetroxide is used in organic chemistry as an oxidizing agent for preparing
$-\underset{OH}{\overset{|}{C}}-\underset{OH}{\overset{|}{C}}-$ from olefins.

The oxidation states found for the platinum metals are summarized in Table 24-9. Oxidation states which can definitely be prepared but are very unstable are shown in parentheses.

QUESTIONS
1. Write balanced chemical equations for:
   (a) the decomposition of arsine
   (b) dissolving As_2S_3 in nitric acid
   (c) dissolving As_2O_3 in HCl; in NaOH; in $(NH_4)_2S$
   (d) the reduction of the antimonyl ion by tin
   (e) the reactions of $SiCl_4$, $SnCl_4$, $PbCl_2$, $ZnCl_2$ and $AsCl_3$ with H_2O.

(f) the oxidation of $Sn(OH)_4^{2-}$ by $Bi(OH)_3$
(g) dissolving gold in aqua regia, $HNO_3 + HCl$
(h) the reaction of Ag_2CrO_4 with Cl^-
(i) the air oxidation of gold in the presence of cyanide
(j) the reactions of ReO_4^- with H_2S and Na_2S

2. From the following sets of four or five species, pick the one specified at the left.
    (a) the softest acid: P^{3+}, P^{5+}, As^{3+}, As^{5+}
    (b) the least soluble: As_2O_3, $AsCl_3$, Bi_2O_3, Bi_2S_3, As_2S_3
    (c) the most reactive with H_2O: SO_3, SeO_3, TeO_3, TeO_2
    (d) the most soluble in H_2O: P_4O_{10}, As_2O_5, Sb_2O_5, Sb_2O_3, P_4O_6
    (e) the strongest oxidizing agent: P(V), Sb(III), Sb(V), Sn(IV), Si(IV)
    (f) the most acidic: Bi_2O_3, PbO_2, SnO_2, Sb_2O_3, SnO
    (g) the most deeply colored: $SiCl_4$, $SnCl_4$, $PbCl_4$, $TlCl_3$
    (h) the softest acid: Pt(II), Au(I), Au(III), Hg(II), Tl(I)
    (i) the least soluble: $CuCl_2$, $InCl_3$, $HgCl_2$, $TlCl_3$, $PtCl_2$
    (j) the most noble: Re, Os, Ir, Pt, Au
    (k) the softest acid: Ag^+, Au^+, Cd^{2+}, Hg^{2+}, Au^{3+}

3. Name two ores of arsenic.
4. Write the formulas of the thio complex anions of As, Sb, and Sn.
5. What are the colors of Sb_2S_3, As_2S_3, Bi_2S_3, CuS?
6. List six elements with common allotropes.
7. Why is Bi(V) a much stronger oxidizing agent than As(V) or Sb(V)?
8. Arrange each set of three species in the order specified at the left.
    (a) increasing softness: As^{3+}, Sb^{3+}, Bi^{3+}
    (b) ease of reduction: As(III), Sb(III), Bi(III)
    (c) strength as an acid: As_4O_6, Sb_2O_3, Bi_2O_3
    (d) strength as a base: Sb_2O_5, SnO_2, PbO_2
    (e) strength as oxidizing agent: PbO_2, SnO_2, Sb_2O_5
    (f) darker color: Sb_2O_3, Sb_2O_5, Bi_2O_5

9. What oxidizing agents in groups IVA and VA will oxidize Mn^{2+} to MnO_4^- in acid solution? $MnO_4^- + 8H^+ + 5e^- = Mn^{2+} + 4H_2O \quad E^0 = 1.51$ volts

10. List one use for each of the IB and IIB elements.
11. $PtCl_2$ does not precipitate with 0.10 M HCl. Why not?
12. Why are the noble metals such poor reducing agents?
13. Why is CuO more easily reduced than ZnO?
14. Complete and balance the following equations.
    $\_Au + \_CN^- + \_O_2 \longrightarrow$
    $\_Pt + \_S_2^{2-} \longrightarrow$
    $\_SnCl_6^{4-} + \_Hg^{2+} \longrightarrow$
    $\_AgBr + \_S_2O_3^{2-} \longrightarrow$
15. In what oxidation states will iridium form robust complexes?

PROBLEMS I

1. H^0 for $AsCl_3$ is -73 kcal mole^{-1}. Calculate α_{As-Cl}.
2. Predict H^0 for $AsCl_5$.
3. Calculate the pH of a solution prepared by adding 0.10 mole of HCl to 1 liter of a 0.15 M Na_2HAsO_4 solution.

Sec. VIII. The Platinum Metals

4. The solubility product of $Sn(OH)_2$ is about 10^{-26}. What pH is needed to dissolve 0.15 mole of $Sn(OH)_2$ in a liter of water?
5. Calculate the formula of the Cu-Sn alloy which is 59% Sn (see Fig. 17-3).
6. The solubility product of BiOCl is 4×10^{-8}. Calculate the concentration of BiO^+ present in a saturated solution in 2 M HCl. What is the chemical form of the bismuth dissolved in this solution?
7. E^0 for the reaction $BiO^+ + 3e^- + 2H^+ \longrightarrow Bi + H_2O$ is -0.32 volt. Calculate E^0 for the half-reaction $BiOCl + 3e^- + 2H^+ \longrightarrow Bi + H_2O + Cl^-$.
8. Calculate K_{sp} for PbI_2 from the G^0 values: PbI_2, -41.53; Pb^{2+}, -5.81; I^-, -12.35.
9. From the data in the text, calculate the pH of a 0.010 M solution of TlOH.
10. From the data in Table 24-7 calculate the solubility of AgCl in 0.10 M NH_3.
11. What concentration of CN^- is needed to dissolve Ag_2O if the pH is 13?
12. The entropy of Pt is 9.95 eu. The thermodynamic values for $Pt(OH)_2$ are $H^0 = -87.2$ kcal mole^{-1} and $S^0 = 26.5$ eu. Calculate the equilibrium constant for the reaction $2Pt(OH)_2 \longrightarrow 2Pt + 2H_2O + O_2$. What pressure of O_2 is required to form $Pt(OH)_2$ in the presence of Pt and pure liquid H_2O at 25°C?
13. The electrode potential of a Pt electrode at a pH of 6.0 in the presence of $Pt(OH)_2$ is $+0.64$ volt. Calculate E^0 for the electrode reaction $Pt(OH)_2 + 2H^+ + 2e^- \longrightarrow Pt + 2H_2O$.
14. If $E^0 = +1.2$ volts for $Pt^{2+} + 2e^- \longrightarrow Pt$, calculate K_{sp} for $Pt(OH)_2$.
15. Calculate K_{sp} for TlCl from the data on p. 567.

PROBLEMS II

1. Given the H^0 values: $ZnCl_2$, -99.20; Zn^{2+}, -36.8; Cl^-, -39.95 kcal mole^{-1}, calculate ΔH^0 for the reaction $\frac{1}{2}ZnCl_2(s) \longrightarrow \frac{1}{2}Zn^{2+} + Cl^-$.
2. H^0 for $CdCl_2$ is -93.6 and $H^0 = -18.14$ for Cd^{2+}. Calculate ΔH^0 for the corresponding reaction of $CdCl_2$.
3. Why is the dissolving of $CdCl_2$ less exothermic than that of $ZnCl_2$?
4. Using data from Appendix IV calculate ΔH for the reaction $Zn(OH)_2(s) + H_2S(aq) \longrightarrow ZnS(s) + 2H_2O(l)$.
5. Calculate ΔH for the reaction $Cu_2O + H_2S(aq) \longrightarrow Cu_2S + H_2O(l)$.
6. Which is softer, Zn^{2+} or Cu^+?
7. Calculate ΔH^0 for the reaction $ZnS \longrightarrow Zn^{2+} + S^{2-}$.
8. ΔS^0 for the reaction in Problem 7 is -44.1 eu. Calculate K_{sp} for ZnS.
9. What concentration of Zn^{2+} will be in equilibrium with ZnS if $a_{S^{2-}} = 0.1$?
10. The equilibrium constant for the reaction $ZnCl_4^{2-} \longrightarrow Zn^{2+} + 4Cl^-$ is 130. Calculate $a_{Zn^{2+}}$ if $[Cl^-] = 1$ M and $[ZnCl_4^{2-}] = 0.10$ M.
11. Using the equilibrium constant from Table 24-8 calculate $a_{Zn^{2+}}$ in equilibrium with 1 M I^- and 0.10 M ZnI_4^{2-}.
12. For the complex ion $Zn(NH_3)_4^{2+}$, $H^0 = -127.5$ kcal mole^{-1} and $S^0 = 72$ eu. Using also data from Appendix IV calculate the equilibrium constant for the reaction $Zn(NH_3)_4^{2+} = Zn^{2+} + 4NH_3$.
13. Calculate $a_{Zn^{2+}}$ in a 1 M NH_3 solution if $a_{Zn(NH_3)_4^{2+}} = 0.10$.
14. Construct a table like Table 24-7 for Zn^{2+} with the ligands NH_3, Cl^-, I^-, and S^{2-}.

> In the earth's core certain elements—gold, platinum, and nickel are examples—may be highly concentrated. Others again appear to concentrate in silicate rocks of the crust, whereas yet others may tend to combine with sulfur and concentrate in sulfur minerals. It has been noted also that many meteorites are characterized by the presence of three principal solid phases, iron, sulfide (troilite), and silicate, in each of which different elements tend to concentrate.*
>
> L. H. AHRENS

Chapter 25 Nonmetals

The chemistry of the negative oxidation states of the nonmetals is covered in the earlier chapters in connection with hydrogen and the metals. Also the bonding in diatomic molecules has been treated in detail in Chapters 7 and 12.

I. Elemental Sulfur

The name polycatenasulfur is used for the long sulfur chains.

Sulfur exists in a variety of elemental forms. At room temperature sulfur can exist either as long chains ($S-(S)_n-S$) or as rings of various sizes (S_6, S_8, and so forth). There are also several different ways of packing S_8 rings into crystals.

The thermodynamically stable form of the element at 25°C and 1 atm pressure is rhombic sulfur. The orthorhombic crystals of rhombic sulfur contain S_8 rings. Between 95.5°C and the melting point of sulfur (114.3°C) a monoclinic packing of the S_8 molecules is stable. The equilibrium relationship between orthorhombic and monoclinic sulfur is shown in the temperature-pressure phase diagram in Fig. 25-1.

The rings in the S_8 molecules of both rhombic and monoclinic sulfur are puckered as shown in Fig. 25-2. The S_8 molecules persist when monoclinic sulfur is melted rapidly. The liquid at the equilibrium melting point (114.3°C) consists of 96% S_8 molecules. In addition to the S_8 rings there are some S_6, S_7, and S_9 rings present.

---

* Distribution of the Elements in Our Planet (New York: McGraw Hill, 1965).

Sec. 1 Elemental Sulfur

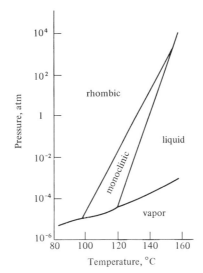

Figure 25-1 *The phase diagram for elemental sulfur.*

Sulfur that consists of only S_6 rings crystallizes in rhombohedral crystals. It readily decomposes to give a mixture of rhombic S_8 and polycatenasulfur. The S_7 and S_9 ring systems have not yet been isolated in a pure form.

As long as liquid sulfur is predominantly S_8 molecules (114.5–159°C) it flows freely and has a light yellow color. When heated above 160°C it becomes dark reddish brown and very viscous. The high viscosity is due to the presence of long chains of sulfur atoms. The dark color is due to electronic transitions that involve the unpaired electrons at the ends of the chains. If the viscous liquid is cooled rapidly the long chains persist. This material is known as plastic sulfur or polycatenasulfur. In this form sulfur is insoluble in organic solvents such as carbon disulfide (CS_2). On standing plastic sulfur converts very slowly to rhombic sulfur. Sulfur as it is normally found in the laboratory is a mixture of rhombic sulfur and polycatenasulfur.

S_6 *sulfur may be prepared according to the reaction* $H_2S_4 + S_2Cl_2 \longrightarrow S_6 + 2HCl$

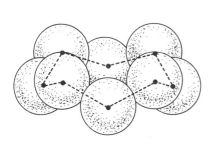
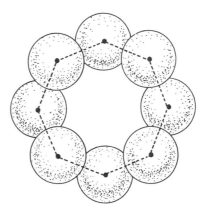

Figure 25-2 *The structure of the S_8 molecule.*

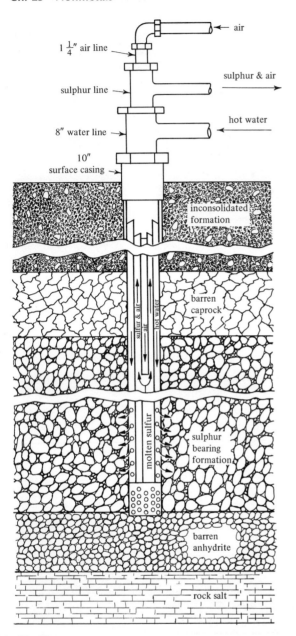

Figure 25-3 The Frasch process for mining sulfur. (*After R. E. Kirk and D. F. Othmer, eds., Encyclopedia of Chemical Technology, New York: Interscience, 1954.*)

Large amounts of sulfur are present in sulfide minerals, and substantial amounts of sulfuric acid are produced from the sulfur dioxide obtained from roasting sulfur ores

$$ZnS + \tfrac{3}{2}O_2 \longrightarrow ZnO + SO_2$$

$$SO_2 + \tfrac{1}{2}O_2 \xrightarrow{Pt} SO_3$$

$$SO_3 + H_2O \xrightarrow{H_2SO_4} H_2SO_4$$

Sulfur is also found as $CaSO_4 \cdot 2H_2O$ (gypsum), $BaSO_4$, and so forth. The principal source is the free sulfur found in large underground deposits associated with salt domes.

In about 1894, Herman Frasch developed a process for extracting sulfur from underground deposits by melting it with superheated steam. As the process was developed it was found that compressed air would make a froth of sulfur light enough to float on water, and this would rise naturally in a third concentric pipe (Fig. 25-3). This process accounts for about 70% of all the sulfur currently being produced.

II. Compounds and Reactions of Sulfur

A. SULFUR(VI)

When sulfur is burned in air sulfur dioxide is produced. The oxidation of sulfur(IV) to sulfur(VI) is the critical step in the production of sulfuric acid and it is very slow. The oxides of nitrogen will catalyze the reaction

$$2SO_2 + O_2 \longrightarrow 2SO_3$$

in the gas phase. The use of nitrogen oxides as catalysts for the oxidation reaction is known as the lead chamber process when it is conducted in lead vessels. This process has been largely superceded by the more efficient "contact process," which uses solid catalysts. Platinum was first proposed as a catalyst for the production of sulfur trioxide (SO_3) as early as 1831, but there were serious problems with impurities in the sulfur dioxide which destroyed the catalytic activity. These problems were not successfully solved for about 60 years. Currently both platinum and vanadium pentoxide (V_2O_5) are widely used as catalysts in the contact process.

The next step in the production of sulfuric acid, the reaction of water with sulfur trioxide is extremely exothermic. To avoid boiling the water, sulfur trioxide is bubbled into a concentrated solution of sulfuric acid which boils at a much higher temperature than water.

The sulfur-oxygen bonding in sulfuric acid is similar to the phosphorus-oxygen bonding in phosphates. The bonding is exceptionally strong and the strength of sulfur-oxygen bonds is dependent on the environment of the oxygen atoms. Compounds containing the S-O-S linkage can be prepared by the addition of excess sulfur trioxide to sulfuric acid

$$SO_3 + H_2SO_4 \longrightarrow H_2S_2O_7$$

The hydrolysis of pyrosulfuric acid ($H_2S_2O_7$) is more exothermic than that of high-energy phosphate bonds

$$H_2S_2O_7 + H_2O \longrightarrow 2H_2SO_4 \qquad \Delta H = -16.4 \text{ kcal mole}^{-1}$$

B. REDOX REACTIONS OF SULFUR

Tabulated G^0 values can be used to calculate the equilibrium constants for any particular sulfur redox reaction. Where there are a number of different oxidation states to consider, the calculations can be simplified by using a consistent stoichiometry for each oxidation state under consideration.

In order to compare the free energy of HSO_4^- with H_2S, oxygen atoms must be added in the form of water. The combination $H_2S + 4H_2O$ has the same stoichiometry as $HSO_4^- + 9H^+ + 8e^-$ and the free energies are directly comparable. With the four G^0 values

Reaction	G^0, kcal mole^{-1}
$H_2S + 4H_2O$	-234.77
$S + 4H_2O + 2H^+ + 2e^-$	-226.748
$SO_2 + 2H_2O + 6H^+ + 6e^-$	-185.112
$HSO_4^- + 9H^+ + 8e^-$	-180.69

the ΔG^0 value for any of six half-reactions can be obtained by subtraction. If the free energy of the sulfur species is plotted against the oxidation state of the sulfur, it is immediately apparent which disproportionation reactions are spontaneous. A species under consideration can disproportionate if the adjusted free energy of the species lies above a straight line connecting the free energies of the products. The disproportionation of SO_2 into S and HSO_4^- is therefore spontaneous (Figs. 25-4 and 25-5). A thermodynamically stable species such as free sulfur lies below the line connecting species of lower and higher oxidation states (Fig. 25-6). Under these conditions sulfur cannot disproportionate into H_2S and HSO_4^-.

This type of diagram can also be used to test for disproportionation possibilities under different conditions. If we choose to ignore the possibility that SO_4^{2-}, HSO_4^-, and H_2SO_4 are products of the reaction, since they are formed slowly, we find that dithionic acid ($H_2S_2O_6$) should not disproportionate to sulfur and sulfur trioxide (Fig. 25-7).

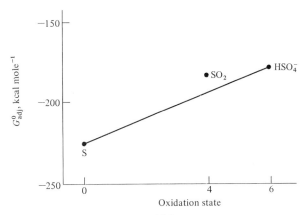

Figure 25-4 *The disproportionation of* SO_2.

Similarly, Fig. 25-8 shows that in the absence of water SO_2 cannot disproportionate to S and SO_3. In basic solutions (Figs. 25-9 and 25-10) elemental sulfur can disproportionate into S^{2-} and SO_4^{2-}. Elemental sulfur also reacts with both S^{2-} and SO_3^{2-} to form species which contain sulfur in an intermediate oxidation state. These are the polysulfide ions

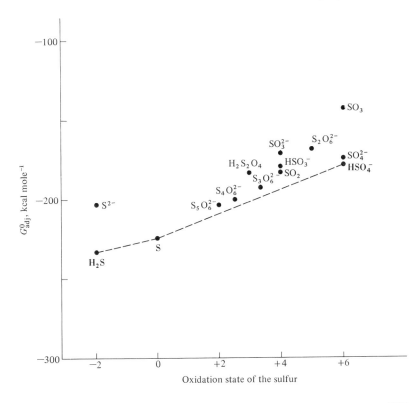

Figure 25-5 G^0 *values for sulfur species in acid solution adjusted to the stoichiometry of* $H_2S + 4H_2O$.

Ch. 25 Nonmetals

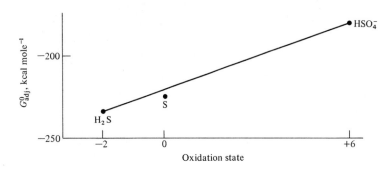

Figure 25-6 *The stability of sulfur(0) in the presence of acid.*

(S_n^{2-}) and thiosulfate ion ($S_2O_3^{2-}$) respectively. Both of these reactions take place through a nucleophilic displacement at an S-S bond.

$$Z-S\colon + X-S-S-Y \longrightarrow Z-S-S-Y + X-S\colon$$

This type of reaction is relatively fast and equilibrium is established in a few minutes.

The product of the reaction between SO_3^{2-} and S_8 would be $S_9O_3^{2-}$

$$SO_3^{2-} + S_8 \longrightarrow O-\underset{O}{\overset{O}{\underset{\|}{\overset{\|}{S}}}}-S-S-S-S-S-S-S-S^{2-}$$

but this ion cannot be isolated because it reacts rapidly with more SO_3^{2-} until finally thiosulfate

$$O-\underset{O}{\overset{O}{\underset{\|}{\overset{\|}{S}}}}-S^{2-}$$

is the predominate product.

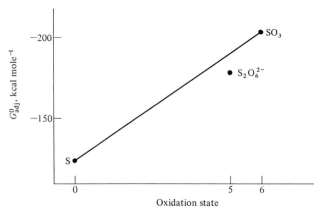

Figure 25-7 *The relative stabilities of $S_2O_6^{2-}$ and $S + SO_3$.*

Sec. II Compounds and Reactions of Sulfur

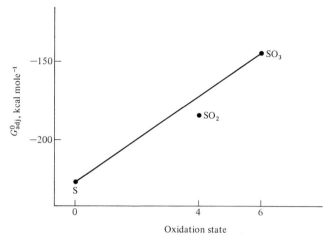

Figure 25-8 *The stability of SO_2 towards disproportionation in the absence of water.*

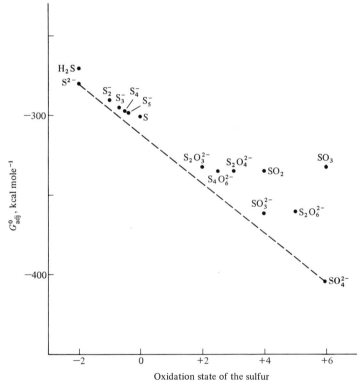

Figure 25-9 *G^0 values for sulfur species in basic solution adjusted to the stoichiometry of $S^{2-} + 8OH^-$.*

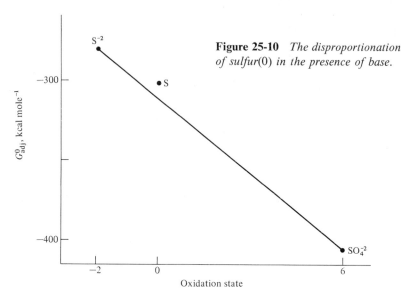

Figure 25-10 *The disproportionation of sulfur(0) in the presence of base.*

The thiosulfate ion has two main uses. It is an excellent complexing agent for soft cations and it is used to dissolve silver bromide in film processing. Thiosulfate ion is also a reducing agent. The stoichiometry of a reaction in which thiosulfate acts as a reducing agent is usually complex and variable because a great variety of products are possible. However, the mechanism of the $S_2O_3^{2-}$–I_2 reaction is quite simple. It may be represented by the equations

$$S_2O_3^{2-} + I_3^- \rightleftharpoons \text{O}-\underset{\underset{\text{O}}{|}}{\overset{\overset{\text{O}}{\|}}{S}}-S-I^- + 2I^-$$

$$S_2O_3^{2-} + S_2O_3I^- \longrightarrow \text{O}-\underset{\underset{\text{O}}{|}}{\overset{\overset{\text{O}}{\|}}{S}}-S-S-\underset{\underset{\text{O}}{|}}{\overset{\overset{\text{O}}{\|}}{S}}-O^{2-} + I^- \quad \text{rate determining step}$$

Other oxidizing agents can be titrated with thiosulfate by first reacting them with iodide and titrating the I_2 produced. When acid is added to a thiosulfate solution a white precipitate of colloidal sulfur forms within a few minutes. The thermodynamically stable products of the reaction are S_8 and SO_4^{2-}. Considerable amounts of these species may be formed; however, other products including S_6 and polythionic acids are produced as well.

The polythionic acids have the formula $H_2S_nO_6$. Compounds with values of n from 2 to at least 10 are known. It appears that the tetrathionic and pentathionic acids are the most stable. Polythionic acids with much larger n values, $n = 20$-80, are present in certain formulations of colloidal sulfur such as that prepared by the reaction of sodium thiosulfate with sulfuric acid. The polythionic acids all have three oxygen atoms on the two sulfur atoms at the ends of the chain.

Other even less stable sulfur-oxygen acids are known. When sulfuric acid or sulfate ion is treated with a very strong oxidizing agent, peroxysulfuric acid or peroxydisulfuric acid is formed (Fig. 25-11). Both of these acids are stronger oxidizing agents than oxygen. When sulfurous acid (H_2SO_3), SO_2, or HSO_3^- is treated with a very strong reducing agent such as zinc or sodium amalgam, salts of dithionous acid may be prepared.

$$2HSO_3^- + Zn \longrightarrow Zn^{2+} + S_2O_4^{2-} + 2OH^-$$

Even though the acid is extremely unstable, the anhydrous sodium salt ($Na_2S_2O_4$) can be stored for reasonable periods of time. The salt is a strong enough reducing agent to reduce copper ion to metallic copper. Since sodium dithionite has a somewhat different reducing action than the metals used in its preparation, the substance has a number of industrial applications. For example, sodium dithionite is used to reduce disulfide linkages in wool and to bleach wood pulp and molasses.

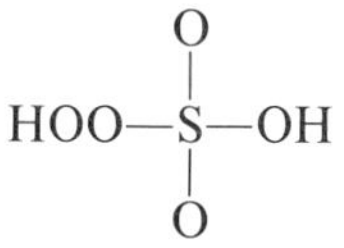

peroxomonosulfuric acid

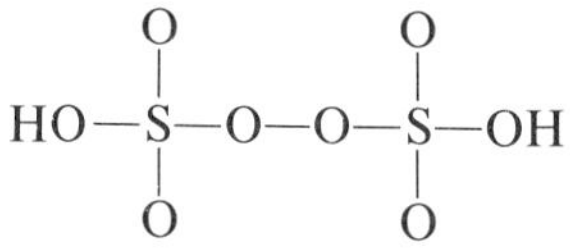

peroxodisulfuric acid

Figure 25-11 *Some sulfur acids with oxygen-oxygen bonds.*

C. SULFUR HALIDES

When fluorine reacts with elemental sulfur, SF_6 is the major product. Because of the slowness of the final steps in the oxidation, substantial amounts of SF_4 and S_2F_{10} (corresponding to SO_2 and dithionate respectively) are also produced. An additional sulfur fluoride (S_2F_2) can be prepared according to the reaction

$$\tfrac{3}{8}S_8 + 2AgF \longrightarrow Ag_2S + S_2F_2$$

S_2F_2 reacts rapidly and completely with water at room temperature

$$2S_2F_2 + 2H_2O \longrightarrow SO_2 + 3S + 4HF$$

SF_4 undergoes a similar hydrolysis

$$SF_4 + 2H_2O \longrightarrow SO_2 + 4HF \qquad \Delta G^0 = -45 \text{ kcal mole}^{-1}$$

SF_6 on the other hand does not react with water at an appreciable rate. It is a colorless gas (sublimes at $63.8°C$) that is widely used for its insulating properties as a high pressure gas. Most fluorides, including SF_4, are poisonous, but SF_6 is so inert chemically that it is odorless, tasteless, and nontoxic.

SCl_4 can be prepared, but it decomposes rapidly above $-20°C$.

$$SCl_4 \longrightarrow SCl_2 + Cl_2$$

Sulfur chlorides can be prepared with chains of from two (S_2Cl_2) up to at least five (S_5Cl_2) sulfur atoms.

The chloride ion is readily displaced by water from all compounds with sulfur-chlorine bonds. The sulfur-oxygen compounds produced in this reaction usually disproportionate to polythionic acids or a mixture of elemental sulfur and sulfurous acid.

III. Chlorine Compounds

Chlorine dioxide (ClO_2) is a yellow gas which shows even less tendency than NO_2 to dimerize. The decomposition of ClO_2 to the elements is exothermic and may occur explosively

$$2ClO_2 \longrightarrow Cl_2 + 2O_2 \quad \Delta H^0 = -49 \text{ kcal mole}^{-1}$$

but it is such a convenient and strong oxidizing agent that it is used industrially in the bleaching of flour and textiles.

Most of the redox reactions of chlorates and chloric acids are moderately slow. Although only Cl^-, Cl_2, and ClO_4^- appear in the E_h-pH diagram (Fig. 5-4, p. 149), the intermediate chloric acids (HClO, $HClO_2$, and $HClO_3$) are all well known. The pK_a's and reduction potentials for all the chloric acids are given in Table 25-1.

Hypochlorous acid (HClO) is readily prepared by the hydrolysis of elemental chlorine

$$OH^- + Cl_2 \longrightarrow HOCl + Cl^-$$

$$HOCl \longrightarrow H^+ + OCl^-$$

Table 25-1 pK_a and E^0 Values for the Chloric Acids

Acid	pK_a	E^0, volts reduction to Cl^-	E^0, volts oxidation to ClO_4^-
HClO	7.2	1.50	1.35
$HClO_2$	2.0	1.54	1.20
$HClO_3$	-1.0	1.44	1.19
$HClO_4$	-10	1.39	—

Over a longer period of time or at a higher temperature the disproportionation of OCl$^-$ proceeds to chlorate(V).

$$3OCl^- \longrightarrow ClO_3^- + 2Cl^-$$

The last step in the oxidation of chlorine to chlorine(VII), like that of sulfur to sulfur(VI), is exceptionally slow. The perchlorate ion is generally produced in high temperature reactions such as the decomposition of KClO$_3$ or by the electrolytic oxidation of ClO$_3^-$

$$2KClO_3 \begin{matrix} \nearrow 2KCl + 3O_2 \\ \searrow KClO_4 + KCl + O_2 \end{matrix}$$

$$2OH^- + ClO_3^- \longrightarrow ClO_4^- + H_2O + 2e^-$$

The most convenient way to prepare chloric(III) acid is the disproportionation of ClO$_2$

$$2ClO_2 + H_2O \longrightarrow HClO_2 + H^+ + ClO_3^-$$

IV. Heavier Elements in Groups VIA and VIIA

A. SELENIUM

The stable form of elemental selenium consists of gray hexagonal crystals. These crystals contain long chains of selenium atoms. They conduct electricity poorly in the dark, but the conductance increases by a factor of about 1000 on exposure to light. Selenium is widely used in photocells. Its properties as a semiconductor are also used to convert alternating current to direct current. There are at least three other allotropic forms of selenium, an amorphous form and two red monoclinic crystalline forms containing Se$_8$ molecules.

The same changes which we noted down the series P, As, Sb are evident on comparing Se and Te with S or Br and I with Cl. In selenates and bromates the d-p π bonding is expected to be weaker than that in sulfates and chlorates. As a consequence SeO$_4^{2-}$ is a stronger oxidizing agent than SO$_4^{2-}$. The ionization constants of H$_2$SeO$_4$ are very close to those of H$_2$SO$_4$. The ion SeO$_4^{2-}$ is labile, and as a result the redox reactions are fast. The E_h-pH diagram (Fig. 25-12) is a good representation of the chemistry of selenium.

Selenium and tellurium compounds are among the most poisonous materials known. Ingested selenium and tellurium are slowly released

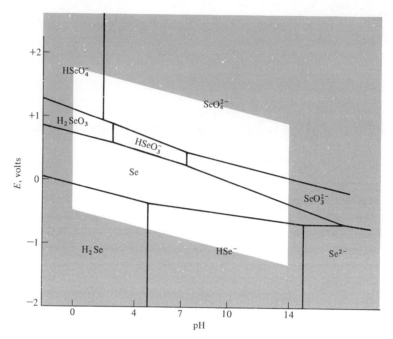

Figure 25-12 E_h-pH *diagram for selenium.*

from the body in the breath as organic compounds with foul odors. The rotten egg odor of hydrogen sulfide is reportedly mild in comparison. Thus it is not surprising that very few chemists have actually worked with selenium or tellurium. The toxicity of selenium compounds is so severe that traces concentrated by plants from the soil will kill grazing cattle.

B. TELLURIUM

Telluric(VI) acid (H_6TeO_6) is not as strong an oxidizing agent as H_2SeO_4, and it is a much weaker acid. The first ionization constant of this acid is only about 2×10^{-8}. Both of these effects appear to be related to the larger size of the fifth row elements over the fourth row and the resulting six-fold coordination that we have already noted for Sb(V). The formula for telluric acid is H_6TeO_6, in contrast to H_2SeO_4, and H_6TeO_6 shows a marked tendency to polymerize. The salts of complex anions such as $K_2H_{12}Te_4O_{17}$ are known, but their structures have not been determined.

Tellurium(IV) is amphoteric. The equilibrium constant for the reaction

$$H^+ + TeO_2 + H_2O \longrightarrow Te(OH)_3^+$$

is 10^{-2}. The coordination number of Te(IV) has never been determined. It may be three as in IO_3^- or four as in Sb(III).

C. IODINE AND BROMINE

The chemistry of the positive oxidation states of iodine and bromine resembles quite closely the pattern we have seen for Se and Te. We expect from the trends shown in Table 25-2 that Br(VII) is a slightly stronger oxidizing agent than I(VII). The E^0 value for the half-reaction

$$H_5IO_6 + 2e^- + H^+ \longrightarrow IO_3^- + 3H_2O$$

is 1.62 volts. BrO_4^- should be on the borderline of stability in water or just beyond. It was first prepared in 1968 in experiments with ^{83}Br. It was then immediately prepared in larger quantities using the reaction

$$H_2O + XeF_2 + BrO_3^- \longrightarrow Xe + 2HF + BrO_4^-$$

BrO_4^- has the same structure as ClO_4^-; whereas I(VII) generally shows six-fold coordination. A number of oxidizing agents, including Cl_2, are strong enough to prepare periodate in basic solution

$$2Na^+ + IO_3^- + 3OH^- + Cl_2 \longrightarrow Na_2H_3IO_6 + 2Cl^-$$

The E_h-pH diagram for iodine (Fig. 5-4, p. 150) indicates that the E^0 for the I(V)-I(VII) couple drops below that for Cl_2/Cl^- at any pH above about 7.

Although the ions BrO^- and IO^- do not appear in the E_h-pH diagrams, they are well known as the initial products of the disproportionation of the elements in basic solutions

Table 25-2 *Electrode Potentials for Changes in Oxidation State for Elements at the Lower Right of the Periodic Table (volts)*

(0)–(III)	(III)–(V)	(0)–(IV)	(IV)–(VI)	(0)–(V)	(V)–(VII)
P	P	S	S	Cl	Cl
−0.49	−0.30	0.45	0.096	1.47	1.12
As	As	Se	Se	Br	Br
0.232	0.60	0.74	1.09	1.47	~1.7
Sb	Sb	Te	Te	I	I
0.091	0.79	0.56	1.02	1.21	1.62
Bi	Bi	Po	Po	At	At
0.32	1.7	0.76	1.9	~1.3	

$$I_2 + 2OH^- \longrightarrow I^- + OI^- + H_2O$$
$$Br_2 + 2OH^- \longrightarrow Br^- + OBr^- + H_2O$$

BrO_2^- and IO_2^- are apparently intermediates in redox reactions of bromine and iodine, but they disproportionate too rapidly to be studied to any great extent.

The fluorides of iodine and bromine have been listed in Table 1-8 (pp. 18–19). Like SF_4 and the inert gas fluorides these compounds are all readily hydrolyzed to HF and the corresponding oxo acids. The only other compounds where bromine or iodine is in a positive oxidation state are the explosive NBr_3 and NI_3 and a few chlorides, $BrCl$, ICl, and ICl_3.

V. The Electrostatic Valence Rule

We have examined the chemistry of a number of heavy elements. The resemblance of the elements in a row (horizontal) and in groups (vertical) has been emphasized. In comparing elements in different rows there are also some important diagonal relations that result from similarities in charge density. Recall that LiOH and $Mg(OH)_2$ are strong bases, whereas $Be(OH)_2$ and $Al(OH)_3$ are amphoteric. At the other extreme, HNO_3 and H_2SO_4 are strong acids, whereas H_2CO_3 and H_3PO_4 are weak acids.

The compounds B_2O_3 and SiO_2 are weakly acidic in that they form a great variety of borates and silicates, but water solutions of these substances are essentially neutral. Boric acid (H_3BO_3) is a mild antiseptic and a weak acid ($K_1 = 6 \times 10^{-10}$) that can be safely used as an eye wash. It is not safe as a mouthwash ingredient because of its poisonous activity.

The fundamental difference between the borates and silicates is due to the difference in the oxidation state most commonly found (+3 and +4 respectively) and the fact that boron usually coordinates three oxygen atoms compared to the four-fold, tetrahedral coordination exhibited by silicon. The exceptional feature of the chemistry of borates and silicates that we wish to emphasize here is the persistence of the B-O-B and Si-O-Si linkage in the presence of water.

The electrostatic requirement of the nearest neighbor interactions in borates and silicates and any other ionic solid can be simply determined. For a particular ion in a complex solid structure, the oxidation number of each neighboring ion is divided by the coordination number of that ion. The sum of the values obtained for all neighboring ions should be nearly the same in magnitude (± 0.25) and opposite in sign to the charge on the ion in question. For example the oxygen atom connecting two silicate tetrahedra

$$\begin{array}{ccc} & O & O \\ & | & | \\ O- & Si-O-Si & -O \\ & | & | \\ & O & O \end{array}$$

has two Si(IV) atoms as nearest neighbors; $\frac{4}{4} + \frac{4}{4} = 2$ is exactly equal to the oxidation number of the oxygen (-2). Similarly, in a borosilicate an oxygen atom can have one Si(IV) atom with a coordination number of four and one B(III) atom with a coordination number of three as nearest neighbors since $\frac{1}{3}(3) + \frac{1}{4}(4) = 2$.

In contrast the same calculation performed on the oxygen shared between two phosphate tetrahedra gives $\frac{1}{4}(5) + \frac{1}{4}(5) = 2.5$. As a result the hydrolysis of $P_2O_7^{4-}$ is exothermic and highly favored thermodynamically.

$$P_2O_7^{4-} + H_2O \longrightarrow 2HPO_4^{2-} \quad \Delta H = -6.6 \text{ kcal mole}^{-1}$$

VI. Boron and Its Compounds

In crystalline B_2O_3 each boron is in the center of a triangle of oxygen atoms and each oxygen is involved in two BO_3 groups. The crystals melt at 450°C to give a very viscous liquid. Cooling the liquid often results in the formation of a glass instead of a true crystalline phase because the crystals form so slowly. Glass is basically a supercooled liquid that is formed very easily from viscous liquids such as B_2O_3, or SiO_2 and polycatenated sulfur. A glass made from B_2O_3 softens over a range of 570–600°C, whereas a temperature around 1500°C is required to soften fused silica (SiO_2).

The compound reacts with water to form boric acid (H_3BO_3).

$$\tfrac{1}{2}B_2O_3 + 1\tfrac{1}{2}H_2O \longrightarrow H_3BO_3$$

Boric acid crystallizes in a layer structure with hydrogen bonds between the layers. This structure leads to some exceptional physical properties. Boric acid is soft like talc, and greasy to the touch. When heated it loses water to form B_2O_3, but if it is heated in the presence of steam, gaseous H_3BO_3 molecules can be obtained.

The sublimation of boric acid in the presence of steam provides a method for concentrating the rare element.

Boric acid is found in the gases emitted from volcanoes and in the waters of hot springs in volcanic regions. Because the acid is quite soluble in water, it is carried out to the sea in most places. However, in certain rift valleys where drainage from a volcano leads to a closed desert floor below sea level, for example in Death Valley, California, vast deposits of borates including borax ($Na_2B_4O_7 \cdot 10H_2O$), kernite ($Na_2B_4O_7 \cdot 4H_2O$), and colemanite ($Ca_2B_6O_{11} \cdot 5H_2O$) are formed.

Elemental boron can be prepared by the reaction of boron oxide or boron halides with active metals such as magnesium. The product of the boron oxide-magnesium reaction always contains either residual oxygen or excess magnesium. Purer and more highly crystalline boron can be prepared by the reduction of boron trichloride with hydrogen at high temperatures.

$$2BCl_3 + 3H_2 \longrightarrow 2B + 6HCl$$

The element is not particularly metallic in its properties; it is a very poor conductor of electricity and almost as hard as diamond.

Boron and many boron-metal compounds, boron-carbon compounds, and boron-hydrogen compounds are electron deficient. The octet rule can be satisfied for these compounds only if some of the electrons are shared by three atoms. In Fig. 3-9 (p. 96) the electronic structure of B_2H_6 was drawn to show electron pairs occupying molecular orbitals on two boron atoms and a hydrogen atom. Three center bonds such as these are quite characteristic of electron deficient compounds. The formation of three center bonds by three boron atoms at the corners of an equilateral triangle permits the formation of a great variety of boranes. The structure of some boranes are shown in Fig. 25-13. Most boranes are very reactive materials that ignite spontaneously in air and undergo hydrolysis at an appreciable rate to give boric acid and hydrogen.

$$B_2H_6 + 6H_2O \longrightarrow 2H_3BO_3 + 6H_2$$

Boranes may be prepared by the hydrolysis of metal borides in an acid solution or by the reaction of boron trichloride with hydrogen at a somewhat lower temperature than that used for the production of boron.

The arrangement of twelve boron atoms at the corners of an icosahedron is particularly stable. This structure is found in three of the crystalline forms of elemental boron as well as in AlB_{12} and B_4C. Fragments of the structure are present in most of the more stable boranes. The icosahedral structure is also found in the $B_{12}H_{12}^{2-}$ anion in the salt $K_2B_{12}H_{12}$.

The substitution of two carbon atoms for two of the boron atoms

Figure 3-9:

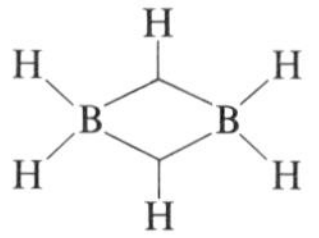

B_2H_6 itself is stable in the presence of cold dry air, but it is extremely reactive toward water.

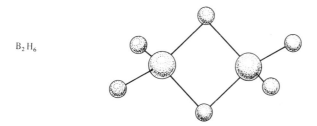

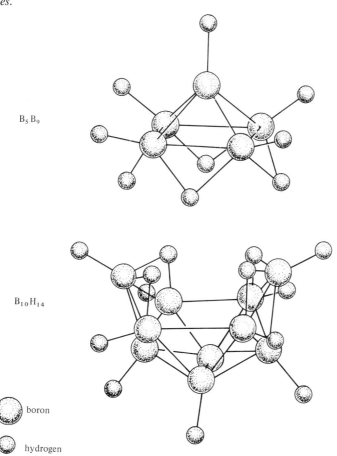

Figure 25-13 *The structures of some boranes.*

in $B_{12}H_{12}^{2-}$ gives carborane ($C_2B_{10}H_{12}$). Carboranes have the stability associated with aromatic rings such as that of benzene (C_6H_6).

VII. Silicon and Its Compounds

A. ELEMENTAL SILICON

Elemental silicon is also a hard substance with a low electrical conductivity. It is an important constituent of some iron alloys. Silicon can be prepared by the reduction of silicon dioxide (SiO_2) with carbon in the presence of iron according to the reaction

$$SiO_2 + 2C \longrightarrow Si + 2CO$$

Pure silicon can be prepared using the sequence of reactions

$$SiO_2 + 2C + 2Cl_2 \longrightarrow SiCl_4 + 2CO$$
$$4Na + SiCl_4 \longrightarrow 4NaCl + Si$$

When still purer material is required, as for use in semiconductors, silicon can be recrystallized from silver or silicon tetraiodide (SiI_4) can be prepared and decomposed.

$$Si + 2I_2 \rightleftharpoons SiI_4$$

B. SILANES

Dozens of different silicon hydrides or silanes can be prepared. The simplest process involves the addition of acid to a silicide.

$$2Li_3Si + 6H^+ \longrightarrow Si_2H_6 + 6Li^+$$

In all silanes each silicon forms four electron pair bonds so the formulas are easily predicted (Si_nH_{2n+2}). Like the boranes, the silanes are very reactive. They ignite spontaneously in air at room temperature or on warming and hydrolyze in basic solution.

$$Si_2H_6 + 4H_2O \longrightarrow 2SiO_2 + 7H_2$$

Although a great variety of silicon compounds can be prepared, those with silicon-silicon, silicon-hydrogen, or silicon-chlorine bonds are generally too reactive to be kept in moist air.

C. SILICONES

Thermodynamically stable silicon-oxygen bonds are used to form the backbone of a group of compounds known as silicones (Fig. 25-14).

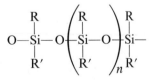

Figure 25-14 *The structure of a typical silicone.*

Sec. VII Silicon and Its Compounds

Silicones with the properties of rubber, lubricating oils, greases, and adhesives can be made by varying the length and branching of the Si-O-Si-O chain and by varying the carbon-containing R groups.

D. SILICATES

The basic structural unit of silicates is the silica tetrahedron in which one Si(IV) atom is surrounded by four oxygen atoms. Silicates may contain discrete SiO_4^{4-} ions (in orthosilicates), or other structures may be built when the tetrahedra share corners in chains, rings, sheets, or three-dimensional structures. In some cases other ions (hydroxide) may occupy some of the sites in the lattice, and aluminum may replace some of the silicon. The structures found in silicates and the equilibria between the different structures may be quite complex. In this chapter we can examine only a few examples in detail. We will start with the structure of the silicates that make up a material called basalt.

Basalt is a dark gray to black igneous rock that consists of orthosilicates, pyroxenes, feldspar, and small amounts of other minerals.

1. THE STRUCTURE OF ORTHOSILICATES

Individual SiO_4 tetrahedra are present in the orthosilicates. There is no sharing of oxygen atoms between tetrahedra. The oxygen atoms of the tetrahedra form close packed layers (Fig. 25-15). Between two successive tetrahedral layers there are metal ions (Mg^{2+} or Fe^{2+}) in octahedral holes. In olivine each oxygen atom is a part of three Mg(II)

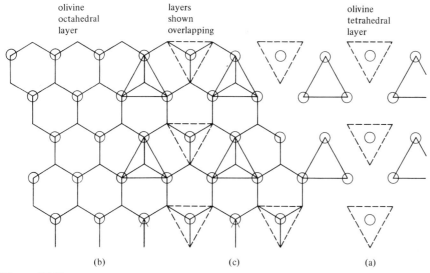

Figure 25-15 *Separate layers of the olivine structure.*

Ch. 25 Nonmetals

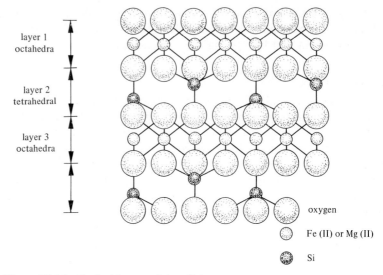

Figure 25-16 *Packed layers of the olivine structure.*

or Fe(II) octahedra (Fig. 25-15b). The octahedral and tetrahedral layers are stacked as shown in Fig. 25-15c or 25-16. The electrostatic valence rule is satisfied; $\frac{4}{4} + 3(\frac{2}{6}) = 2$.

2. The Structure of Metasilicates

The metasilicates (pyroxenes) contain extended chains of SiO_4 tetrahedra in which two of the four oxygen atoms in each SiO_4 tetrahedron are shared. The structure of a single chain is shown in

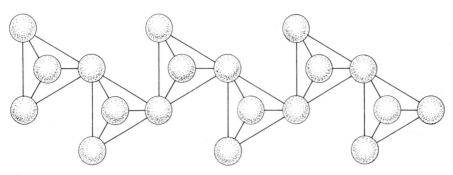

Figure 25-17 *Silicate anion linear chain.*

Fig. 25-17. The metal ions (for example, Mg in $MgSiO_3$) hold the chains together as shown in Fig. 25-18.

The principal pyroxene in basalt is $CaMgSi_2O_6$. In this case the close packing of the oxygen atoms is sufficiently distorted to allow eight-fold coordination of the Ca^{2+} ion.

3. Amphiboles

The amphibole minerals all contain the double chains of silica tetrahedra shown in Fig. 25-19. In this structure the repeating unit is $Si_4O_{11}^{6-}$. The space above each hexagon of Si atoms is occupied by OH^- so the anion is best represented as $Si_8O_{22}(OH)_2^{14-}$. The chains are again held together by cations occupying octahedral holes.

A wide range of compositions is possible for amphiboles since $+1$ cations may occupy some empty holes and there are innumerable possibilities for substitution of Ca, Mg, and Si. Amphiboles that are deficient in calcium are easily cleaved parallel to the silicate chains. Occasionally the fibrous nature of such crystals is so pronounced that the mineral fibers can actually be made into a felt or even woven. Such materials are known by the common name, asbestos. Some forms of asbestos are simple amphiboles; however, the structure of the most common form of asbestos [chrysotile, $Mg_3Si_2O_5(OH)_4$] has not been fully determined.

4. Layer Silicates

If silica tetrahedra share three corners with other tetrahedra, a sheet structure composed of repeating $Si_2O_5^{2-}$ units results (Fig. 25-20). With

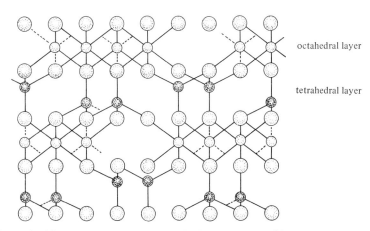

Figure 25-18 *Octahedral and tetrahedral layers in $MgSiO_3$.*

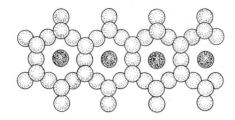

Figure 25-19 *The anion present in amphiboles,* $Si_8O_{22}(OH)_2^{14-}$.

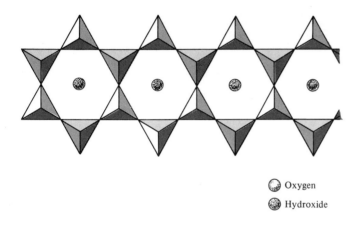

○ Oxygen
● Hydroxide

hydroxide ions present to complete a close packed layer the anion is $Si_4O_{10}(OH)_2^{6-}$.

The sheets of anions are held together by metal ions such as magnesium in $Mg_3Si_4O_{10}(OH)_2$ (Fig. 25-21). This substance is talc. The softness of talc is what is expected when neutral layers are packed together.

In mica as opposed to talc some aluminum ions occupy tetrahedral holes giving a net negative charge that is balanced by cations in the twelve-coordinate holes between the sandwich layers shown in Fig. 25-22. There is one such position for every four tetrahedral positions and the simplest formula for a mica is $KMg_3Si_3AlO_{10}(OH)_2$.

Similar aluminosilicate layers are found in vermiculite [(Ca, Mg)(H_2O)_8(Mg, Fe)_6Si_6Al_2O_{20}(OH)_4] and in the clay minerals montmorillonite [$KMg_{12}Si_{15}AlO_{40}(OH)_8$] and kaolin [$Al_2Si_2O_5(OH)_4$].

Sec. VII Silicon and Its Compounds

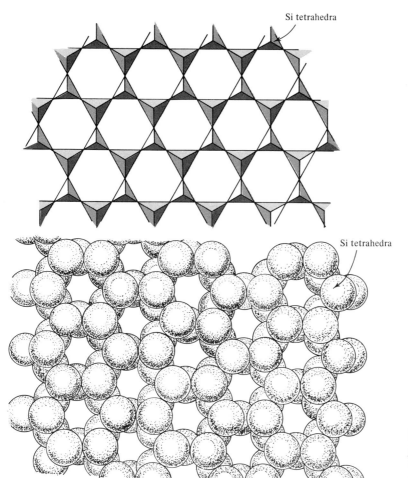

Figure 25-20 *The infinite planar layer of silicate tetrahedra sharing corners found in the mica structure.*

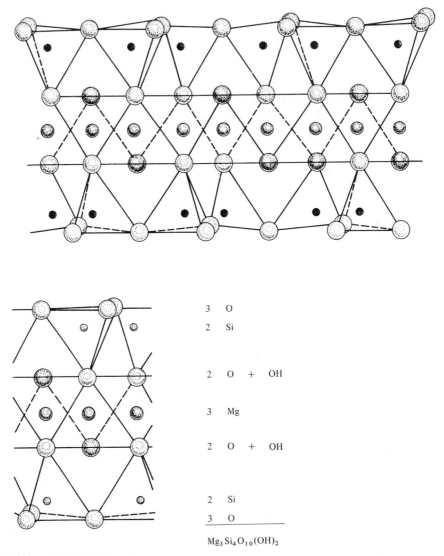

Figure 25-21 *Two representations of the sandwiching of an octahedral layer by two mica tetrahedral layers in the mica structure.*

Sec. VII Silicon and Its Compounds

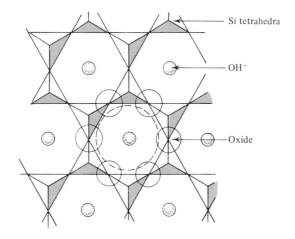

Figure 25-22 *The type A holes between mica layers.*

QUESTIONS

1. SCl_6 cannot be prepared. Why?
2. Describe the bonding in SO_3 and in SO_4^{2-}.
3. In each series of five species, pick the one indicated:
   (a) strongest acid: H_2SO_3, H_2SO_4, $HClO_3$, $HClO$, ClO_4^-
   (b) strongest oxidizing agent: S, SO_3^{2-}, S^{2-}, SO_4^{2-}, H_2S
   (c) strongest base: Cl^-, OCl^-, ClO_2^-, SO_4^{2-}, ClO_4^-
   (d) strongest acid: H_2S, H_2Te, HI, HBr, H_2Se
   (e) weakest acid: H_6TeO_6, H_2SeO_4, H_2SO_4, H_5IO_6, $HBrO_4$
4. What oxidation states of polonium should be most stable?
5. Write balanced chemical equations for the following disproportionation reactions:
   (a) $S_2O_3^{2-}$ into S_8 and SO_3^{2-}
   (b) $S_5O_6^{2-}$ into S_8 and SO_2
   (c) $S_4O_6^{2-}$ into S_8 and SO_4^{2-}
   (d) $S_2O_3^{2-}$ into S^{2-} and SO_3^{2-}
   (e) S into S^{2-} and SO_4^{2-}
   (f) SO_2 into H_2S and SO_3
6. Using the data in Figs. 25-5 and 25-9 predict the direction of spontaneity for each of the reactions in Question 5.

597

Ch. 25 Nonmetals

The formula of a typical feldspar is $K\,AlSi_3O_8$ with the Al^{3+} ions in tetrahedral holes.

7. Why does the melting point of sulfur depend on the rate of heating?
8. Contrast Cl_2, SO_2, $Na_2S_2O_4$, and $Ca(OCl)_2$ as bleaches.
9. What are the shapes of NCl_3 and ClF_3?
10. Write a series of equations for the preparation of ClO_2^- using $ClO_2(g)$ as an intermediate. How is the ClO_2 prepared?
11. Why is NI_3 an explosive?
12. Why does liquid SiO_2 have a high viscosity?
13. How does the viscosity of B_2O_3 compare with that of SiO_2? Relate this to the composition of pyrex glass.
14. Write the structure of a carborane.
15. Why is SiH_4 more reactive than CH_4?
16. Why do amphiboles have a wider range of composition than most other silicates?
17. Granite is composed predominantly of feldspar, quartz, and mica crystals. Which of these can contain the most Fe^{3+} (yellow)? The most Fe^{2+}? Which could contain iron in both valence states?
18. What color would you expect in the last condition described in Question 17?

PROBLEMS I

1. Calculate the pH of a 0.2 M solution of $NaClO_2$, $K_a = 10^{-2}$.
2. Calculate the pH of a 0.3 M solution of $NaHSO_4$. K_2 for H_2SO_4 is 1.2×10^{-2}.
3. The solubility product of $BaSO_4$ is 1.5×10^{-9}. How much $BaSO_4$ will dissolve in a 1 M acid solution?
4. Concentrated H_2SO_4 is 96% H_2SO_4 by weight and has a density of 1.84 g cm^{-3}. Calculate its molarity.
5. How much electricity would be required to prepare 100 g $NaClO_3$ by the electrolytic oxidation of $NaCl$?
6. Given the following equation:

$$Mn_2O_7 + SO_2 + H_2O \longrightarrow H_2SO_4 + MnO_2$$

   (a) Balance the equation.
   (b) What weight of Mn_2O_7 is required to prepare 49.0 g of H_2SO_4?
   (c) What volume of SO_2 gas at 1 atm at 35°C will be required?

7. Solid Na_2S is slowly added to 0.01 M $FeSO_4$ solution. Given the values below, which will be precipitated first, $Fe(OH)_2$ or FeS?

$$K_{sp}Fe(OH)_2 = 2 \times 10^{-15} \qquad K_aHS^- = 1.3 \times 10^{-13}$$
$$K_{sp}FeS\ \ \ \ \ = 4 \times 10^{-17} \qquad K_aH_2S = 1.0 \times 10^{-7}$$

8. Show that the electrostatic valence rule is satisfied in the $NaCl$ and CaF_2 crystal structures.
9. There are four different kinds of oxygen positions in the crystal structure of $MgSiO_3$. Calculate the discrepancy from the electrostatic valence rule for each of them.

10. Calculate the freezing point of a suspension of 10 g kaolin (MW = 8×10^{10}) in 100 g water. The molal freezing point depression constant of water is 1.86°C.
11. What weights of feldspar, kaolin, and flint (SiO_2) should be used to prepare a porcelain which is 75% SiO_2, 10% K_2O, and 15% Al_2O_3?
12. Calculate the pH of a 0.01 M solution of boric acid ($K_a = 6 \times 10^{-10}$). Is the H_3O^+ from the ionization of water negligible?
13. A pH of 13 is required to dissolve SiO_2 and $E^0 = -0.86$ volt for $SiO_2 + 4H^+ + 4e^- \longrightarrow Si + 2H_2O$. Draw the E_h-pH diagram for silicon.

PROBLEMS II

1. Which step of the preparation of sulfuric acid requires a catalyst?
2. What does the reaction of $S + O_2$ give in the absence of a catalyst?
3. Calculate from Appendix IV the equilibrium constant for the reaction
$$2SO_2 + O_2 \longrightarrow 2SO_3$$
4. What is ΔG^0 for the process $3SO_2 \longrightarrow S + 2SO_3$?
5. Calculate ΔG^0 for the half-reaction $HSO_4^- + 3H^+ + 2e^- \longrightarrow SO_2 + 2H_2O$.
6. Balance the half reaction $\_\_HSO_4^- \longrightarrow H_2S$.
7. Calculate ΔG^0 for the half-reaction $HSO_4^- + 9H^+ + 2e^- \longrightarrow SO_2 + 2H_2O + 6H^+$.
8. What advantage is there in including six extra H^+ in the half-reaction in Problem 7?
9. Calculate E^0 for the two half-reactions in Problems 6 and 7.
10. Will SO_2 disproportionate in acid solution?
11. G^0 for $S_3O_6^{2-}$ is -246 kcal mole^{-1}. Calculate G^0 for $\frac{1}{3}S_3O_6^{2-} + 2H_2O + 6H^+ + 5\frac{1}{3}e^-$.
12. What is the significance of the stoichiometry in Problem 11?
13. Will $S_3O_6^{2-}$ disproportionate to S and SO_2?
14. Will $S_3O_6^{2-}$ disproportionate to S and HSO_4^-?
15. Define G_{adj}^0.
16. MgO has the NaCl crystal structure. How many oxide ions are there around each Mg^{2+} ion?
17. What charge does a Mg^{2+} ion contribute to each oxide ion in MgO?
18. How many Mg^{2+} ions should be around each oxide ion in MgO to satisfy the electrostatic valence rule?
19. How many Si^{4+} ions should be around each oxide ion to satisfy the electrostatic valence rule?
20. Sketch a SiO_4 tetrahedron.
21. Sketch a MgO_6 octahedron.
22. In the spinel structure (spinel is $MgAl_2O_4$) there is a cubic close packed arrangement of oxide ions with cations in both tetrahedral and octahedral holes. Each oxygen is part of three octahedra and one tetrahedron. Calculate the positive contribution by the electrostatic valence rule made to each oxygen ion in the structure if Mg^{2+} ions are in the tetrahedral holes and Al^{3+} ions in all the octahedral holes.

23. In the inverse spinel structure, +3 ions are in the tetrahedral holes. Calculate the positive contribution made to an oxygen atom in one AlO_4 tetrahedron, two AlO_6 octahedra, and one MgO_6 octahedron.
24. Calculate the positive contribution made to an oxygen atom in one AlO_4 tetrahedron, one AlO_6 octahedron, and two MgO_6 octahedra.
25. Which structure, that in Question 22 or Question 23, is favored electrostatically?
26. Mn_3O_4 has the spinel structure. In which positions will Mn(II) be located?
27. Which coordination number, four or six, is lower in energy for a d^8 ion such as Ni^{2+}?
28. Why does $NiFe_2O_4$ have the inverse spinel structure?
29. At very high pressures Si(IV) can adopt an octahedral configuration. Describe the high pressure spinel form of Mg_2SiO_4.

SUGGESTED READING

AHRENS, L. H., *Distribution of the Elements in Our Planet*. New York: McGraw-Hill, Inc., 1965.

It need occasion little surprise, though it is perhaps little appreciated, that before 1850 organic chemistry could boast of virtually no synthetic achievements—the synthesis of acetic acid by Kolbe in 1845 being the sole example, if we except the partial synthesis of urea in 1828 by Wohler. Until organic chemistry had passed through its first great revolution, synthetic planning in any realistic sense was simply not possible. But in the decade after 1855, the structure theory was proposed, sharpened, and accepted into the body of science. This massive advance provided the basis for a phenomenal flowering of organic chemistry, mirrored and measured by outstanding advances in synthesis. Alizarin, indigo, coniine, and nicotine, to cite a few examples, fell, and the period may be considered to have culminated in the brilliant synthetic researches of Emil Fischer on the sugars, purines, polypeptides, and tannins.*

ROBERT B. WOODWARD

Chapter 26 Organic Chemistry

The chemistry of carbon is more complex than that of any other element. Over 1.5×10^6 different compounds of carbon have been purified and characterized. The only other element that comes close to this number is hydrogen, and this is only because most organic compounds also contain hydrogen.

Carbon is a second row element that forms p-p π bonds readily. This accounts for the existence of two crystalline allotropic forms of the element, diamond and graphite (Fig. 26-1); for the fact that CO_2 is a gas in contrast to $SiO_2(s)$; and for the formula of the carbonate ion, CO_3^{2-}. Because carbon(IV) has a coordination number quite different from its oxidation number in carbonates there is no tendency for carbonates to polymerize. Thus carbonates are much simpler than silicates. Nevertheless, on the whole carbon chemistry is more complex than that of silicon, and this is owing mainly to the robust character of carbon compounds.

I. Hydrocarbons

To demonstrate the robust nature of carbon compounds we need only consider the hydrocarbons. Hydrocarbons are compounds of carbon and hydrogen. They have the general formula, C_nH_m. The

*Used by permission from Sir Alexander Todd, *Perspectives in Organic Chemistry*, (New York: John Wiley & Sons, 1956)

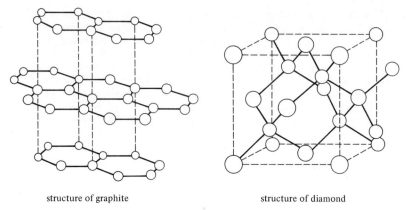

structure of graphite   structure of diamond

Figure 26-1 *The allotropes of carbon.*

reaction of hydrocarbons with oxygen is very exothermic. For example, propane (C_3H_8) is an excellent fuel

$$C_3H_8(g) + 5O_2 \rightleftharpoons 3CO_2(g) + 4H_2O(g)$$
$$\Delta H = -488.52 \text{ kcal mole}^{-1}$$

Propane will not ignite in air below a temperature of about 450°C. This means that hydrocarbons can normally be handled without taking special precautions to exclude air.

The reaction of the hydrocarbons may be compared with that of the silanes. The reaction of Si_3H_8 is also very exothermic.

$$Si_3H_8(l) + 5O_2 \rightleftharpoons 3SiO_2(s) + 4H_2O(g) \qquad \Delta H = -862.2 \text{ kcal mole}^{-1}$$

The big difference between C_3H_8 and Si_3H_8 is that Si_3H_8 will ignite spontaneously on contact with air.

There is no reason why $Si_{15}H_{32}$ cannot be prepared, purified, and characterized; however, it would not be easy and it has not yet been accomplished. In the corresponding hydrocarbon series, compounds with the formulas C_nH_{2n+2} are well known up to at least $n = 30$, and any particular member of the series with a larger number of carbon atoms, say normal $C_{37}H_{76}$, could be made. The hydrocarbons whose formulas are C_nH_{2n+2} are known as alkanes.

A. ALKANES

The nomenclature for the alkane series starts off with four arbitrary names, methane through butane, and then continues with numerical prefixes as shown in Table 26-1.

With appropriate allowance for the fact that alkanes can be formed with branched chains as well as straight chains of carbon atoms, (Fig. 26-2) it is possible to calculate that there are 4,111,847,763 different structural isomers of $C_{30}H_{62}$. Obviously they have not all been prepared, purified, and characterized because, if all the organic chemists ever trained had worked on nothing but the $C_{30}H_{62}$ isomers and were each able to prepare 1000 a year, they would not have been able to make much of a dent on the four billion possibilities.

Normal butane is the first of the normal alkanes to have a branched chain isomer (Fig. 26-2). This compound is known as isobutane, but a better name is obtained from the systematic nomenclature given in Appendix V; according to this system it would be called 2-methylpropane. In more complex cases, for example the nine isomers of C_7H_{16}, the systematic nomenclature is a necessity. The nine isomers are different chemical compounds with different physical and thermodynamic properties (illustrated in Table 26-2).

There are many interesting correlations between properties and structure of alkanes, but we will be satisfied with pointing out the higher volatility of roughly spherical molecules (2,2,3-trimethylbutane) as opposed to molecules with longer chains of carbon atoms. In addition the thermodynamic instability induced by large substituents around a single carbon atom (steric hindrance) is evident, especially for 3-ethylpentane.

Table 26-1 *Some Normal Alkanes*

Formula	Name	Melting point, °C	Boiling point, °C	Density of liquid,‡ g cm^{-3}	H^0, kcal mole^{-1}	S^0, eu
CH_4	methane	−182	−161	0.415	−17.889	44.50
C_2H_6	ethane	−183	−89	0.572	−20.236	54.85
C_3H_8	propane	−190	−44	0.585	−24.820	64.51
C_4H_{10}	butane	−138	−0.5	0.6012	−29.812	74.10
C_5H_{12}	pentane	−130	36	0.6262	−41.40*	62.79*
C_6H_{14}	hexane	−95	68	0.6594	−47.52	70.76
C_7H_{16}	heptane	−91	98	0.6838	−53.63	78.42
C_8H_{18}	octane	−56	126	0.7025	−59.74	85.5
C_9H_{20}	nonane	−51	151	0.7176	−65.84	94
$C_{10}H_{22}$	decane	−30	174	0.7300	−79.95	102
$C_{11}H_{24}$	hendecane	−25	195	0.7402		111
$C_{12}H_{26}$	dodecane	−9.6	216	0.7487	−90.4	118
$C_{13}H_{28}$	tridecane	−6	234	0.7559		
$C_{14}H_{30}$	tetradecane	5.5	254	0.7627		

*Thermodynamic values are for the form stable at 25° and 1 atm pressure; for pentane through tetradecane this is the liquid phase.

‡Density values are for various temperatures up to 20°C.

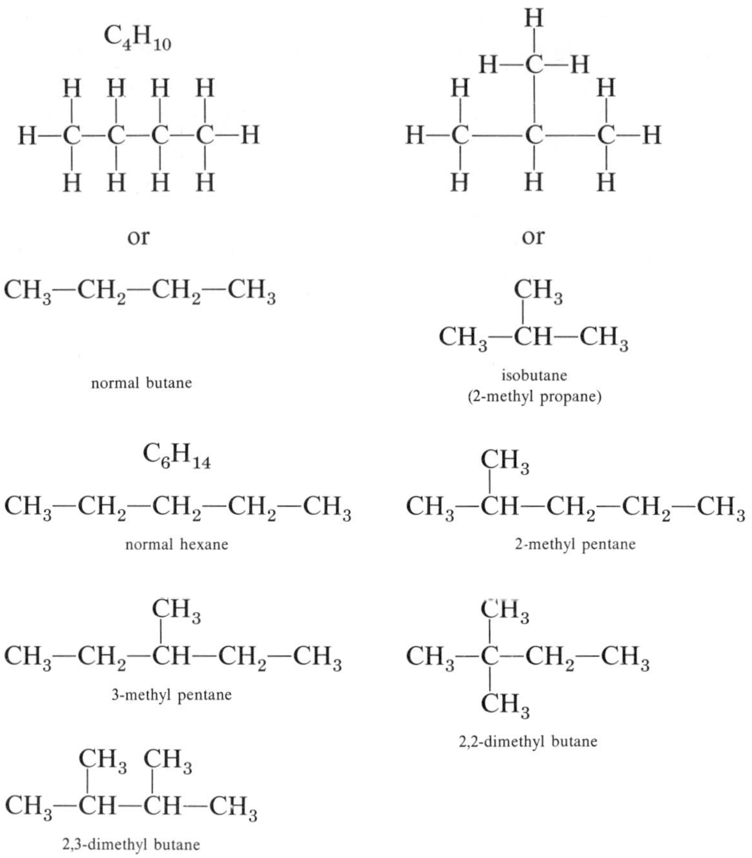

C_4H_{10}

H H H H
H—C—C—C—C—H
H H H H

or

$CH_3—CH_2—CH_2—CH_3$

normal butane

H
H—C—H
H   H
H—C—C—C—H
H   H   H

or

CH_3
|
$CH_3—CH—CH_3$

isobutane
(2-methyl propane)

C_6H_{14}

$CH_3—CH_2—CH_2—CH_2—CH_3$
normal hexane

CH_3
|
$CH_3—CH—CH_2—CH_2—CH_3$
2-methyl pentane

CH_3
|
$CH_3—CH_2—CH—CH_2—CH_3$
3-methyl pentane

CH_3
|
$CH_3—C—CH_2—CH_3$
|
CH_3
2,2-dimethyl butane

CH_3 CH_3
| |
$CH_3—CH—CH—CH_3$
2,3-dimethyl butane

Figure 26-2 *Isomers of C_4H_{10} and C_6H_{14}.*

B. CYCLOALKANES AND ALKENES

The cycloalkanes and alkenes are series of hydrocarbons with the formulas C_nH_{2n}. The simplest member of the alkene series, ethylene (C_2H_4), has no isomers. The six isomers of C_4H_8 are illustrated in Fig. 26-3. In 2-butene the rotation about carbon-carbon double bonds is extremely slow (activation energy $\approx$ 30 kcal mole^{-1}) and this permits the separation of cis and trans isomers.

The thermodynamic values for the lower cycloalkanes are shown in Table 26-3 along with the fraction (X) that would be present if equilibrium among these particular molecules could be established at 25°C. The smaller as well as the larger rings are unstable with respect to cyclohexane. The smaller rings are unstable because the C-C-C bond angles are forced to take on values considerably smaller than normal. The bond angles in cyclopentane (C_5H_{10}) are normal (109°), but the

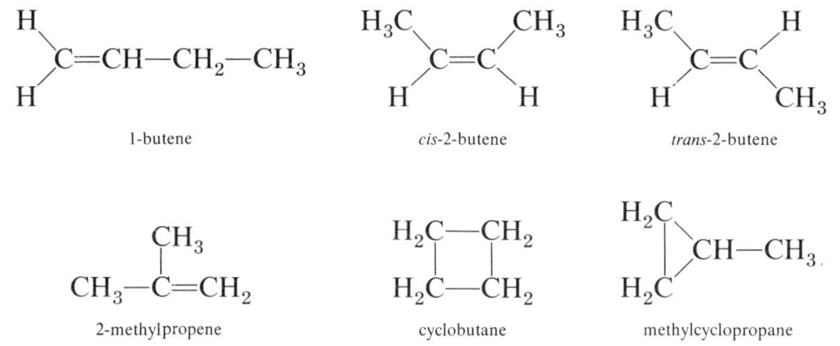

Sec. I Hydrocarbons

$H_2C=CH-CH_2-CH_3$ 1-butene

H_3C \ C=C / CH_3 with H, H   cis-2-butene

H_3C \ C=C / H with H, CH_3 trans-2-butene

$CH_3-C(CH_3)=CH_2$ 2-methylpropene

cyclobutane (four-membered ring of CH_2 groups)

methylcyclopropane (three-membered ring with $CH-CH_3$)

Figure 26-3 *Structures of the six C_4H_8 isomers.*

compound is somewhat unstable because of repulsions between the hydrogen atoms. In cyclohexane (C_6H_{12}) (Fig. 26-4) the hydrogen atoms of adjacent carbon atoms are a little further apart. Two of the chair forms of cyclohexane (see Fig. 26-4a and b) are equivalent, but there is a third form (Fig. 26-4c) in which the C-C bonds are arranged in a way that resembles a boat. The boat form has one rather close pair of hydrogen atoms and it is accordingly 6 kcal mole^{-1} less stable than the chair forms. The forms of cyclohexane are readily interconverted; therefore they do not qualify as different isomers. It is not possible to isolate the different conformations of cyclohexane.

For a quantitative comparison of the energy of the cyclohexane conformations, all the H-H interactions should be considered.

Larger rings are energetically satisfactory, but the entropy term tends

Table 26-2 *Some Properties of the Nine C_7H_{16} Isomers**

Compound	Boiling point, °C	Melting point, °C	Liquid			Gas	
			Density,‡ g cm^{-3}	H^0, kcal mole^{-1}	S^0, eu	H^0, kcal mole^{-1}	S^0, eu
n-Heptane	98.43	−90.61	0.6795	−53.63	78.42	−44.89	102.14
2-Methylhexane	90.05	−118.28	0.6744	−54.93	77.28	−46.60	100.34
3-Methylhexane	91.85		0.6830	−54.35	78.23	−45.96	101.37
3-Ethylpentane	93.48	−118.60	0.6940	−53.77	75.33	−45.34	98.47
2,2-Dimethylpentane	79.20	−123.81	0.6695	−57.05	71.73	−49.29	93.83
2,3-Dimethylpentane	89.78		0.6909	−55.81	76.27	−47.62	98.96
2,4-Dimethylpentane	80.50	−119.24	0.6683	−56.17	72.47	−48.30	94.80
3,3-Dimethylpentane	86.06	−134.46	0.6891	−56.07	73.44	−48.17	95.53
2,2,3-Trimethylbutane	80.88	−24.91	0.6859	−56.63	70.43	−48.96	93.33

*The data in this table are from American Society for Testing and Materials, *Physical Constants of Hydrocarbons C_1 to C_{10}*, (Philadelphia: American Society for Testing and Materials, 1963), and U. S. National Bureau of Standards, *Circular 461, Selected Values of Properties of Hydrocarbons*, (Washington, D. C.: U.S. Government Printing Office, 1947).

‡At 25°C

Table 26-3 *Properties of Some Cycloalkanes*

Formula	Name	Melting point, °C	Boiling point, °C	Density 25 °C, g cm^{-3}	H^0, kcal $mole^{-1}$	S^0, eu	X*	G^0, kcal $mole^{-1}$
C_3H_6	cyclopropane	−127.42	−32.80					
C_4H_8	cyclobutane (g)	−90.73	12.51	0.6890	+0.8	~70	10^{-11}	~19
C_5H_{10}	cyclopentane (l)	−93.87	49.26	0.7404	−25.30	+49	0.0033	8.70
C_6H_{12}	cyclohexane	6.554	80.738	0.7739	−37.34	+48	0.9966	6.37
C_7H_{14}	cycloheptane	−8.0	118.79	0.8066	−37.73	57.97	8×10^{-5}	12.96
C_8H_{16}	cyclooctane	+14.8	151.14	0.8320	−40.42	62.62	4×10^{-8}	18.60

*Mole fraction of the cycloalkane in an equilibrium cycloalkane liquid.

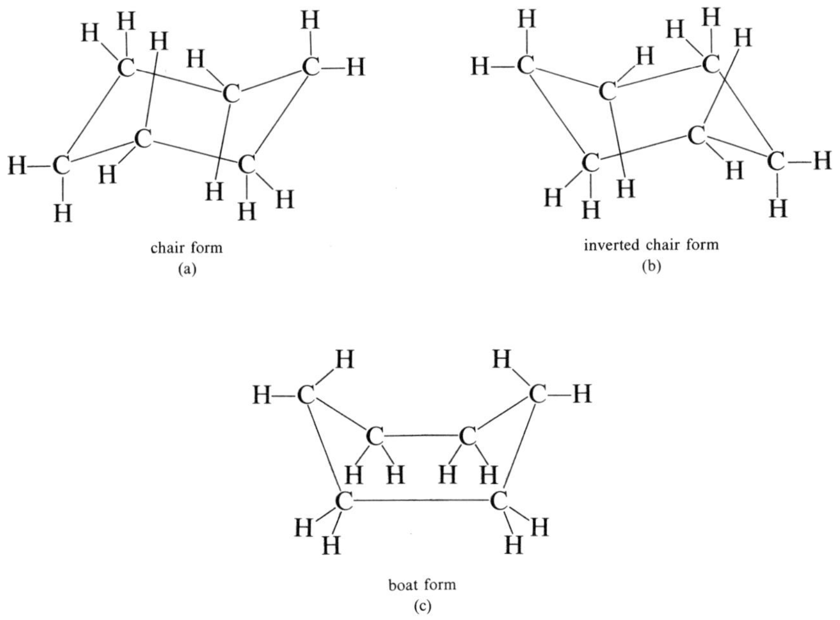

Figure 26-4 *Conformations of cyclohexane.*

to favor the six-member ring. This situation is very similar to the predominance of S_8 rings in liquid sulfur at its melting point (Chapter 25) except in the case of sulfur the equilibrium is readily attainable and for this reason it has not been possible to prepare and study pure S_7 and S_9.

Alkanes and cycloalkanes are not very reactive except at high temperatures. One typical free radical reaction of alkanes, chlorination, was discussed in Chapter 20 (p. 489). The three-member cyclopropane ring has sufficient strain to be quite reactive. Reaction with sulfuric acid produces a primary sulfate

$$(CH_2)_3 + H_2SO_4 \longrightarrow CH_3-CH_2-CH_2-O-SO_3H$$

The word primary is used to designate a substituent on a terminal carbon atom.

II. Reactions of C=C and C=O Bonds

A. POLYMERIZATIONS

The polymerization of alkenes is exothermic because β_{C-C} is substantially smaller than α_{C-C} (65 and 90 kcal mole^{-1} respectively). At high pressures and in the presence of catalysts, ethylene and propylene give good yields of high molecular weight alkanes. The products of the polymerization, polyethylene and polypropylene, soften below 200°C and are thus easily molded into large complex shapes. This together with the cheapness of the starting materials makes these substances two of the most widely used plastics. At the current time about 2×10^6 tons of polyethylene are produced each year.

Many other useful plastics are polymers and copolymers of substituted alkenes such as $CH_2=CHCl$, $CH_2=CH-C_6H_5$, and

$$CH_2=\underset{CH_3}{\underset{|}{C}}-\overset{O}{\overset{\|}{C}}-OCH_3 \text{ (methylmethacrylate).}$$ Some of the polymers formed from these substances have exceptional properties. Polymethylmethacrylate (Lucite or Plexiglass) is transparent.

B. HYDROGENATION

Alkenes can be converted to the corresponding alkanes with the same number of carbon atoms by treatment with hydrogen (hydrogenation under pressure in the presence of a catalyst, Chapter 20). The reaction occurs on the catalyst surface. Both hydrogen atoms are added to the same side of the molecule. Thus hydrogenation of 1,2-dimethylcyclohexene, [structure], yields a molecule of 1,2-dimethylcyclohexane with both hydrogen atoms and both methyl groups on the same side of the ring (cis). The other isomer, *trans*-1,2-dimethylcyclohexane, is more stable since the cyclohexane ring can adopt a chair form with both methyl groups in positions where the steric hindrance is the least. The most stable arrangements (conformations) of these two molecules are shown in Fig. 26-5.

C. ACID CATALYZED ADDITIONS

The π electrons of an alkene are susceptible to chemical attack by an acid. If the reactant is a proton acid, a carbonium ion intermediate is formed. Propene can react with sulfuric acid in two ways:

$$CH_2=CH-CH_3 + H^+ \longrightarrow \begin{cases} (CH_3-\overset{+}{C}H-CH_3)^{\ddagger} \xrightarrow{HSO_4^-} CH_3-\underset{}{\overset{O-SO_3H}{\underset{|}{C}H}}-CH_3 \\ \text{secondary carbonium ion} \\ \\ (^+CH_2-CH_2-CH_3)^{\ddagger} \xrightarrow{HSO_4^-} CH_3-CH_2-CH_2-O-SO_3H \\ \text{primary carbonium ion} \end{cases}$$

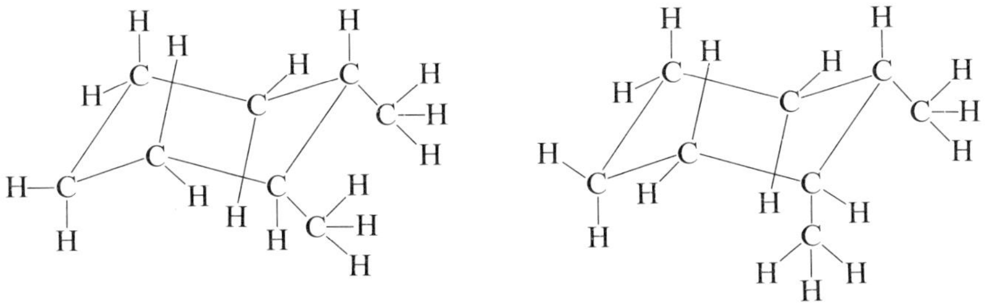

Figure 26-5 *The cis and trans isomers of 1,2-dimethylcyclohexane.*

but the reaction goes predominantly through the more stable secondary carbonium ion (the charge is not on the terminal carbon atom).

In general the direction of addition in these reactions is determined by the relative stabilities of the possible carbonium ions, so that generally secondary or tertiary sulfates, halides, or alcohols are produced

$$CH_2=CH-CH_2-CH_3 + HBr \longrightarrow CH_3-CHBr-CH_2-CH_3$$
a secondary halide

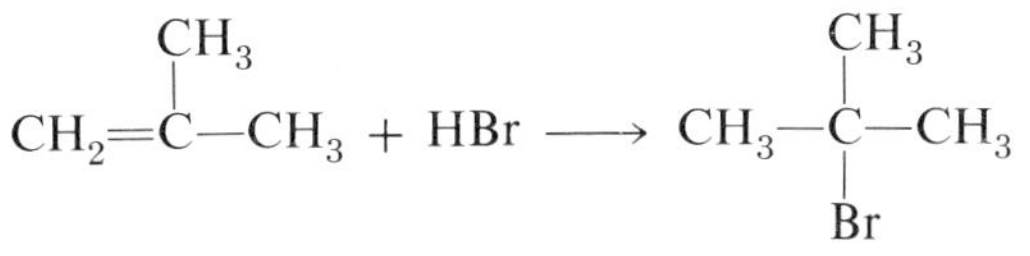

$$CH_2=\underset{CH_3}{\overset{CH_3}{C}}-CH_3 + HBr \longrightarrow CH_3-\underset{Br}{\overset{CH_3}{C}}-CH_3$$
a tertiary halide

The empirical statement of this fact, the acid sulfate, halide, or hydroxide group adds to the carbon with the least number of hydrogen atoms on it, is known as Markovnikov's rule.

In the reaction of ethylene with sulfuric acid, there are four different major products as shown in Fig. 26-6. These products can also all be interconverted by S_N2 type substitutions on carbon or by substitutions on the sulfur atom. Sulfur trioxide may be involved as an intermediate in the substitution

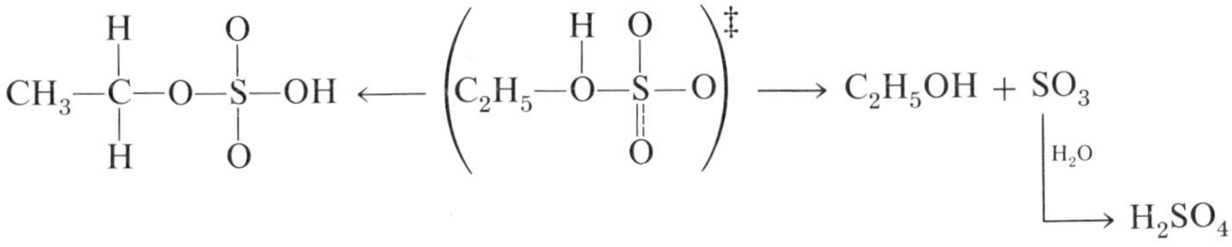

$$CH_3-\underset{H}{\overset{H}{C}}-O-\underset{O}{\overset{O}{S}}-OH \longleftarrow \left(C_2H_5-\underset{O}{\overset{H\ \ O}{O-S-O}} \right)^{\ddagger} \longrightarrow C_2H_5OH + SO_3$$

$$\qquad\qquad\qquad\qquad\qquad\qquad\qquad\qquad\qquad\qquad\downarrow H_2O$$
$$\qquad\qquad\qquad\qquad\qquad\qquad\qquad\qquad\qquad\qquad H_2SO_4$$

The relative amounts of the products can be controlled by the rate of the reactions and by shifting the equilibria involved. This can be done by varying the temperature, the concentration of sulfuric acid,

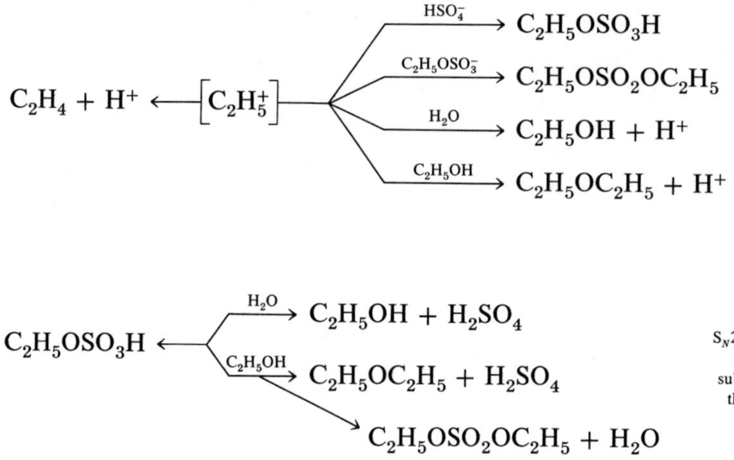

Ch. 26 Organic Chemistry

$$C_2H_4 + H^+ \longleftrightarrow [C_2H_5^+] \begin{array}{l} \xrightarrow{HSO_4^-} C_2H_5OSO_3H \\ \xrightarrow{C_2H_5OSO_3^-} C_2H_5OSO_2OC_2H_5 \\ \xrightarrow{H_2O} C_2H_5OH + H^+ \\ \xrightarrow{C_2H_5OH} C_2H_5OC_2H_5 + H^+ \end{array} \quad \text{carbonium ion mechanisms}$$

$$C_2H_5OSO_3H \begin{array}{l} \xrightarrow{H_2O} C_2H_5OH + H_2SO_4 \\ \xrightarrow{C_2H_5OH} C_2H_5OC_2H_5 + H_2SO_4 \\ \phantom{\xrightarrow{xxx}} C_2H_5OSO_2OC_2H_5 + H_2O \end{array} \quad \begin{array}{l} S_N2 \text{ substitutions on C} \\ \text{or} \\ \text{substitutions through three coordinate S} \end{array}$$

Figure 26-6 *The major pathways in the reactions of C_2H_4 and C_2H_5OH with H_2SO_4.*

and the proportions of carbon and sulfur. At high temperatures (155–175°C) with concentrated sulfuric acid, $C_2H_4(g)$ is the major product from ethyl alcohol. At lower temperatures (140–145°C) with alcohol added continuously, the activation energy for the carbonium ion mechanism is a substantial barrier so that good yields of diethyl ether ($CH_3CH_2OCH_2CH_3$) can be obtained. Ethanol can be prepared from ethylene and concentrated sulfuric acid at a low temperature. The product of this reaction is $CH_3CH_2OSO_3H$. This solution is then diluted and heated to about 100°C where ethanol distills off as it is formed.

The process for preparing ethanol cheaply from ethylene was developed commercially about 1931. At current prices the ethanol produced by the traditional fermentation of grain is about twice as expensive as that produced from ethylene. The fermentation process is

Maltase and zymase are catalysts for the reactions. They are enzymes present in the yeast.

$$C_{12}H_{22}O_{11} + H_2O \xrightarrow{\text{maltase from yeast}} 2C_6H_{12}O_6$$

$$C_6H_{12}O_6 \xrightarrow{\text{zymase from yeast}} 2CO_2 + 2C_2H_5OH$$

About 11% of the 3.50×10^8 gal of ethanol produced in the United States in 1965 was prepared by the fermentation process.

The reaction of ethylene with an acid is not typical. This is the only alkene for which the reaction must proceed through a primary carbonium ion (the terminal carbon atom in the chain). Other alkenes will react at lower temperatures and with more dilute acids. 2-methylpropene reacts with cold 50% sulfuric acid, and it can be prepared from tertiary butanol (2-methylpropan-2-ol) with 20% sulfuric acid at

90°C. The ease with which alkenes are produced must be kept in mind in considering all reactions of tertiary alcohols and alkyl halides.

D. PREPARATION OF ALDEHYDES AND KETONES

Hydrogen can be removed from alcohols to give aldehydes and ketones, respectively

$$R-\underset{H}{\overset{H}{C}}-OH \longrightarrow H_2 + R-\overset{O}{\underset{}{C}}-H$$

primary alcohol   an aldehyde

The formula for an aldehyde may also be written RCH=O.

$$R-\underset{R'}{\overset{H}{C}}-OH \longrightarrow H_2 + R-\overset{O}{\underset{}{C}}-R'$$

secondary alcohol   a ketone

The dehydrogenation reaction is characteristic of primary and secondary alcohols. Acetaldehyde can also be prepared by the hydration of acetylene (ethyne) with Hg^{2+} as a catalyst

$$HC\equiv CH + H_2O \xrightarrow{H^+, Hg^{2+}} CH_2=\underset{}{\overset{H}{C}}-OH \rightleftharpoons CH_3\overset{O}{\underset{}{C}}-H$$

The postulated intermediate is an enol. Enols are normally present in small quantities in equilibrium with compounds having C=O bonds if there is a hydrogen on the next carbon atom of the chain,

$$CH_3CH=O \rightleftharpoons CH_2=CHOH$$

The small amount of enol is important in certain reactions of acetaldehyde such as the base catalyzed polymerization

$$CH_2=CHOH + OH^- \rightleftharpoons CH_2=CHO^- + H_2O$$

$$^-O-CH=CH_2 + CH_3CH=O \xrightarrow{\text{rate determining step}} O=CH-CH_2\underset{CH_3}{\overset{}{C}}H-O^-$$

$$O=CHCH_2\underset{CH_3}{\overset{}{C}}H-O^- + H_2O \longrightarrow O=CHCH_2\underset{CH_3}{\overset{}{C}}H-OH + OH^-$$

The anion of the enol, $CH_2=CHO^-$, has some negative charge on the oxygen and on the end carbon atom. This can be explained by considering the resonance between the two forms

$$CH_2{=}CH{-}O^- \qquad \text{and} \qquad \bar{:}CH_2{-}CH{=}O$$

or from the molecular orbital diagram.

The largest single industrial use of ethanol (1×10^8 gallons per year) is in the production of acetaldehyde.

$$C_2H_5OH \xrightarrow{\text{Cu, 250°C}} CH_3\overset{\overset{O}{\|}}{C}H + H_2$$

E. REACTIONS OF ALDEHYDES

The C=O bond is reactive toward both acids and bases. Because there is a substantial difference in electronegativity between carbon and oxygen there is even more selectivity in the direction of addition to the C=O bond than for addition to the C=C bond. Strong bases attack the carbon atom exclusively. The reaction of aldehydes with carbanions or metal alkyl compounds provides a method of synthesizing C-C bonds.

$$(CH_3)_2\overset{\overset{H}{|}}{C}Br + Mg \longrightarrow (CH_3)_2\overset{\overset{H}{|}}{C}MgBr$$

$$(CH_3)_2\overset{\overset{H}{|}}{C}MgBr + CH_3\overset{\overset{O}{\|}}{C}H \longrightarrow (CH_3)_2CH{-}\underset{\underset{H}{|}}{\overset{\overset{CH_3}{|}}{C}}{-}OMgBr$$

$$(CH_3)_2CH{-}\underset{\underset{H}{|}}{\overset{\overset{CH_3}{|}}{C}}{-}OMgBr + H_2O \longrightarrow (CH_3)_2CH{-}\underset{\underset{H}{|}}{\overset{\overset{CH_3}{|}}{C}}{-}OH + \tfrac{1}{2}Mg(OH)_2 + \tfrac{1}{2}Mg^{2+} + Br^-$$

The corresponding reaction of an aldehyde with a base containing oxygen (water or alcohol) gives a hydrate or a hemiacetal

$$H_2O + CH_3\overset{\overset{O}{\|}}{C}{-}H \xrightarrow{OH^-} CH_3CH\begin{matrix} \diagup OH \\ \diagdown OH \end{matrix}$$

hydrate of acetaldehyde

$$C_2H_5OH + CH_3\overset{\overset{O}{\|}}{C}H \xrightarrow{OH^-} CH_3CH\begin{matrix} \diagup OH \\ \diagdown OC_2H_5 \end{matrix}$$

a hemiacetal

When the reaction of an aldehyde with an alcohol is catalyzed by strong acids the reaction can go one stage further to give an acetal

$$H^+ + C_2H_5OH + CH_3\overset{O}{\overset{\|}{C}}H \longrightarrow \left(CH_3CH\underset{OC_2H_5}{\overset{OH_2}{\diagup}}\right)^+$$

$$\left(CH_3-CH\underset{OC_2H_5}{\overset{OH_2}{\diagup}}\right)^+ \longrightarrow CH_3-CH=\overset{+}{O}-C_2H_5 + H_2O$$

$$C_2H_5OH + CH_3CH=\overset{+}{O}-C_2H_5 \xrightarrow{\text{rate determining step}} \left(CH_3CH\underset{\underset{H}{\overset{|}{O}}-C_2H_5}{\overset{OC_2H_5}{\diagup}}\right)^+$$

$$\longrightarrow H^+ + CH_3CH\underset{OC_2H_5}{\overset{OC_2H_5}{\diagup}}$$

an acetal

The energy of the activated complex in this reaction is higher than that of the intermediate, $CH_3-CH=O^+-C_2H_5$. The intermediate is present at a significant concentration only in strong acid solutions. Therefore, once an acetal is formed it is quite stable in neutral or basic solutions.

Glucose contains an aldehyde group at one end of a six-carbon chain (Fig. 26-7). Although it exists primarily as a mixture of the two hemi-

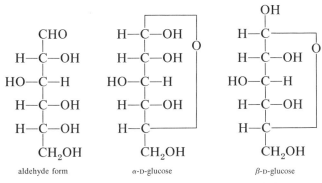

Figure 26-7 *Schematic representations of the structure of glucose.*

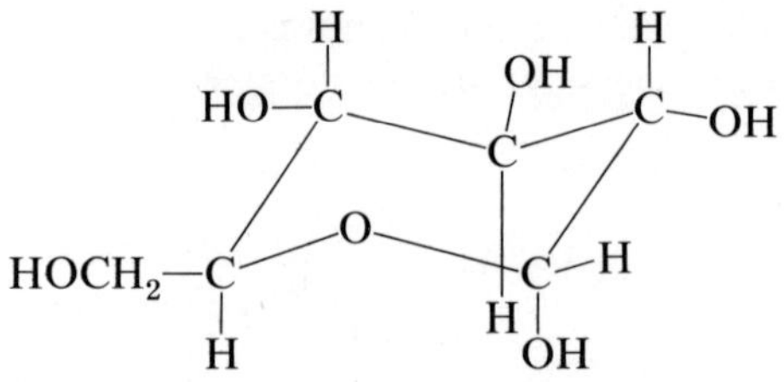

Figure 26-8 *The two hemiacetal forms of glucose.*

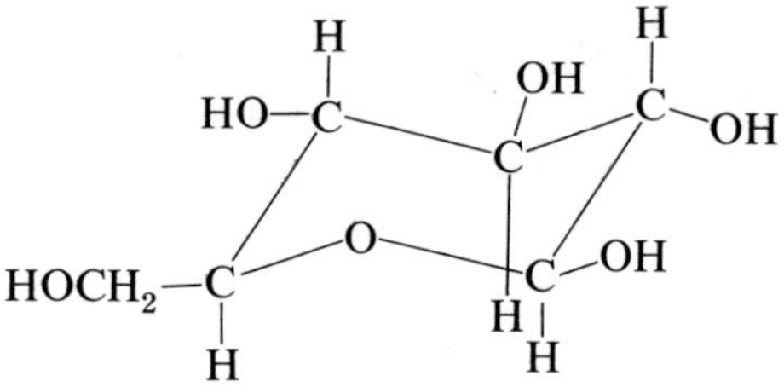

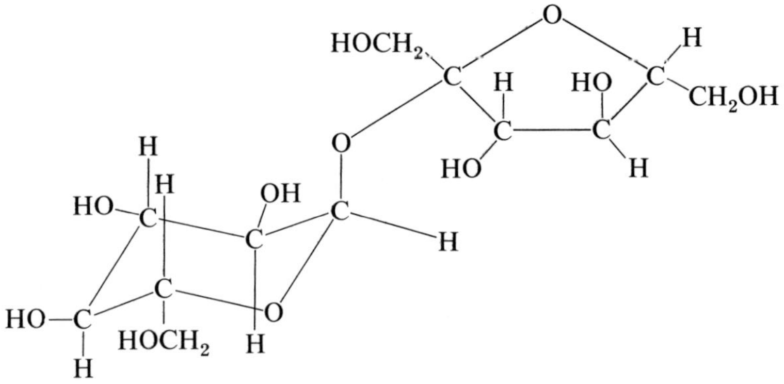

Figure 26-9 *The structure of sucrose.*

acetals shown in Fig. 26-8, glucose is easily oxidized by silver ion or Fehling solution. In sucrose the glucose part of the molecule is present as an acetal (Fig. 26-9) and sucrose will not react with either of these oxidizing agents.

III. Optical Isomerism

Optical isomerism (described on p. 361) was first noted for compounds of carbon. One of the simplest optically active organic molecules is 2-butanol,

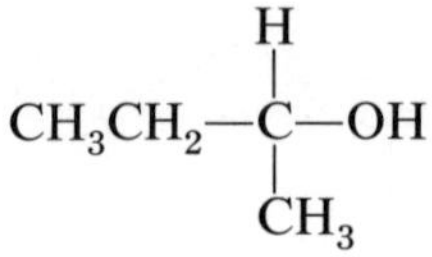

$$CH_3CH_2-\underset{CH_3}{\overset{H}{C}}-OH$$

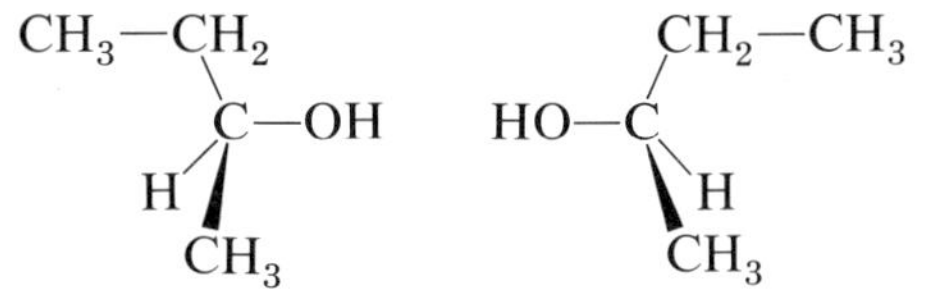

CH₃—CH₂   CH₂—CH₃
      \\C—OH   HO—C
   H /   \\ H
     CH₃   CH₃
   s-2-butanol   r-2-butanol

Figure 26-10 R *and* S *forms of 2-butanol.*

The pairs of letters (D, L) *and* (R, S) *are used to specify the configuration of asymmetric carbon atoms.* D *and* L *are very useful for sugars, but the more general* R *and* S *nomenclature is best for the alcohols shown here. In either case the assignment is made by a set of conventional rules which we need not go into here. See Morrison, R. T. and Boyd, R. N., Organic Chemistry, (Boston: Allyn and Bacon, 1966) pages 86–90 and 956.*

2-Butanol can exist in two forms (Fig. 26-10) that are mirror images of each other but which are not superimposable. The physical properties of the isomers are identical. They differ in their effect on polarized light and in their reactivity toward other optical isomers.

Almost all optically active organic molecules have a carbon atom with four different groups attached (H, OH, CH₃, and C₂H₅ in 2-butanol). Such carbon atoms are called asymmetric. A carbon atom with a π bond generally lies in a plane of symmetry and so cannot contribute to optical activity.

In general the total number of optical isomers of an organic molecule is given by 2^n where n is the number of asymmetric carbon atoms in the molecule. There are exceptions to this rule, as in the case of 2,3-butanediol which has only three optical isomers.

The hemiacetal form of glucose has five asymmetric carbon atoms, so the two forms in Fig. 26-8 are 2 of 32 possible optical isomers. The other 30 are all known.

There are many other sugars that are not optical isomers of glucose. Fructose ($C_6H_{12}O_6$) has a keto group but no aldehyde group. It is a functional isomer of glucose but not a geometrical isomer. Other common sugars are sucrose ($C_{12}H_{22}O_{11}$) and ribose ($C_5H_{10}O_5$).

IV. Acids and Esters

The mild oxidation of an aldehyde will give the corresponding acid. The name acetaldehyde comes from its relationship with acetic acid (ethanoic acid). Organic acids can also be prepared by the oxidation of the corresponding primary alcohols.

$$3C_2H_5OH + 2Cr_2O_7^{2-} + 16H^+ \longrightarrow 3CH_3COOH + 4Cr^{3+} + 11H_2O$$

The corresponding oxidation of secondary alcohols normally stops at a ketone.

$$3CH_3CHOH-CH_3 + Cr_2O_7^{2-} + 8H^+ \longrightarrow 3CH_3\overset{O}{\overset{\|}{C}}CH_3 + 2Cr^{3+} + 7H_2O$$

An organic acid has a C=O group which is susceptible to acid and base catalyzed reactions that are analogous to the reactions of aldehydes. The products of the reaction with alcohols are esters. The reaction is called an esterification reaction.

$$CH_3C(=O)OH + OH^- \longrightarrow CH_3-\overset{O^-}{\underset{OH}{C}}-OH \longrightarrow CH_3\overset{O}{\underset{OH}{C}} + OH^-$$

$$CH_3C(=O)OH + CH_3OH + H^+ \rightleftharpoons CH_3\overset{OH}{\underset{H-O^+-CH_3}{C-OH}} \rightleftharpoons CH_3-C(=O)OCH_3 + H^+ + H_2O$$

an ester

The acid catalyzed esterification equilibrium was the first chemical equilibrium studied in detail (see p. 485). The rate of both the forward and reverse reaction can be easily measured. Since the esterification reaction is an equilibrium, it is often desirable to prepare esters by reactions of alcohols with acyl halides under anhydrous conditions.

Acyl halides are prepared by the reaction of organic acids with PCl_3 (p. 539), S_2Cl_2 (p. 582), or $SOCl_2$.

$$CH_3C(=O)Cl + R-OH \longrightarrow CH_3-C(=O)-OR + HCl$$

an acyl halide

$\alpha_{C-O} = 96$ kcal mole^{-1}
$\alpha_{H-Cl} = 103$ kcal mole^{-1}
$\alpha_{C-CL} = 80$ kcal mole^{-1}
$\alpha_{O-H} = 119$ kcal mole^{-1}

$\beta_{C-O} = 100$ kcal mole^{-1}
$\sqrt{1.5}\,\beta_{C-O} - \beta_{C-O} = 22.5$ kcal mole^{-1}

In this reaction the C—Cl and H—O single bonds are replaced by C—O and H—Cl bonds. The value α_{C-Cl} is 80 kcal mole^{-1} so that $\alpha_{C-O} + \alpha_{H-Cl} - \alpha_{C-Cl} - \alpha_{O-H} = 96 + 103 - 80 - 119 = 0$. From this estimate of ΔH the reaction should be easily reversed. However there is, in addition, a substantial increase in the π bond resonance (about 22.5 kcal mole^{-1}) which is sufficient to drive the reaction well to the right. The reaction

$$CH_3CCl(g) + CH_3OH(g) \longrightarrow CH_3COCH_3(g) + HCl(g)$$

is exothermic. ΔH for the reaction is about -16 kcal mole^{-1}.

The substitution of OR for Cl on a carbon atom can take place through an S_N2 type mechanism. In basic solution, reaction of the HCl produced with base will again drive the reaction to the right. In this way ortho esters (relatively high energy compounds analogous to ethers and acetals), can be prepared.

$$CH_3CCl_3 + 3NaOC_2H_5 \longrightarrow CH_3-\overset{OC_2H_5}{\underset{OC_2H_5}{C-OC_2H_5}} + 3NaCl$$

an ortho ester

616

Esters of orthocarbonic acid can be made if a reactive starting material is used.

$$CCl_3NO_2 + 4NaOR \longrightarrow C(OR)_4 + NaNO_2 + 3NaCl$$

Ortho esters, like acetals, are hydrolyzed in acid solution. There is a considerable amount of π bond resonance in $H\overset{\overset{\displaystyle O}{\|}}{C}OH$ and H_2CO_3, and the hydrolysis of ortho esters is highly exothermic.

On the other hand the hydrolysis of a normal ester is most easily accomplished in base (saponification) since the reaction of the free organic acids with base shifts the equilibrium in this direction.

$$CH_3COOC_2H_5 + OH^- \longrightarrow CH_3COO^- + C_2H_5OH$$

$$\begin{array}{l} C_{17}H_{35}COO-CH_2 \\ \phantom{C_{17}H_{35}COO-}| \\ C_{17}H_{35}COO-CH + 3OH^- \longrightarrow 3C_{17}H_{35}COO^- + C_3H_5(OH)_3 \\ \phantom{C_{17}H_{35}COO-}| \\ C_{17}H_{35}COO-CH_2 \end{array}$$

a fat   octadecanoate ion or stearate ion   glycerine

The sodium salt of a long-chain acid produced in this reaction is known as a soap. The hydrocarbon portion of the molecule is of course hydrophobic, and soap molecules tend to form aggregates that dissolve grease and other materials soluble in hydrocarbons (Fig. 26-11). Syn-

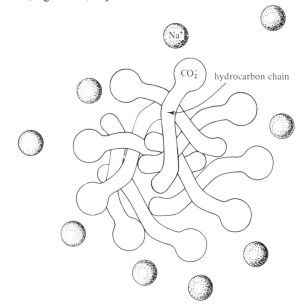

Figure 26-11 *Schematic diagram of a soap micelle.*

Table 26-4 *Examples of Some Typical Detergent Molecules*

$CH_3(CH_2)_{14}COO^-Na^+$

$CH_3(CH_2)_{10}OSO_3^-Na^+$

$CH_3(CH_2)_{12}C_6H_4OSO_3^-Na^+$

$CH_3(CH_2)_{14}N(CH_3)_3^+Cl^-$

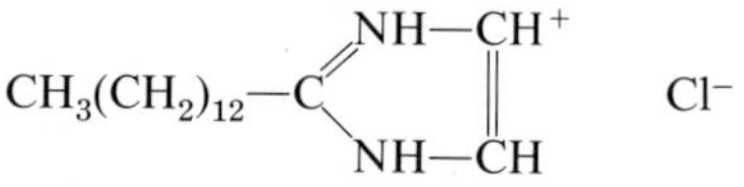

$$CH_3(CH_2)_{12}-C\begin{matrix}\diagup NH-CH^+ \\ \diagdown NH-CH \end{matrix} \quad Cl^-$$

$$CH_3(CH_2)_{12}-\overset{O}{\overset{\|}{C}}-(O-CH_2-CH_2)_5-OH$$

thetic detergents are similar molecules with a long hydrocarbon part and a polar end. Some examples are shown in Table 26-4.

The attack of a base on the carbon of the C=O group of an ester is probably the most widely used synthetic reaction in which C-C bonds are formed. Metal alkyls and enolate ions will react with esters to give easily predicted products

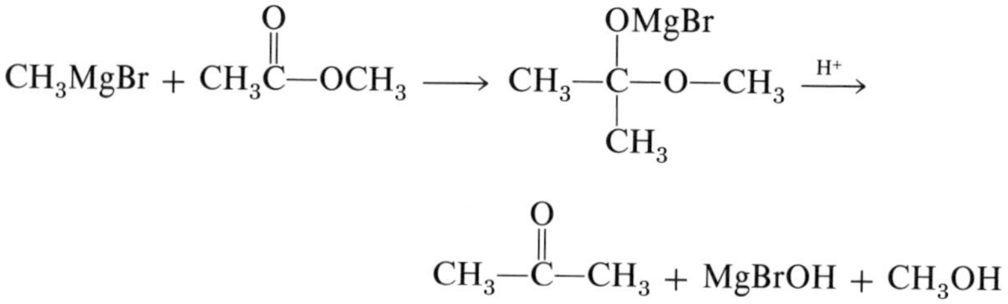

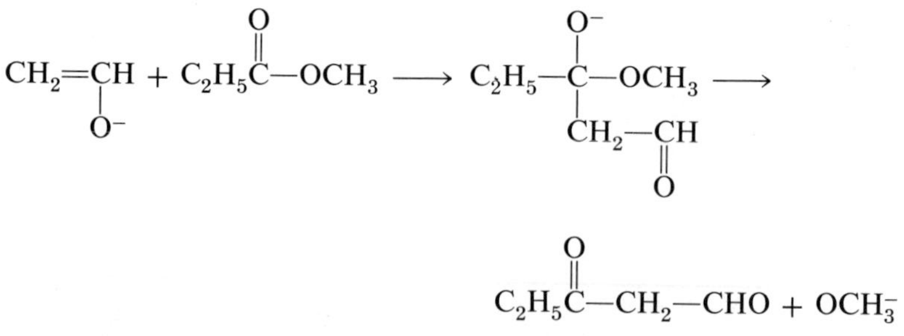

$$CH_3MgBr + CH_3\overset{O}{\overset{\|}{C}}-OCH_3 \longrightarrow CH_3-\underset{CH_3}{\overset{OMgBr}{\underset{|}{C}}}-O-CH_3 \xrightarrow{H^+}$$

$$CH_3-\overset{O}{\overset{\|}{C}}-CH_3 + MgBrOH + CH_3OH$$

$$CH_2=\underset{O^-}{\overset{}{C}}H + C_2H_5\overset{O}{\overset{\|}{C}}-OCH_3 \longrightarrow C_2H_5-\underset{\underset{\overset{\|}{O}}{CH_2-CH}}{\overset{O^-}{\underset{|}{C}}}-OCH_3 \longrightarrow$$

$$C_2H_5\overset{O}{\overset{\|}{C}}-CH_2-CHO + OCH_3^-$$

In many cases the enol form of one ester will react with another

$$C_2H_5-O-\overset{O}{\overset{\|}{C}}-CH_2CH_2CH_2CH_2-\overset{O}{\overset{\|}{C}}-OC_2H_5 + {}^-OC_2H_5 \longrightarrow$$

$$C_2H_5-O-\overset{O^-}{\underset{\parallel}{C}}\overset{}{\underset{CHCH_2CH_2CH_2-\overset{O}{\underset{\parallel}{C}}-OC_2H_5}{|}} \quad + C_2H_5OH$$

$$C_2H_5O-\overset{O}{\underset{\parallel}{C}}\underset{\underset{O}{\underset{\parallel}{C}}}{\overset{H_2C-CH_2}{\underset{CHCH_2}{||}}} + C_2H_5O^- \longleftarrow C_2H_5-O-\overset{O}{\underset{\parallel}{C}}\underset{\underset{O^-}{\underset{|}{C}}OC_2H_5}{\overset{H_2C-CH_2}{\underset{CHCH_2}{||}}}$$

Acids do not give corresponding reactions since the acid will react to protonate any carbanion.

V. Organic Nitrogen Compounds

Esters, acyl halides, and acids will react with ammonia to form C-N bonds in compounds known as amides

$$CH_3\overset{O}{\underset{\parallel}{C}}OC_2H_5 + NH_3 \longrightarrow CH_3-\overset{O}{\underset{\parallel}{C}}-NH_2 + C_2H_5OH$$
an amide

$$C_2H_5\overset{O}{\underset{\parallel}{C}}-Cl + 2NH_3 \longrightarrow C_2H_5-\overset{O}{\underset{\parallel}{C}}-NH_2 + NH_4Cl$$

$$CH_3\overset{O}{\underset{\parallel}{C}}-OH + NH_3 \longrightarrow NH_4CH_3CO_2 \overset{\Delta}{\longrightarrow} H_2O + CH_3\overset{O}{\underset{\parallel}{C}}-NH_2$$

In an amide there is considerable $p\text{-}p$ π overlap between carbon and nitrogen (resonance); thus in acetamide ($CH_3\overset{O}{\underset{\parallel}{C}}-NH_2$) the oxygen, nitrogen and the two carbon atoms all lie in one plane.

Amides can also be prepared from alkyl amines.

Note that a single m is used in the word amine to distinguish it from the term ammine used to designate ammonia as a ligand in complex ions.

$$CH_3Cl + NH_3 \longrightarrow CH_3NH_2 + H^+ + Cl^-$$

$$CH_3NH_2 + CH_3\overset{O}{\underset{\parallel}{C}}-OCH_3 \longrightarrow CH_3\overset{O}{\underset{\parallel}{C}}-NHCH_3 + CH_3OH$$

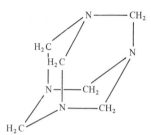

Figure 26-12 *Hexamethylene tetramine.*

High molecular weight polymers such as nylon can be formed from the similar reactions of molecules with two functional groups (two acids, two NH_2 groups, or one of each).

$$n\ ^+NH_3-(CH_2)_6-NH_3^+ + n\ ^-OC-(CH_2)_4-CO^- \longrightarrow$$

$$NH_3-(CH_2)_6-NH-\left(\underset{\parallel}{C}-(CH_2)_4-\underset{\parallel}{C}-\underset{|}{N}\ (CH_2)_6\ \underset{|}{N}\right)_{n-1}-\underset{\parallel}{C}-(CH_2)_4-\underset{\parallel}{C}-O^-$$

$$+ (2n-1)H_2O$$

Ammonia will also react with other molecules that contain a carbonyl group. The reactions with simple aldehydes and ketones generally give a complicated mixture of products. Formaldehyde and ammonia give a compound with a polycyclic structure known as hexamethylene tetramine (Fig. 26-12) which is the starting material for the preparation of the explosive, cyclonite, p. 530.

The reaction of substituted ammonia molecules with aldehydes generally results in simpler products. The reactions with hydroxylamine, hydrazine, and phenylhydrazine are commonly used to identify aldehydes and ketones

$$NH_2OH + CH_3CHO \longrightarrow H_2O + CH_3CH=NOH$$

hydroxylamine   acetaldehyde

cyclopentanone + hydrazine $\longrightarrow H_2O$ + [cyclopentanone hydrazone]

$$C_6H_5CHO + NH_2NHC_6H_5 \longrightarrow H_2O + C_6H_5CH=N-NH-C_6H_5$$
benzaldehyde   phenylhydrazine

The solid products are identified by their melting points.

The reaction of a secondary amine (a dialkyl amine) with an aldehyde or ketone results in a slightly different product because in this case a C=N bond cannot be easily formed. Compounds known as enamines are formed.

Note that for amines the designation primary, secondary, or tertiary reflects the number of alkyl groups on nitrogen, not those on carbon, so the nomenclature for amines and for alcohols is quite different. Sec-butylamine, $C_2H_5\underset{\underset{CH_3}{|}}{C}HNH_2$, is a primary amine.

cyclopentanone + NH(CH$_3$)$_2$ ⟶ H$_2$O + enamine (N,N-dimethyl-1-cyclopentenylamine)

Enamines are structurally similar to enols. They are almost as reactive as enolate ions. They are finding considerable application in current synthetic problems as illustrated in Fig. 26-13.

A. THE CARBON-NITROGEN TRIPLE BOND

One of the simplest carbanions that can be used to form C-C bonds in reactions with carbonyl compounds is CN^-. CN^- reacts very readily with aldehydes. It can be used to add an additional carbon to a sugar.

$$CH_2OH-(CHOH)_4-CHO + HCN \longrightarrow$$

$$CH_2OH-(CHOH)_4-\underset{\underset{}{|}}{\overset{\overset{OH}{|}}{C}}H-CN \xrightarrow[H_2O]{H^+}$$
a nitrile

$$CH_2(OH)-(CHOH)_4-CHOH-COOH \longrightarrow$$

$$CH_2OH-\underset{\underset{O\text{———}}{|}}{CH}-(CHOH)_4-\overset{O}{\underset{}{C}} + H_2O \xrightarrow{Na(Hg)}$$

$$CH_2OH-\underset{\underset{O\text{———}}{|}}{CH}-(CHOH)_4-\overset{H}{\underset{\underset{OH}{|}}{C}}$$

The equilibria in the hydrolysis of nitriles

$$R-CN + 2H_2O \rightleftharpoons R-\overset{O}{\underset{}{\overset{\|}{C}}}-NH_2 + H_2O \rightleftharpoons RCOOH + NH_3$$

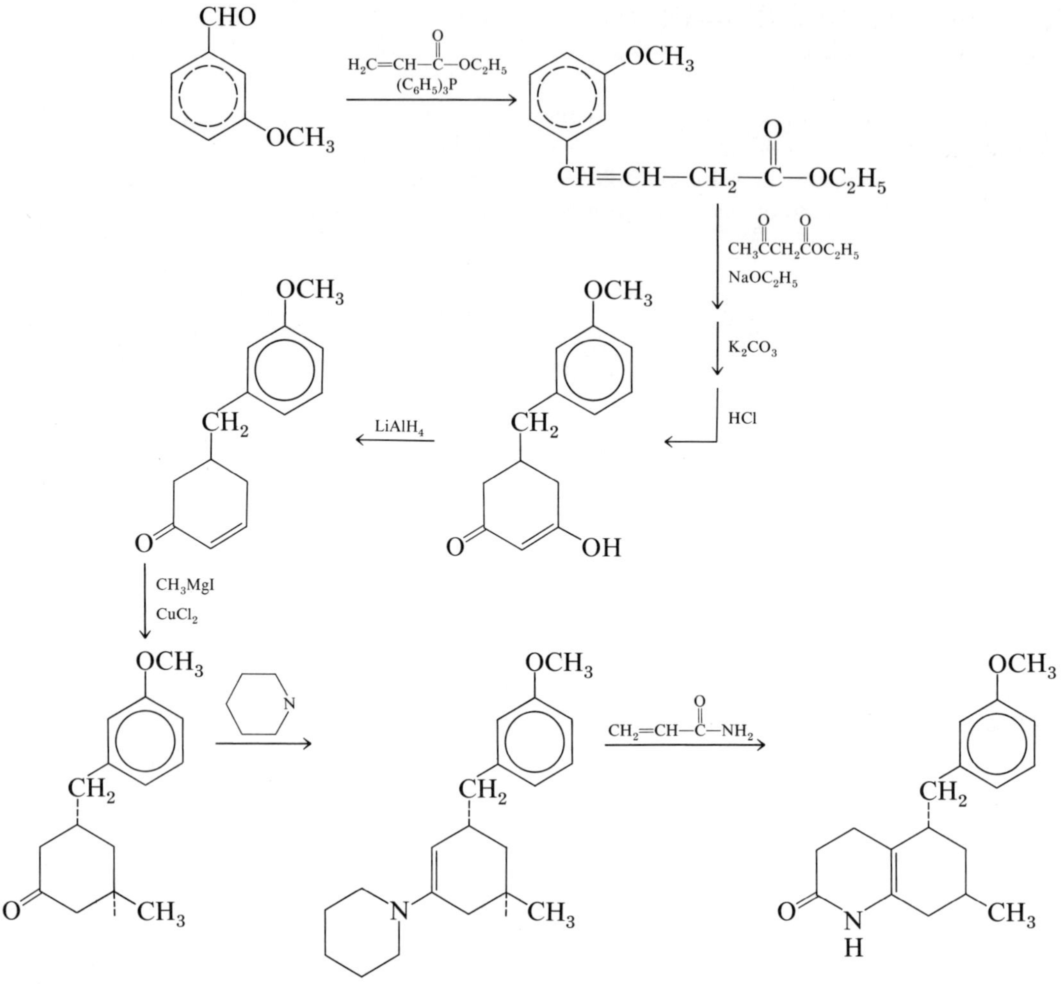

the enamine steps

Figure 26-13 *The total synthesis of the alkaloid lycopodine, including the condensation of an enamine with acrylamide.*

can be shifted in either direction. Under acid conditions the ammonia is fully protonated and the equilibrium shown lies well over to the right so that RCOOH and NH_4^+ are produced

$$CH_3CN + 2H_2O \rightleftharpoons CH_3\overset{\overset{\displaystyle O}{\|}}{C}{-}NH_2 + H_2O \rightleftharpoons CH_3\overset{\overset{\displaystyle O}{\|}}{C}{-}OH + NH_3$$

$$\rightleftharpoons CH_3\overset{\overset{\displaystyle O}{\|}}{C}O^- + NH_4^+$$

Figure 26-13 (*Continued*)

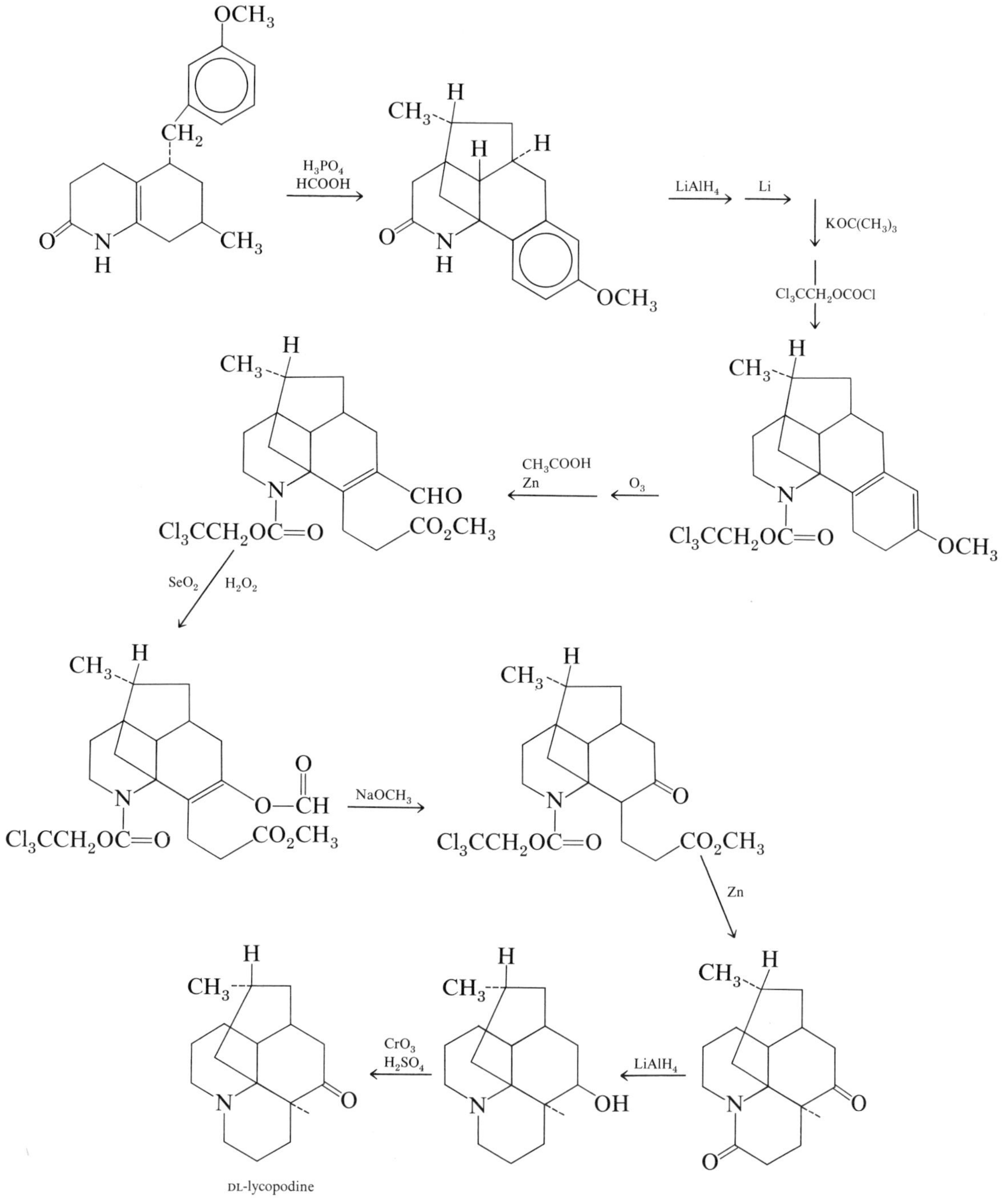

DL-lycopodine

In addition, because of the basicity of NH_3, $NH_3(g)$ is not easily produced under acidic conditions. Consequently heating an ammonium salt will drive off water

$$RC(=O)-O^-NH_3R'^+ \xrightarrow{\Delta} RC(=O)-NHR' + H_2O$$

$$RC(=O)-O^-NH_4^+ \xrightarrow{\Delta} RC(=O)-NH_2 + H_2O$$

More rigorous conditions are required to remove another water molecule from an amide. The conversion to a nitrile can be accomplished by heating in the presence of P_2O_5 or Al_2O_3

$$RC(=O)-NH_2 \xrightarrow[425°C]{\Delta, Al_2O_3} RCN + H_2O$$

$$RC(=O)-NH_2 + P_2O_5 \xrightarrow{\Delta} RCN + 2HPO_3$$

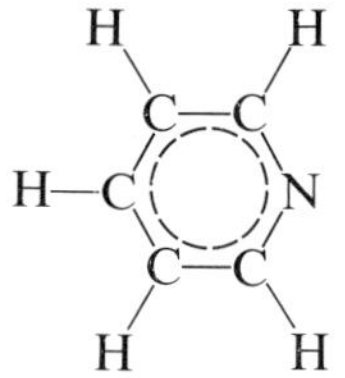

Figure 26-14 *The structure of pyridine.*

When the amide and nitrile are not appreciably volatile because of a high molecular weight, the equilibrium can be shifted by heating in a continuous stream of dry ammonia gas

$$C_{17}H_{35}CONH_2 \xrightarrow[NH_3]{300°C} C_{17}H_{35}CN + H_2O$$

Because of the extreme conditions required for dehydration of an amide, a nitrile is more easily prepared from an alcohol with one less carbon atom

$$C_3H_7CH_2OH \xrightarrow{HBr} C_3H_7CH_2Br \xrightarrow{CN^-} C_3H_7CH_2CN$$

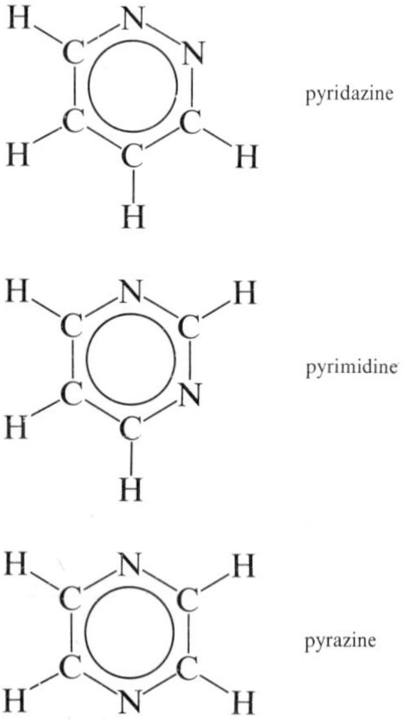

B. HETEROCYCLIC COMPOUNDS

Nitrogen can also substitute for carbon in the aromatic ring. Pyridine, C_5H_5N (Fig. 26-14), is a much weaker base ($K_b = 2.3 \times 10^{-9}$) than ammonia or the aliphatic amines. Other heterocyclic compounds that contain nitrogen in the ring are pyridazine, pyrimidine, and pyrazine (Fig. 26-15). These three compounds are all isomers of $C_4H_4N_2$.

Figure 26-15 *The structures of some additional heterocyclic nitrogen compounds.*

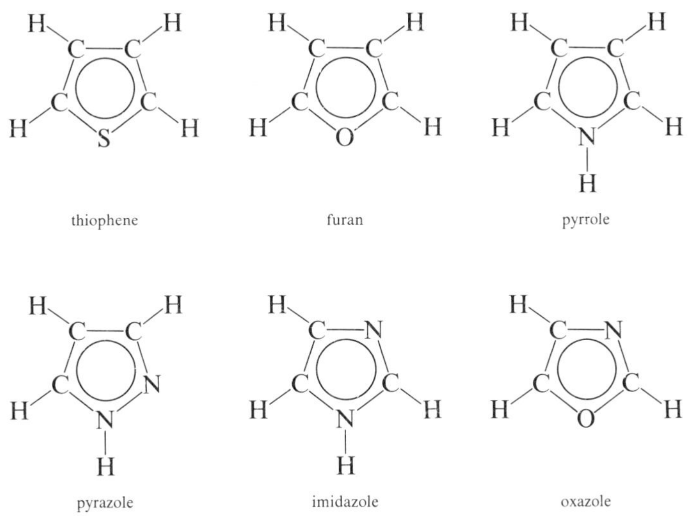

Sec. V Organic Nitrogen Compounds

Figure 26-16 *Some five-member aromatic heterocyclic ring compounds.*

The exceptional stability of aromatic molecules is associated with the presence of six π electrons in the ring. Six π electrons also give strong π bonding in five-member rings, as in $C_5H_5^-$. In some cases a nitrogen or oxygen atom can contribute two electrons to the π system of a five-member ring as illustrated by the structures in Fig. 26-16.

Obviously a tremendous variety of such compounds can be prepared. Because the π bonding is exceptionally strong, the compounds are expected to be quite stable. A few of them are quite important biologically. The imidazole ($C_3N_2H_4$) ring appears in the amino acid histidine and in adenosine triphosphate.

C. DYES

William Henry Perkin entered the Royal College of Chemistry in London in 1853 at the age of 15 to study chemistry under A. W. Hofmann. Hofmann had done outstanding work on the nature of aromatic amines. At that time the structure of quinine (Fig. 26-17) was completely unknown, but in 1856 Perkin attempted the oxidation of allyl toluidine (Fig. 26-18) with the hope that it might lead to quinine or a similar material. He obtained an intractible red-brown precipitate, and attempts with other amines were equally discouraging. Aniline gave a black tarry mess (polymeric), but Perkin found that a bright purple compound could be extracted from the tar with alcohol. The structure of this product, mauve, is shown in Fig. 26-19. It cannot be prepared from pure aniline, and Perkin soon found that the toluidine impurity in his original aniline was necessary. Not all colored compounds can be satisfactorily attached to textile fabrics or are

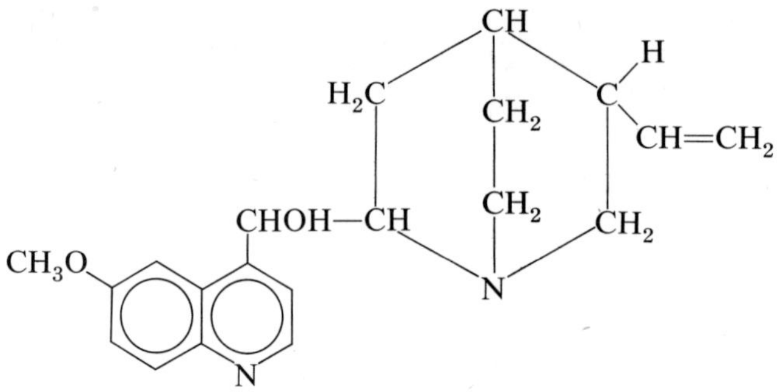

Figure 26-17 *The structure of quinine, $C_{20}H_{24}O_2N_2$.*

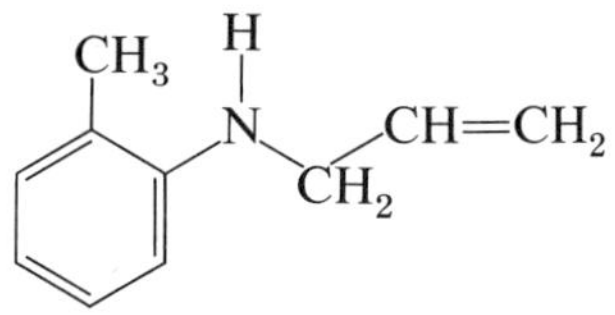

Figure 26-18 *Allyl toluidine, $C_{10}H_{13}N$.*

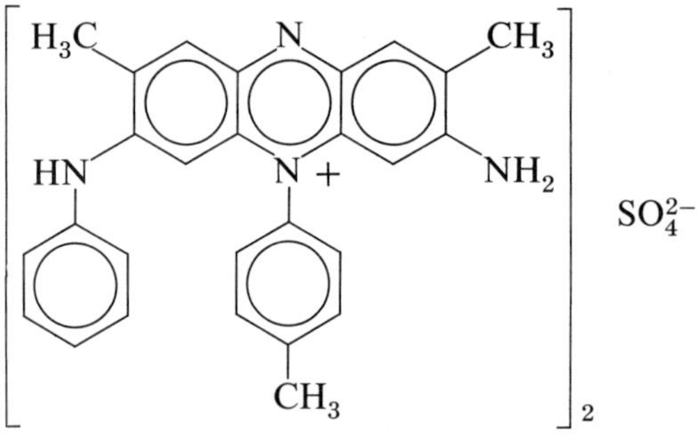

Figure 26-19 *The dye, mauve.*

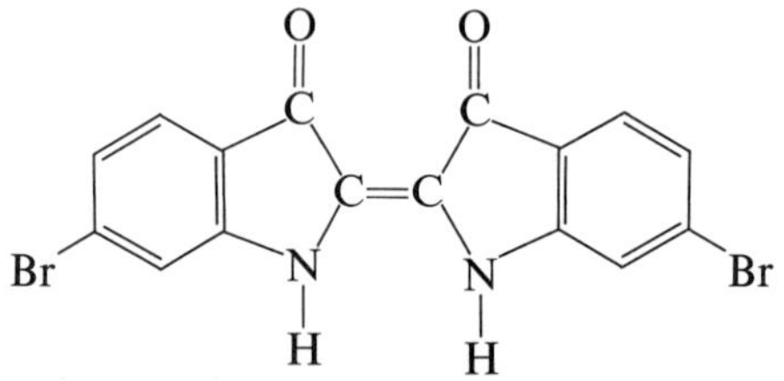

Figure 26-20 *The structure of Tyrian purple.*

sufficiently light-stable for use as dyes, but mauve is quite satisfactory. Perkin undertook the manufacture of mauve in 1857. This was the first commercial synthesis of a dye. The enterprise was highly successful, and Perkin continued work on basic organic chemistry and the synthesis of other dyes (alizarin).

At the present time essentially all the important dyes are synthetic as opposed to natural dyes. Such historically important natural dyes as Tyrian purple (Fig. 26-20) can now be synthesized, but very few natural products can compete with the wide variety of purely synthetic dyes now available. Figure 26-21 shows the structures of four of the dyes most widely used in current practice.

Sec. V Organic Nitrogen Compounds

Name	Chemical constitution	Production in 1962, lb	Price, $/lb
Direct Black 38	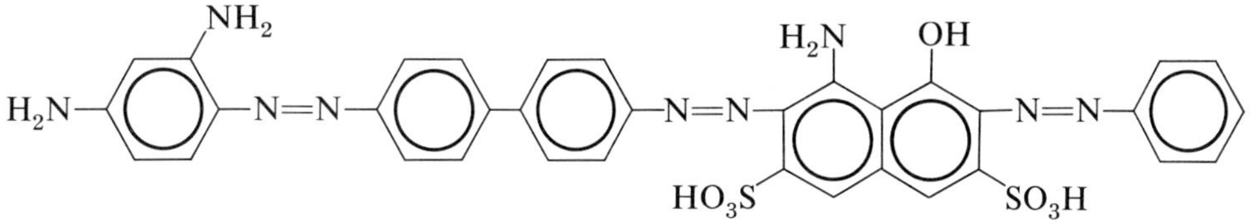	6,439,000	0.88
Vat Green 1	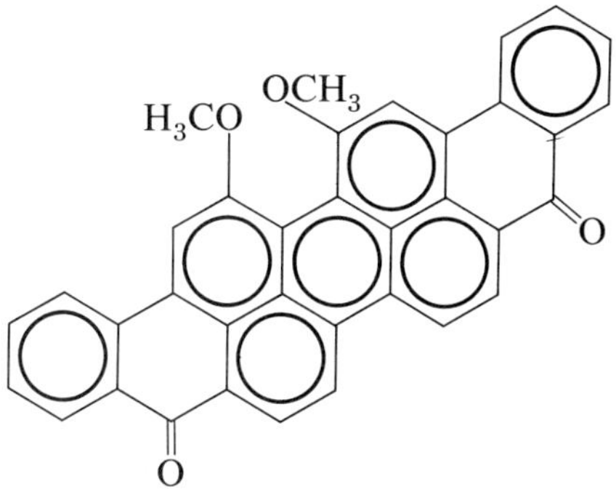	5,325,000	0.67
Basic Violet 1 (Methyl Violet)	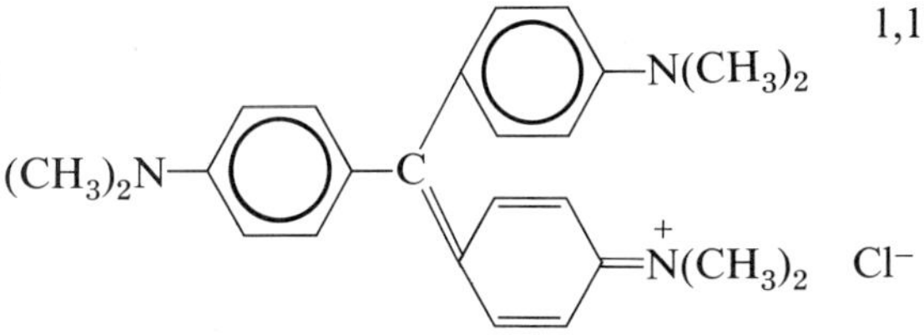	1,174,000	1.29
Acid Orange 7 (Orange II)	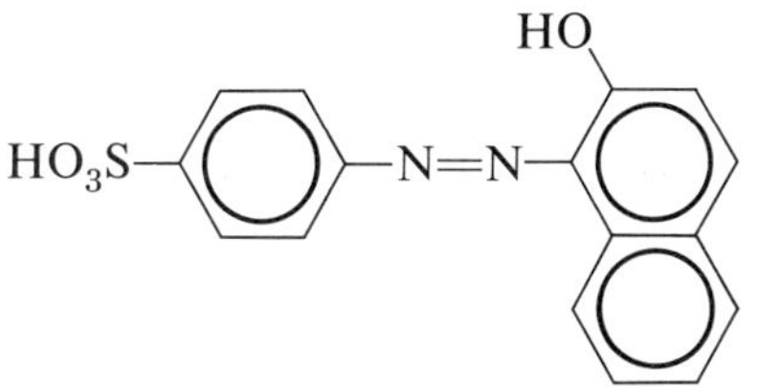	819,000	0.81

Figure 26-21 *The structures of four common dyes.*

QUESTIONS

1. List and name nine isomers of C_7H_{16}.
2. List and name six isomers of C_4H_8O.
3. Write three reactions for the preparation of 2-butanone (methylethyl ketone).

4. Predict the products of the following reactions:

(a) $CH_3CHCH(=O)$ with CH_3 substituent $\xrightarrow{KMnO_4}$

$$CH_3-\underset{CH_3}{\underset{|}{CH}}-\overset{O}{\overset{\|}{C}}-H \xrightarrow{KMnO_4}$$

(b) $CH_3-CH_2-\overset{O}{\overset{\|}{C}}-OH \xrightarrow{SOCl_2}$

(c) $CH_3CBr_2CH_3 \xrightarrow{KOH}$

(d) $C_6H_5CHO \xrightarrow{C_2H_5MgBr}$

(e) 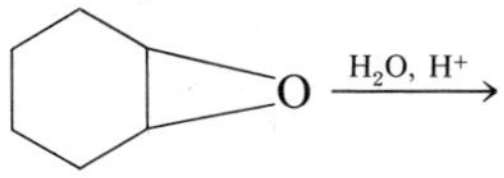 [bicyclic epoxide] $\xrightarrow{H_2O, H^+}$

(f) 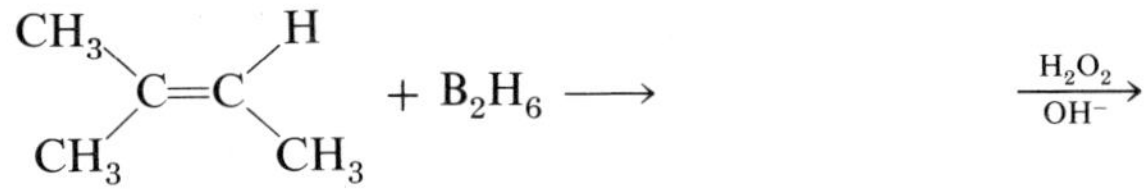 $\underset{CH_3}{\overset{CH_3}{}}C=C\underset{CH_3}{\overset{H}{}} + B_2H_6 \longrightarrow \xrightarrow{H_2O_2, OH^-}$

(g) $CH_3-\underset{OH}{\underset{|}{\overset{CH_3}{\overset{|}{C}}}}-CH_2-CH_3 \xrightarrow[H^+]{K_2CrO_4}$

(h) 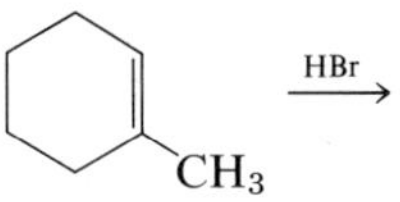 [cyclohexene with CH₃ substituent] $\xrightarrow{HBr}$

(i) 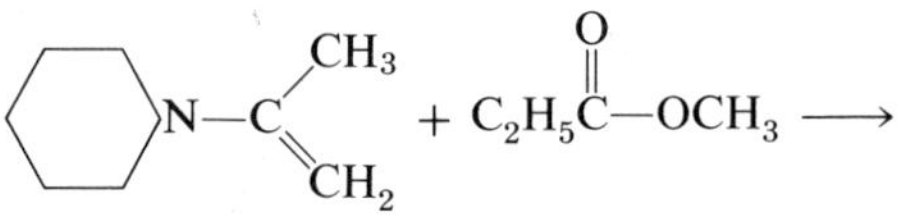 [piperidine]$N-C\underset{CH_2}{\overset{CH_3}{}} + C_2H_5\overset{O}{\overset{\|}{C}}-OCH_3 \longrightarrow$

5. Suggest a sequence of reactions for the following transformations:

(a) 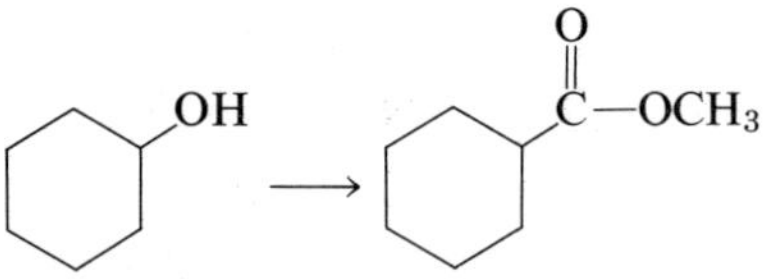 [cyclohexanol] $\longrightarrow$ [methyl cyclohexanecarboxylate]

(b) 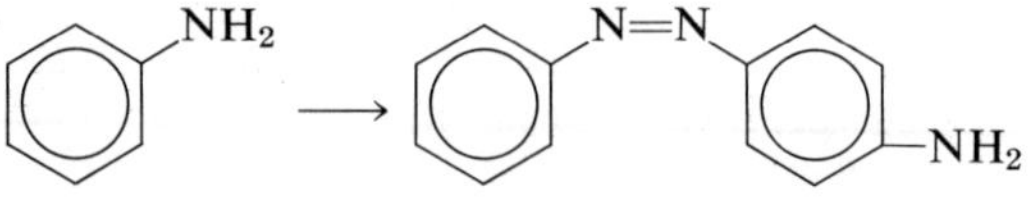 [aniline] $\longrightarrow$ [4-aminoazobenzene]

(c) $CH_3CH_2Br \longrightarrow CH_3-CH_2-\underset{CH_3}{\underset{|}{CHOH}}$

6. Define geometrical isomerism and list four examples.
7. Four drops of each of the following alcohols were shaken with a solution

of $ZnCl_2$ in concentrated HCl. Assume a mechanism for the reactions taking place and explain the observations below.

(a) 1-butanol: dissolves and remains clear
(b) 2-butanol: dissolves, but after 2 min the solution becomes cloudy and eventually a second liquid layer forms
(c) 2 methyl-2-propanol: does not appear to dissolve

8. Suggest a mechanism for the reaction of acetyl chloride with ammonia.
9. Suggest a mechanism for each step in the following synthesis of the dye Orange II.

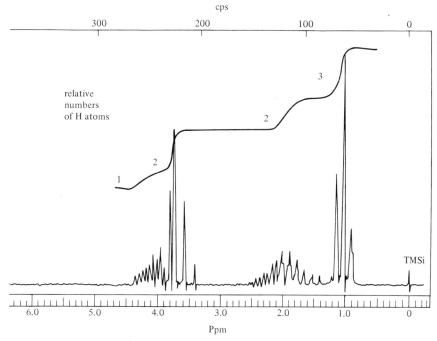

10. The NMR spectrum of one isomer of $C_4H_8Br_2$ is given below. Which isomer is this? How does the spectrum differ from that expected for the other isomers?

11. How many optical isomers are there for each of the following structures?

(a) $CH_2OH—CHOH—CHOH—CHO$
(b) $COOH—CHOH—CHOH—COOH$
(c) cis-$Co(NH_2CH_2CH_2NH_2)_2(H_2O)_2^{3+}$
(d) 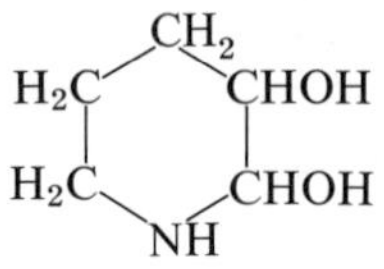
```
        CH_2
   H_2C/   \CHOH
    |       |
   H_2C\   /CHOH
        NH
```

12. Treatment of the two butene-2 isomers with Br_2 gave the results given below. Explain these observations. (Hint: a cyclic intermediate such as

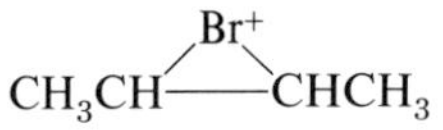

$$CH_3CH \overset{Br^+}{-\!\!\!-\!\!\!-} CHCH_3$$

has been proposed.)

> cis- and trans-2-Butenes. Samples of the pure diastereomeric dibromides were prepared by adding bromine to each olefin in dichloromethane at $-60°$. The dibromides prepared in this manner contained less than 1% of the other diastereomer as determined by glpc (Tide column at $85°$), the cis olefin giving only dl-2,3-dibromobutane and the trans olefin only meso-2,3-dibromobutane. The reaction products in acetic acid showed that only one dibromide was formed from each olefin. Small amounts of acetoxy bromides were formed in each case, as shown by the presence of carbonyl absorption in the infrared spectra of the crude reaction mixture, but in insufficient quantities to allow definite characterization.

PROBLEMS I

1. A compound was found to contain both carbon and hydrogen. It may contain oxygen as well. A sample weighing 0.1085 g gave 0.3411 g CO_2 and 0.1394 g H_2O on combustion. Calculate the empirical formula.
2. What volume of air at standard temperature and pressure would be required for the complete combustion of 1 gal of normal C_7H_{16}(2490 g)?
3. A sample of 5.00 g ethyl alcohol was oxidized with excess K_2CrO_4 in H_2SO_4. This reaction produced 6.2 g acetic acid. What was the percent yield?
4. Calculate the percent composition of 5 chloro-2-cis-4-trans pentadienoic acid. This was prepared and analyzed recently. The experimental values [J. Am. Chem. Soc., 91: 1136 (1969)] are 45.28%C, 3.86%H, and 26.30%Cl.
5. In the synthesis of lycopodine (Fig. 26-13) the yields given below were obtained. What weight of m-methoxybenzaldehyde must be used to prepare 500 mg of lycopodine by this sequence of reactions?

m-methoxybenzaldehyde
↓ 36%
trans-3-(m-methoxybenzyl)-5-methylcyclohexanone

25% in the enamine steps
↓
16%
↓
dihydrolycopodine
↓
80%
↓
lycopodine

PROBLEMS II

1. Draw the structural formula of 2-methylpropane.
2. How many hydrogen atoms are present in 2-methylpropane?
3. How many hydrogen atoms of 2-methylpropane are in methyl groups?
4. Name the straight chain isomer of 2-methylpropane.
5. There are two isomers of C_4H_9Cl with a branched chain of carbon atoms. Write their structures.
6. Draw the structures of 1-chlorobutane and 2-chlorobutane.
7. What is the correct name for 3-chlorobutane?
8. Name four isomers of C_4H_9Cl.
9. Tertiary chlorides react in aqueous acid solutions to form alcohols. Which isomer of C_4H_9Cl is a tertiary chloride?
10. Write an equation for the formation of 2-methyl-2-propanol from the chloride.
11. Which isomer of C_4H_9Cl is optically active?
12. What are the five isomers of C_4H_9Cl?
13. R-2-Chlorobutane reacts slowly with hydroxide ions to form S-2-butanol. Write an equation for the reaction.
14. Sketch the activated complex for the reaction in Problem 13.
15. Which of the C_4H_9OH alcohols is easily oxidized to a ketone?
16. What product would you expect from the oxidation of 2-methyl-1-propanol?
17. Write a series of equations for the preparation of methyl 2-methyl propionate from 1-chloro-2-methylpropane and methanol. Use any inorganic reagents you want.

And it should be clear to the reader that given these elemental facts of life, given growth and death and reproduction with individual variation in a world that changes, life must change in this way, modification and differentiation must occur, old species must disappear, and new ones appear. We have chosen for our instance here a familiar sort of animal, but what is true of furry beasts in snow and ice is true of all life, and equally true of the soft jellies and simple beginnings that flowed and crawled for hundreds of millions of years between the tidal levels and in the shallow, warm waters of the Proterozoic seas.*

H. G. WELLS

Chapter 27 Biochemistry

Biochemistry is the study of the chemical reactions involved in living organisms. An incredibly complex system of interacting chemical reactions is involved in the growth and reproduction of a cell, and even as simple a system as the membrane enclosing a cell is not yet understood in any detail. Many individual reaction sequences in cells have been worked out in detail, and even the most rudimentary knowledge of the chemistry of cells has paid dividends in the development of drugs and hormones.

I. Iatrochemistry

The simplest application of chemistry to biological systems is the testing of chemical preparations for their biological effect. This was first put on a reasonably systematic basis in the sixteenth century by Paracelsus (T. B. von Hohenheim), and his followers. Although Paracelsus was successful in many cases with opium and mercury compounds, he was important primarily because he popularized the idea that the study of chemistry could make major contributions to medicine. One of his immediate followers discovered tartar emetic, (potassium antimonyl tartrate, $KSbOC_4H_4O_6$) which is still used in medicine

---

*Used by permission of Professor G. P. Wells from H. G. Wells, *The Outline of History* (New York: Macmillan, 1920).

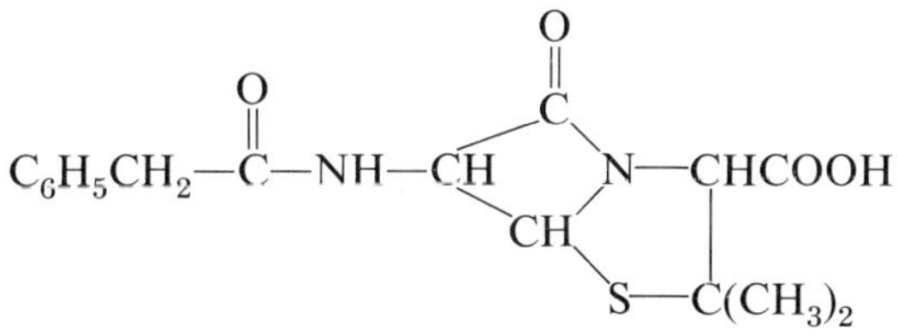

Figure 27-1 *The structure of penicillin G.*

to a limited extent. Another significant advance was the preparation of alcoholic solutions of a medicine (a tincture). This approach to the development of knowledge is exemplified today in the routine testing of over 20,000 compounds each year for anticancer activity.

The more systematic application of chemistry to medical problems required detailed information on the cause of diseases. A breakthrough was provided by Louis Pasteur, a chemist, who proved that bacteria and other microorganisms are responsible for the spoiling of food. Just a few years later (1867) Lister established that phenol (C_6H_5OH) was useful in preventing the infection of surgical wounds, and the chemists were ready at that time with the anaesthetics, ether and nitrous oxide, to make surgery a reality. A great variety of antiseptics (or antibacterials) and anaesthetics have since been developed.

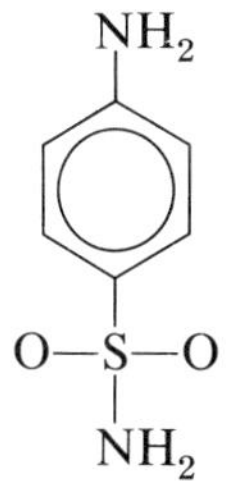

The recent trend in the development of antibiotics has been the use of reasonably complex organic molecules; these are accepted as a substitute for some natural material in the metabolism of a microorganism but later block some critical metabolic step. The material may be originally of biological origin such as penicillin (Fig. 27-1) or purely synthetic like sulfanilamide (Fig. 27-2).

The sulfanilamides were discovered when a synthetic dye was found to be effective against streptococcal infections (1935). Within a few years it was found that sulfanilamide, a degradation product of the dye, was equally effective. By 1940 a whole array of useful drugs had been developed from the hundreds of derivatives of sulfanilamide that had been synthesized and tested.

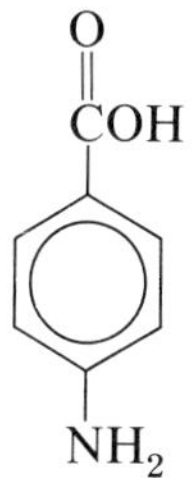

Sulfanilamide is effective because of its resemblance to *para*-aminobenzoic acid ($C_7H_7NO_2$, Fig. 27-3). This substance is an essential part of the folic acid molecule, shown in Fig. 27-4, which is involved

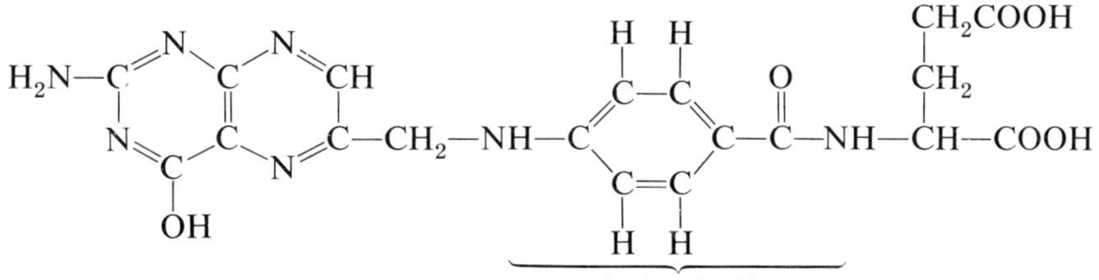

Figure 27-2 *The structure of sulfanilamide.*

Figure 27-3 *Para-aminobenzoic acid.*

Figure 27-4 *Folic acid.*

in the enzymatic synthesis of a number of essential parts of a cell. Humans and most other mammals cannot synthesize folic acid; therefore they are dependent on the presence of the substance in their diet. It is one of the B complex of vitamins. On the other hand most bacteria make folic acid from *para*-aminobenzoic acid.

Sulfanilamide present at a concentration as low as 10^{-4} M blocks the synthesis of folic acid, and without folic acid the bacteria cannot grow. The block is due to the competition between sulfanilamide and *para*-aminobenzoic acid in bonding to an enzyme. An enzyme molecule that is bound to sulfanilamide is no longer available for the synthesis of folic acid. If the concentration of *para*-aminobenzoic acid is increased, the effect of sulfanilamide is cancelled. Some of the experimental data on this point are summarized in Table 27-1.

The data indicate that the enzyme system involved shows a strong preference for the natural material, but sulfanilamide can block bacterial growth provided it is 5000 times more concentrated than *para*-aminobenzoic acid.

II. Proteins and Enzymes

The biological organization of cells is possible only if the materials that catalyze the reactions taking place in the cell are highly specific. Specificity can be partially understood by visualizing an enzyme as a lock and the molecule that is undergoing the reaction (the substrate) as the key. The two molecules fit together more or less perfectly. A substance that is chemically similar to the substrate might by accident also fit the active site. Sulfanilamide just happens to be absorbed by the enzyme in competition with *para*-aminobenzoic acid. It is much less strongly held.

Another example of enzyme selectivity is found in the digestive enzyme trypsin. This enzyme catalyzes the hydrolysis of protein chains

Table 27-1 *Competition Between Sulfanilamide and Para Amino Benzoic Acid*

Sulfanilamide concentration, M	Concentration of *para*-aminobenzoic acid required to restore the growth of bacteria, M	Ratio of the two concentrations
0.3×10^{-3}	0.6×10^{-7}	5000
1.5×10^{-3}	3×10^{-7}	5000
7.5×10^{-3}	15×10^{-7}	5000

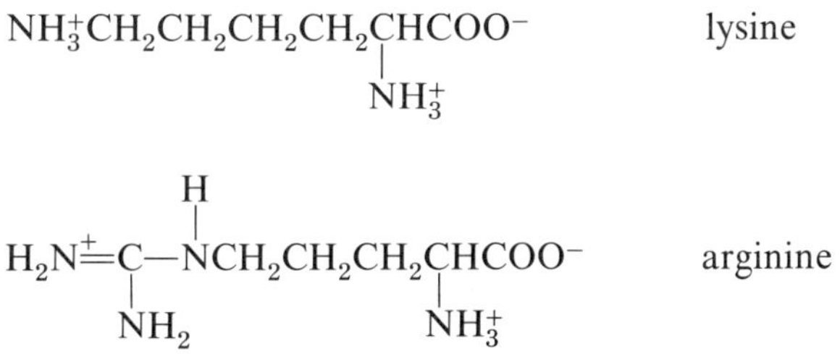

$$\mathrm{NH_3^+CH_2CH_2CH_2CH_2CHCOO^-} \qquad \text{lysine}$$
$$\underset{\mathrm{NH_3^+}}{|}$$

$$\mathrm{H_2N}\overset{+}{=}\underset{\mathrm{NH_2}}{\overset{}{\underset{|}{C}}}-\overset{\mathrm{H}}{\underset{}{\overset{|}{N}}}\mathrm{CH_2CH_2CH_2}\underset{\mathrm{NH_3^+}}{\overset{}{\underset{|}{C}}}\mathrm{HCOO^-} \qquad \text{arginine}$$

Figure 27-5 *The structure of lysine and arginine. Trypsin will hydrolyze the amide linkages in a protein chain at the carboxyl end of these two amino acids.*

only at the carboxyl end of the two amino acids lysine and arginine. The structural similarity of these two amino acids is evident from Fig. 27-5. The surprising fact is that trypsin does not catalyze the hydrolysis of the amide linkage of any of the other eighteen common amino acids in a protein chain.

The structure of proteins, which are made up of an ordered series of amino acids, is ideal for constructing molecules with the detailed configuration required of an enzyme. All known enzymes are primarily protein in nature. There are some twenty different amino acids found in proteins, each of which has been assigned a conventional three letter symbol. These symbols can be used to write the primary structure (if it is known) of a protein or an enzyme in a fairly concise form. By convention the first amino acid in the chain contains the free amine group. Figure 27-6 shows the structure of insulin. Since an error in a single amino acid in a critical part of the chain can lead to a biologically inactive enzyme, it is clear that the synthesis of proteins is one of the most critical parts of the metabolism of cells.

III. Adenosine Triphosphate and Glucose

ATP is involved in the initial activation of amino acids for the synthesis of proteins. The energy required to form ATP from ADP and inorganic phosphate is supplied primarily by the absorption of light by the chloroplasts of plants.

$$\mathrm{ADP^{3-} + H_2PO_4^- \xrightarrow{h\nu} ATP^{4-} + H_2O} \qquad \Delta G° = +8 \text{ kcal mole}^{-1}$$

$$\mathrm{C_{10}N_5H_{12}O_{10}P_2^{3-} + H_2PO_4^- \xrightarrow{h\nu} C_{10}N_5H_{12}O_{13}P_3^{4-} + H_2O}$$

The structure of ATP is shown in Fig. 27-7.

Ch. 27 Biochemistry

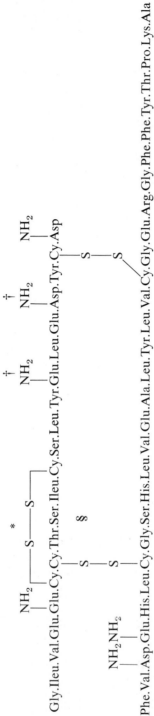

*The sulfur atoms of cysteine are bonded together in pairs as shown by the solid lines.

†The second carboxyl groups in the amino acids asparagine and glutamine are found as amides in some of the positions of insulin. This is indicated by writing an NH_2 group bonded to the amino acid.

§The three amino acids at this position between the two cysteines can be replaced by others without loss of biological activity. Horse insulin has glycine in place of serine, and in beef and sheep insulins these sequences are Ala.Ser.Val. and Ala.Gly.Val., respectively.

Figure 27-6 *The structure of insulin as isolated from pigs or sperm whales.*

Sec. III Adenosine Triphosphate and Glucose

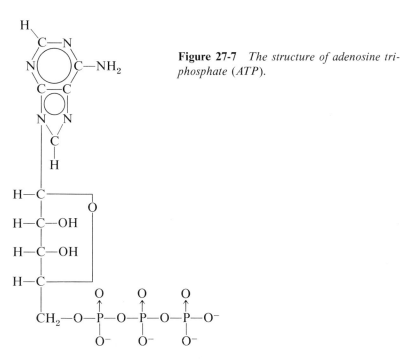

Figure 27-7 *The structure of adenosine triphosphate (ATP).*

A. GLUCOSE FORMATION AND OXIDATION

In the process of photosynthesis ATP is used for the formation of glucose and oxygen from carbon dioxide and water. In the metabolism of animals these reactions can be reversed to produce ATP as it is needed. For still longer term energy conversion, glucose is commonly converted to a polymeric material such as starch or glycogen. ΔG^0 for the reaction

$$C_6H_{12}O_6 + 6O_2 \longrightarrow 6CO_2 + 6H_2O$$

is -690 kcal mole^{-1}. Since the free energy of formation of ATP is 8 kcal mole^{-1} the oxidation of one molecule of glucose could theoretically lead to the preparation of $690/8 = 86$ molecules of ATP.

For biological purposes the rate of the reaction is important. In the actual system an enzymatic mechanism is set up so that only about 36 molecules of ATP are produced for each glucose molecule oxidized. In addition, this provides a substantial rate for each step. The equilibrium for the overall reaction

$$C_6H_{12}O_6 + 6O_2 + 36ADP^{3-} + 36H_2PO_4^- \longrightarrow 36ATP^{4-} + 42H_2O + 6CO_2$$

is such that the concentration of ATP in cells is substantially higher

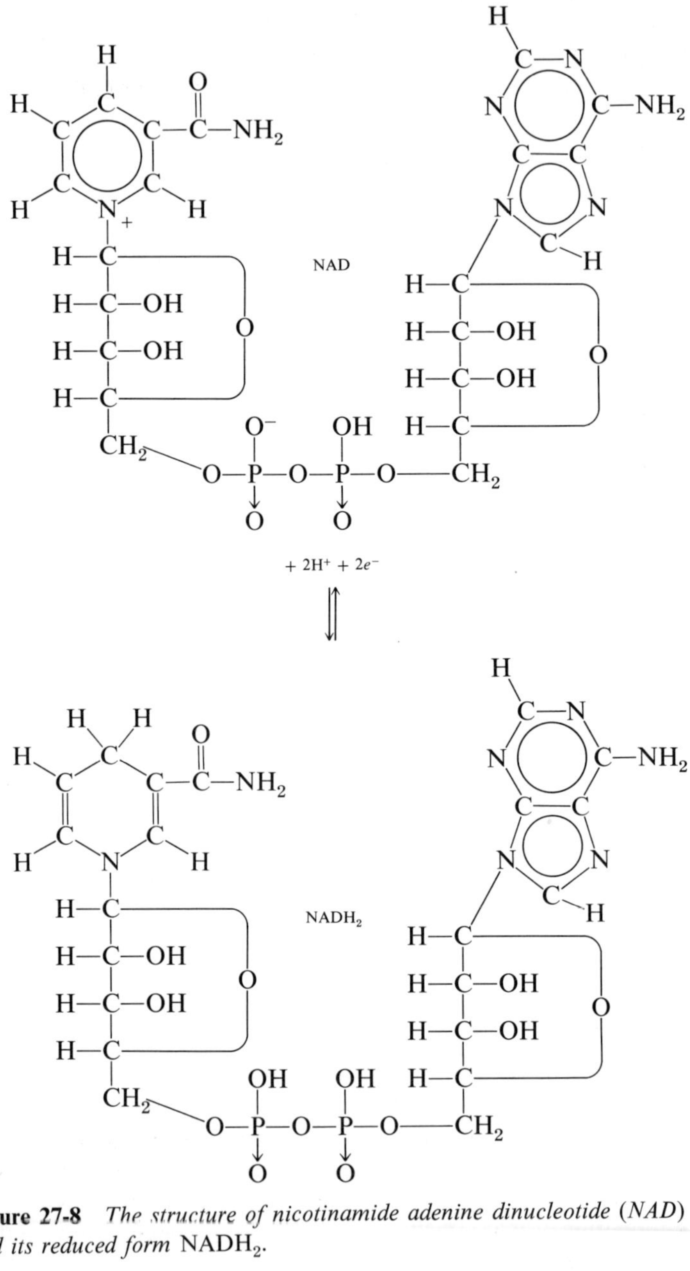

Figure 27-8 *The structure of nicotinamide adenine dinucleotide (NAD) and its reduced form* $NADH_2$.

than the concentration of ADP and inorganic phosphate. Under physiological conditions ΔG for the hydrolysis of ATP is about -12 kcal mole^{-1}, substantially shifted from the ΔG^0 value of -8 kcal mole^{-1}. Thus, the 36 molecules of ATP store about 432 kcal mole^{-1}.

The aerobic oxidation of glucose has an overall efficiency of about $(432/690)100 = 63\%$.

Most of the ATP available from the oxidation of glucose is formed during the oxidation of reduced nicotinamide adenine dinucleotide, ($NADH_2$), (Fig. 27-8). The standard potential for the half-reaction $NAD + 2H^+ + 2e^- \longrightarrow NADH_2$ is -0.32 volt, and ΔH of the oxidation with molecular oxygen

$$NADH_2 + \tfrac{1}{2}O_2 \longrightarrow NAD + H_2O$$

is -51 kcal mole^{-1}. There is sufficient energy for the formation of at least three molecules of ATP. The redox steps actually coupled to ATP synthesis occur in a highly organized system of oxidizing and reducing agents, mainly cytochromes, in a mitochondrial membrane, and the details are exceedingly difficult to unravel. A very similar redox system, but with somewhat different cytochromes and enzymes, is found in the chloroplasts which are the basic structures involved in photosynthesis. The individual steps in ATP synthesis are best illustrated by a third system, anaerobic glycolysis, since in this case all the principal reactants are present in solution where they are more easily studied.

B. ANAEROBIC GLYCOLYSIS—FERMENTATION

This sequence of reactions begins with the consumption of two moles of ATP and the conversion of glucose to fructose-1,6-diphosphate. This substance then splits into two molecules of glyceraldehyde-3-phosphate.

$$C_6H_{12}O_6 + 2ATP^{4-} \longrightarrow \begin{array}{c} H_2C-O-PO_3^{2-} \\ | \\ C=O \\ | \\ HO-C-H \\ | \\ H-C-OH \\ | \\ H-C-OH \\ | \\ H_2C-O-PO_3^{2-} \end{array} + 2ADP^{3-} + 2H^+$$

$$\longrightarrow 2\begin{array}{c} H-C=O \\ | \\ H-C-OH \\ | \\ H_2C-O-PO_3^{2-} \end{array}$$

A phosphate group can add to the aldehyde and, if the aldehyde is oxidized at this stage, a new high-energy phosphate bond is formed.

$$\begin{array}{c} H-C=O \\ H-C-OH \\ H_2C-O-PO_3^{2-} \end{array} + HPO_4^{2-} \longrightarrow \begin{array}{c} OH \\ H-C-O-PO_3^{2-} \\ H-C-OH \\ H_2C-O-PO_3^{2-} \end{array}$$

$$\begin{array}{c} OH \\ H-C-O-PO_3^{2-} \\ H-C-OH \\ H_2C-O-PO_3^{2-} \end{array} + NAD + ADP^{3-} \longrightarrow \begin{array}{c} O \\ \parallel \\ C-O-PO_3^{2-} \\ H-C-OH \\ H_2C-O-PO_3^{2-} \end{array} + NADH_2 + ADP^{3-}$$

$$\longrightarrow \begin{array}{c} O \\ \parallel \\ C-O^- \\ H-C-OH \\ H_2C-O-PO_3^{2-} \end{array} + NADH_2 + ATP^{4-}$$

In a series of rearrangements that introduce more and more C=O π bonding to the oxygen shared by phosphorus, the remaining phosphate group acquires a high energy phosphate bond.

$$\begin{array}{c} O \\ \parallel \\ C-O^- \\ H-C-OH \\ H_2C-O-PO_3^{2-} \end{array} \longrightarrow \begin{array}{c} O \\ \parallel \\ C-O^- \\ H-C-O-PO_3^{2-} \\ H_2C-OH \end{array} \longrightarrow \begin{array}{c} O \\ \parallel \\ C-O^- \\ C-O-PO_3^{2-} \\ \parallel \\ CH_2 \end{array} + H_2O$$

$$\begin{array}{c} O \\ \parallel \\ C-O^- \\ H-C-O-PO_3^{2-} \\ \parallel \\ CH_2 \end{array} + ADP^{3-} + H^+ \longrightarrow \begin{array}{c} O \\ \parallel \\ C-O^- \\ C=O \\ CH_3 \end{array} + ATP^{4-}$$

pyruvic acid

Thus far 4 moles of ATP have been formed from 1 mole of glucose. There is a net gain of 2 moles of ATP. For the reaction to continue the $NADH_2$ must be reoxidized. In muscle tissue lactic acid is produced.

$$\underset{\underset{\underset{CH_3}{|}}{\overset{C=O}{|}}}{\overset{O}{\overset{\|}{C}-O^-}} + NADH_2 \longrightarrow \underset{\underset{\underset{CH_3}{|}}{\overset{H-C-OH}{|}}}{\overset{O}{\overset{\|}{C}-O^-}} + NAD$$

The enzymes of yeast catalyze the decarboxylation of pyruvic acid prior to the reduction step

$$\underset{\underset{\underset{CH_3}{|}}{\overset{C=O}{|}}}{\overset{O}{\overset{\|}{C}-O^-}} + H^+ + NADH_2 \longrightarrow CO_2 + \underset{\underset{CH_3}{|}}{H-C=O} + NADH_2$$

$$\longrightarrow CO_2 + CH_3CH_2OH + NAD$$

The overall net reactions are

$$C_6H_{12}O_6 + 2ADP^{3-} + 2HPO_4^{2-} \longrightarrow 2C_3H_5O_3^- + 2ATP^{4-} + 2H_2O$$

and

$$C_6H_{12}O_6 + 2ADP^{3-} + 2H_2PO_4^- \longrightarrow$$
$$2CO_2 + 2C_2H_5OH + 2ATP^{4-} + 2H_2O$$

IV. Deoxyribonucleic Acid (DNA) and Ribonucleic Acid (RNA)

In both NAD and ATP the heterocyclic ring system of adenine is bonded to the sugar ribose which is bonded in turn to a phosphate. The use of a particular grouping of molecules in several different molecules is quite characteristic of biological systems. The adenine ring in NAD and ATP does not take part directly in the redox reaction or the phosphorylation. It appears to be present as a label for the recognition of the molecule by the enzymes involved. The protein-adenine recognition system was apparently developed first in connection with the mechanism of protein synthesis and then was adapted to the energy transfer and redox reactions of ATP and NAD. The significance of the heterocyclic ring system of adenine becomes apparent only when we examine the structure of nucleic acids.

Unravelling the chemical steps by which the ordered sequence of purine and pyrimidine bases in the genetic material DNA is transcribed into a similar sequence in RNA and the further steps in which RNA serves as a template for the synthesis of an ordered chain of amino acids are exciting areas of current biochemical research. Some points in this sequence are known now in detail.

The basic structure of DNA consists of a backbone of the sugar deoxyribose and phosphate ions. These groups are arranged as shown in Fig. 27-9. The various heterocyclic bases are strung onto the backbone as shown in Fig. 27-10. The representation in this figure is single-stranded. In most cases DNA is actually a double stranded-polymer. The two strands are bonded together through hydrogen bonding between nucleotide bases.

The pairing of bases, guanine with cytocine, adenine with thymine, and uracil with adenine is shown in Fig. 27-11. Pairing in just this way is dictated by both the size of the bases and the number of possible hydrogen bonds. The pair thymine-adenine can form two hydrogen bonds whereas the pair cytosine-guanine forms three. Several simplified pictures of the structure of DNA are shown in Fig. 27-12. RNA differs from DNA in that uracil is substituted for thymine and ribose takes the place of deoxyribose.

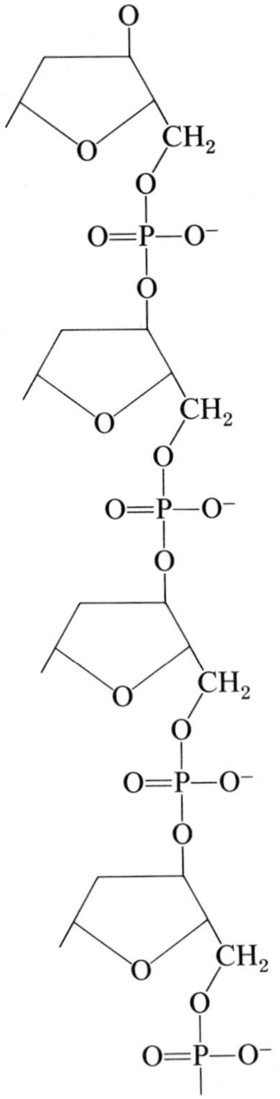

The double-stranded structure of DNA and RNA is highly stable. If broken by ionizing radiation, ultraviolet light, or chemicals, the other chain may hold it in position and it may be properly repaired.

The base sequence in DNA differs from one gene to another. The arrangement of bases in the DNA *is* the genetic information. All of the information required to direct the production of red hair or blue eyes, the growth of whiskers, the production of thousands of enzymes involved in the metabolism of cells, the differentiation of cells, and the production of hormones that control growth (pituitary gland) and activity (adrenal glands) is coded in the base sequence of the DNA molecules of the genes for these traits.

In short, biological activity is controlled by DNA.

A sequence of three bases in DNA corresponds to an amino acid in a protein. For example, the three bases guanine (G), adenine (A), and adenine (A), arranged in that order *is* a code for the amino acid glutamic acid. The three bases guanine (G), thymine (T), and adenine (A), arranged in that order is a code for valine. The code for glutamic acid differs from that for valine by only one base. The code for glutamic acid is GAA, that for valine is GTA.

A protein that contains valine in place of glutamic acid will result if thymine is substituted for adenine. This simple change in a DNA molecule is a mutation. The result is an alteration in the protein.

A mutation such as this may have no overall effect on the organism involved. If the affected gene is involved in the production of a protein, only one amino acid in the protein would be altered. If this amino acid is not important to the functioning of the protein, there will be

Figure 27-9 *The sugar-phosphate backbone of DNA.*

Sec. IV Deoxyribonucleic Acid (DNA) and Ribonucleic Acid (RNA)

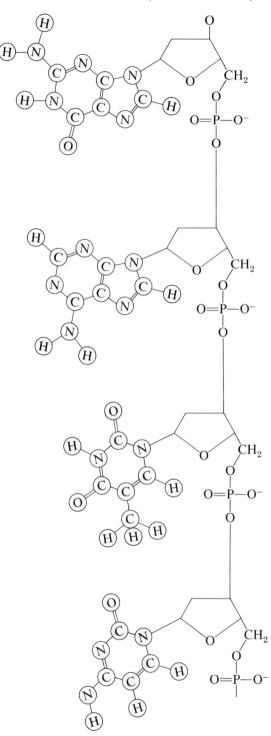

Figure 27-10 The chemical bonding in single-stranded DNA.

Ch. 27 Biochemistry

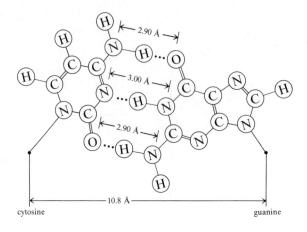

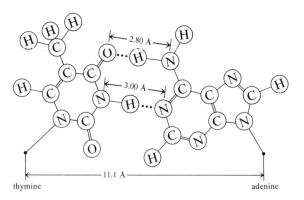

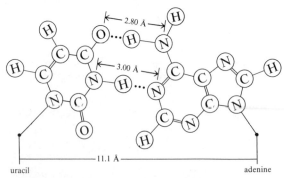

Figure 27-11 *Base pairing in DNA and RNA.*

644

Sec. IV Deoxyribonucleic Acid (DNA) and Ribonucleic Acid (RNA)

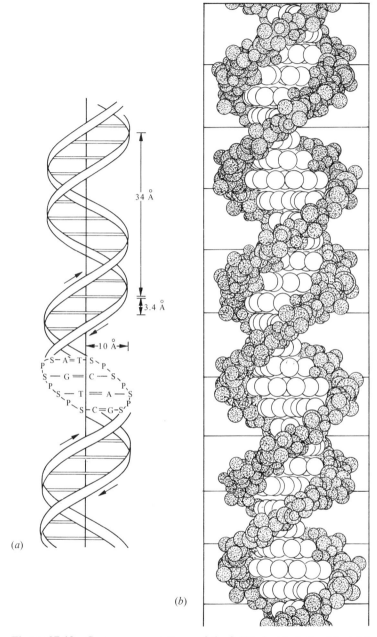

Figure 27-12 *Some representations of the DNA double helix. [Part (a) of this figure appeared first in Steiner, R. F., The Chemical Foundations of Molecular Biology, 1965. Used with permission from Van Nostrand Reinhold Co., New York. Part (b) is from Scientific American, 216, 1967. Used with permission from W. H. Freeman and Co., San Francisco, California.]*

no noticeable effect. This type of mutation can be seen in many enzymes.

On the other hand, any alteration of the base sequence can manifest itself in a startling manner. A striking example involves an inherited disorder called sickle cell anemia. This disease is common in some areas of the United States and in Equatorial Africa. Besides anemia, individuals with this disease suffer from acute attacks of abdominal pain, ulceration of the lower extremities, and pain in the joints. People with sickle cell anemia have an abnormal hemoglobin in their red blood cells. Hemoglobin is involved in the transport of oxygen from the lungs to the cells. It contains a protein that is made up of four chains of amino acids linked together. Each chain contains about 145 amino acids for a total of about 580 amino acid molecules.

The total amino acid composition of the protein in normal hemoglobin can vary to some extent without abnormal effects. For example, the hemoglobin present in a human fetus is different from that in an adult. This is a completely normal situation.

However, the hemoglobin in a person who has sickle cell anemia is the result of an "error" in the base sequence. Anemia results because of a change in the hemoglobin molecule. The hemoglobin present in these individuals differs from normal hemoglobin *by only one amino acid*. The amino acid valine has been substituted for glutamic acid. A change of only one base (thymine for adenine) out of a total of about 1740 bases in the DNA sequence for the hemoglobin gene results in severe anemia for hundreds of thousands of people.

This is a single example that illustrates the exceedingly fine line that differentiates between normal and abnormal function. The majority of individuals are normal in this respect, but a tiny change that might appear to be of no consequence can be very important. Even a tiny change can result in disaster because the chemistry of a cell at the molecular level is so closely regulated.

The mutation that causes sickle cell anemia is known as a point mutation. At most, point mutations involve three bases in the sequence. Group mutations can involve larger segments. For example, the deletion of a single base from the sequence throws off the whole sequence. The result may be a nonsense protein that has no function or, if it has a function, may be pathological.

Another type of group mutation might involve a gene that does not direct the production of a specific protein (a structural gene) but a gene that controls the activity of a whole sequence of genes. A gene of this type is called a regulator gene. Some regulator genes must be involved in the process of cell differentiation.

Sec. IV Deoxyribonucleic Acid (DNA) and Ribonucleic Acid (RNA)

QUESTIONS
1. List the properties desirable in an antibiotic.
2. Why are fluorine-containing compounds (5-fluorouracil, and so forth) worth testing as antibiotics or anticancer agents?
3. Suggest a mechanism for the reaction of *para*-aminobenzoic acid with aspartic acid.
4. Describe the structure of a protein.
5. Where will the insulin molecule be hydrolyzed by the enzyme trypsin?
6. How many molecules of ATP are required for the synthesis of the peptide bonds in insulin?
7. What is a mitochondrion? In what sense can it be called a chemical factory?
8. In coronary cases if the heart is not started again within 1 min, a base is added intravenously to the blood to maintain the pH of blood. What acid is present and where does it come from?
9. What is a nucleic acid?
10. Which of the amino acids have side chains that could provide the nonpolar environment for lipids in a cell wall or for the oxygen molecule in hemoglobin?
11. Can you suggest a function for the iron present in chloroplasts and mitochondria?
12. The ferrodoxins are iron-containing proteins with redox potentials of about -0.42 volt. Comment on the suggestion that they are primitive electron acceptors serving the same function as cytochromes but in an atmosphere without oxygen.

PROBLEMS I
1. What is the efficiency of energy transfer in the formation of ATP from the oxidation of glucose to carbon dioxide? (See the text for the thermodynamic data.)
2. What weight of ATP is formed in the fermentation of 20 g of sugar?
3. Given the G^0 values for glucose, -219.22; ethanol, -43.39; and CO_2, -92.31 kcal mole^{-1}; calculate ΔG^0 for the reaction $C_6H_{12}O_6 \longrightarrow 2CO_2 + 2CH_3CH_2OH$. What weight of ATP could be formed from 20 g of sugar if the energy transfer were 100% efficient?
4. Calculate the relative amounts of the various anions of phosphoenolpyruvic acid, (CH_2=C—COOH), $pK_1 = 1.4$, $pK_2 = 3.5$, $pK_3 = 6.4$,
   $\phantom{(CH_2\text{=C—COOH),}}|$
   $\phantom{(CH_2\text{=C—COOH),}}OPO_3H_2$
   present at a pH of 7.50.

PROBLEMS II
1. What is the common denominator of energy in biological systems?
2. How is ATP synthesized in animals?
3. What is the primary source of glucose in the diet of animals?
4. What is the source of the energy used for photosynthesis in plants?
5. How is ATP synthesis coupled to the oxidation of glucose?
6. What is a mitochondrion?

Ch. 27 Biochemistry

7. Write a chemical equation for the oxidation of glucose.
8. Calculate ΔH^0 and ΔG^0 for the reaction in Problem 7. For glucose $H^0 = -304.6$ and $G^0 = -217.6$ kcal mole^{-1}.
9. Why does the oxidation of glucose coupled to ATP synthesis occur more rapidly in a cell than the direct reaction of glucose with molecular oxygen?
10. Describe the structure of enzymes.
11. What kinds of bonds hold the amino acids in an enzyme in the proper position?
12. What determines the sequence of amino acids in a protein?
13. Briefly describe the structure of DNA.
14. What forces are involved in the pairing of adenine with thymine?
15. What forces are involved in the pairing of adenine with uracil?
16. List the various times the pairing of bases occurs in the course of protein synthesis.

APPENDIX I
Wave Functions for the Hydrogen Atom

n	l	m_l	$= [R_n^l(r)][\Theta_l^{m_l}(\theta)][\Phi^{m_l}(\varphi)]$
1	0	0	$1s = [2(Z/a_0)^{3/2}e^{-Zr/a_0}][(1/4\pi)^{1/2}][1]$
2	0	0	$2s = [(Z/2a_0)^{3/2}(2 - Zr/a_0)e^{-Zr/2a_0}][(1/4\pi)^{1/2}][1]$
2	1	0	$2p_z = [3^{-1/2}(Z/2a_0)^{3/2}(Zr/a_0)e^{-Zr/2a_0}][(3/4\pi)^{1/2}\cos\theta][1]$
2	1	± 1	$2p_x = [3^{-1/2}(Z/2a_0)^{3/2}(Zr/a_0)e^{-Zr/2a_0}][(3/4\pi)^{1/2}\sin\theta][\cos\varphi]$
			$2p_y = [3^{-1/2}(Z/2a_0)^{3/2}(Zr/a_0)e^{-Zr/2a_0}][(3/4\pi)^{1/2}\sin\theta][\sin\varphi]$
3	0	0	$3s = [(2/27)(Z/3a_0)^{3/2}(27 - 18Zr/a_0 + 2(Zr/a_0)^2)e^{-Zr/3a_0}][(1/4\pi)^{1/2}][1]$
3	1	0	$3p_z = [(1/81)3^{-1/2}(2Z/a_0)^{3/2}(6 - Zr/a_0)(Zr/a_0)e^{-Zr/3a_0}][(3/4\pi)^{1/2}\cos\theta][1]$
3	1	± 1	$3p_x = [(1/81)3^{-1/2}(2Z/a_0)^{3/2}(6 - Zr/a_0)(Zr/a_0)e^{-Zr/3a_0}][(3/4\pi)^{1/2}\sin\theta][\cos\varphi]$
			$3p_y = [(1/81)3^{-1/2}(2Z/a_0)^{3/2}(6 - Zr/a_0)(Zr/a_0)e^{-Zr/3a_0}][(3/4\pi)^{1/2}\sin\theta][\sin\varphi]$
3	2	0	$3d_{z^2} = [(1/81)15^{-1/2}(2Z/a_0)^{3/2}(Zr/a_0)^2 e^{-Zr/3a_0}][(5/16\pi)^{1/2}(3\cos^2\theta - 1)][1]$
3	2	± 1	$3d_{xz} = [(1/81)15^{-1/2}(2Z/a_0)^{3/2}(Zr/a_0)^2 e^{-Zr/3a_0}][(15/4\pi)^{1/2}\sin\theta\cos\theta][\cos\varphi]$
			$3d_{yz} = [(1/81)15^{-1/2}(2Z/a_0)^{3/2}(Zr/a_0)^2 e^{-Zr/3a_0}][(15/4\pi)^{1/2}\sin\theta\cos\theta][\sin\varphi]$
3	2	± 2	$3d_{xy} = [(1/81)15^{-1/2}(2Z/a_0)^{3/2}(Zr/a_0)^2 e^{-Zr/3a_0}][(15/16\pi)^{1/2}\sin^2\theta][\sin 2\varphi]$
			$3d_{x^2-y^2} = [(1/81)15^{-1/2}(2Z/a_0)^{3/2}(Zr/a_0)^2 e^{-Zr/3a_0}][(15/16\pi)^{1/2}\sin^2\theta][\cos 2\varphi]$

$a_0 = 0.529 \times 10^{-8}$ cm

APPENDIX II
Symbols and Abbreviations

Greek alphabet

α	Alpha	η	Eta	ν	Nu	υ	Upsilon
β	Beta	θ	Theta	ξ	Xi	φ	Phi
γ	Gamma	ι	Iota	π	Pi	χ	Chi
δ	Delta	κ	Kappa	ρ	Rho	ψ	Psi
ϵ	Epsilon	λ	Lambda	σ	Sigma	ω	Omega
ζ	Zeta	μ	Mu	τ	Tau		

Other symbols

Δ	Delta	$\leftarrow$	Is formed from
$<$	Less than	$\rightarrow$	Yields
$>$	Greater than	$\rightleftharpoons$	Formed from or yields
∞	Infinity	Å	Angstrom

Abbreviations

A	work function	bp	boiling point
Å	Angstrom	(c)	crystalline
amp	ampere	°C	degrees centigrade (or Celsius)
amu	atomic mass unit		
(aq)	aqueous	cal	calorie
atm	atmosphere	ccp	cubic close packed
bcc	body centered cubic	cm	centimeter
BDE	bond dissociation energy	cm^2	square centimeter
		cm^3	cubic centimeter

Abbreviations (cont.)

cos	cosine	ln	natural logarithm base e
d	density	log	logarithm to base 10
debye	Debye units	M	molar (moles liter^{-1})
dis sec^{-1}	disintegrations per second	μ	micron
		mμ	millimicron
DNA	deoxyribonucleic acid	mg	milligram
		min	minute
dyne	dyne	ml	milliliter
E	energy	mm	millimeter
EA	electron affinity	mm^2	square millimeter
emf	electromotive force	mp	melting point
erg	erg	MW	molecular weight
esu	electrostatic units	N_0	Avogadro's number
eu	entropy unit		
ev	electron volt	P	pressure
°F	degrees Fahrenheit	PA	polarizing ability
$\mathcal{F}$	Faraday	PE	potential energy
fcc	face centered cubic	pH	negative log [H$^+$]
		pK	negative log equilibrium constant
g	grams		
(g)	gas	ppm	parts per million
G	free energy		
gal	gallon	R	gas constant
		R	organic radical
H	enthalpy	r	radius
hcp	hexagonal close packed	RNA	ribonucleic acid
$h\nu$	light energy	(s)	solid
Hz	hertz (cycles per second)	S	entropy
in.	inches	SCSE	spin correlation stabilization energy
IP	ionization potential		
K	degrees Kelvin or Absolute	sec	second
K	various ionization constants	sin	sine
kcal	kilocalorie	STP	standard temperature and pressure
lb	pound		
LCAO	linear combination of atomic orbitals		

APPENDIX III
Fundamental Constants

Name	Symbol	Value
Avogadro's number	N_0	6.02257×10^{23} mole^{-1}
Gas constant	R	0.08205 liter atm mole^{-1} deg^{-1}
		1.98717 cal mole^{-1} deg^{-1}
Planck's constant	h	6.62554×10^{-27} erg sec
		9.5369×10^{-14} kcal mole^{-1} sec
Boltzmann constant	k	1.38053×10^{-16} erg deg^{-1}
Velocity of light	c	2.997925×10^{10} cm sec^{-1}
Charge on electron	e^-	4.80296×10^{-10} esu
		1.602095×10^{-19} coulomb
Faraday Constant	$\mathcal{F}$	96,500 coulombs mole^{-1}
		23.061 kcal mole^{-1} volt^{-1}
Atomic mass unit	amu	9.31×10^8 ev
		1.6604×10^{-24} g
		2.148×10^{10} kcal mole^{-1}
1000 mμ photon	—	10,000 cm^{-1}
		28.5912 kcal mole^{-1}
1 atm	—	1.013×10^6 dynes cm^{-2}
		760 mm Hg
Mass of an electron	—	9.10904×10^{-28} g
		5.486×10^{-4} amu
		1.1784×10^7 kcal mole^{-1}
Acceleration of gravity	—	980 cm sec^{-2}
ln 10	—	2.302585
$2.3026R$	—	4.57563
0°C	—	273.15 K
$298.15R/\mathcal{F}$	—	0.059157

APPENDIX IV
Thermodynamic Properties of Selected Materials*

Formula	State	H^0, kcal mole^{-1}	G^0, kcal mole^{-1}	S^0, eu	C_p^0, eu
Aluminum					
Al	(s)			6.77	5.82
Al	(g)	78.0	68.3	39.30	5.11
AlF$_3$	(s)	−359.5	−340.6	15.88	17.95
AlCl$_3$	(s)	−168.3	−150.3	26.45	21.95
Al$_2$O$_3$	(s)	−400.5	−378.2	12.17	18.89
Al$_2$S$_3$	(s)	−173	−117.7	23	
Al^{3+}	(aq)	−127	−116	−76.9	
AlO$_2^-$	(aq)	−219.6	−196.8	−5	
Antimony					
Sb	(s)			10.92	6.03
Sb	(g)	60.8	51.1	43.06	4.97
SbCl$_3$	(s)	−91.34	−77.37	44.0	25.8
SbCl$_5$	(l)	−105.2	−83.7	72.0	
Sb$_2$O$_4$	(s)	−216.9	−190.2	30.4	27.39
Sb$_2$O$_5$	(s)	−232.3	−198.2	29.9	28.1
Sb$_2$S$_3$	(s)	−41.8	−41.5	43.5	28.65
Sb(OH)$_3$	(s)		−163.8		
SbH$_3$	(g)	34.68	35.31	55.61	9.81
H$_3$SbO$_3$	(aq)	−184.9	−154.1	27.8	

*Most of the data in Appendix IV is taken from the Technical Notes 270-3 and 270-4 and Circular 500 (Washington, D.C.: National Bureau of Standards). However, the data from Circular 500 (1952) has been extended and revised in accordance with a variety of more recent compilations including the JANAF Thermochemical Tables (Joint Army Navy Air Force) (Midland, Mich.: Dow Chemical Co., 1965–1967; Springfield, Virginia: Clearinghouse for Federal Scientific and Technical Information, 1965–1967); Kubaschewski, O., E. Ll. Evans, and C. B. Alcock, *Metallurgical Thermodynamics*, (Oxford: Pergamon Press, 1967); and Brewer, Somayajulu, and Brackett, *Chem. Revs.*, **63**: 111 (1963). Estimated values from the various sources are shown in parentheses.

App. IV Thermodynamic Properties of Selected Materials

Formula	State	H^0, kcal mole^{-1}	G^0, kcal mole^{-1}	S^0, eu	C_p^0, eu
Arsenic					
As	(s)			8.4	5.89
As	(g)	60.64	50.74	41.62	4.97
AsF_3	(g)	−220.04	−216.46	69.07	15.68
$AsCl_3$	(g)	−61.80	−58.77	78.17	18.10
As_2O_5	(s)	−221.05	−187.0	25.2	27.85
As_2S_3	(s)	−40.4	−40.3	39.1	27.8
$AsCl_3$	(aq)	−72.9	−61.37	49.6	
AsO_2^-	(aq)	−102.54	−83.66	9.9	
AsO_4^{3-}	(aq)	−212.27	−155.00	−38.9	
$HAsO_2$	(aq)	−109.1	−96.25	30.1	
$HAsO_4^{2-}$	(aq)	−216.62	−170.82	−0.4	
$H_2AsO_3^-$	(aq)	−170.84	−140.35	26.4	
$H_2AsO_4^-$	(aq)	−217.39	−180.04	28	
H_3AsO_3	(aq)	−177.4	−152.94	46.6	
Barium					
Ba	(s)			16	6.30
Ba	(g)	41.96	34.60	40.70	4.97
BaF_2	(s)	−286.9	−274.5	23.0	17.02
$BaCl_2$	(s)	−205.56	−193.8	30	18.0
BaO	(s)	−133.4	−126.3	16.8	11.34
$Ba(OH)_2$	(s)	−226.2			
BaH_2	(s)	−40.9			
$BaCO_3$	(s)	−291.3	−272.2	26.8	20.40
$BaSO_4$	(s)	−350.2	−323.4	31.6	24.32
Ba^{2+}	(aq)	−128.67	−134.0	3	
Beryllium					
Be	(s)			2.28	3.93
Be	(g)	78.26	69.23	32.55	4.97
BeF_2	(s)	−242.3	−234.7	12.75	12.38
$BeCl_2$	(s)	−119	−107	18.12	14.92
BeO	(s)	−143.1	−136.1	3.38	6.10
$Be(OH)_2$	(s)	−216.6	(195)	(11)	(16)
Be_3N_2	(s)	−140.6	−127.3	8.16	15.38
$BeSO_4$	(s)	−287	−260	18.64	20.48
Be^{2+}	(aq)	−93			
BeO_2^{2-}	(aq)	−188			
Bismuth					
Bi	(s)			13.56	6.10
Bi	(g)	49.5	40.2	44.67	4.97
$BiCl_3$	(s)	−90.6	−75.3	42.3	25

App. IV Thermodynamic Properties of Selected Materials

Formula	State	H^0, kcal mole^{-1}	G^0, kcal mole^{-1}	S^0, eu	C_p^0, eu
Bismuth (cont.)					
$BiCl_3$	(g)	−63.5	−61.2	85.74	19.04
Bi_2O_3	(s)	−137.16	−118.0	36.2	27.13
BiS	(g)	43	29	68	
Bi_2S_3	(s)	−34.2	−33.6	47.9	29.2
Bi^{3+}	(aq)		19.8		
Boron					
B	(s)			1.40	2.65
B	(g)	134.5	124.0	36.65	4.971
BF	(g)	−29.2	−35.8	47.89	7.07
BF_3	(g)	−271.75	−267.77	60.71	12.06
BCl_3	(l)	−102.1	−92.6	49.3	25.5
BCl_3	(g)	−96.50	−92.91	69.31	14.99
B_2Cl_4	(l)	−125.0	−111.1	62.7	32.9
B_2Cl_4	(g)	−117.2	−110.0	85.4	22.80
BO	(g)	6	−1	48.62	6.98
BO_2	(g)	−71.8	−73.1	54.84	10.28
B_2O_2	(g)	−108.7	−110.5	57.93	13.69
B_2O_3	(s)	−304.20	−285.30	12.90	15.04
BS	(g)	81.74	69.02	51.65	7.18
H_3BO_3	(aq)	−256.29	−231.56	38.8	
BO_2^-	(aq)	−184.60	−162.27	−8.9	
BH_4^-	(aq)	11.51	27.31	26.4	
Bromine					
Br_2	(l)			36.384	18.090
Br_2	(g)	7.387	0.751	58.641	8.61
Br	(g)	26.741	19.701	41.805	4.97
BrF	(g)	−22.43	−26.09	54.70	7.88
BrF_3	(g)	−61.09	−54.84	69.89	15.92
BrF_3	(l)	−71.9	−57.5	42.6	29.78
BrF_5	(g)	−102.5	−83.8	76.50	23.81
BrF_5	(l)	−109.6	−84.1	53.8	
BrCl	(g)	3.50	−0.23	57.36	8.36
BrO	(g)	30.06	25.87	56.75	7.67
HBr	(g)	−8.70	−12.77	47.463	6.97
Br_2	(aq)	−0.62	0.62	31.2	
Br^-	(aq)	−29.05	−24.85	19.7	−33.9
BrO^-	(aq)	−22.5	−8.0	10	
BrO_3^-	(aq)	−20.0	0.4	39.0	
HBrO	(aq)	−27.7	−19.7	34	

App. IV Thermodynamic Properties of Selected Materials

Formula	State	H^0, kcal mole^{-1}	G^0, kcal mole^{-1}	S^0, eu	C_p^0, eu
Cadmium					
Cd	(s)			12.37	6.21
Cd	(s)(α)	−0.14	−0.14	12.37	
Cd	(g)	26.77	18.51	40.07	4.97
CdF_2	(s)	−167.4	−154.8	18.5	
$CdCl_2$	(s)	−93.57	−82.21	27.55	17.85
CdO	(s)	−61.7	−54.6	13.1	10.38
CdS	(s)	−38.7	−37.4	15.5	
$Cd(OH)_2$	(s)	−134.0	−113.2	23	
Cd^{2+}	(aq)	−18.14	−18.542	−17.5	
Calcium					
Ca	(s)			9.95	6.28
Ca	(g)	46.04	37.98	36.993	4.968
CaF_2	(s)	−290.3	−277.7	16.46	16.02
$CaCl_2$	(s)	−190.0	−179.3	27.2	17.36
CaI_2	(s)	−127.8	−126.6	34	
CaO	(s)	−151.9	−144.4	9.5	10.23
CaO_2	(s)	−157.5			
CaS	(s)	−115.3	−114.1	13.5	11.33
$Ca(OH)_2$	(s)	−235.80	−214.33	18.2	20.2
CaH_2	(s)	−45.1	−35.8	10	
$CaSO_4$	(s)	−342.42	−315.56	25.5	23.8
$Ca(NO_3)_2$	(s)	−224.00	−177.34	46.2	35.69
$CaCO_3$	(s)	−288.45	−269.78	22.2	19.57
Ca^{2+}	(aq)	−129.77	−132.18	−13.2	
Carbon					
C	(s)(graphite)			1.372	2.038
C	(s)(diamond)	0.4533	0.6930	0.568	1.4615
C	(g)	171.291	160.442	37.7597	4.9805
CF_4	(g)	−221	−210	62.50	14.60
CCl_4	(l)	−32.37	−15.60	51.72	31.49
CCl_4	(g)	−24.6	−14.49	74.03	19.91
CO	(g)	−26.416	−32.780	47.219	6.959
CO_2	(g)	−94.051	−94.254	51.06	8.87
CS	(g)	56	44	50.30	7.12
CS_2	(l)	21.44	15.60	36.17	18.1
CS_2	(g)	28.05	16.05	58.82	10.85
CH_4	(g)	−17.88	−12.13	44.49	8.439
HCN	(g)	32.3	29.8	48.20	8.57
HCN	(aq)	25.7	28.6	29.8	
CO	(aq)	−28.91	−28.66	25.0	
CO_2	(aq)	−98.90	−92.26	28.1	

App. IV Thermodynamic Properties of Selected Materials

Formula	State	H^0, kcal mole^{-1}	G^0, kcal mole^{-1}	S^0, eu	C_p^0, eu
Carbon (cont.)					
CO_3^{2-}	(aq)	−161.84	−126.17	−13.6	
CH_4	(aq)	−21.28	−8.22	20.0	
HCO_3^-	(aq)	−165.39	−140.26	21.8	
H_2CO_3	(aq)	−167.22	−148.94	44.8	
CN^-	(aq)	36.0	41.2	22.5	
$C_2O_4^{2-}$	(aq)	−197.2	−161.1	10.9	
CH_3COO^-	(aq)	−116.16	−88.29	20.7	−1.5
CH_3COOH	(aq)	−116.10	−94.78	42.7	
Cesium					
Cs	(s)			19.8	7.42
Cs	(g)	18.83	12.24	41.944	4.968
CsF	(s)	−126.9			
CsCl	(s)	−103.5			
CsBr	(s)	−94.3	−91.6	29	12.4
CsI	(s)	−80.5	−79.7	31	12.4
Cs_2O	(s)	−75.9			
CsOH	(s)	−97.2			
Cs^+	(aq)	−59.2	−67.41	31.8	
Chlorine					
Cl_2	(g)			53.288	8.104
Cl	(g)	29.082	25.262	39.457	5.220
ClF	(g)	−13.02	−13.37	52.05	7.66
ClF_3	(g)	−39.0	−29.4	67.28	15.26
ClO	(g)	24.34	23.45	54.14	7.52
ClO_2	(g)	24.5	28.8	61.36	10.03
Cl_2O	(g)	19.2	23.4	63.60	10.85
HCl	(g)	−22.062	−22.777	44.646	6.96
Cl^-	(aq)	−39.952	−31.372	13.5	−32.6
Cl_2	(aq)	−5.6	1.65	29	
ClO_4^-	(aq)	−30.91	−2.06	43.5	
Chromium					
Cr	(s)			5.68	5.58
Cr	(g)	94.8	84.1	41.68	4.97
CrF_3	(s)	−277	−260	22.44	18.82
$CrCl_2$	(s)	−94.5	−85.1	27.56	17.01
$CrCl_3$	(s)	−133.0	−116.2	29.4	21.94
Cr_2O_3	(s)	−272.4	−252.9	19.4	28.38
CrO_2Cl_2	(l)	−138.5	−122.1	53.0	
Ag_2CrO_4	(s)	−174.89	−153.40	52.0	34.00
CrO_4^{2-}	(aq)	−210.60	−173.96	12.00	

App. IV Thermodynamic Properties of Selected Materials

Formula	State	H^0, kcal mole^{-1}	G^0, kcal mole^{-1}	S^0, eu	C_p^0, eu
Chromium (cont.)					
$Cr_2O_7^{2-}$	(aq)	−356.2	−311.0	62.6	
$HCrO_4^-$	(aq)	−209.9	−182.8	44.0	
Cobalt					
Co	(s)			7.18	5.93
Co	(g)	101.5	90.9	42.88	5.502
CoF_2	(s)	−165.4	−154.7	19.59	16.44
$CoCl_2$	(s)	−74.7	−64.5	26.09	18.8
CoO	(s)	−56.87	−51.20	12.66	13.20
Co_3O_4	(s)	−213	−185	24.5	29.5
$Co(OH)_2$	(s)(pink)	−129	−109.5	16	
$CoSO_4$	(s)	−212.3	−187.0	28.2	
$[Co(NH_3)_5Cl]Cl_2$	(s)	−243.1	−139.3	87.5	57.2
Co^{2+}	(aq)	−13.9	−13.0	−27	
Co^{3+}	(aq)	22	32	−73	
$Co(NH_3)_6^{3+}$	(aq)	−139.8	−38.9	40	
$Co(NH_3)_5Cl^{2+}$	(aq)	−150.1	−69.8	81.6	
Copper					
Cu	(s)			7.92	5.84
Cu	(g)	80.86	71.37	39.744	4.968
CuCl	(s)	−32.8	−28.6	20.6	11.6
$CuCl_2$	(s)	−52.6	−42.0	25.83	13.82
CuI	(s)	−16.2	−16.6	23.1	12.92
CuO	(s)	−37.6	−31.0	10.19	10.11
Cu_2O	(s)	−40.3	−34.9	22.3	15.2
CuS	(s)	−12.7	−12.8	15.9	11.43
Cu_2S	(s)	−19.0	−20.6	28.9	18.24
$CuSO_4$	(s)	−184.36	−158.2	26	23.9
$CuSO_4 \cdot 5H_2O$	(s)	−544.85	−499.34	71.8	67
Cu^+	(aq)	17.1	12.0	9.7	
Cu^{2+}	(aq)	15.48	15.66	−23.8	
$Cu(NH_3)_4^{2+}$	(aq)	−83.3	−26.60	65.4	
Fluorine					
F_2	(g)			48.44	7.48
F	(g)	18.88	14.80	37.917	5.436
F_2O	(g)	−5.2	−1.1	59.11	10.35
HF	(g)	−64.8	−65.3	41.508	6.963
F^-	(aq)	−79.50	−66.64	−3.3	−25.5
HF	(aq)	−76.50	−70.95	21.2	

App. IV   Thermodynamic Properties of Selected Materials

Formula	State	H^0, kcal mole^{-1}	G^0, kcal mole^{-1}	S^0, eu	C_p^0, eu
Hydrogen					
H_2	(g)			31.208	6.889
H	(g)	52.095	48.581	27.391	4.9679
H_2O	(g)	−57.796	−54.634	45.104	8.025
H_2O	(l)	−68.315	−56.687	16.71	17.995
H_2O_2	(g)	−32.58	−25.24	55.6	10.3
H_2O_2	(l)	−44.83	−28.78	26.2	21.3
OH	(g)	9.31	8.18	43.89	7.14
H_2O_2	(aq)	−45.69	−32.05	34.4	
OH^-	(aq)	−54.970	−37.594	−2.57	−35.5
H^+	(aq)	0.000	0.000	0.000	
H_3O^+	(aq)	−68.315	−56.687	16.71	
HO_2^-	(aq)	−38.32	−16.1	5.7	
Iodine					
I_2	(s)			27.757	
I_2	(g)	14.923	4.627	62.28	8.82
I	(g)	25.535	16.798	43.184	4.968
IF	(g)	−22.86	−28.32	56.42	7.99
IF_7	(g)	−225.6	−195.6	82.8	32.6
ICl	(l)	−5.71	−3.25	32.3	
ICl	(g)	4.25	−1.30	59.14	8.50
ICl_3	(s)	−21.4	−5.34	40.0	
IO	(g)	41.84	35.80	58.65	7.86
HI	(g)	6.33	0.41	49.351	6.969
I^-	(aq)	−13.19	−12.33	26.6	−34.0
I_2	(aq)	5.4	3.92	32.8	
Iron					
Fe	(s)			6.52	6.00
Fe	(g)	99.5	88.6	43.11	6.14
$FeCl_2$	(s)	−81.69	−72.26	28.19	18.32
$FeCl_3$	(s)	−95.48	−79.84	34.0	23.10
Fe_2O_3	(s)	−197.0	−177.4	20.89	24.82
Fe_3O_4	(s)	−267.3	−242.7	35.0	34.28
FeS	(s)	−23.9	−24.0	14.41	12.08
FeS_2	(s)	−42.6	−39.9	12.65	14.86
$Fe(OH)_2$	(s)	−136.0	−116.3	21	
$Fe(OH)_3$	(s)	−196.7	−166.5	25.5	
Fe_3C	(s)	6.0	4.8	25.0	25.3
$Fe(CO)_5$	(g)	−175.4	−166.65	106.4	
Fe_2SiO_4	(s)	−353.7	−329.6	34.7	31.76
$FeAl_2O_4$	(s)	−470	−442	25.4	29.5
Fe^{2+}	(aq)	−21.3	−18.85	−32.9	

App. IV Thermodynamic Properties of Selected Materials

Formula	State	H^0, kcal mole^{-1}	G^0, kcal mole^{-1}	S^0, eu	C_p^0, eu
Iron (cont.)					
Fe^{3+}	(aq)	−11.6	−1.1	−75.5	
$FeBr^{2+}$	(aq)	−34.6	−26.8	−33	
$FeNO^{2+}$	(aq)	−2.2	19.8	−51	
$FeOH^{2+}$	(aq)	−69.5	−54.83	−34	
$Fe(CN)_6^{3-}$	(aq)	134.3	174.3	64.6	
$Fe(CN)_6^{4-}$	(aq)	108.9	166.09	22.7	
Lead					
Pb	(s)			15.49	6.32
Pb	(g)	46.6	38.7	41.889	4.968
PbF_2	(s)	−158.7	−147.5	26.4	
$PbCl_2$	(s)	−85.90	−75.08	32.5	
PbO	(s)(yellow)	−51.94	−44.91	16.42	10.94
PbO	(s)(red)	−52.34	−45.16	15.9	10.95
PbO_2	(s)	−66.3	−51.95	16.4	15.45
Pb_2O_3	(s)			36.3	25.74
Pb_3O_4	(s)	−171.7	−143.7	50.5	35.1
PbS	(s)	−24.0	−23.6	21.8	11.83
$PbSO_4$	(s)	−219.87	−194.36	35.51	24.667
Pb^{2+}	(aq)	−0.4	−5.83	2.5	
Lithium					
Li	(s)			6.70	5.65
Li	(g)	38.4	30.6	33.143	4.968
LiF	(s)	−146.3	−139.6	8.57	10.04
LiCl	(s)	−97.58	−91.79	14.17	11.48
LiCl	(g)	−53	−58	51.01	7.88
LiBr	(s)	−83.9	(−81.6)	(18)	(11.7)
LiI	(s)	−64.6	(−64.5)	(20.5)	(11.97)
Li_2O	(s)	−143.1	−134.3	9.056	12.93
LiOH	(s)	−115.8	−104.8	10.23	10.22
LiH	(g)	30.7	25.2	40.77	7.06
LiH	(s)	−21.67	−16.37	4.79	6.69
Li_2CO_3	(s)	−290.6	−270.6	21.55	23.0
Li_3N	(s)	−47	−37	9	18.48
Li^+	(aq)	−66.554	−70.22	3.4	
Magnesium					
Mg	(s)			7.77	5.71
Mg	(g)	35.3	27.0	35.504	4.968
MgF	(g)	−20	−27	54.85	7.82
MgF_2	(s)	−268.7	−256.0	13.68	14.72
$MgCl_2$	(s)	−153.4	−141.6	21.42	17.06
$MgBr_2$	(s)	−124	−119	28.50	17.52

App. IV Thermodynamic Properties of Selected Materials

Formula	State	H^0, kcal mole^{-1}	G^0, kcal mole^{-1}	S^0, eu	C_p^0, eu
Magnesium (cont.)					
MgI_2	(s)	−86.0	(−84.4)	(30)	
MgO	(s)	−143.7	−136.0	6.440	8.88
$Mg(OH)_2$	(s)	−221.00	−199.25	15.10	18.41
MgS	(s)	−83	(−82)	(11)	
MgH_2	(s)	−18.2	−8.8	7.43	8.45
MgH	(g)	41	34	47.61	7.050
$MgCO_3$	(s)	−266	−246	15.7	18.21
$MgSO_4$	(s)	−305.5	−278.2	21.9	23.02
Mg_3N_2	(s)	−110	(−96)	(21)	24.98
Mg_2SiO_4	(s)	−520	−492	22.73	28.24
$MgSiO_3$	(s)	−370	−349	16.19	19.55
Mg^{2+}	(aq)	−110.41	−108.99	−28.2	
Manganese					
Mn	(s)			7.65	6.29
Mn	(s)(γ)	0.37	0.34	7.75	6.59
Mn	(g)	67.1	57.0	41.49	4.968
MnF_2	(s)	−189	−179	22.05	15.96
$MnCl_2$	(s)	−115.03	−105.29	28.26	17.43
MnO	(s)	−92.07	−86.74	14.27	10.86
MnO_2	(s)	−124.29	−111.18	12.68	12.94
Mn_3O_4	(s)	−331.7	−306.7	37.2	33.38
MnS	(s)	−51.2	−52.2	18.7	11.94
$Mn(OH)_2$	(s)	−166.2	−147.0	23.7	
$MnCO_3$	(s)	−213.7	−195.2	20.5	19.48
$MnSO_4$	(s)	−254.60	−228.83	26.8	24.02
Mn^{2+}	(aq)	−52.76	−54.5	−17.6	12
$Mn(OH)^+$	(aq)	−107.7	−96.8	−4	
MnO_4^-	(aq)	−129.4	−106.9	45.7	
MnO_4^{2-}	(aq)	−156	−119.7	14	
Mercury					
Hg	(l)			18.17	6.69
Hg	(g)	14.65	7.61	41.79	4.968
Hg_2Cl_2	(s)	−63.39	−50.377	46.0	24.3
$HgCl_2$	(s)	−53.6	−42.7	34.9	
$HgBr_2$	(s)	−40.8	−36.6	41	
Hg_2Br_2	(s)	−49.45	−43.28	52	
HgI_2	(s)	−25.2	−24.3	43	
Hg_2I_2	(s)	−29.00	−26.53	55.8	
HgO	(s)(red)	−21.7	−13.99	16.8	10.53
HgO	(s)(yellow)	−21.62	−13.96	17.0	
HgS	(s)(red)	−13.90	−12.1	19.7	

App. IV Thermodynamic Properties of Selected Materials

Formula	State	H^0, kcal mole^{-1}	G^0, kcal mole^{-1}	S^0, eu	C_p^0, eu
Mercury (cont.)					
HgS	(s)	−12.8	−11.4	21.1	
HgH	(g)	57.2	51.6	52.46	7.16
Hg_2SO_4	(s)	−177.61	−149.59	47.96	31.44
Hg^{2+}	(aq)	40.9	39.30	−7.7	
Hg_2^{2+}	(aq)	41.2	36.70	20.2	
$Hg(CN)_4^{2-}$	(aq)	125.8	147.8	71	
Nickel					
Ni	(s)			7.14	6.23
Ni	(g)	102.7	91.9	43.519	5.583
NiF_2	(s)	−155.7	−144.4	17.59	15.31
$NiCl_2$	(s)	−72.98	−61.92	23.34	17.13
NiO	(s)	−57.3	−50.6	9.08	10.59
NiS	(s)	−19.6	−19.6	12.66	11.26
$Ni(OH)_2$	(s)	−126.6	−106.9	21	
$NiSO_4$	(s)	−208.63	−181.6	22	33
$Ni(CO)_4$	(g)	−144.10	−140.36	98.1	34.70
Ni^{2+}	(aq)	−12.9	−10.9	−30.8	
$Ni(NH_3)_6^{2+}$	(aq)	−150.6	−61.2	94.3	
$Ni(CN)_4^{2-}$	(aq)	87.9	112.8	52	
Nitrogen					
N_2	(g)			45.77	6.961
N	(g)	112.979	108.886	36.613	4.968
NF_2	(g)	10.3	13.8	59.71	9.80
NF_3	(g)	−29.8	−19.9	62.29	12.7
N_2F_4	(g)	−1.7	19.4	71.96	18.9
NO	(g)	21.57	20.69	50.347	7.133
NO_2	(g)	7.93	12.26	57.35	8.89
N_2O	(g)	19.61	24.90	52.52	9.19
N_2O_3	(g)	20.01	33.32	74.61	15.68
N_2O_4	(l)	−4.66	23.29	50.0	34.1
N_2O_4	(g)	2.19	23.38	72.70	18.47
NH_3	(g)	−11.02	−3.94	45.97	8.38
N_3^-	(aq)	65.76	83.2	25.8	
NO_2^-	(aq)	−25.0	−8.9	33.5	−23.3
NO_3^-	(aq)	−49.56	−26.61	35.0	−20.7
NH_3	(aq)	−19.19	−6.35	26.6	
NH_4^+	(aq)	−31.67	−18.97	27.1	19.1
HNO_2	(aq)	−28.5	−13.3	36.5	
Oxygen					
O_2	(g)			49.003	7.016
O_3	(g)	34.1	39.0	57.08	9.37

App. IV Thermodynamic Properties of Selected Materials

Formula	State	H^0, kcal mole^{-1}	G^0, kcal mole^{-1}	S^0, eu	C_p^0, eu
Oxygen (cont.)					
O	(g)	59.553	55.389	38.467	5.237
O_2	(aq)	−2.8	3.9	26.5	
H_2O	(g)	−57.796	−54.634	45.104	8.025
H_2O	(l)	−68.315	−56.687	16.71	17.995
H_2O_2	(g)	−32.58	−25.24	55.6	10.3
H_2O_2	(l)	−44.83	−28.78	26.2	21.3
OH	(g)	9.31	8.18	43.89	7.14
H_2O_2	(aq)	−45.69	−32.05	34.4	
OH$^-$	(aq)	−54.970	−37.594	−2.57	−35.5
H_3O^+	(aq)	−68.315	−56.687	16.71	
HO_2^-	(aq)	−38.32	−16.1	5.7	
Phosphorus					
P_4	(s)			39.28	22.79
P	(s)	−4.2	−2.9	5.45	5.07
P	(g)	75.20	66.51	38.978	4.968
P_2	(g)	34.5	24.8	52.108	7.66
P_4	(g)	14.08	5.85	66.89	16.05
PF_3	(g)	−219.6	−214.5	65.28	14.03
PCl_3	(l)	−76.4	−65.1	51.9	
PCl_3	(g)	−68.6	−64.0	74.49	17.17
PCl_5	(g)	−89.6	−72.9	87.11	26.96
P_4O_{10}	(s)	−713.2	−644.8	54.70	50.60
PH_3	(g)	1.3	3.2	50.22	8.87
PH_3	(aq)	−2.16	0.35	48.2	
PO_4^{3-}	(aq)	−305.3	−243.5	−53	
HPO_4^{2-}	(aq)	−308.83	−260.34	−8.0	
$H_2PO_4^-$	(aq)	−309.82	−270.17	21.6	
H_3PO_4	(aq)	−307.92	−273.10	37.8	
Potassium					
K	(s)			15.34	7.07
K	(g)	21.31	14.5	38.296	4.968
KF	(s)	−135.6	−128.5	15.92	11.79
KCl	(s)	−104.37	−97.71	19.73	12.26
KCl	(g)	−51.6	−56.2	57.24	8.66
KBr	(s)	−94.1	−90.9	22.93	12.52
KI	(s)	−78.31	−77.46	26.48	12.60
K_2O	(s)	−86.8			
KOH	(s)	−101.5	−90.6	18.95	15.75
KII	(s)	−13.82	(−8)	(12)	
KH	(g)	30.0	25.1	47.3	
KNO_3	(s)	−117.76	−93.96	31.77	23.01

App. IV Thermodynamic Properties of Selected Materials

Formula	State	H^0, kcal mole^{-1}	G^0, kcal mole^{-1}	S^0, eu	C_p^0, eu
Potassium (cont.)					
K_2SO_4	(s)	−342.66	−314.62	42.0	31.1
K_2CO_3	(s)	−274.9	−254.4	37.17	27.35
K^+	(aq)	−60.32	−67.70	24.5	5.2
Rubidium					
Rb	(s)			16.6	7.27
Rb	(g)	20.51	13.35	40.628	4.968
RbF	(s)	−131.28			12.2
RbCl	(s)	−102.91			12.3
RbBr	(s)	−93.03	−90.38	25.88	12.68
RbI	(s)	−78.5	−77.8	28.21	12.50
Rb_2O	(s)	−78.9			
Rb_2S	(s)	−83.2			
RbOH	(s)	−98.9			
RbH	(s)				
Rb^+	(aq)	−58.9	−67.45	29.7	
Silicon					
Si	(s)			4.50	4.78
Si	(g)	108.9	98.3	40.12	5.318
Si_2	(g)	142	128	54.92	8.22
SiF	(g)	1.7	−5.8	53.94	7.80
SiF_2	(g)	−148	−150	60.38	10.49
SiF_4	(g)	−385.98	−375.88	67.49	17.60
$SiCl_2$	(g)	−39.59	−42.35	67.0	12.16
$SiCl_4$	(l)	−164.2	−148.16	57.3	34.17
$SiCl_4$	(g)	−157.03	−147.47	79.02	21.57
SiO	(g)	−23.8	−30.2	50.55	7.15
SiO_2	(s)(quartz)	−217.72	−204.75	10.00	10.62
SiO_2	(s)(cristobalite)	−217.37	−204.56	10.20	10.56
SiO_2	(s)(tridymite)	−217.27	−204.42	10.4	10.66
SiS	(g)	26.88	14.56	53.43	7.71
SiH_4	(g)	8.2	13.6	48.88	10.24
Si_2H_6	(g)	19.2	30.4	65.14	19.31
H_2SiO_3	(aq)	−282.7	−258.0	26	
H_4SiO_4	(aq)	−351.0	−314.7	43	
Silver					
Ag	(s)			10.17	6.06
Ag	(g)	68.01	58.72	41.32	4.968
AgF	(s)	−48.9	−44.6	20	
AgCl	(s)	−30.370	−26.244	23.0	12.14

App. IV Thermodynamic Properties of Selected Materials

Formula	State	H^0, kcal mole^{-1}	G^0, kcal mole^{-1}	S^0, eu	C_p^0, eu
Silver (cont.)					
AgBr	(s)	−23.99	−23.16	25.6	12.52
AgI	(s)	−14.78	−15.82	27.6	13.58
Ag$_2$O	(s)	−7.42	−2.68	29.0	15.74
Ag$_2$O$_2$	(s)	−5.8	6.6	28	21
Ag$_2$S	(s)	−7.79	−9.72	34.42	18.29
AgNO$_3$	(s)	−29.73	−8.00	33.68	22.24
Ag$^+$	(aq)	25.234	18.433	17.37	5.2
AgCl$_2^-$	(aq)	−58.6	−51.5	55.3	
Ag(NH$_3$)$_2^+$	(aq)	−26.60	−4.12	58.6	
Ag(CN)$_2^-$	(aq)	64.6	73.0	46	
Sodium					
Na	(s)			12.24	6.75
Na	(g)	25.75	18.48	36.715	4.968
NaF	(s)	−137.1	−129.9	12.26	11.20
NaCl	(s)	−98.26	−91.788	17.24	12.07
NaBr	(s)	−81.11	−79.45	24.94	14.9
NaI	(s)	−68.8	−68.0	23.5	12.48
Na$_2$O	(s)	−99.4	−90.1	17.99	17.44
Na$_2$O$_2$	(s)	−120.6			
NaO$_2$	(s)	−61.9			
Na$_2$S	(s)	−89	−86	23.4	
NaOH	(s)	−102.24	−91.19	15.34	14.26
NaH	(s)	−13.49	−8.02	9.56	8.70
NaH	(g)	29.88	24.78	44.93	7.002
Na$_2$CO$_3$	(s)	−270.26	−250.50	33.17	26.53
Na$_2$SO$_4$	(s)	−331.6	−303.4	35.76	30.55
Na$_2$SiO$_3$	(s)	−372.1	−349.6	27.21	26.90
NaNO$_3$	(s)	−111.54	−87.45	27.8	22.24
Na$^+$	(aq)	−57.39	−62.593	14.1	11.1
Strontium					
Sr	(s)			13.0	6.0
Sr	(g)	39.2	26.3	39.25	4.9680
SrF$_2$	(s)	−290.3	(−277.8)	(19.5)	
SrCl$_2$	(s)	−198.0	−186.7	28	18.9
SrI$_2$	(s)	−135.5	(−134.7)	(38.0)	19.5
SrO	(s)	−141.1	−133.8	13.0	10.76
Sr(OH)$_2$	(s)	−229.3			
SrH	(g)	52.4	45.8	49.43	7.179
SrH$_2$	(s)	−42.3			
SrCO$_3$	(s)	−291.2	−271.9	23.2	19.46
Sr^{2+}	(aq)	−130.38	−133.2	−9.4	

App. IV Thermodynamic Properties of Selected Materials

Formula	State	H^0, kcal mole^{-1}	G^0, kcal mole^{-1}	S^0, eu	C_p^0, eu
Sulfur					
S_8	(s)			60.8	43.3
S	(g)	66.636	56.951	40.084	5.658
S_2	(g)	30.68	18.96	54.51	7.76
S_8	(g)	24.45	11.87	102.98	37.39
SF_4	(g)	−185.2	−174.8	69.77	17.45
SF_6	(g)	−289	−264.2	69.72	23.25
S_2Cl_2	(g)	−4.4	−7.6	79.2	17.6
SO	(g)	1.496	−4.741	53.02	7.21
SO_2	(g)	−70.994	−71.748	59.30	9.53
SO_2	(l)	−76.6			
SO_3	(s)	−108.63	−88.19	12.5	
SO_3	(l)	−105.41	−88.04	22.85	
SO_3	(g)	−94.58	−88.69	61.34	12.11
HS	(g)	34.10	27.08	46.74	7.72
H_2S	(g)	−4.93	−8.02	49.16	8.18
S^{2-}	(aq)	7.9	20.5	−3.5	
SO_2	(aq)	−77.194	−71.871	38.7	
SO_4^{2-}	(aq)	−217.32	−177.97	4.8	−70
HS^-	(aq)	−4.2	2.88	15.0	
H_2S	(aq)	−9.5	−6.66	29	
HSO_4^-	(aq)	−212.08	−180.69	31.5	−20
Tin					
Sn	(s)			12.32	6.45
Sn	(s)(II)	−0.50	0.03	10.55	6.16
Sn	(g)	72.2	63.9	40.243	5.081
$SnCl_4$	(l)				
$SnCl_4$	(g)	−112.7	−103.3	87.4	23.5
SnO	(s)	−68.3	−61.4	13.5	10.59
SnO	(g)			55.45	7.55
SnO_2	(s)	−138.8	−124.2	12.5	12.57
$Sn(OH)_2$	(s)	−134.1	−117.5	37	
Sn^{2+}	(aq)	−2.1	−6.5	−4	
Sn^{4+}	(aq)	7.3	0.6	−28	
$SnOH^+$	(aq)	−68.4	−60.9	12	
Titanium					
Ti	(s)			7.325	5.987
Ti	(g)	112.4	101	43.066	5.839
TiF_3	(s)	−337.5	−321.0	24.5	21.99
TiF_4	(s)	−394.19	−372.66	32.02	27.31
$TiCl_4$	(g)			84.4	22.88
$TiCl_4$	(l)	−179.3	−161.2	60.4	37.5

App. IV Thermodynamic Properties of Selected Materials

Formula	State	H^0, kcal mole^{-1}	G^0, kcal mole^{-1}	S^0, eu	C_p^0, eu
Titanium (cont.)					
TiI_2	(s)	−64	−64	35	
TiI_4	(s)	−92	−91	59	30.03
TiO	(s)	−124.2	−118.3	12.2	9.55
TiO_2	(s)(rutile)	−225.8	−212.6	12.03	13.15
TiO_2	(s)(anatase)	−223	−210	11.93	13.26
Ti_2O_3	(s)	−363	−342	18.83	23.27
Ti_3O_5	(s)	−584	−550	30.92	37.00
Vanadium					
V	(s)			7.05	5.85
V	(g)	120	109	43.546	6.2166
VCl_2	(s)	−108	−97	23.2	17.26
VCl_3	(s)	−137	−120	31.3	22.27
V_2O_3	(s)	−290	−271	23.58	24.83
V_2O_4	(s)	−344	−318	24.65	28.30
V_2O_5	(s)	−373	−344	31.3	31.00
Zinc					
Zn	(s)			9.95	6.07
Zn	(g)	31.245	22.748	38.450	4.968
ZnF_2	(s)	−182.7	−170.5	17.61	15.69
$ZnCl_2$	(s)	−99.20	−88.296	26.64	17.05
ZnO	(s)	−83.24	−76.08	10.43	9.62
ZnS	(s)	−49.23	48.11	13.8	11.0
$Zn(OH)_2$	(s)	−153.74	−132.68	19.5	17.3
Zn^{2+}	(aq)	−36.78	−35.14	−26.8	11

APPENDIX V
Nomenclature in Organic Chemistry

In this appendix the systematic rules that have been developed for naming the compounds usually classified under the broad field of organic chemistry will be introduced.

The basic rules considered in this appendix are those approved by International Union of Pure and Applied Chemistry (IUPAC). These are the result of periodic revision of the first international proposals (Geneva, 1892) on the nomenclature of organic substances. Note that a number of trivial names (propane) have been retained or included as a part of systematic names (2-chloro-3-propylhexane; 1,4-benzene-dicarboxylic acid).

I. In general, organic compounds are named as derivatives of hydrocarbons. Thus we must consider first the naming of hydrocarbons.
   A. Saturated hydrocarbons.
      1. The trivial names methane, ethane, propane, and butane are used for the first four saturated straight chain hydrocarbons. The names of the higher members of this series consist of a numerical prefix and the termination "ane."
      2. A univalent radical that is obtained by removing a hydrogen atom from the end carbon atom of a saturated straight chain hydrocarbon is named by replacing the ending "ane" by "yl." For example, $CH_3-CH_2-CH_2-$ is propyl.
      3. Branched saturated hydrocarbons are named by prefixing the name of the longest chain present with the name of all side chains. If there are two chains of equal length the one with the most side chains is selected.
      4. The position of each side chain is designated by the number of the carbon atom on the main chain to which the side chain is attached. The direction of numbering is chosen to give the

lowest possible numbers to the side chains. For example,

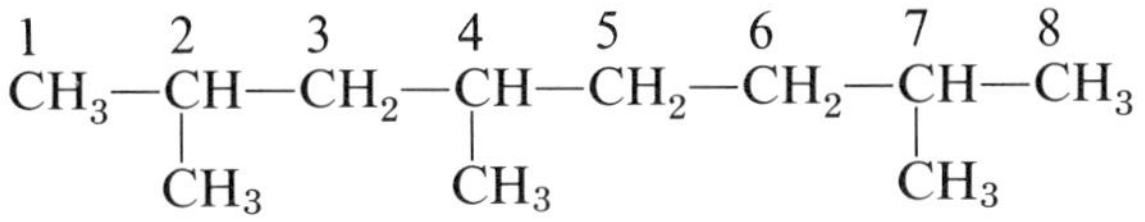

$$\underset{1}{CH_3}-\underset{2}{\underset{|}{CH}}-\underset{3}{CH_2}-\underset{4}{\underset{|}{CH}}-\underset{5}{CH_2}-\underset{6}{CH_2}-\underset{7}{\underset{|}{CH}}-\underset{8}{CH_3}$$
$$\qquad\;\;CH_3\quad\;\;CH_3\quad\quad\quad\;CH_3$$

5. The branches of side chains are specified similarly. In this case the numbering starts at the point of attachment to the main chain. A branched side chain is enclosed in parentheses.

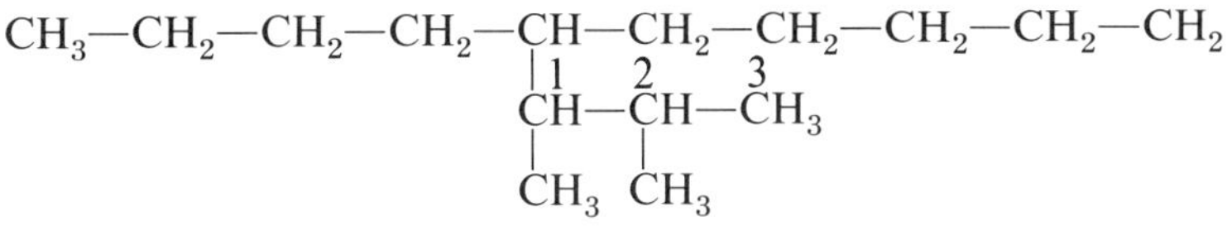

$$CH_3-CH_2-CH_2-CH_2-\underset{\underset{\underset{CH_3}{|}}{\underset{1}{CH}}}{\overset{}{|}}-CH_2-CH_2-CH_2-CH_2-CH_2$$
$$\qquad\qquad\qquad\qquad\quad\underset{2}{\underset{|}{CH}}-\underset{3}{CH_3}$$
$$\qquad\qquad\qquad\qquad\quad CH_3$$

is 5-(1,2-dimethylpropyl)decane.

Some trivial names such as isopropyl, *sec*-butyl, and *tert*-butyl are permitted as names of branched side chains.

6. The side chains are cited in alphabetical order. 4-ethyl-2,2-dimethyl hexane is

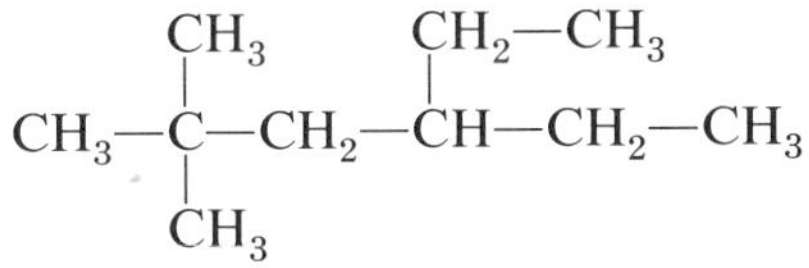

$$CH_3-\underset{\underset{CH_3}{|}}{\overset{\overset{CH_3}{|}}{C}}-CH_2-\underset{\underset{}{|}}{\overset{\overset{CH_2-CH_3}{|}}{CH}}-CH_2-CH_3$$

7-(1,2-dimethylpentyl)-5-ethyl tridecane is

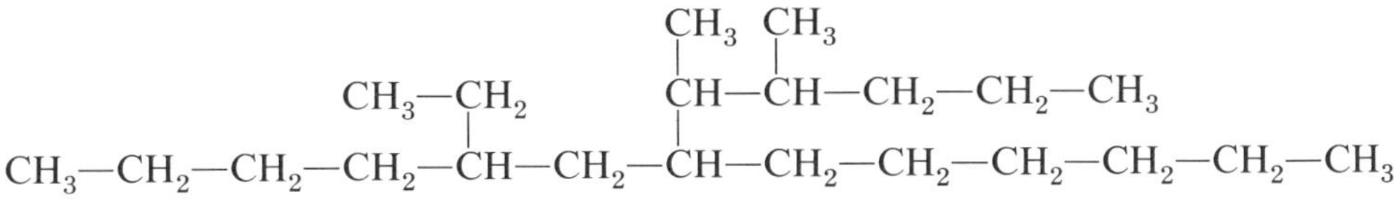

$$CH_3-CH_2-CH_2-CH_2-\underset{\underset{}{|}}{\overset{\overset{CH_3-CH_2}{|}}{CH}}-CH_2-\underset{\underset{}{|}}{\overset{\overset{CH_3\;\;CH_3}{\overset{|\quad|}{CH-CH}-CH_2-CH_2-CH_3}}{CH}}-CH_2-CH_2-CH_2-CH_2-CH_2-CH_3$$

An alternative order according to increasing complexity of the branch is permitted.

B. Unsaturated hydrocarbons are named as derivatives of the hydrocarbon chain with the maximum number of double and triple bonds.
1. The ending "ane" is replaced with:

"ene" for compounds that contain one double bond
"a(numerical prefix)ene" for compounds that contain more than one double bond
"yne" for compounds that contain one triple bond

669

"a(numerical prefix)yne" for compounds that contain more than one triple bond

"enyne," "adienyne," "enediyne," and so forth, are used for combinations of triple and double bonds.

The position of the multiple bonds is indicated by using the number of the smallest number of the carbon atoms involved in the bond. The numbers are placed before the name if this is unambiguous; otherwise it is placed just before "ene," "yne," or the numerical prefix.

Examples

2-butene	$CH_3-CH=CH-CH_3$
1-pentene-4-yne	$CH_2=CH-CH_2-C\equiv CH$
3-pentyl-1,3-butadiene	$CH_2=CH-C=CH_2$
	$\quad\quad\quad\quad\quad\quad\;	$
	$\quad\quad\quad\quad\; CH_2-CH_2-CH_2-CH_2-CH_3$
3-hexene-1,5-diyne	$CH\equiv C-CH=CH-C\equiv CH$

2. If an unsaturated side chain must be cited, the endings "enyl," "ynyl," "trienyl," and so forth are used.

5-ethynyl-1,3,6-heptatriene is

$$CH_2=CH-CH=CH-\underset{\underset{CH}{\overset{\|}{C}}}{CH}-CH=CH_2$$

3. Monocyclic hydrocarbons are named with the word "cyclo" used as a prefix to the name of the hydrocarbon that forms the ring.

Examples

cyclobutane

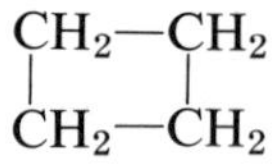

$$\begin{array}{c} CH_2-CH_2 \\ |\quad\quad\; | \\ CH_2-CH_2 \end{array}$$

1,4-cyclohexadiene

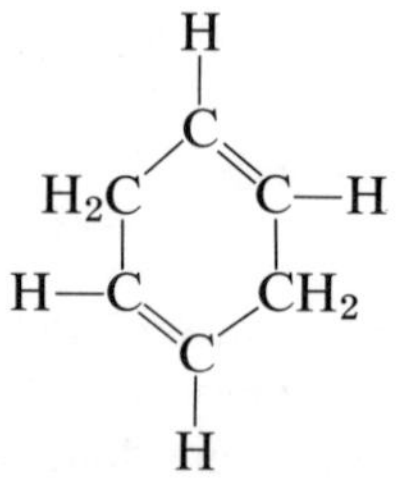

4-methylcyclohexene

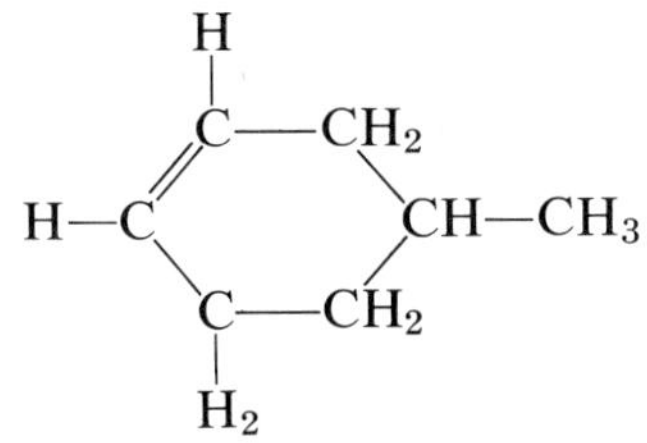

$$\begin{array}{c} H \\ | \\ C-CH_2 \\ H-C \diagup \diagdown CH-CH_3 \\ \diagdown C-CH_2 \diagup \\ | \\ H_2 \end{array}$$

Systematic rules are available for naming polycyclic hydrocarbons, but they need not be given here.

II. The systematic names of most organic compounds consist of the name of a straight hydrocarbon chain with the ending modified to specify one type of substituent. Chain branching and all other substituents are specified as prefixes.

A. Some characteristic groups including chloro (—Cl), bromo (—Br), fluoro (—F), iodo (—I), nitroso (—N=O), nitro (—NO$_2$), alkoxy (—OR), alkdioxy (—OOR), alkathio (—SR) are cited only as prefixes.

Examples

2-chlorobutane	CH_3—CHCl—CH_2—CH_3
dinitromethane	O_2N—CH_2—NO_2
1,1,2-trichloroethane	$CHCl_2$—CH_2Cl
ethoxyethane	CH_3—CH_2—O—CH_2—CH_3

B. Prefixes are arranged alphabetically.
1-bromo-3-ethyl-2,5-diiodopentane

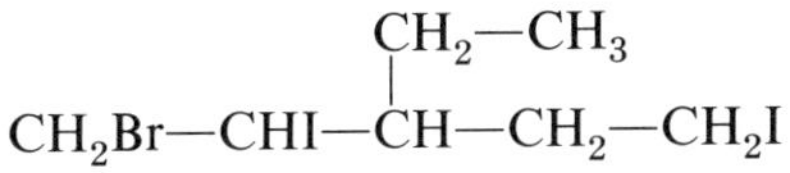

$$\begin{array}{c} CH_2-CH_3 \\ | \\ CH_2Br-CHI-CH-CH_2-CH_2I \end{array}$$

C. If any characteristic groups that may be cited as a suffix are present, one must be cited as a suffix, but only one. The group cited as a suffix is called the principal group. The principal group should be the highest type in the list in Table I. All other groups are listed as prefixes.

For purposes of illustration we will list in Table II one or two examples of functional groups in Table I. Note that the systematic rules for substitution may be applied to trivial names for a hydrocarbon or heterocyclic framework.

App. V Nomenclature in Organic Chemistry

Table I

Class	Formula	Group used as prefix	Group used as suffix
Cations		onio	onium
Carboxylic acids	—COOH OOH	carboxy	carboxylic acid oic acid*
Sulfonic acids	SO_3H	sulfo	sulfonic acid
Salts	COOM OOM	†	metal . . . carboxylate metal . . . oate*
Esters	COOR OOR	R-oxycarbonyl	R . . . carboxylate R . . . oate*
Acid halides	COX OX	haloformyl	carbonyl halide oyl halide*
Amides	$CONH_2$ ONH_2	carbamoyl	carboxamide amide*
Nitriles	—C≡N ≡N	cyano	carbonitrile nitrile*
Aldehydes	—CHO =O	formyl oxo	carbaldehyde al*
Ketones	=O	oxo	one
Alcohols	—OH	hydroxy	ol
Thiols	—SH	mercapto	thiol
Hydroperoxides	—OOH	hydroperoxy	§
Amines	$—NH_2$‡	amino	amine
Imines	=NH	imino	imine

*The groups COOH, COOM, COOR, COX, $CONH_2$, CN, and CHO are considered as a group when treated as prefixes and when substituted directly on a ring. Otherwise the carbon atom is considered as part of the parent chain as in hexanoic acid, $CH_3CH_2CH_2CH_2COOH$.

†When there are both positive and negative ionic centers in a single structure, the cationic group is designated as a prefix and acid salts are named as hydrogen salts so there is no need for a prefix to designate the carboxylate anion. HOOC—$(CH_2)_5$—COOK is potassium hydrogen heptanedioate.

‡Secondary and tertiary amines are named by citing each R group as a substituent on ammonia. Both systems are used for primary amines. For example, we have ethanamine or ethylamine ($C_2H_5NH_2$), diethylamine [$(C_2H_5)_2NH$], and triethylamine [$(C_2H_5)_3N$].

§When the principal group is hydroperoxide, the name of the radical R in ROOH is used in front of the word hydroperoxide. For example, CH_3OOH is methylhydroperoxide.

Table II

Compound	Name
$^+H_2OCH_2CH_2NH_3^+$	oxonioethanammonium ion
HOOC—CH$_2$—CH(CH$_2$COOH)—CH$_2$—CH$_2$—CH$_2$—COOH	3-(carboxymethyl)heptanedioic acid
C$_6$H$_{11}$—COOH (cyclohexane-COOH)	cyclohexanecarboxylic acid
toluene with SO$_3$H at 2 and 4 positions	2,4-toluenedisulfonic acid
NaOOC—CH$_2$—CH(CH$_3$)—CH$_2$Cl	sodium 4-chloro-3-methylbutanoate
CH$_3$—(CH$_2$)$_4$—CH$_2$—C(=O)—O—CH$_2$—CH(CH$_3$)—CH$_2$—CH$_3$	2-methylbutyl heptanoate
CH$_3$—C(=O)—Cl	ethanoyl chloride
CH$_3$—CH$_2$—CH$_2$—CH$_2$—CH$_2$—C(=O)—NH$_2$	hexanamide
furan with NC at 5 and COOH at 2	5-cyano-2-furoic acid
NC—CH$_2$—CH$_2$—CH$_2$—CN	pentanedinitrile
CH$_3$—CH$_2$—CHO	propanal
CH$_3$—C(=O)—CH$_2$—CH=CH$_2$	4-pentene-2-one
CH$_3$—C(=O)—CH$_2$—CHO	3-oxobutanal

673

App. V Nomenclature in Organic Chemistry

Table II (Continued)

Compound	Name
$CH_2OH-CH_2-CH_2-CH_2OH$	1,4-butanediol
cyclohexene with H and SH at position 1	2-cyclohexen-1-thiol
2,5-cyclohexadiene with HO-O and H at C4 and $-COC_2H_5$ at C1	ethyl 4-hydroperoxy-2,5-cyclohexadiene-1-carboxylate
pyridine with NH_2 groups at 2 and 4	2,4-pyridinediamine
$CH_3-CH_2-CH_2-CH_2-CH_2-CH=NH$	1-hexanimine

APPENDIX VI
Answers to Problems II

Chapter 1
1. 32.0 g.   **2.** 44.0 g.   **3.** 44.0   **4.** 660 g.   **5.** 901 g.   **6.** 98.0   **7.** 310.2   **8.** 0.323
9. 0.969   **10.** 38.9%.   **11.** 1.28 moles.   **12.** 2.56   **13.** CaF_2.   **14.** Mn_3O_4.   **15.** H_3PO_4.

Chapter 2
1. Ne, neon.   **2.** 10.   **3.** ^{22}Ne.   **4.** 10.   **5.** H, hydrogen.   **6.** 1.   **7.** $l = 0$. $m_l = 0$. $m_s = \frac{1}{2}$ or $-\frac{1}{2}$.
8. 1, 0, 0, $\frac{1}{2}$ or 1, 0, 0, $-\frac{1}{2}$.   **9.** 3, 1, 1, $\frac{1}{2}$ or 3, 1, 1, $-\frac{1}{2}$.   **10.** $m_l = 0$ and $m_l = -1$.   **11.** 6.
12. B, boron.   **13.** $1s^2$, $2s^2$, $2p^6$, $3s^2$, $3p$.
14. Either, depending on the atom. For scandium the $3d$ is lower in energy than the $4p$.
15. $1s^2$, $2s^2$, $2p^6$, $3s^2$, $3p^6$, $4s^2$, $3d$.   **16.** $1s^2$, $2s^2$, $2p^6$, $3s^2$, $3p^6$, $4s^2$, $3d^6$.   **17.** 4.   **18.** 5.
19. 14 electrons.   **20.** $1s^2$, $2s^2$, $2p^6$, $3s^2$, $3p^6$, $4s^2$, $3d^{10}$, $4p^3$.

Chapter 3
1. The ionization potential, 300.01 kcal mole^{-1} (Table 3-2).
2. The X-ray K absorption edge, 65050 kcal mole^{-1} (Table 3-4).   **3.** 64750 kcal mole^{-1}.
4. 1.98 Å (twice the covalent radius, Table 3-6).   **5.** $0.99 + 0.77 = 1.76$ Å.
6. 3.94 Å (twice the metallic radius, Table 3-5).   **7.** $0.99 + 1.81 = 2.80$ Å (Table 3-7).
8. It is closer to the value 2.10 Å predicted from covalent radii than to 2.22 Å, the sum of the ionic radii.
9. Most electronegative F, Br, Ge, Be, Ca least electronegative.   **10.** 0.   **11.** -1 in all three cases.
12. +7.   **13.** BrF_5.   **14.** -1.   **15.** -2.   **16.** +6 or VI.   **17.** Fe_2O_3.   **18.** 5.   **19.** P_2O_5.
20. P_2O_3 and possibly P_2O.   **21.** -3.   **22.** Na_3P and Ca_3P_2.   **23.** -2.   **24.** PH_3.   **25.** H:$\overset{..}{\underset{\underset{H}{..}}{P}}$:H.
26. $4 + 6 + 6 = 16$.   **27.** $24 - 16 = 8$ to be shared.   **28.** 10.   **29.** 0 or none.   **30.** 2.   **31.** 6.

App. VI   Answers to Problems II

Chapter 4
1. $\Delta H = -71$ kcal mole^{-1}.   2. Negative.   3. Endothermic.   4. 800 kcal.   5. 400 kcal mole^{-1}.
6. 0.00 kcal mole^{-1}, as it is for any element in its standard state.   7. Solid Al at 25°C and 1.00 atm pressure.
8. $H^0 = -400$ kcal mole^{-1}.   9. $267/3.0 = 89$ kcal.   10. $-1600 - (-801) = -799$ kcal mole^{-1}.
11. Endothermic.   12. $-400 - (+59) = -459$ kcal mole^{-1}.   13. 48.7 kcal.   14. 6 moles or 162 g.
15. $\Delta H^0 = -260.6 - (-71 - 99.4) = -90.2$ kcal mole^{-1}.   16. 3.2 g (0.050 moles).
17. $0.050 \times 90.2 = 4.51$ kcal.   18. 0.00832 moles or 0.515 g.   19. -120.6 kcal mole^{-1}.
20. $+42.4$ kcal mole^{-1}.   21. -118.8 kcal mole^{-1}.   22. 104 kcal mole^{-1} or 1.72×10^{-22} kcal per bond.
23. The H-Cl bond is stronger.   24. 103 kcal mole^{-1}.   25. $52 + 29 - 103 = -22$ kcal mole^{-1}.
26. $+36$ kcal mole^{-1}.   27. 119.2 kcal mole^{-1}.   28. 9 kcal mole^{-1}.   29. $\Delta H^0 = -103$ kcal mole^{-1}.
30. $\Delta H^0 = -119$ kcal mole^{-1}.   31. 119 kcal mole^{-1} (see Problems 27 and 30).
32. $\Delta H^0 = -289$ kcal mole^{-1}.   33. 51 kcal mole^{-1}.   34. $H^0 = +18$ kcal mole^{-1}.
35. $H^0 = -20$ kcal mole^{-1}.   36. $-20 - 22 - (-58) = +16$ kcal mole^{-1}.   37. 85 kcal mole^{-1}.
38. $H^0 = -23$ kcal mole^{-1}.   39. $-22 + 0.4 - (-23) = +1.0$ kcal mole^{-1}.

Chapter 5
1. $Cl_2 + 2e^- \longrightarrow 2Cl^-$.   2. 1.   3. 6.   4. 6.   5. $ClO_3^- + 6H_3O^+ + 6e^- \longrightarrow 9H_2O + Cl^-$.
6. Yes.   7. $PbCl_2 + 2e^- \longrightarrow Pb + 2Cl^-$.   8. $3Pb + 5Cl^- + ClO_3^- + 6H_3O^+ \longrightarrow 3PbCl_2 + 9H_2O$.
9. No, Pb will react with H^+ under these conditions.   10. $H_2 + Pb^{2+} \longrightarrow 2H^+ + Pb$.

11. $K = \dfrac{a_{Pb} a_{H^+}^2}{a_{Pb^{2+}} a_{H_2}} = \dfrac{[H^+]^2}{[Pb^{2+}] P_{H_2}}$.

12. $\Delta E^0 = -0.13$

$\Delta G^0 = -n\mathcal{F}\Delta E^0 = +6.0$ kcal mole^{-1}

$K = 4 \times 10^{-5}$.

13. 1.8×10^{-5}.   14. 1.8×10^{-5}.   15. 0.81 volt.   16. 2.7

Chapter 6
1. 41.1%.   2. $1s^2 2s^2 2p^6 3s^2 3p^6 4s^2 3d^{10} 4p^6 5s^1$.   3. $+1$.   4. 48 neutrons, 37 protons, 36 electrons.
5. $Cs > Rb > Rb^+ > K^+$.   6. $Te^{2-} > Se^{2-} > Br^- > Br^+ > K^+$.   7. LiI.   8. K^+ [4N-4-N-4-N4]$^-$.
9. Not as negative as -78 because of the size mismatch in LiI.   10. $+13$ kcal mole^{-1}.
11. 1.05×10^{-10}.   12. 1.8×10^{-12}.   13. 1.8×10^{-12} atm.   14. $Li > Rb > Na$.   15. 10^{75} atm.

Chapter 7
1. 1.86 Å.   2. 8.92×10^{-18} esu cm or 8.92 debye.   3. 14.5%.   4. Molecular.   5. 8.   6. 6.
7. F_2, it has a lower bond order.
8. The electrons in antibonding orbitals have higher energy than the $2p$ orbital of F atoms.
9. 38 kcal mole^{-1}.   10. $47 + 30 = 77$.   11. -29 kcal mole^{-1}.

Chapter 8
1. sp^3.   **2.** To minimize electron-electron repulsions, the lone pair on nitrogen should be assigned more room than the bond pairs.
3. Pyramid (tetrahedral with one of the positions a lone pair) H-Bi-H bond angle much less than 109°, probably $94 \pm 4°$.
4. Single bonds, sp^3 hybridization.   **5.** Bond order = 2, sp hybridization.
6. NO_2: bond order $1\frac{3}{4}$, bent, angle > 120°. O_3: bond order $1\frac{1}{2}$, bent, angle < 120°.
7. NO_3^-, BF_3, SO_3.   **8.** Because they have different numbers of valence electrons.
9. Bond order $1\frac{1}{2}$, bent, angle close to 120° (slightly less).   **10.** Trigonal bipyramid.
11. SF_4: trigonal bipyramid with a lone pair in one equatorial position. ClF_3: trigonal bipyramid with a lone pair in two of the equatorial positions.

Chapter 9
1. Pressure increases.   **2.** 1.44 atm.   **3.** 14.4 liters.   **4.** 10.6 liters.   **5.** 0.473 moles.   **6.** 20.9 g.
7. 80 liters.   **8.** 138 atm.   **9.** 138 + 0.023 atm.   **10.** 0.12 g.   **11.** 6.12 eu.   **12.** 105.   **13.** 106.4.
14. 14.2 eu or 0.103 cal g^{-1} deg^{-1}.   **15.** 3 translational, 3 rotational, and 3 vibrational.   **16.** 7.38 eu.
17. 1.41 eu.

Chapter 10
1. The maximum occurs for H_2S which has the greatest hydrogen bonding capability.
2. $K = 62$ at 25°C.   **3.** $K = 1.00$.   **4.** 0.00 kcal $mole^{-1}$.   **5.** 20.9 eu.
6. Hydrogen bonding in liquid H_2S introduces considerable order in the liquid. Also the entropy of SiH_4 liquid is particularly high because rotation occurs quite freely for these nonpolar, nearly spherical molecules even in the liquid phase.
7. 175 K.   **8.** +29 kcal $mole^{-1}$.   **9.** +43 kcal $mole^{-1}$.   **10.** 1720 K.
11. The bonding is ionic, KH is an ionic solid.   **12.** 737 K.
13. KH decomposes below its melting point.

Chapter 11
1. 1.70   **2.** 3.22

3. $K = 10^{-5} = \dfrac{[ScOH^{2+}][H^+]}{[Sc^{3+}]}$.

4. 3.35
5. Fe^{3+} because it is smaller.   **6.** $FeOH^{2+}$.   **7.** $K = 1.0 \times 10^{-2}$.
8. $2.0 \times 10^{-4} M = [FeOH^{2+}]$.   **9.** 4.57   **10.** 4.75   **11.** 11.7

Chapter 12
1. 76 kcal $mole^{-1}$.   **2.** 134 kcal $mole^{-1}$.   **3.** +204 kcal $mole^{-1}$.   **4.** 187 kcal $mole^{-1}$.
5. 289 kcal $mole^{-1}$.   **6.** There are more bonding electrons in CO.   **7.** 215 kcal $mole^{-1}$.

App. VI Answers to Problems II

8. The additional electron in NO is an antibonding electron. **9.** Weakened by 63 kcal mole^{-1}.
10. 1 bonding molecular orbital for each Ca atom. **11.** 1 bond for each Ca atom.
12. 46 kcal mole^{-1}. **13.** 35 kcal mole^{-1}.
14. The low energy d orbitals of Ca interact to give low energy bonding molecular orbitals which are not present at similar energies in solid Mg.
15. 52 kcal mole^{-1}. **16.** 3×10^{-14} atm. **17.** Cs in IA; Mg in IIA.
18. Mixing with low energy d orbitals lowers the electrical conductivity.

Chapter 13

1. $K = \dfrac{a_{Ca^{2+}} a_{F^-}^2}{a_{CaF_2}} = [Ca^{2+}][F^-]^2$. **2.** 1.00. **3.** 1.0×10^{-8}. **4.** 3.16×10^{-5}. **5.** -0.35. **6.** 1.75.
7. 5. **8.** $5 \times .15 = 0.75$. **9.** 2.89. **10.** 0.81. **11.** Sc^{3+}.

Chapter 14

1. 4. **2.** Chloro. **3.** Tetrachlorocuprate(II). **4.** p or d orbitals.
5. π^* made up of p orbitals on C and O. **6.** π^*. **7.** The $4s$, the three $4p$ orbitals, and $3d_{z^2}$. **8.** sp^3d.
9. Trigonal bipyramid. **10.** $d_{xy}, d_{yz}, d_{xz}, d_{x^2-y^2}$. **11.** 11 instead of 10. **12.** Mn.
13. Trigonal bipyramid. **14.** Two (the NO can be equatorial or axial).
15.

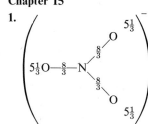

  16. 2.62 **17.** Basic. **18.** Yes.

Chapter 15

1.

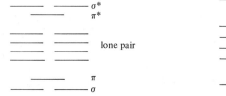

  2. Planar. **3.** sp^2. **4.** p_z. **5.**

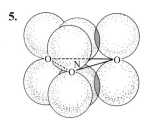

6.

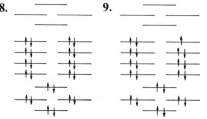

  7. 16. **8.**

9.

10. On oxygen (in one of the lone pair orbitals, probably on all three oxygen atoms).
11. Yes, π bonding on overlap with p_z on nitrogen.   12. None.
13. A combination of ligand orbitals which does not overlap significantly with any orbitals of the central atom.
14. The antibonding character of orbitals (a node perpendicular to the bond) gives lower electron density between the nuclei and hence a higher energy than an orbital without significant overlap.

Chapter 16
1. 24 kcal mole^{-1}.   2. Much less than 24 kcal mole^{-1}.   3. 12 kcal mole^{-1}.
4.

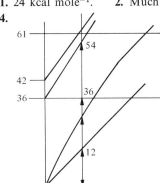

$12 \approx \Delta$
$36 \approx 36$
$\approx 21 + \Delta = 33$
$54 \approx 36 + 3/2 \Delta = 54$

5. 12 kcal mole^{-1}, infrared. 36 kcal mole^{-1}, red. 54 kcal mole^{-1}, green.
6. Reasonable window between 36 and 54, about 46 kcal mole^{-1}, 6200 Å, orange or yellow light.
7. Orange or yellow.   8. Blue, window is shifted to shorter wavelength from green.   9. ~3500 cm^{-1}.
10. 10 kcal mole^{-1}.   11. The molecule is bent.
12. Charges of $+1.23$ esu on each hydrogen, -2.46 esu or 0.51 electron on the oxygen.
13. 18$^+$, 17$^+$, 16$^+$, 1$^+$ for the isotopes ^{1}H and ^{16}O.   14. One.

Chapter 17
1. Cu_5Sn_4.   2. About 2%.   3. Cu_5Sn_4, solid Sn, and liquid of 2% Cu, 98% Sn.   4. 91% Sn, 9% Cu.
5. Cu_5Sn_4 or 60% Sn, 40% Cu.   6. The liquid is depleted in Cu, the percentage of Sn increases.
7. 1.8 g of Cu and 85.2 g Sn or 2.1% Cu.   8. Cu_5Sn_4 and solid Sn.   9. 1153°C.
10. 4.1% C, 95.9% Fe.
11. Add wrought iron; remove carbon by oxidation with O_2; remove carbon by oxidation with Fe_2O_3.

Chapter 18
1. 1.01   2. No, it is not amphoteric or acidic.   3. $La_2O_3 + 3H_2O + 6e^- \longrightarrow 2La + 6OH^-$.
4. $+351.6$ kcal mole^{-1}.   5. -2.54 volts.   6. $E = -2.54 - \dfrac{0.059}{6} \log a_{OH^-}^6$.
7. It will increase 0.059 volts each time the pH decreases by one unit.   8. 9.8.   9. $+176$ kcal mole^{-1}.
10. $+60$ eu.   11. $+158$ kcal mole^{-1}.   12. -2.29 volts.   13.

$$\begin{array}{c|c} La^{3+} & La_2O_3 \\ \hline & La \end{array}$$

14. -2.29 volts, pH = 9.8.   15. 2.5×10^{29}.

App. VI Answers to Problems II

Chapter 19
1. $^{221}_{87}Fr \longrightarrow ^{4}_{2}He + ^{217}_{85}At$.
2. The ^{26}Na nucleus has too many neutrons (15) for the number of protons present (11).
3. $^{26}_{11}Na \longrightarrow ^{26}_{12}Mg + _{-1}e^- + \nu$.   4. Positron emission.   5. $^{25}_{12}Mg + ^{1}_{1}p \longrightarrow ^{26}_{13}Al + \gamma$.   6. 0.999497.
7. 9.2×10^7 kcal mole^{-1}.   8. 9.36×10^{-8} year^{-1} or 2.98×10^{-15} sec^{-1}.   9. 1.79×10^9 atoms sec^{-1}.
10. 2.73×10^{-7} kcal.   11. One quarter of it, 1.5×10^{23} atoms.   12. 4.5×10^8 atoms sec^{-1}.
13. Probably none, decay formula gives 1.2×10^{-17} atoms left.

Chapter 20
1. Second order.   2. Third order.   3. Rate = $k[C_2H_5Cl]$.   4. 20 min at 800 K.   5. $\frac{1}{32}$.
6. 0.0347 min^{-1} or 5.77×10^{-4} sec^{-1} at 800 K.   7. $(C_2H_5Cl)^{\ddagger}$.   8. $(N_2O_4)^{\ddagger}$.   9. Rate = $k[H_2O_2][I^-]$.
10. $(FeTl^{5+})^{\ddagger}$.   11. The first one, $Fe^{2+} + Tl^{3+}$.

Chapter 21
1. Second order.   2. 6.93×10^6 sec.   3. 71 liter moles^{-1} sec^{-1}.
4. $(Co(NH_3)_4NH_2Br^+)^{\ddagger}$ and $(Co(NH_3)_5OHBr^+)^{\ddagger}$.   5. 2.   6. 4.3   7. About 0.62 Å.   8. About 3.4
9. Amphoteric.   10. 2.26   11. 6.   12. 1, 5.15, 0.57, 4.5, acidic, 2.46, 4.   13. $(HVCrO_4^{2+})^{\ddagger}$.
14. $HCrO_3$ or H_3CrO_4.

Chapter 22
1. 25 kcal mole^{-1}.   2. +5 eu.   3. +26.2 kcal mole^{-1}.
4. +22.2 kcal mole^{-1}, not quite as large as $\Delta H^{\ddagger}$.   5. 109 kcal mole^{-1}.   6. +24 kcal mole^{-1}.
7. +12 kcal mole^{-1}.   8. −12 kcal mole^{-1}.   9. 3 N-5-O 3.   10. +21 kcal mole^{-1}.
11. 177 kcal mole^{-1} to fit $H^0 = +17$.   12. $175 \approx 177$.

Chapter 23
1. 75 kcal mole^{-1}.   2. $6\frac{2}{5}$ $6\frac{2}{5}$ 3. 4.   4. −66 kcal mole^{-1}.   5. 6

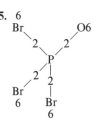

6. 329 kcal mole^{-1}.   7. 37 kcal mole^{-1}.   8. 0.40 M.   9. 0.20 M.
10. $[Na^+] + [H^+] = [OH^-] + [H_2PO_4^-] + 2[HPO_4^{2-}] + 3[PO_4^{3-}]$.
11. Subtract; $0.40 = 2[H_2PO_4^-] + 2[HPO_4^{2-}] + 2[PO_4^{3-}]$; from answer to 10.
12. $\dfrac{[HPO_4^{2-}][H^+]}{[H_2PO_4^-]} = K_2$.   13. $\dfrac{[PO_4^{3-}][H^+]}{[HPO_4^{2-}]} = K_3$.   14. $K_2K_3 = [H^+]^2$.   15. 9.8   16. 5.7

App. VI Answers to Problems II

Chapter 24
1. -8.8 kcal mole^{-1}.   2. -2.2 kcal mole^{-1}.   3. Cd^{2+} is a softer acid than Zn^{2+}.
4. -22.6 kcal mole^{-1}.   5. -37.5 kcal mole^{-1}.   6. Cu^+.   7. $+20.3$ kcal mole^{-1}.   8. 3×10^{-25}.
9. 3×10^{-24}.   10. 13.   11. 5×10^{-4}.   12. 3×10^{-9}.   13. 3×10^{-10}.

14. Ligand		K	$[Zn^{2+}]$, M
Cl^-	$ZnCl_4^{2-}$	130	13
I^-	ZnI_4^{2-}	5×10^{-3}	5×10^{-4}
NH_3	$Zn(NH_3)_4^{2+}$	3×10^{-9}	3×10^{-10}
S^{2-}	ZnS	3×10^{-25}	3×10^{-24}

Chapter 25
1. $2SO_2 + O_2 \longrightarrow 2SO_3$ and $SO_3 + H_2O \longrightarrow H_2SO_4$.   2. SO_2.   3. 6.8×10^{24}.
4. $+37.86$ kcal mole^{-1}.   5. $+4.43$ kcal mole^{-1}.   6. $HSO_4^- + 9H^+ + 8e^- \longrightarrow H_2S + 4H_2O$.
7. -4.43 kcal mole^{-1}.   8. It makes the stoichiometry the same as that for the half-reaction in Problem 6.
9. $HSO_4^- | H_2S$, $+0.293$ volt; $HSO_4^- | SO_2$, -0.096 volt.   10. Yes.   11. -195 kcal mole^{-1}.
12. Same as in Problems 6 and 7.   13. No.   14. Yes.
15. The G^0 value for a species plus the values for the amounts of standard materials (H_2O, H^+, and so forth) to bring it to a standard stoichiometry.
16. 6.   17. $+\frac{1}{3}$.   18. 6.   19. 2.   20.   21.

22. 2.00.   23. 2.083.   24. 1.917.   25. Spinel.   26. Tetrahedral.   27. 6.
28. Ni^{2+} is a high spin d^8 ion, favoring octahedral coordination. This effect overbalances the electrostatic preference.
29. The octahedral holes are filled with Si^{4+} and half of the Mg^{2+}; the rest of the Mg^{2+} is in the tetrahedral holes.

Chapter 26
1. CH_3—CH(—CH_3)—CH_3 with CH_3 branch.   2. 10.   3. 9.   4. Butane.   5. CH_3—CH(CH_3)—CH_2Cl and CH_3—CCl(CH_3)—CH_3.
6. CH_3—CH_2—CH_2—CH_2Cl and CH_3—CH_2—$CHCl$—CH_3.   7. 2-chlorobutane.
8. 1-chlorobutane, 2-chlorobutane, 1-chloro-2-methylpropane, and 2-chloro-2-methylpropane.
9. 2-chloro-2-methylpropane.   10. CH_3—CCl(CH_3)—CH_3 + H_2O $\longrightarrow$ CH_3—COH(CH_3)—CH_3 + Cl^- + H^+.
11. 2-chlorobutane.
12. R-2-chlorobutane, S-2-chlorobutane, 1-chlorobutane, 1-chloro-2-methylpropane, and 2-chloro-2-methylpropane

681

App. VI Answers to Problems II

13. $CH_3—CH_2CHCl—CH_3 + OH^- \longrightarrow CH_3—CH_2—CHOH—CH_3 + Cl^-$.

14.
$$\left(\begin{array}{c} CH_3 \quad CH_2—CH_3 \\ \diagdown \; C \; \diagup \\ HO \; | \; Cl \\ H \end{array} \right)^-$$

15. 2-butanol.   **16.** 2-methylpropanoic acid.

17. $CH_3—\underset{CH_3}{\underset{|}{CH}}—CH_2Cl + OH^- \longrightarrow CH_3—\underset{CH_3}{\underset{|}{CH}}—CH_2OH \xrightarrow{CrO_3}$

$CH_3—\underset{CH_3}{\underset{|}{CH}}—COOH \xrightarrow{CH_3OH} CH_3—\underset{CH_3}{\underset{|}{CH}}—COOCH_3$

Chapter 27

1. Adenosine triphosphate.   **2.** By the oxidation of food.   **3.** Photosynthesis in plants.   **4.** The sun.
5. By enzyme systems.   **6.** A membrane structure containing redox systems and enzymes found in cells for the oxidation of materials with the formation of ATP.
7. $C_6H_{12}O_6 + 6O_2 \longrightarrow 6CO_2 + 6H_2O$.   **8.** -669.2 and -688.0 kcal mole^{-1}.
9. The cell does not contain any enzymes to catalyze the direct reaction.
10. High molecular weight proteins with a specific order of amino acids.
11. Covalent bonds; amide linkages.   **12.** The sequence of bases in DNA.
13. A double helix with paired bases in the center.   **14.** Hydrogen bonds.   **15.** Hydrogen bonds.
16. (a) In the DNA; (b) between one strand of DNA and messenger RNA being synthesized;
(c) between messenger RNA and transfer RNA as the protein is synthesized.

Index

Absolute temperature, 218
Absorbance, 403
Absorption edges, 75–76
Acetaldehyde, 611
Acetic acid, 130–31, 272, 284, 288–90, 615
Acetylene, 611, 208–209
Acid, 271–94, 363–67
   electron acceptor, 351
   equilibria, 283–94
   hard and soft, 279–80, 326, 363–68, 555–62
   organic, 615–19
   pK, 273, 275, 279, 289, 291
   polarizing ability, 327–28
   strength, 271–94
Actinium series, 436, 441–44
Activated complex, 477–87
   bridged, 507–509, 513–14
   outer sphere, 507–509
Activation energy, 478–84
Activity, 128–32, 142
Activity coefficient, 237–40, 348
Adenine, 635–46
Adenosine triphosphate, 542–44, 625, 635–41
Adrenal gland, 642
Ahrens, L. H., 572
Ahrland, S., 363
Algae, 348
Alizarin, 626
Alkali metals, 153–63
Alkane, 601–605
Alkenes, 604–608
Alpha particles, 28–31, 464
Allotropes, 105, 534, 550, 572, 601

Allred, E. L., 84
Aluminum, 432–33, 438–39, 653
   amphoteric, 367–68
   atomic structure, 45–58
   chloride, 438
   complexes, 364–65
   fluoride, 191–92
   hydroxide, 505–506
   ionization potential, 69, 78–79
   metallurgy, 432–33
   oxide, 438
Alyea, H. N., 153
America, 431
Amides, 517–19, 528, 619–24
Amines, 619–24
Amino acids, 520–22, 634–36, 642
Ammonia, 91–92, 107, 258, 260–64, 519–20
   base, 271–74
   bonding, 202, 203, 285–88
   complex ammines, 327–29
   oxidation, 262–63
   solvent, 261–62
Amphibole, 593–94
Amphoterism, 276, 367–68, 437–41
Anaesthetics, 633
Anemia, 646
Anion, 157
Anode, 138, 427
Antibiotics, 633
Antibonding molecular orbitals, 185–90
Antifluorite structure, 163–65
Antimony, 552–58, 653
   chloride, 550
   oxidation states, 552–54

Antimony (cont.)
   stibine, 260, 264
Antiseptics, 633
Apatite, 533–34, 544
Arginine, 520–21, 635
Argon, 57, 156, 231
Aristotle, 1–3
Aromatic molecules
   bonding, 307–308
Arrhenius, S. A., 153, 478
Arsenic, 550–52, 654
   arsine, 260–64
Asbestos, 593
Asymmetric carbon, 615
Atmosphere, 15, 220, 516
Atom, 1–7
   electronic structure, 28–66
   enthalpy, 106–108, 112
Atomic number, 77
Atomic orbitals, 28–66, 280–81
Atomic radius, 77, 79–84, 208
Atomic spectra, 40–44
Atomic volume, 6–7
Atomic weight, 3–5, 28
ATP, 542–44, 635–41
Avogadro, A., 100, 169, 219, 227, 482, 651–52
Azide ion, 528

Bacon, F., 215
Bacteria, 633
Balanced equations, 106, 140, 152
Band theory, 308–17
Barium, 318–19
   halides, 172, 339–40
   sulfate, solubility, 344–45

683

Index

Barium (*cont.*)
  thermodynamics, 654
Basalt, 591
Bases, 271–94, 351
Benzene, 354, 529
  molecular orbitals, 307–308
Bergman, T., 318
Beryllium, 45, 318, 438–40, 654
  Be$_2$, 300–302
  chloride, bonding, 198–99
  complexes, 364
  hydride, 249–52, 376–78
  ionization potential, 69, 79
Berzelius, J. J., 554
Bessemer, H., 431
Beta decay, 462–65
Biochemistry, 632–48
Bismuth, 330, 555–58, 654–55
Blast furnace, 430
Boat form, 605–607
Bohr, N., 39–44
Boiler scale, 347
Boiling point, 8–10, 215–17
Boltzmann, L., 219, 226–30, 652
Bond angles, 200–203, 209–11
Bond energy, 107–18, 192–95, 252
Bond length, 207, 211
Bonding orbitals, 183–90
Bonds, 79–80, 157, 176–95, 295–317
  d-p π bonding, 209, 538–41
  multiple, 207–24, 523–25
Born, M., 38, 116, 168, 170, 333–34, 524
Boron, 586–90
  atomic structure, 45–58
  B$_2$, 189–90, 297–301
  BH$_3$, 476–77
  complex ions, 365
  fluoride, 271, 378–83
  halides, bonding, 199–200, 203, 378–83
  ionization potential, 69, 78–79
  thermodynamics, 655
Boundary condition, 48–49
Boyle, R., 1
Bragg, W. H., 34–36
Breeder reactors, 470
Bromine, 585–86
  bromides, 25
  dissociation, 490
  E_h-pH diagram, 149
  fluorides, 18–19, 90
  hydrobromic acid, 265, 273, 283
  reaction with hydrogen, 487
  thermodynamics, 655
  vapor pressure, 216
Brønsted, J. N., 271–72
Bronze, 423–28

Brown, R., 223, 224
Buffer solutions, 292, 545–48
Butadiene, 304–307
Butane, 603–604

Cadmium, 163, 449, 536, 656
Cadmium chloride structure, 163
Calcium, 2–4, 318–49
  biochemistry, 319, 348–49
  bonding in metal, 312–13
  carbonate, 318, 348–49
  fluoride, 442
  halides, 172, 339–40, 344
  occurrence, 318, 346–49
  oxide, 347
  phosphide, 536
  sulfate, 344
  thermodynamics, 656
Calorimeter, 105
Carbene, 212
Carbon, 601–31
  aldehydes and ketones, 611–14
  alkenes, 304–308, 604–608
  amino acids, 520–22, 634–36, 642
  benzene, 307–308, 354, 529
  biochemistry, 632–48
  bond energies, 192, 616
  bond to hydrogen, 115–18
  C$_2$, 189–90
  carbonate, 334–49, 382–83, 601
  carbon monoxide, 216–17, 485–87
  carbon monoxide as ligand, 355–59, 397, 411
  cyanide, 352, 361, 365
  dioxide, 191, 232–33, 238–39, 243–44
  dioxide bonding, 209
  ester, 485, 615–19
  ethanol, 417–19, 609–11
  ether, 633
  ethylenimine, 405–409
  glucose, 637–41
  glyceraldehyde, 639–40
  heterocyclic compounds, 624–27
  hydrocarbons, 601–608
  ionization potential, 109–14
  lactic acid, 640–41
  methane, 383–84, 489, 602–603
  nitriles, 621–24
  nucleophilic substitution, 479, 491–94, 498
  organic acids, 615–19
  organic chemistry, 601–31, 668–74
  organic nitrogen compounds, 619–27
  phenanthroline, 509
  phenol, 633
  proteins, 520–22, 634–39

Carbon (*cont.*)
  pyruvic acid, 640
  reactions at pi bonds, 607–14
  ribose, 641–46
  tartrate, 632–33
  thermodynamics, 656–57
Carbonium ion, 608–10
Carborane, 590
Catalysis, 485–87, 495–97
Cathode, 138
Cations, 157
Caves, 346–47
Cell, 632, 634–35, 646
Cementation, 430
Cerium, 436, 442–43, 509
Cesium, 153–73, 657
  cesium chloride structure, 163
  electronic structure, 155–56
  halides, 170–73, 180–81
  solubility of salts, 335–39
Chain reactions, 487–90
Chair form, 605–609
Charge transfer, 557–58
Chatt, J., 351, 363
Chelates, 354–55
Chemical kinetics, 474–515, 525–29
Chemiluminescence, 533
China, 430–31
Chlorine, 2, 582–83
  bond to hydrogen, 193
  chlorates, 87–91, 137, 148–49, 582–83
  chloric acid, 277–79, 582–83
  chlorides, 25, 157–73, 331–32, 559–60
  chloro complexes, 364, 369
  dioxide, 88, 210–11, 582–83
  E_h-pH diagram, 148–49
  fluoride, 419–20
  hydrochloric acid, 265, 273, 281–83, 286
  oxidation of chlorides, 148–49
  perchlorate, 87, 137, 148–49, 331, 336, 343, 502–503, 582–83
  reaction with hydrogen, 489
  thermodynamics, 559–62, 657
Chloroplasts, 635, 639
Chromium, 5, 439, 501–12, 657–58
  atomic structure, 57–62
  chromates, 509–12
  chromium II, 508
  chromium III hydroxide, 505–507
  chromium IV and V, 509–11
  complex ions, 200, 365, 501–508
Chrysotile, 593
Cis-trans isomerism, 360, 604–605
Clapeyron, B. P. E., 242
Clausius, R. J. E., 119, 242

Clay, 594
Closest packing, 157-60
Cloud chamber, 29-31
Coal, 468
Cobalt, 512-14, 658
  amine complex, 205
  biochemistry, 353
  crystal structure, 159
  oxidizing agent, 508, 513-14
  phosphide, 536
  solubility of halides, 341-42
Colloid, 224
Combustion, 100-102
Comets, 15
Common ion effect, 289-92
Complexes, 327-29, 345, 351-400
  spectra, 401-17
Compressibility factor, 237-41
Conjugate acid-base pairs, 272-73, 287
Contact process, 575
Coordinate covalent bond, 351
Coordination chemistry, 351-400
  biochemistry, 353
  carbonyl complexes, 397
  isomerism, 361-66, 412-14
  molecular orbitals, 388-99
  nomenclature, 361-62
  π bonding, 355-56
  polarizing ability, 365-71
  rate of substitution, 500-15
  spectra, 401-409
Coordination number, 158-62, 352, 367-71
Copper, 561-63, 658
  alloys, 423-28
  atomic structure, 58-64
  catalyst, 495
  copper III, 445-46
  halides, 172-73, 332, 341-42
  hydroxide, 332, 343
  metallurgy, 422-27
  occurrence, 422, 426
  sulfide, 561-62
Corals, 348
Correlation diagram, 300
Cosmic abundance, 454-56
Coulomb, C. A. de, 39-44
Covalent bond, 80, 181-92
Covalent radius, 80-82
Covalent solids, 191
Critical point, 237-44
Cryolite, 432
Crystal, 32-39, 157-73
Curie, M. S., 153
Cyanogen, 97
Cycloalkanes, 604-609
Cyclonite, 620

Cytochromes, 639
Cytocine, 642-46

d orbitals, 53-66, 73, 323-26
Dalton, J., 2-4
Davies, N. R., 363
Davisson, C. J., 30-35
Davy, H., 2, 432
deBroglie, L. V., 36-38
Debye, P. J. W., 414, 651
deChancourtois, A. E. B., 5
Degeneracy, 308
Degree of freedom, 233, 425-26
Democritus, 2-3
Deoxyribonucleic acid, 268, 641-46
Detergents, 617-18
Deuterium, 26, 247-49
Diamagnetism, 409-10
Diamond, 312-15
Diatomic molecules, 177-90
  dissociation, 490-91
  heat capacity, 232-33
  molecular orbitals, 295-304
Diborane, 96-97, 588-89
Diffraction, 31-35
Diffusion, 224-25
Dimethyl glyoxime, 408
Dipole moment, 177-78, 413-14
DNA, 268, 641-46
Drugs, 632-35
Dulong, P. L., 234-35
Dynamite, 529

Earth, 468
Ease of polarization, 320-21, 326-27, 555-62
Eddington, A. S., 99
Edison, T. A., 449
Edta, 406-407
Effusion, 225-26
Egypt, 423
e_g, 388
E_h-pH diagram, 145-51
Einstein, A., 37, 235
Electrode potential, 137-52, 445-46
Electrolysis, 137-45, 432
Electron, 28-29, 77, 652
  repulsion, 68-72, 77, 203-205
  sharing, 91-98
Electron affinity, 166, 280
Electron-volt, 36
Electronegativity, 68, 84-86, 280-81
  acid strength, 273-74
  bond strength, 178-79, 192-95
Electrostatic valence rule, 586-87, 592
Elements, 1-4, 8-14
  cosmic abundance, 454-56
  formation, 454-60

emf, 138
Enamines, 621-23
Endothermic, 101
Energy, 39, 99-135, 155, 215-46
  levels, 228-36
  transfer, 491
Enol, 611
Enthalpy, 100-135, 218
  of activation, 480
  and heat capacity, 231-32
  heat of vaporization, 216-17, 239-44
  hydration, 276, 333-40
  hydrogen compounds, 251
  of ions, 329-30
  of nitrides, 518
  periodicity, 99-112
  standard, 105
Entropy, 99, 118-35
  of activation, 481-85
  in cycloalkane equilibria, 605
  hydrogen compounds, 254-57
  of ions, 330, 337-38
  standard, 119
  of vaporization, 217
Enzyme, 634-41, 646
Equilibrium constant, 124-35, 144
Equipartition of energy, 234
Erbium, 441-43
Esters, 485, 615-19
Ethane, 97, 208, 603
Ethanol, 417-19, 609-11
Ether, 610, 633
Ethylene, 208-209, 607
Ethylenediamine, 354, 355, 361
Europe, 431
Europium, 443-44
Exclusion principle, 62-63
Exothermic, 101
Explosives, 489-90, 529-30, 620
Exponential function, 127

f orbitals, 53-66, 73
Fahrenheit, G. D., 651
Faraday, M., 142, 651, 652
Fat, 497, 617
Fatty acid, 544
Feldspar, 591, 598
Fermentation, 610, 639-41
Fertilizer, 155, 544-45
Fischer, H., 601
Fission, 469-70
Flame tests, 155-56
Flotation, 426
Fluorine, 106, 302-304, 658
  E_h-pH diagram, 145-47
  fluorides, 15-25, 103, 126, 331-42
  fluorination, 265

Index

Fluorine (cont.)
  fluoro complexes, 369, 371
  hydrofluoric acid, 264–69, 273–74
Fluorite structure, 162–65
Folic acid, 633
Fossil fuels, 468
Foucault, J. B. L., 37
Francium, 173
Frasch, H., 574–75
Free energy, 122–35, 218, 232
  hydrogen compounds, 255
  of ions, 274–77, 329–30
  periodicity, 124–26
  standard potential, 140–45
Free radicals, 17, 96–97, 487–90
Fructose, 543, 615, 639
Fusion, 454–57, 471, 480

Gadolin, J., 441
Galen, 1
Gallium, 5–6, 8, 364, 439–40
Gamma rays, 75
Gamow, G., 295
Gas, 215–46
  activity coefficients, 240
  collisions, 224–26
  compressibility factor, 241
  critical constants, 240
  diffusion, 224
  gas constant, 124, 143, 219–20, 652
  heat capacity, 231–34
  real, 236–41
Gene, 641–46
Gerlach, W., 44–50, 323
Germanium, 5–6, 212, 315
  germane, 259–60
Germer, L. H., 30–35
Gibbs, J. W., 122, 424–26
Gillespie, R. J., 183, 203–14, 360
Glucinium, 319
Glucose, 543, 637–41
Glutamic acid, 521, 642–46
Glyceryl trinitrate, 529
Glycine, 520–22, 636
Glycogen, 637
Gold, 442, 505, 565, 572
Gouy, G., 410
Gravity, 2–6, 230, 652
Greece, 427
Grignard reaction, 618
Guanine, 642–46
Gypsum, 575

Haber, F., 116, 170, 333–34, 524
Hafnium, 444–45
Half-cell, 138–45
Half-filled shell, 79
Half-life, 465–67, 475

Half-reactions, 138–45
Hall, C. M., 432
Hamilton, W. R., 50
Hard water, 347
Heart, 319
Heat, 100–107, 215
Heat capacity, 230–36
Heisenberg, W., 38
Helium, 45, 57, 216, 231, 455–56
Hemiacetal, 612–15
Hemoglobin, 646
Heraclitus, 474
Heroult, P. L. T., 432
Hertz, H., 37, 651
Hess, G. H., 104
Heterogeneous reactions, 494–97
Heterolytic dissociation, 281–82
Heteronuclear, 177
Hexamethylene tetramine, 620
High-energy phosphate bond, 542–44
High spin, 395
Hippocrates, 1
Histadine, 521, 625
Hofmann, A. W., 625
Holmes, H. N., 516
Homer, 427
Homonuclear, 177
Hormones, 632, 635–36, 642
Hund, F., 62–63, 109, 187
Huxley, T. H., 247
Huygens, C., 36
Hybridization, 190, 198–203, 298–300
Hydrates, 274–77, 344, 364–65
Hydrazine, 528–29
Hydrocarbons, 601–608
Hydroelectric power, 468
Hydrogen, 247–70
  abundance, 454–56
  atomic structure, 45–58
  boiling point, 216
  bond energies, 252, 259
  bonding, 181–86, 261–70
  combustion of hydrides, 259–60
  compounds, 15–22
  diffusion, 224–26
  dissociation, 106
  fusion reactions, 454–57, 471
  heat capacity, 232
  hydrogenation, 496–97, 608
  hydrolysis of hydrides, 260
  isotopes, 247–49
  metallic hydrides, 201, 254
  peroxide, 90
  reaction with halogens, 248, 483–90
  saline hydrides, 248–54
  spectrum, 41–44
  thermodynamics, 251, 255, 659
  volatile hydrides, 251–70

Hydrogen (cont.)
  wave functions, 51–57, 649
Hydronium ion, 121
Hydroxides, solubility, 332, 342–43
Hydroxyl radical, 90, 96
Hydroxylamine, 528
Hypo, 327, 564, 578–80

Ice, 118–28
Ideal gas, 217–30
Ideal gas law, 218
  calculation, 220–22, 245–46
  derivation, 218–26
Imidazole, 625
Imides, 517–19
India, 430–31
Indium, 364
Induced polarizing ability, 323–26, 559–62
Induced reactions, 510–11
Infrared spectra, 358, 401, 403, 410–13
Insulin, 635–36
Interference, 33–39
Inversion, 492
Iodine, 585–86, 659
  chloride, 203–205, 586
  E_h-pH diagram, 150
  fluorides, 18, 19, 23, 88, 200, 586
  heat capacity, 232
  hydrogen iodide, 265, 273–74, 283
  reaction with hydrogen, 483–84, 488–89
Ion, 28, 30, 80–84
Ionic bonding, 165–69, 176–81
  bond energies, 178–81
  crystal structures, 157–65
  properties of solids, 171, 191–92
Ionic equilibrium, 286–94
Ionic radius, 79–84, 207–208
Ionization constant, 130–31
Ionization potential, 68–79, 109–14, 155
Ionizing radiation, 642
Iridium, 567–69
Iron, 445–46, 448–49
  biochemistry, 353
  bonding in complexes, 206
  coordination number, 352
  Fe_3C, 429–31
  ferrocyanide, 504, 508
  iron age, 427–31
  metallurgy, 428–31
  pyrite, 562
  solubility of halides, 341–42
  thermodynamics, 656–60
  thiocyanate, 500–501
Isobaric spin, 463
Isomers, 501, 360–61, 492, 602–605, 614–15

686

Isomers (*cont.*)
  optical, 361, 492, 614–15
Isotopes, 3, 247–49
IUPAC, 361, 668

Jørgensen, C. K., 110

K shell, 75–78
Kaolin, 594
Kcal mole^{-1}, 36
Keats, J., 533
Kelvin, W. T., 651
Kinetic theory, 222–30
Kolbe, A. W. H., 601
Kopp, H. F. M., 234–36

L shell, 75–78
Labile complexes, 500–505
Lanthanum, 442–43
  series, 436, 441–44
Lattice energy, 167–75, 334–40
Lavoisier, A. L., 1, 516
LCAO, 308
Lead, 555–61, 568, 660
  lead chamber process, 575
  metallic character, 315
  oxide structure, 212
  pigments, 558
  sulfide, 212
Leaving group, 492–93
Leibniz, C. W., 2
Lemery, N., 271
Lewis, G. N., 91, 116, 272–73, 529
Ligand, 351, 354, 361–62
  exchange reactions, 500–505
  field, 374–400
Light, 31–39, 642, 652
Lime, 2, 347
Limestone, 346–49
Lister, J., 633
Lithium, 153–173, 660
  atomic structure, 45–58, 73–74
  halides, 150–64, 172–73, 181
  hydride, 249–52
  oxide, 165
  solubility of salts, 335–39
Lowry, T. M., 271
Low-spin, 395
Lutetium, 443–44
Lysine, 521, 635

M shell, 75–78
Madelung, E., 169–71
Magnesium, 2, 318–49
  biochemistry, 319
  bonding in metal, 308–13
  halides, 171–73, 339–40
  hydride, 249–52

Magnesium (*cont.*)
  ionization potential, 67–79
  metallic radius, 311
  metallurgy, 432–33
  Mg$_2$, 296
  occurrence, 318, 348–49
  sulfate, 344
  thermodynamics, 660–61
  uses, 318–19
Magnetism, 44–47, 409–10
Main sequence, 454–57
Maltase, 610
Manganese, 445–48, 661
  permanganate, 508
  solubility of salts, 340–43
Marignac, J. C. G. de, 5
Markovnikov, V. V., 609
Martensite, 430
Mass, 28
  number, 455
  spectra, 415–17
Mauve, 625–26
Maxwell, J. C., 99, 229
Mean free path, 224
Membrane, 632–39
Mendeleev, D. I., 1–15, 77
Mendeleevium, 15
Mercury, 8, 216, 566–67, 611, 661–62
Metal, 8–11, 308–15
Metallic radius, 80–81
Metallurgy, 422–35
Metasilicates, 591–93
Methane, 115, 258, 383, 603
  bonding, 196–98
  crystal structure, 158
Methyl radical, 96–97, 247
Mica, 594–97
Mitochondria, 639
Mixing, 190, 299–311, 378
Mole, 100
Molecular orbitals, 97, 176–96, 295–317
  band theory, 308–15
  for complexes, 374–400
  degenerate, 382–85
  diatomic molecules, 295–304
  π bonding, 304–308
Molecular solids, 191–92
Molecules, 15, 86–97
  energy, 215–46
  shapes, 196–214
Monazite, 442, 444
Montaigne, M. E. de, 67
Montmorillonite, 594
Moseley, H. G. J., 77
Mu, 362
Mulliken, R. S., 84, 280–81
Muscle, 543
Mutation, 646

NAD, 638–39
Neon, 190
Nernst, W. H., 143–47
Neutrino, 462–64
Neutron, 28, 457–71
  capture, 460, 463
Newlands, J. A. R., 4–5
Newton, I., 1–6, 36, 223, 234
Nickel, 445–46, 449–50
  coordination complexes, 206, 323–26, 352–53
  nickel(III), 413–14
  nickel-arsenic structure, 163–64
  phosphide, 535–36
  robust complexes, 504
  spectra, 403–409
  thermodynamics, 662
Niobium, 5, 445
Nitrile, 621–24
Nitrogen, 158, 516–52
  amides, 517, 528–29
  ammonia, 260–64, 519–20
  ammonia as ligand, 355, 360–61, 364–65, 369
  atomic structure, 58–62
  bond energies, 107–18, 517, 523–25
  dioxide, 88, 210–11, 262–63
  dissociation, 107
  E_h-pH diagram, 522
  explosives, 529–30
  hydrazine, 528–29, 620
  hydroxylamine, 528, 620
  hyponitrous acid, 528
  ionization potential, 69, 79, 109–14
  molecular orbital diagram, 187–90
  nitrate, 92–95, 331, 343, 382–83, 516
  nitration, 529–30
  nitric acid, 262, 263, 279, 529–30
  nitric oxide, 262–63, 359, 525–27
  nitrides, 303–304, 319, 517–19
  nitrites, 529
  nitrous oxide, 528, 633
  oxide, bonding, 522–26
  oxide, reactions, 477, 485–87, 525–27
  proteins, 520–22, 634–39
  thermodynamics, 518, 662
Node, 53–56, 185
Nonmetal, 8–11, 315, 572–97
Novae, 460
Nuclear magnetic resonance, 417–20
Nuclear reactions, 444
Nuclei, 29, 454–73
  fission, 464, 469–70
  fusion, 454–57, 471, 480
  spin, 458, 464
  subshells, 460–62

Index

Nucleon 463
Nucleophile, 492–93
Nucleosynthesis, 454–60
Nucleotide, 638–46
Nyholm, R. S., 183, 203–14, 360
Nylon, 620

Octahedral, 160–65, 388–94
Octet rule, 91–98
Odd-odd nuclei, 464–65
Operator, 50–51
Oppenheimer, J. R., 38
Optical isomers, 361, 492, 614–15
Orbitals, 57–61, 375–99
  e_g and t_{2g}, 391
  LCAO, 375–77
Order of reaction, 475
Organic chemistry 601–31, 668–74, (*See also* Carbon)
Organometallic compounds, 354
Orgel, L. E., 403–404
Orthocarbonic acid, 617
Orthosilicates, 591–92
Osmium, 567–69
Overlap, 183–90, 197–98, 298–99
Overvoltage, 147–48, 494
Oxidation-reduction reactions, 136–52
Oxidation states, 86–91, 136, 145–51
Oxidizing agent, 136–45
Oxygen, 2–4, 15–24, 58–63
  bond to hydrogen, 115–18
  deviations from ideality, 237–38
  fluoride, 87
  gaseous oxides, 303–304
  ionization potential, 68–72, 79, 109–14
  molecular orbital diagram, 186–89
  ozone, 210
  thermodynamics, 102, 125, 232, 662–63
  vaporization, 216–17
  velocity distribution, 230
  water, 86–91, 115–26, 364–65, 387–89

p orbitals, 53–66, 73
Paint, 558
Palladium, 254, 497, 505, 567–69
Palmitic acid, 544
Paramagnetism, 409–10
Partial pressures, 221
Pasteur, L., 633
Pauli, W., 62–63, 182
Pauling, L. C., 84, 115, 192, 198, 281
Penetration, 61–62, 73–79
Pentagonal bipyramid, 199–200, 397
Penicillin, 633
Periodic Table, 1–27, 63–64, 78–86, 99

Perkin, W. H., 625–26
Petit, J. L., 234–35
Pfeifer, P., 374
pH, 131–32, 144–52
Phase, 33, 243–44, 424–26
Phenanthroline, 507
Phenol, 633
Phoenicia, 427
Phosphorescence, 533
Phosphorus, 4, 533–49
  allotropes, 534, 535, 548
  bonding, 192–93, 538–44
  E_h-pH diagram, 534–35
  halides, 92–95, 200, 537–40
  high energy phosphate bond, 537–44, 576, 587
  ionization potential, 69, 79
  oxides, 542–43
  oxyacids, 278–79, 541–42
  phosphides, 535–36
  phosphine, 259–64, 536–37
  production, 533–34
  solubility of phosphates, 344
  thermodynamics, 536, 663
Photocells, 583
Photoelectric effect, 37
Photography, 563–64
Photon, 37–38
Photosynthesis, 468, 635, 639
π bonding, 184–90, 302, 355, 381, 496–97, 604–14
Pitchblende, 444
Pituitary gland, 642
Planck, M. K. E. L., 36–37, 482, 652
Plaster of Paris, 344
Platinum, 139, 442, 505, 567–72, 575
  catalyst, 139, 495–97, 575
Plato, 2
Polarizing ability, 320–50
  coordination complexes, 365–71
  induced, 559–62
Polarography, 494
Polybius, 422
Polycatenasulfur, 572
Polymerization, 520–22, 607–608
Polyprotic acids, 287–88, 290–91
Potassium, 153–73, 663
  atomic structure, 57–61, 73–74, 155–56
  biochemistry, 154–55
  dihydrogen phosphate, 547–48
  halides, 162–72
  in fertilizer, 155, 544
  solubility of salts, 335–39
Potential, 138–45
  energy, 39, 230
Powder metallurgy, 433
Precambrian, 348

Pressure, 219
Primary, 607–10, 621
Promethium, 442–44
Protein, 268, 529–32, 634–39
Proterozoic, 632
Proton, 28, 454–71
Purine, 641–46
Pyridine, 624
Pyrimidine, 624, 641–46
Pyrite, 562
Pyroxenes, 591–93

Quantum number, 43–66
Quinine, 625–26

Radius ratio, 160–62, 334
Rare earths, 436, 441–44
Rate constant, 466–67, 474–82
Reaction mechanism, 474, 476–82
Real gases, 236–41
Red giant, 455
Redox reactions, 136–52, 507–11
Reefs, 348–49
Resonance, 93–94
Rhenium, 5–6
Rhodium, 567–69
Ribose, 615
Ring strain, 604–607
RNA, 641–46
Robinson, R., 401
Robust complexes, 500–505, 542
Rochow, E. G., 84
Rocket, 248
Rotation of molecules, 232–36
Rubidium, 153–73, 249, 335–39
Ruthenium, 567–69
Rutherford, E., 28–39, 516
Rutile structure, 162–65
Rydberg, 77

s orbitals, 53–66, 73
Salt domes, 574
Sanderson, R. T., 84
Sandwich compounds, 362, 397
Saponification, 617
Scandium, 5–6, 443
Scheele, K. W., 2
Schrödinger, E., 50–52
Screening, 323
Secondary, 609–11, 621
Selenium, 258, 264, 583–84
Semiconductor, 315, 583
Shakespeare, W., 196
Shapes of molecules, 196–214
Shielding, 72–79, 323
Sickle cell anemia, 646
Sigma bonding, 184–90

Index

Silicon, 586–87, 590–97, 664
  dioxide, 191
  fluoride, 191–92
  as semiconductor, 315
  silanes, 259–60, 590
  silicates, 422, 572, 591–97
  silicones, 590–91
Silver, 422, 563–65, 664–65
  chloride, 172–73, 331–32, 366–67, 559–63
Simplicius, 176
Slater, J. C., 323
S_N1 reactions, 498
S_N2 reactions, 479, 491–94
Soap, 617–18
Sodium, 73–74, 153–73, 665
  biochemistry, 154
  bond to oxygen, 179–80
  chloride, 90, 162–81
  E_h-pH diagram, 153–54
  halides, 162–64, 170–72, 181
  hydride, 249–52
  oxide, 165
  phosphates, 545–48
  solubility of salts, 335–39
Solubility, 320–50, 423–33
  calculations, 344–50
Sound, 224
Spectra, 40–42, 401–21, 434
Spectrochemical series, 398–99, 403–406
Spedding, F. H., 436
Spherical coordinates, 51–52
Spin, 44, 417–20, 458–64
Spin correlation stabilization energy, 79, 109–18, 314
Square planar complexes, 398
Standard potentials, 139–42
Standard state, 105, 139
Starch, 637
Stars, 454–60
Steel, 429–31
Stern, O., 44–50, 323
Stevenson, A. E., 454
Streptococci, 633
Strontium, 318–22, 665
  biochemistry, 319
  solubility of halides, 339–40
Sucrose, 543, 614
Sulfanilamide, 633
Sulfur, 572–82
  allotropes, 572–74
  boiling point, 216
  chloride, 25, 89–90, 193, 581, 582
  dithionate, 576–79
  dithionite, 581

Sulfur (cont.)
  fluorides, 18–25, 88, 200, 205, 581–82
  hydrogen sulfide, 258, 264
  ionization potential, 69, 79
  occurrence, 574–75
  peroxyacids, 581
  polysulfide, 577–79
  polythionic acids, 577–81
  redox reactions, 576–82
  ring equilibria, 572–73, 607
  solubility of sulfides, 331–32, 343
  sulfate, 331, 343–44, 411–12
  sulfuric acid, 278–79, 285, 575–76
  thermodynamics, 559–62, 666
  thiocyanate, 500–502
  thiosulfate, 327, 564, 578–80
Sun, 454, 467, 468
Supernovae, 460

t_{2g} orbitals, 388
Talc, 594
Tantalum, 445
Tartrate, 632
Taube, H., 500
Tellurium, 258–59, 264, 565, 583–85
Terbium, 441–44
Tertiary, 609–11, 621
Tetrahedral, 160–65, 197–98, 394
Thallium, 364, 566–68
Thermal conductance, 12–14
Thermodynamics, 100, 104, 119, 653–67
Thölde, J., 550
Thompson, G. P., 35
Thompson, J. J., 28, 136
Thorium, 442, 444, 470
Three center bonds, 96–97
Thulium, 443–44
Thymine, 642–46
Tin, 212, 315, 554–55, 558, 666
  bronze, 423–28
  stannane, 259–60
  stannic chloride, 216, 554, 558
Tincture, 633
Titanium, 5, 433, 444–46, 666–67
Tool steel, 431
Translational motion of molecules, 228
Transportation, 468
Trigonal bipyramid, 199–200, 204
Trimethyl amine, 212
Trisilyl amine, 211–12
Tritium, 247–48
Trouton, F. T., 217
Trypsin, 634
Troilite, 572
Tyrian purple, 626

Unsaturation, 184–90, 302, 355, 381, 496–97, 604–14
Uracil, 642–46
Uranium, 444, 469–70

Valence bond, 183, 196–203, 207, 357–58, 378
Valence shell expansion, 93
Valentine, B., 550
Valine, 521, 642–46
Vanadium, 5, 45, 445–47, 575, 667
van der Waals, J. D., 79–80, 182, 238, 267–68
Vapor pressure, 216–17
Vector, 44
Velocity distribution, 226–30
Vermiculite, 594
Volatile hydrides, 251–70
Voltage, 138
von Hohenheim, T. B., 632

Water, 86–87, 91, 115, 387–89
  amphoterism, 271–73
  heat capacity, 232
  hybridization, 200–202
  ionization, 271–87
  as ligand, 361–65
  phase equilibria, 118–26, 216–22, 243
Waves, 3, 34–36, 48–49
Wave function, 47–66, 649
Wells, H. G., 632
Werner, A., 374
Wilson, C. T. R., 29–31
Wöhler, F., 601
Woodward, R. B., 601
Work function, 218, 232
Wrought iron, 428–29

Xenon, 96, 210–11, 415–16, 585
X-rays, 31–35, 74–78

Yeast, 610, 641
Young, T., 36
Ytterbium, 441–44
Yttrium, 436, 441–43

Zinc, 439–41, 566, 667
  metallurgy, 431–32
  reaction with acids, 494–97
  solubility of halides, 341–42
  zinc blende structure, 162–65
Zirconium, 444–45
Zwitterion, 520
Zymase, 610

IA									
1 **H** 1.01	IIA								
3 **Li** 6.94	4 **Be** 9.01								
11 **Na** 22.99	12 **Mg** 24.31	IIIB	IVB	VB	VIB	VIIB	VIII		
19 **K** 39.10	20 **Ca** 40.08	21 **Sc** 44.96	22 **Ti** 47.90	23 **V** 50.94	24 **Cr** 52.00	25 **Mn** 54.94	26 **Fe** 55.85	27 **Co** 58.93	28 **Ni** 58.71
37 **Rb** 85.47	38 **Sr** 87.62	39 **Y** 88.91	40 **Zr** 91.22	41 **Nb** 92.91	42 **Mo** 95.94	43 **Tc** (99)	44 **Ru** 101.07	45 **Rh** 102.91	46 **Pd** 106.4
55 **Cs** 132.91	56 **Ba** 137.34	57 **La** 138.91	72 **Hf** 178.49	73 **Ta** 180.95	74 **W** 183.85	75 **Re** 186.2	76 **Os** 190.2	77 **Ir** 192.2	78 **Pt** 195.09
87 **Fr** (223)	88 **Ra** (226)	89 **Ac** (227)	104 (257)						

- ▨ METALS
- ▨ NONMETALS
- ☐ RARE GASES

58 **Ce** 140.12	59 **Pr** 140.91	60 **Nd** 144.24	61 **Pm** (147)	62 **Sm** 150.35	63 **Eu** 151.96
90 **Th** 232.04	91 **Pa** (231)	92 **U** 238.03	93 **Np** (237)	94 **Pu** (244)	95 **Am** (243)